K. Uyemura, K. Ka
T. Yazaki (Eds.)

Neural Development

With 186 Figures, Including 21 in Color

Springer

Keiichi Uyemura, M.D., Ph.D.
Professor and Chair, Department of Physiology, School of Medicine, Keio University, 35 Shinanomachi, Shinjuku-ku, Tokyo 160-8582, Japan

Koki Kawamura, M.D., Ph.D.
Professor and Chair, Department of Anatomy, School of Medicine, Keio University, 35 Shinanomachi, Shinjuku-ku, Tokyo 160-8582, Japan

Takahito Yazaki, M.D., Ph.D.
Assistant Professor, Department of Physiology, School of Medicine, Keio University, 35 Shinanomachi, Shinjuku-ku, Tokyo 160-8582, Japan

ISBN 4-431-70232-6 Springer-Verlag Tokyo Berlin Heidelberg New York

Library of Congress Cataloging-in-Publication Data

Neural development / K. Uyemura, K. Kawamura, T. Yazaki (eds.).
p. cm.
"Proceedings of the second symposium of the Keio University International Symposia for Life Sciences and Medicine . . . "—Foreword.
Includes bibliographical references and index.
ISBN 4-431-70232-6 (hbk. : alk. paper)
1. Neurons—Growth—Congresses. 2. Molecular neurobiology—Congresses. 3. Nervous system—Regeneration—Congresses. 4. Developmental neurobiology—Congresses. I. Uyemura, K. (Keiichi), 1935- II. Kawamura, Koki, 1934- III. Yazaki, T. (Takahito), 1958- IV. Keio University International Symosium for Life Sciences and Medicine (2nd : 1997 : Tokyo, Japan)
QP356.25 . N46 1998
573.8′536—dc21

98-45347
CIP

Printed on acid-free paper

Typesetting, printing, and binding: Best-set Typesetter Ltd., Hong Kong
SPIN: 10731271 9 8 7 6 5 4 3 2

Foreword

This volume contains the proceedings of the second symposium of the Keio University International Symposia for Life Sciences and Medicine under the sponsorship of the Keio University Medical Science Fund. As stated in the address by the President of Keio University at the opening of the symposium, the fund was established by the generous donation of Dr. Mitsunada Sakaguchi. The Keio University International Symposia for Life Sciences and Medicine constitute one of the core activities of the fund. The objective is to contribute to the international community by developing human resources, promoting scientific knowledge, and encouraging mutual exchange. Every year, the Committee of the International Symposia for Life Sciences and Medicine selects the most interesting topics for the symposium from applications received in response to a call for papers to the Keio medical community. The publication of these proceedings is intended to publicize and distribute information arising from the lively discussions of the most exciting and current issues during the symposium. We are grateful to Dr. Mitsunada Sakaguchi, who made the symposium possible, the members of the program committee, and the office staff whose support guaranteed the success of the symposium. Finally, we thank Springer-Verlag, Tokyo, for their assistance in publishing this work.

Akimichi Kaneko, M.D., Ph.D.
Chairman
Committee of the International Symposia for Life Sciences and Medicine

Preface

Neuroscience is one of the most attractive and important research fields in life science and medicine. Remarkable progress and new discoveries have emerged in recent years, while new questions arise, waiting to be solved. Therefore, now is an excellent time to hold this symposium on neuroscience at Keio University.

Our proposal on "Neuroscience: Frontiers of Neural Development" was accepted as the theme for the second of the Keio University International Symposia for Life Sciences and Medicine: 1997 Conference by the committee of the Keio University Medical Science Fund in the autumn of 1996. The symposium was organized to present, exchange, and develop the latest results and ideas in the field of neural development. For these purposes, our organizing and program committees invited some 50 neuroscientists from all over the world as speakers and more than 50 young scientists as poster presenters. The symposium was held at the New North Building (Kita-shinkan) on the Mita campus of Keio University, December 1–4, 1997. All the invited speakers accepted our invitation, presented interesting reports, and joined in vigorous discussions. Owing to the cooperation of all participants, the symposium was successful not only academically but also socially.

A total of 94 papers have been included in this volume, arranged according to the subject area: 1) Early Neurogenesis and Pattern Formation, 2) Regional and Cellular Specification, 3) Trophic Factors, 4) Adhesion Molecules, 5) Glial Cell Lineage and Myelination, 6) Circuit Formation, 7) Synaptic Refinement and Neurotransmitter Release, 8) Neural Regeneration, and 9) Human Genome Project. In this way, our intention has been to cover various aspects of recent research progress in neural development.

The editors are deeply grateful to Dr. Mitsunada Sakaguchi, who was an ardent researcher of neuroscience in the Departments of Physiology and Anatomy, School of Medicine, Keio University. We would like to express our thanks to all participants for their contributions and to the members of related committees and conference secretariats. We also thank the staff of Springer-Verlag, Tokyo, for their assistance.

Keiichi Uyemura
Koki Kawamura
Takahito Yazaki
EDITORS

Keio International Symposium on Neuroscience: Frontiers of Neural Development

First row, from right: Dr. Asai, Dr. Spassky, Dr. Kawamura, Dr. Lemmon, Dr. Kaneko, Dr. McConnell, Dr. C. Campagnoni, Dr. Jan, Dr. Gould, Dr. Zalc, Dr. Eisen, Dr. Lindsay, Dr. Davies, Dr. Uyemura, Dr. Sotelo, Dr. Colman, Dr. Lemke, Dr. A. Campagnoni, Dr. Bregman, Dr. Lansford, Dr. Filbin, Dr. Walsh, Dr. Sakaguchi, Dr. Tsukada, Dr. Aguayo, Dr. Scheller;

Second row: Dr. Murase, Dr. Yano, Dr. Nishiyama, Dr. Ito, Dr. Yamauchi, Dr. Taniguchi, Dr. Kumakura, Dr. Okano, Dr. Nakatsuji, Dr. Shirao, Dr. Mikoshiba, Dr. Pfeiffer, Dr. Trapp, Dr. Ikenaka, Dr. Nagahama, Dr. Takai, Dr. Raper, Dr. Barde, Dr. Augustine, Dr. Kohsaka, Dr. Kawaguchi, Dr. Ide, Dr. H. Hatanaka, Dr. Akagawa, Dr. Baba, Ms. Okazeri, Dr. Ivanova;

Third row: Dr. R. Yoshida, —,* Dr. Shimamura, Dr. Nishiwaki, Dr. Torii, Dr. Ikeya, —, Dr. M. Yoshida, Dr. Iwasaki, Dr. Nakahira, Dr. Miura, Dr. Takeda, Dr. Yamaguchi, —, Dr. Murakami, Dr. T. Inoue, Dr. Ohara-Imaizumi, Dr. Maeda, —, Dr. Okada, Dr. Kuroda, —, —, —, Ms. Tatsumi;

Fourth row: Ms. Hayashi, Mr. Endo, Ms. Nakamura, Dr. Kawano, Dr. Toda, Dr. Nitta, —, —, Dr. Kitamura, Dr. Asou, Dr. Yoshihara, Dr. Inoue, Dr. Takahashi, Dr. Noda, Dr. Kageyama, Dr. Nawa, Dr. Okamoto, Dr. Nakafuku, —, —, Dr. Kano, —, —, —, Dr. Y. Hatanaka, Dr. Nakai, Dr. Toba;

Fifth row: Dr. Uemura, Dr. Terao, Dr. Segawa, Dr. Kikuchi, Dr. Hanaoka, Dr. Hoshi, Dr. Mieda, Dr. Han, —, Dr. Nakamura, Dr. Kamiguchi, Dr. Ooyama, Dr. Hayashi, Dr. Takamatsu, —, —, —, —, Dr. Nishimura, Dr. Tanaka, Dr. Yazaki, Dr. Miyashita, Dr. Yu, Dr. Horie, Dr. Iizuka, Dr. Matsuo.

**—: The editors regret that some participants shown in the photograph could not be identified at the time this volume went to press.*

Contents

Part 1 Early Neurogenesis and Pattern Formation

Part 3 Trophic Factors

Part 4 Adhesion Molecules

Part 5 Glial Cell Lineage and Myelination

Part 6 Circuit Formation

Part 7 Synaptic Refinement and Neurotransmitter Release

Part 8 Neural Regeneration

Part 9 Human Genome Project

List of Contributors

Opening Remarks*

PROFESSOR YASUHIKO TORII
PRESIDENT, KEIO UNIVERSITY
CHAIRMAN, KEIO UNIVERSITY MEDICAL SCIENCE FUND

Distinguished Guests, Ladies and Gentlemen:

On behalf of Keio University, I take great pleasure in welcoming all the guests and participants to the second of the Keio University International Symposia for Life Sciences and Medicine. I am particularly grateful to those scientists who traveled such long distances to participate in this symposium.

The topic of our symposium this year is "Neuroscience: Frontiers of Neural Development." Neuroscience is one of the most attractive and important research fields. Remarkable progress and breakthroughs are reported almost every day. Therefore, it seems an auspicious time to hold a symposium on neuroscience at Keio University. All invited speakers kindly accepted our invitation to contribute to the symposium program, and we are sure that this unique meeting will be both exciting and successful.

There are several reasons for us at Keio University to host this international symposium for the life sciences and medicine. I would like to explain the reasons by giving you a brief history of Keio University and of the Keio University Medical Science Fund.

Keio University is unique in being one component in an educational corporation named Keio Gijuku, which was founded in 1858 by Yukichi Fukuzawa, a pioneer in the modernization of Japan. I assume some of you are already familiar with Mr. Fukuzawa's personal appearance, because his portrait is printed on the Japanese ten-thousand-yen note. We are proud that, in the more than 139 years since our establishment, Keio, as the oldest of Japan's 587 universities, has played a major role in developing human resources including academic, business, and political leaders. Prime Minister Ryutaro Hashimoto is one of our alumni. At Keio University we now have eight faculties and nine graduate schools, and among the faculties, the school of medicine is one of the most highly regarded in Japan and throughout the world.

---

* This opening address was given by Professor Yasuhiko Torii, President of Keio University, at the opening session of the Keio University International Symposium for Life Sciences and Medicine, "Neuroscience: Frontiers of Neural Development," on the morning of Monday, December 1, 1997, in the conference hall of the New North Building on the Mita campus of Keio University.

For many years we have also been carrying out a wide range of international exchanges with people from various countries. In this connection, I would like to tell you that our founder, Mr. Fukuzawa, was a member of the very first mission of the Tokugawa Shogunate government to the United States in 1860, after Japan was opened to the outside world by Commodore Perry knocking on our door.

During his visits to the United States and Europe as a member of the official Japanese mission, Fukuzawa realized that education was critically important to the future of Japan. After coming back to Japan, he reorganized his school to create a modern institution of learning. Thus, Keio has its origin in international exchanges and has long aimed for international exchanges of culture and science with many countries. Please understand that international exchanges, such as this occasion, have been one of the most important academic and social missions of Keio University since its foundation.

In the fall of 1994, Dr. Mitsunada Sakaguchi, a 1940 alumnus of our medical school, donated 5 billion yen to the university, expressing his wish that it should be used to encourage research in life sciences and medicine and, in addition, to promote worldwide advances in medical sciences. I, myself, being a political economist especially interested in the nation's health-care policy, totally sympathized and agreed with his wishes, and thus launched the Keio University Medical Science Fund in April of 1995. The International Symposia for Life Sciences and Medicine have thus been organized as part of several projects supported by the fund, the objective of which is to contribute to the international community by developing human resources, promoting scientific knowledge, and encouraging mutual exchange.

Time passes very quickly. The year 1997 is nearly over, and as we witness the dawn of the third millennium, we realize that our society faces many problems from this century which will carry over into the next. In the field of life sciences and medicine alone, we are still unable to cure many kinds of cancer, AIDS, and neurological and psychiatric disorders. In addition, many new and unknown problems await us in the new century. We will have to overcome numerous obstacles, including diseases and problems that arise with the over-sophistication of civilization and the aging of our population.

I believe that exploring new horizons in life sciences is one of the most vital tasks that we face at the dawn of the 21st century. It is equally important to ensure that the knowledge obtained through these pursuits is used in ways that bring genuine happiness to humankind. Conceived in the belief and philosophy I have described, Keio University has organized the Keio University International Symposia for Life Sciences and Medicine. It is therefore more than a pleasure, indeed it is an honor, for me to meet you, distinguished medical researches from world-renowned institutions, and to share and exchange views and opinions in the fields of medicine and the life sciences. I am also grateful for the efforts of the organizing committee, chaired by Professor Keiichi Uyemura, who devoted themselves to making this symposium a high-quality and enjoyable event.

Finally, I do hope that this symposium will be both fruitful and productive for all of you. Let me close now by wishing you great success in this year's symposium and further success in your research. Thank you very much for your attention.

Part 1
Early Neurogenesis and Pattern Formation

Regulation of Asymmetric Cell Divisions by Wnt and Frizzled in *Caenorhabditis elegans*

Hitoshi Sawa and Hideyuki Okano

Summary. Asymmetric cell divisions are fundamental to the differentiation of a variety of cells including neurons and glias. In *Caenorhabditis elegans*, asymmetric cell divisions are regulated by extrinsic signals and receptors. Both signaling molecules encoded by *lin-44* and *mom-2* belong to the Wnt family. Both receptors encoded by *lin-17* and *mom-5* are related to Frizzled. We describe how these molecules regulate asymmetric cell divisions.

Key words. *wnt*—*frizzled*—Cell polarity—Asymmetric cell division—*C. elegans*

Introduction

Cells can divide asymmetrically to produce daughter cells with different characters or fates. Asymmetric cell division is a fundamental mechanism to produce cellular diversity during development [1]. To produce distinct daughter cells by cell divisions, the mother cells should have polarities when they divide. In *Drosophila*, asymmetric cell divisions of sense organ precursor cells and neuroblasts involve the polarized localization of the Numb and Prospero proteins in the mother cells [2]. It is not known how the mother cells establish polarities required for the asymmetry of the division. Here, we describe examples of how extrinsic signals determine the polarities of asymmetric divisions in the nematode *Caenorhabditis elegans*.

Postembryonic Asymmetric Cell Divisions Are Regulated by *lin-17/frizzled* and *lin-44/wnt*

In *lin-17* and *lin-44* mutants of the nematode *C. elegans*, certain asymmetric cell divisions that occur during postembryonic development are abnormal [3,4]. The divisions often become symmetric in *lin-17* and reversed in *lin-44* mutants (Fig. 1). For example, in the wild type, posterior daughters of the T cells in the tail of the animals produce neural cells and the anterior ones produce hypodermal cells. In *lin-17* mutants, both daughter cells produce hypodermal cells. In *lin-44* mutants, anterior

Department of Neuroanatomy, Osaka University Medical School, JST CREST, 2-2 Yamadaoka, Suita, Osaka 565-0871, Japan

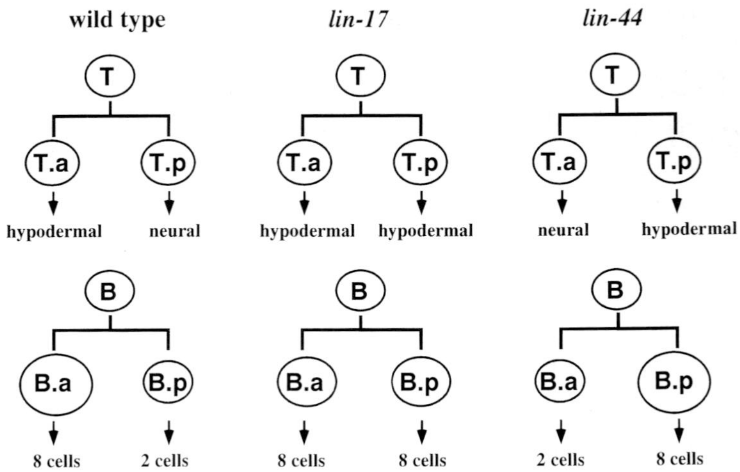

FIG. 1. T- and B-cell divisions in wild type, *lin-17*, and *lin-44* mutants

rather than posterior daughters produce neural cells and the posterior daughters generate hypodermal cells. In the case of the B cell, in the wild type its daughter cells differ in size as well as in subsequent fate. The anterior daughter B.a cell is larger than the posterior daughter B.p cell. Subsequently, B.a and B.p produce eight and two cells, respectively, in the L2 stage. In *lin-17* mutants, B.a and B.p are similar in size as well as in fate. In *lin-44* mutants, the size and the fates of B.a and B.p are reciprocally altered. This suggested that both *lin-17* and *lin-44* activities are required in the mother cells before or at the time of their divisions.

Both *lin-17* and *lin-44* have been cloned. *lin-44* encodes a putative signaling protein that is a member of the *Wnt* gene family [5]. Wnt proteins are secreted glycoproteins that can function as morphogens [6]. *lin-17* encodes a receptor protein that belongs to the Frizzled family [7]. It was shown that some of the Frizzled proteins can act as receptors of Wingless, which is a Wnt protein of *Drosophila* [8]. Because *lin-17* and *lin-44* mutations affect divisions of the same cells, it is likely that LIN-17/Frizzled functions as a receptor of LIN-44/Wnt. The LIN-44 proteins are expressed at the tip of the tail in cells posterior to the T and B cells [5]. Thereby, the LIN-44 signal might activate the LIN-17 receptor on the posterior side of the T and B cells to polarize these cells. Polarity reversal in *lin-44* was proposed to be caused by an unidentified signal "X" expressed anterior to the T and B cells [7]. In that model, the LIN-17 receptor on the anterior rather than posterior side of the T and B cell is activated by the X signal in *lin-44* mutants.

Asymmetric Division of the EMS Blastomere Requires *mom* Genes

In the four-cell-stage embryo of *C. elegans*, the EMS, blastomere divides to produce the MS cell, which is a mesodermal precursor, and the E cell, which is an endodermal precursor [9]. Here signals from a neighboring blastomere, P2, are essential for EMS

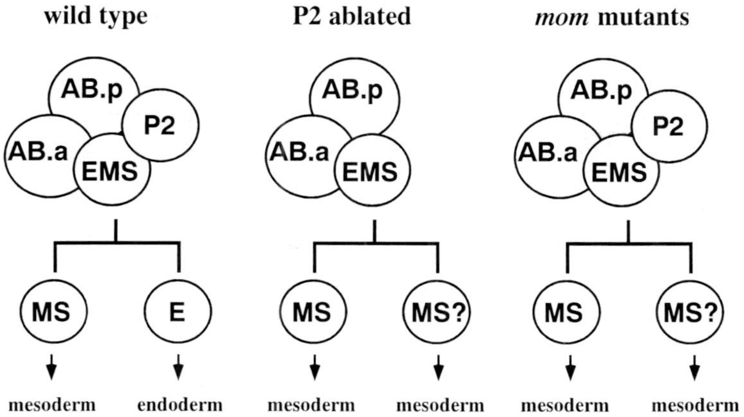

FIG. 2. Four-cell-stage embryo and division of EMS shown schematically

to divide asymmetrically (Fig. 2). Removing the P2 cell results in a gutless embryo because both daughters of EMS take the MS-like cell fate. Similarly, mutations in five *mom* genes (*mom-1* through *mom-5*) disrupt the asymmetry of the EMS division [10,11]. Mosaic analyses have shown that *mom-1*, *mom-2*, and *mom-3* act in the signaling cell P2, whereas *mom-4* is required in the receiving cell EMS. Molecular analyses have shown that *mom-2* and *mom-5* encode a Wnt protein and a Frizzled protein, respectively. *mom-5* was predicted to act in EMS because it encodes a receptor-like protein. Therefore, as in the case of LIN-44 and LIN-17, MOM-2/Wnt produced in P2 might polarize EMS by activating MOM-5/Frizzled.

Wnt Signaling and Cell Polarity

wnt genes are involved in many developmental processes in various organisms [6]. Among them, the *wingless* gene, the best studied *wnt* gene of *Drosophila*, is required for the specification of cell fates in many developmental stages. However, involvement of *wingless* in cell polarity or asymmetric cell divisions has not been demonstrated. In contrast, the *frizzled* gene, the first identified *frizzled* family gene, is required for polarities of many cells and tissues [12]. Because signal transduction pathways of *wingless* and *frizzled* involve different components, such as Armadillo and Pangolin/DTcf for *wingless* and Pricked-spiny-legs for *frizzled*, one might imagine that different signaling components produce different effects, cell-fate specification for *wingless* and cell polarity for *frizzled*. However, regulation of the EMS polarity in *C. elegans* involves WRM-1 and POP-1, which are related to Armadillo and Pangolin/DTcf, respectively [11,13]. Therefore, a signaling pathway similar to that of *wingless* seems to be involved in the regulation of cell polarity in *C. elegans*.

Although it is not understood how polarity is regulated by these components, one clue might be the finding that the MS cell has a higher amount of the POP-1 protein than its sister E [13]. Such difference is disrupted in the mutants of *mom-2/wnt*, *mom-5/frizzled*, and *wrm-1/armadillo* [10,11]. One possibility might be that the difference in the intracellular signaling in MS and E results in that in the amount of POP-1 in the nuclei. However, considering the fact that the cell–cell interaction occurs between P2 and EMS rather than between their daughters [9], it could be that the consequence of the signaling is the difference in the inheritance of the POP-1 protein between two daughter cells, as in the case of the Numb and Prospero proteins in the asymmetric divisions of neuronal precursor cells in *Drosophila* [2].

Involvement of *wrm-1* and *pop-1* in LIN-44-LIN-17 signaling remains to be elucidated. Molecular genetic analyses of these genes will help to elucidate the mechanism for asymmetric cell divisions.

References

1. Horvitz HR, Herskowitz I (1992) Mechanism of asymmetric cell division: two Bs or not two Bs, that is the question. Cell 68:237–255
2. Doe CQ, Spana EP (1995) A collection of cortical crescents: asymmetric protein localization in CNS precursor cells. Neuron 15:991–995
3. Herman MA, Horvitz HR (1994) The *Caenorhabditis elegans* gene *lin-44* controls the polarity of asymmetric cell divisions. Development (Camb) 120:1035–1047
4. Sternberg PW, Horvitz HR (1988) *lin-17* mutations of *Caenorhabditis elegans* disrupt certain asymmetric cell divisions. Dev Biol 130:67–73
5. Herman MA, Vassilieva LL, Horvitz HR, et al (1995) The *C. elegans* gene *lin-44*, which controls the polarity of certain asymmetric cell divisions, encodes a Wnt protein and acts cell nonautonomously. Cell 83:101–110
6. Cadigan KM, Nusse R (1997) Wnt signaling: a common theme in animal development. Genes Dev 11:3286–3305
7. Sawa H, Lobel L, Horvitz HR (1996) The *Caenorhabditis elegans* gene *lin-17*, which is required for certain asymmetric cell divisions, encodes a putative seven-transmembrane protein similar to the *Drosophila* Frizzled protein. Genes Dev 10: 2189–2197
8. Bhanot P, Brink M, Samos CH, et al (1996) A new member of the *frizzled* family from *Drosophila* functions as a Wingless receptor. Nature (Lond) 382:225–230
9. Goldstein B (1992) Induction of gut in *Caenorhabditis elegans* embryos. Nature (Lond) 357:255–227
10. Thorpe CJ, Schlesinger A, Carter JC, et al (1997) Wnt signaling polarizes an early *C. elegans* blastomere to distinguish endoderm from mesoderm. Cell 90:695–705
11. Rocheleau CE, Downs WD, Lin R, et al (1997) Wnt signaling and an APC-related gene specify endoderm in early C. elegans embryos. Cell 90:707–716
12. Adler PN (1992) The genetic control of tissue polarity in *Drosophila*. BioEssays 14:735–741
13. Lin R, Thompson S, Priess JR (1995) *pop-1* encodes an HMG box protein required for the specification of a mesoderm precursor in early *C. elegans* embryos. Cell 83:599–609

The Regulatory Mechanisms of Neural Development: Roles of Cell-Autonomous and Non-Cell-Autonomous Cues in Cell-Fate Decisions

HIDEYUKI OKANO[1,2], YUKIKO KANEKO[1], YUKI NAKAMURA[1], WADO AKAMATSU[1,3], YOSHIHIRO YUASA[1], TAKAO IMAI[1], YUKI HIROTA[1], MING-HAO JIN[1,2], KATSUHIKO TABUCHI[1], AKIKO TAGUCHI[1], MASATAKA OKABE[1,2,4], SHIN-ICHI SAKAKIBARA[1,2], TAKAHIRO GOTO[5], YASUO UCHIYAMA[5], and KAZUNOBU SAWAMOTO[1,2]

Summary. By using the powerful genetic strategies of *Drosophila*, we have identified new gene products required in neural development, which we review here. Based on their functions, these gene products can be classified into two major categories: (1) gene products regulating cell–cell interactions and (2) cell-autonomous intrinsic cues. An example of the first category is Argos, a lateral or remote inhibitor that acts during neural development to inhibit EGF-R signaling and its downstream RAS/MAPK pathway. A neural RNA-binding protein, Musashi, is an example of the second category, which is required for the asymmetric cell division of a neural precursor. The characterization of the mammalian homolog of Musashi is also discussed.

Key words. Cell autonomous cues—Non-cell-autonomous cues—Argos—Musashi—Hu—Neural RNA-binding proteins

Introduction

During development in the vertebrate central nervous systems (CNS), stem cells proliferate in the ventricular zone of the neural tube. Cells then migrate away from the ventricular zone toward the pia mater, where they differentiate into a wide variety of neurons and glia. These diverse types of cells provide the cellular basis for the higher functions of the nervous system. We are interested in how this cellular diversity is

[1] Department of Neuroanatomy, Biomedical Research Center, Osaka University Medical School, Suita, Osaka 565-0871, Japan
[2] Core Research for Evolutional Science and Technology (CREST), Japan Science and Technology Corporation (JST), Suita, Osaka 565-0871, Japan
[3] Department of Pediatrics, Keio University School of Medicine, 35 Shinanomachi, Shinjuku-ku, Tokyo 160-8582, Japan
[4] National Institute of Genetics, Mishima, Shizuoka 411-8540, Japan
[5] Department of Cell Biology and Anatomy I, Osaka University Medical School, Suita, Osaka 565-0871, Japan

generated and how the number of cells of a given type is regulated during neural development. It is known that cellular diversity is generated by the actions of both cell-autonomous intrinsic cues and microenvironmental cues including cell–cell interactions. Furthermore, recent studies have indicated that the gene products and mechanisms regulating the early steps of neural development are surprisingly well conserved between invertebrates and vertebrates [1–3].

To identify new gene products regulating neural development, we have performed transposon P-element insertional mutagenesis; 8300 *Drosophila* lines with a single P-element insertion were generated. Homozygote mutants were screened for phenotypes affecting their nervous systems. Consequently, a number of novel mutants affecting neural development were isolated. Some of the corresponding genes have been isolated and their functions characterized by analyzing in detail the mutant phenotypes and genetic interactions. The gene products identified in this screening can be classified into two major categories. The first are the non-cell-autonomous cues, namely, environmental cues including cell–cell interactions. The second type of determinants that can influence neural cell diversity are intrinsic, such as cytoplasmic factors or patterns of gene expression inherited by a cell from its progenitors that can restrict or determine cell fates.

In this review we describe work from our laboratory to identify and characterize gene products that regulate neural development. These components include both non-cell-autonomous and intrinsic determinants.

Regulator of Cell–Cell Interactions: Argos, a Secreted Protein with an EGF Motif, Is Required for Lateral Inhibition During *Drosophila* Neural Development

To identify and characterize the function of the new gene products required for cell–cell interactions during neural development, we have been studying *Drosophila* eyes. During retinal development, cell-fate decisions are mainly regulated by cell–cell interactions rather than by cell lineage restrictions [4]. If a gene that is required for cell–cell interaction is mutated, a developmental defect occurs that results in a rough eye phenotype. Thus, we screened mutants affecting *Drosophila* eye development by selecting flies with rough eye phenotypes.

A *Drosophila* compound eye is composed of 800 repeating units called ommatidia. Each ommatidium consists of exactly the same cellular components, i.e., eight photoreceptor neurons, four lens-forming cells, and eight accessory cells (pigment cells) (Fig. 1). Thus, it is easy to imagine that there should be a strict regulatory mechanism for each cell type. For the recruitment of each type of cell within an ommatidium, sequential activations of the epidermal growth factor (EGF) receptor signaling pathway are known to be required [5]. In the *Drosophila* eye, the EGF receptor is called DER (*Drosophila* EGFR) and its ligand is Spitz, a *Drosophila* homolog of TGF-α.

One gene required for *Drosophila* eye development is *argos*, a mutant of which was identified in this screening [6]. This gene was identified and reported almost simultaneously by three groups including ours [6–8]. In the loss-of-function mutants of *argos*, the numbers of each retinal cell are increased by overrecruitment [6] rather

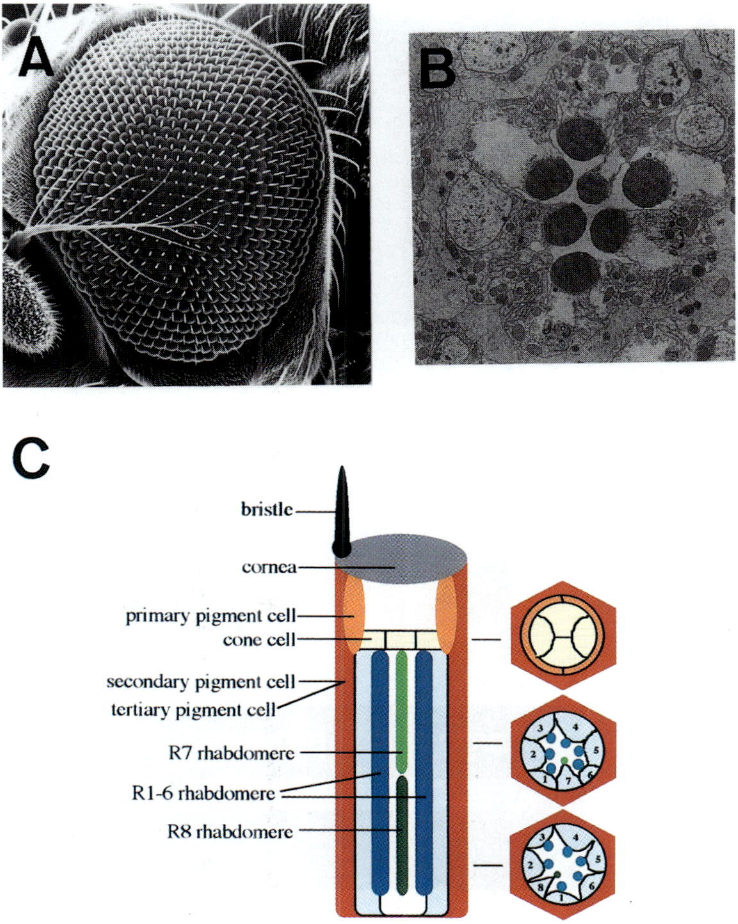

FIG. 1A–C. Structure of the *Drosophila* compound eye. **A** Scanning electron microscopic view of a wild-type compound eye. **B** Tangential section through a single adult ommatidium. The rhabdomeres in R1–R6 are larger and arrayed around the smaller rhabdomere of R7. **C** Schematic view of *Drosophila* adult ommatidium. (From [4], with permission)

than enhanced mitosis [7] (Fig. 2A). *argos* expression begins when each ommatidial cell starts to be recruited into an ommatidium (Fig. 2B). Mosaic analysis showed that *argos* acts non-cell-autonomously to inhibit the differentiation of cells that are adjacent to *argos*-expressing cells [7]. Molecular analysis indicates that the *argos* gene encodes a secreted protein with an EGF motif, which is generally known to be involved in cell–cell interactions [6–8] (Fig. 2C).

On the basis of these facts and work from another group [5], we are proposing that Argos acts as a lateral or remote inhibitor during *Drosophila* eye development. We define lateral inhibition as occurring between cells that have direct contact and remote inhibition as occurring between cells that do not contact one another directly. The Argos protein is produced in differentiated retinal cells and acts on the neighbor-

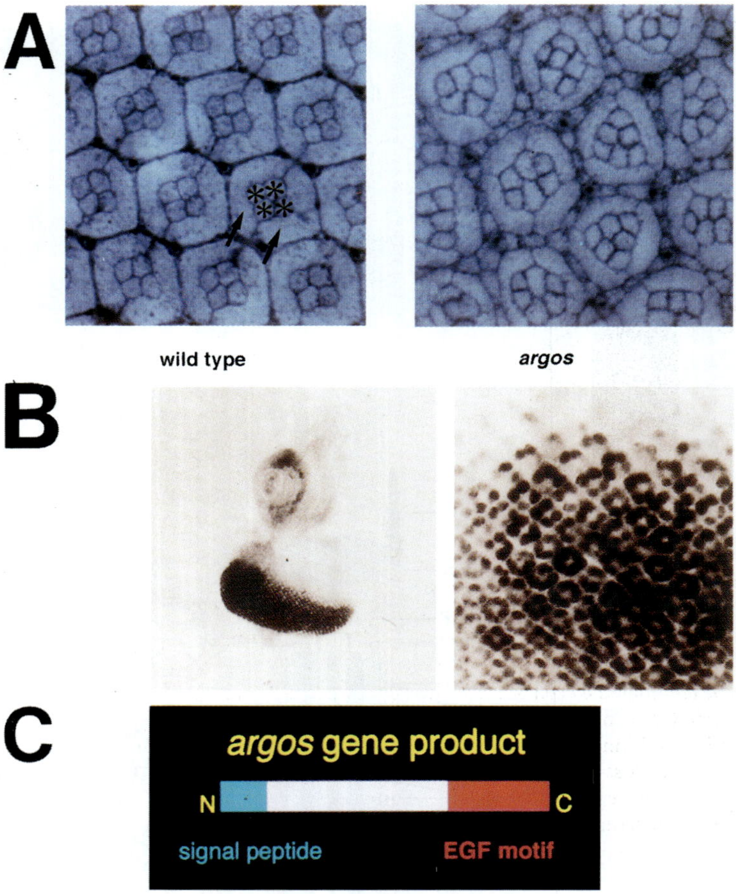

FIG. 2A–C. The expression pattern and structure of the *argos* gene product. **A** Cobalt sulfide staining of pupal retinae at 50 h APF from wild type (*left*) and an *argos* loss-of-function mutant (*argos*^{styp2}/*argos*^{styp2}) (*right*). Each wild-type ommatidium has four cone cells and two primary pigment cells. Note that ommatidia in *argos*^{styp2}/*argos*^{styp2} flies have increased numbers of cone and primary pigment cells. **B** The *argos* expression pattern in the developing eye disk (from [6]). The expression of *argos* is monitored using the β-galactosidase protein in *argos*^{styp1} heterozygotes used as an enhancer trap line [30]. *argos* appears to be expressed in each ommatidial cell as it begins to be differentiated. **C** The deduced Argos protein. The *argos* gene encodes an open reading frame of 444 amino acids [6–8]. Based on the hydropathy profile and functional analyses [6–8,14], Argos is likely to be a secreted protein with an EGF motif

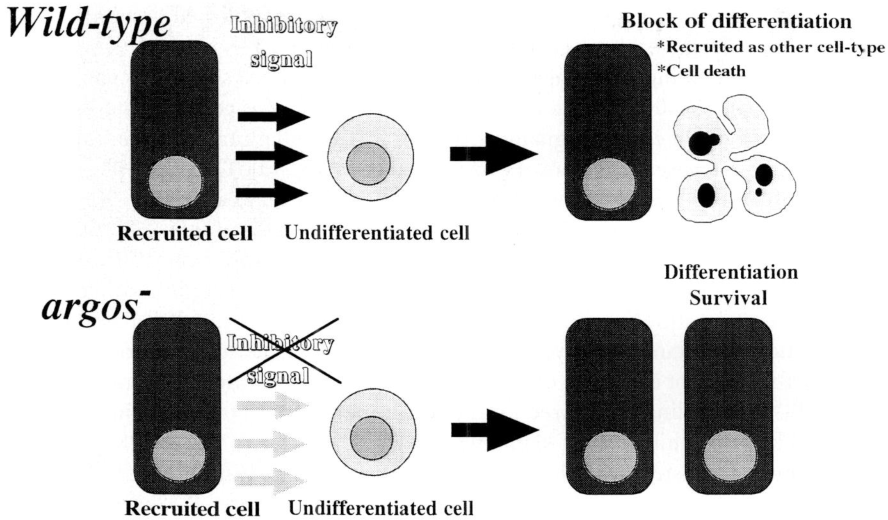

FIG. 3. Proposed role of *argos* gene product as a lateral inhibitor (adapted from [2]). *Upper panel:* The *argos* gene product is produced in differentiated cells, including photoreceptor cells, cone cells, and pigment cells, and is putatively secreted from these cells. In the developing eye disk at least, the *argos* gene product is proposed to inhibit neighboring undifferentiated cells from terminal differentiation. Once differentiation is blocked, these undifferentiated cells are likely to be eliminated by cell death at the pupal stage in "*Wild-type*". *Lower panel:* The loss-of-function phenotype of *argos* indicate that the Argos protein acts on neighboring undifferentiated cells by inhibiting their differentiation. Such a function is referred to as lateral or remote inhibition. In the loss-of-function mutants, the inhibitory signal is removed. Consequently, the number of photoreceptor cells increases in *argos*. When this inhibitory signal is too strong, the number of photoreceptor cells decreases

ing undifferentiated cells non-cell-autonomously, as a secreted protein (Fig. 3). Argos prevents these undifferentiated cells from differentiating in response to immediate environmental cues. Thus, they are likely to take on alternate fates or to be eliminated by apoptosis. In the mutants lacking Argos, the neigboring undifferentiated cells are likely to be recruited into the same as the *argos*-expressing cells simultaneously by the action of an inductive signal, Spitz. Once recruited, they are protected from apoptosis. In this way, the number of each type of cell in an ommatidium can be increased without increasing the mitotic index.

To determine the feasibility of this model, we tried to make the inhibitory signal stronger by overexpressing Argos in transgenic flies [9,10]. To overexpress Argos, we used the GMR vector as well as an heat shock vector. The GMR vector carries multiple binding sites for the eye-specific transcription factor Glass, which is used for eye-specific overexpression of transgenes [11]. In each ommatidium of adult *GMR-argos* flies, we observed that the number of each type of retinal cell was reduced. This is quite the opposite phenotype from the *argos* loss-of-function mutant. Based on a detailed analysis of *GMR-argos* flies, we concluded that this *GMR-argos* phenotype is mainly the result of enhanced apoptosis [10] and partly that of the suppressed recruitment of each retinal cell type (Sawamoto and Okano, unpublished results). These results are consistent with the current model shown in Fig. 3.

What is the signaling mechanism of Argos? In *Drosophila*, a signal cascade can be explored by genetically characterizing the interactions of a gene of interest with mutations in genes of various signaling cascades. In the null mutant of *argos* (*argos*^{257}) [12], there are increased numbers of each retinal cell type [13]. This *argos* phenotype, interestingly, was suppressed by reducing the activity of genes encoding *Drosophila* Ras (*Ras1*), Raf (*D-raf*), and MEK (*Dsor1*) [13]. Heterozygotes of the null mutants of these genes do not show abnormal eye phenotypes. Thus, components of *Drosophila* Ras/MAPK signaling act downstream of Argos, although the possibility that Argos acts in a parallel pathway to Ras/MAPK signaling cannot be excluded. Based on its loss-of-function mutant phenotype, we suggest that *argos* gene activity is required for the proper number of each type of retinal cell to be recruited into each ommatidium. Because EGFR signaling is required for the recruitment of each type of retinal cell [5], Argos is likely to inhibit EGFR signaling and its downstream Ras/MAPK pathway. This scheme is consistent with the biochemical functional analysis of Argos in the cell culture system reported by the laboratory of M. Freeman and B. Shilo [14]; i.e., recombinant Argos inhibits the activation of EGFR induced by Spitz. Even though Argos specifically inhibits EGFR signaling, the receptor for Argos is still unknown. Argos may bind to DER as a competitive inhibitor. Alternatively, Argos may inhibit the dimerizartion of DER. Another possibility is that Argos has its own receptor, the activation of which may inhibit EGFR signaling and its downstream Ras/MAPK pathway in trans.

The framework of the Argos-mediated signaling cascade is shown in Fig. 4. Some black boxes have not yet been elucidated: (1) we do not know the receptor of Argos;

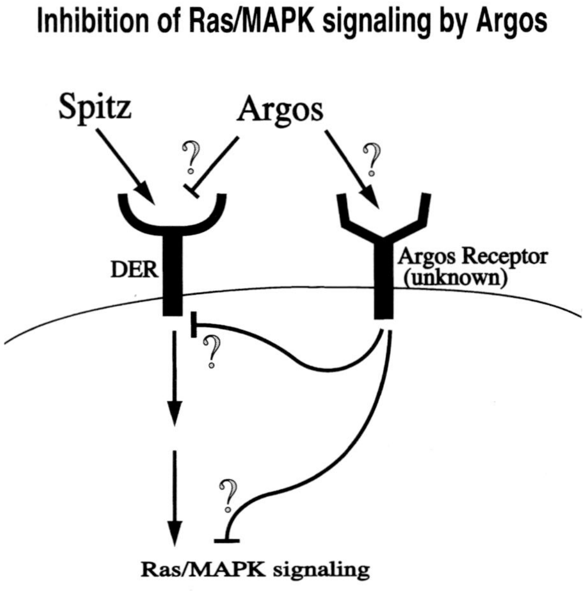

FIG. 4. Argos-mediated signaling cascade

(2) we do not know how the activation of DER is inhibited by Argos; and (3) we do not know how the repression of the Ras/MAPK pathway results in the induction of apoptosis in the eye. For a complete understanding of the Argos signaling pathway, we are currently performing a genetic screen for modifiers of the gain-of-function *argos* phenotype, together with biochemical experiments to examine the direct binding of Argos to the EGF receptor.

Argos inhibits EGFR signaling and its downstream RAS pathway. Overexpression of Argos can induce programmed cell death. Thus, it will be very interesting to identify vetebrate Argos homologs from clinical aspects. Interestingly, lateral or remote inhibition of differentiation by means of blocking EGFR signaling is likely to occur even in vertebrates. In the chick, the endogenous soluble type of EGFR is known to be generated through alternative splicing. This truncated receptor blocks EGFR signaling in a dominant-negative fashion [15]. Thus, extracellular negative regulation of EGFR is likely to be universal among invertebrates and vertebrates.

Musashi, a Neural RNA-Binding Protein Required for the Asymmetric Cell Division of Neural Precursor Cells

Drosophia *Musashi*

Some cell-autonomous intrinsic cues are required for the asymmetric cell division of neural precursor cells or for cell lineage restrictions. Recent studies by McConnell's group have indicated that asymmetric cell division of neural precursor cells occurs in the developing ferret cerebral cortex [16]. Although asymmetric cell division is required for the generation of cellular diversity in the developing nervous system, its regulatory mechanism remains largely unknown, especially in vertebrates. The *Drosophila* external sensory (Es) organ provides an excellent model system for studying the regulatory mechanism of asymmetric cell division. The *Drosophila* adult Es organ is composed of four different cells: a neuron, a thecogen (glia cell), a tormogen (socket cell), and a tricogen (shaft cell) (Fig. 5A). These cells are generated by two successive asymmetric cell divisions of a single sensory organ precursor (SOP) cell. The SOP divides asymmetrically to generate two second-order precursor cells, the IIa precursor (which gives rise to the tricogen and tormogen) and the IIb precursor (which gives rise to the neuron and thecogen).

Finally, there is a second order of asymmetric cell division, resulting in the production of the four progeny cells. There are a number of mutations affecting the asymmetric cell division of the SOP of the *Drosophila* Es organ. The *Drosophila musashi* mutant was isolated on the basis of on the double-bristle phenotype of the adult Es organ [2,17]. The *musashi* bristle phenotype indicated that loss of *musashi* function results in the frequent appearance of mechanosensory bristles containing an extra tricogen and tormogen at the expense of the neuron and thecogen (Fig. 5C). Therefore, we propose that the *musashi* product is required for the asymmetric division of the SOP. In *musashi* loss-of-function mutants, cells of the IIb lineage are likely to be transformed into those of IIa lineage, implying a requirement for *musashi* during or after the formation of IIb precursors (Fig. 5B). *musashi* encodes an RNA-binding protein, with two tandem RNA recognition motifs (RRMs). Interestingly, *musashi* expression is higher in the cells of IIb lineages than in those of IIa lineages, corresponding to its requirement during asymmetric division (Okabe and Okano, unpublished results). Therefore, *musashi* is likely to be required intrinsically (probably in the IIb precursor or its progeny), and presumably acts by regulating expression of downstream genes posttranscriptionally, although a lineage-specific mosaic analysis is still required to reveal its cell autonomy. Our preliminary results have indicated that the RNA-binding activity of the Musashi protein is required for the asymmetric division of neural precursor cells. We are investigating the function of *musashi* further by characterizing the phenotype of *musashi* misexpression, determining the localization of Musashi protein during SOP cell division, through mosaic analysis, and by identifying the downstream target genes. To identify the target genes, we are currently applying two strategies including an in vitro selection method and the examination of in vivo genetic interactions between *musashi* and mutants affecting the development of the Es organ.

Besides Musashi, there is another class of *Drosophila* neural RNA-binding proteins that includes Elav [18]. We propose that Musashi and Elav belong to a novel gene family of neural RNA-binding proteins that are required for neural development and which act by regulating the expression of downstream genes at the posttranscriptional level. *Drosophila* Elav and Rbp9, and the mammalian Hu proteins (HuB, HuC, HuD) [19], are members of the Elav subfamily (based on homology with *Drosophila* Elav [1].

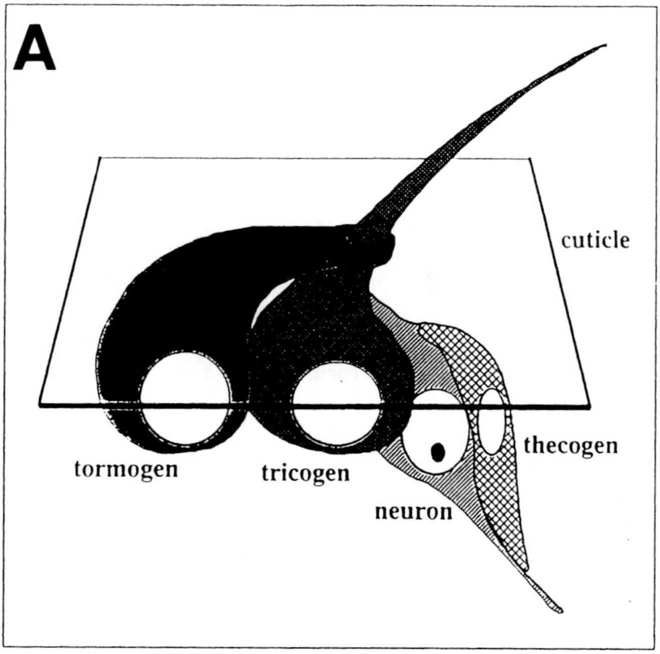

FIG. 5A. Structure of the adult external sensory (Es) organ and altered cell lineage in the *musashi* mutant. **A** Structure of the *Drosophila* adult Es organ: *neuron*, *thecogen* (glial cell), *tricogen* (shaft cell), and *tormogan* (socket cell). (From [2] with permission)

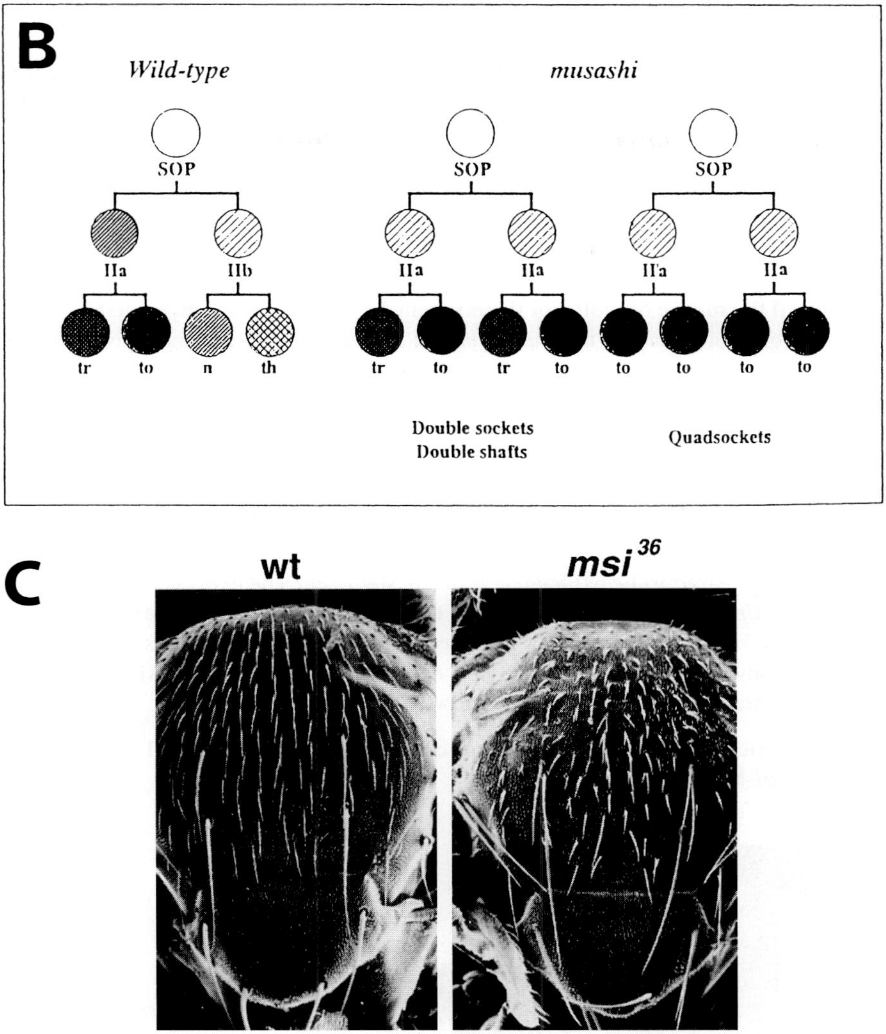

FIG. 5B,C. *Continued.* **B** Cell lineage of the Es organ in the wild type and *msi* mutant. The cell lineage of the Es organ in the wild type is shown on the *left*. Asymmetric cell division of the singled-out sensory organ precursor cell (SOP) gives rise to a IIb and IIa precursor. Subsequently, the IIb precursor gives rise to a neuron (*n*) and thecogen (*th*), and the IIa precursor gives rise to a tricogen (*tr*) and tormogen (*to*). The *musashi* loss-of-function mutant shows an increased number of tricogens and tormogens within each Es organ. The phenotype is variable among sensilla, e.g., double sockets, double shafts, quadruple sockets, etc. [8]. The following cell lineage can be proposed for the *musashi* bristle phenotype. In the sensillum showing the double-sockets:double-shafts phenotype, the IIb precursor is transformed into IIa precursor, yielding two IIa precursors and subsequently two tormogens and two tricogens. In the sensillum showing the quadruple-socket phenotype, the IIb precursor is transformed into IIa precursors, yielding two IIa precursors and subsequently four tormogens and no tricogens. (From [2], with permission). **C** The bristle phenotype of the *musashi* loss-of-function mutant (from [17], with permission). Shown are scanning electron micrographs of the adult nota of wild type (*left*) and *msi*^{1} (*right*) flies

All these proteins have three sets of highly conserved RNA recognition Motifs (RRMs) [20]. Interestingly, Hu proteins are all expressed in postmitotic neurons, not in the mitotic neural precursor cells within various parts of the developing nervous system [21,22].

Mammalian Musashi

To elucidate the function of the Musashi subfamily in mammalian nervous systems, we isolated the mouse homolog of the *msi* gene by low-stringency hybridization. We isolated a clone and named it *mouse-musashi-1* (*m-msi-1*) [23]. In the RRMs, 60% of the amino acid sequence is identical between *Drosophila* and mouse. Even outside the motif, 30% of the amino acids are identical. As is the case with the *Drosophila msi* gene, *m-msi-1* is highly expressed in the developing mouse nervous system (Fig. 6). In the embryonic CNS, m-Msi-1 expression is highly enriched in mitotically active neural precursor cells, in the same way as is Nestin [24]. Based on the results of a single-cell culture experiment, when a CNS stem cell under goes asymmetric cell division to produce a newly formed stem cell and a differentiating cell, m-Msi-1 expression appeared to be associated with the stem cell. The expression of m-Msi-1 showed an almost complementary pattern to that of another mammalian RNA-binding protein, Hu [23,25], which is localized in differentiated neurons in the CNS. Detailed analysis has indicated that the expression patterns of m-Msi-1 and Hu change in a stage-specific manner as follows:

1. The mitotically active neural precursor cells are positive for m-Msi-1 and negative for Hu: m-Msi-1 (+), Hu (−).

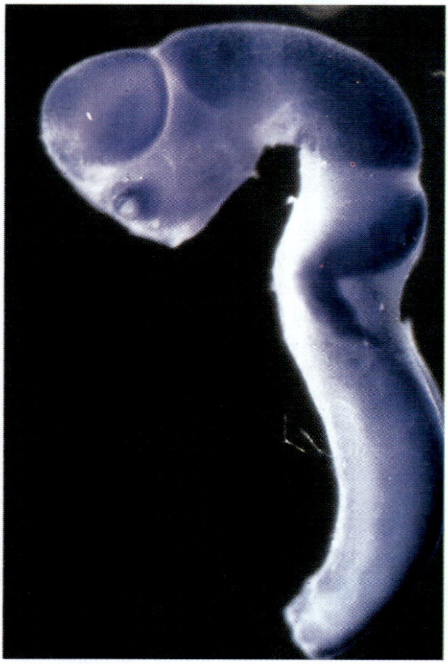

Fig. 6. Expression of the *m-msi-1* gene in the developing neural tube in the E10 mouse CNS was analyzed by whole-mount in situ hybridization

2. When neurogenesis begins, cells become positive for both m-Msi-1 and Hu: m-Msi-1 (+), Hu (+) (Fig. 7).
3. Differentiating neurons are positive for Hu and negative for m-Msi-1: m-Msi-1 (−), Hu (+).

The presence of cells transiently expressing both m-Msi-1 and Hu indicates that m-Msi-1 may repress the expression of Hu at the posttranscriptional level.

Interestingly, the expression pattern of m-Msi-1 is associated with the asymmetric cell divisions of neural precursors or else with cell lineage restrictions at various time points (Fig. 8). First, m-Msi-1 expression is associated with stem cells during their asymmetric cell divisions [23], which occur during embryonic neurogenesis [16] (Fig. 8A). Second, it also shows a restricted pattern in astrocyte and O-2A-oligodendrocyte lineages, which are potentially generated from bipotential glial precursor cells present in the postnatal subventricular zone (SVZ) [26] (Fig. 8B). It remains unclear, however, whether asymmetric cell division of a bipotential glial precursor cell is required for the generation of astrocyte and O-2A-oligodendrocyte lineages, in a way similar to the asymmetric division of the *Drosophila* ES organ SOP cell to produce precursors IIa and IIb (see Fig. 5B). We are currently examining whether m-Msi-1 is involved in these asymmetric divisions of neural precursor cells and in cell lineage restrictions by generating *m-msi-1* knockout mice and characterizing them using a single neuroepithelial cell culture experiment [23].

All members of the Musashi subfamily, including *Drosophila* Musashi [17], *Xenopus* Nrp-1 [27], m-Msi-1 [23], and human Musashi [28] have two tandem sets of RRMs, which are highly conserved within the subfamily. They share similar expression patterns during development, starting with neural precursor cells [17,22,25,29]. Members of the Msi subfamily are distinct from those of the Elav subfamily, based on

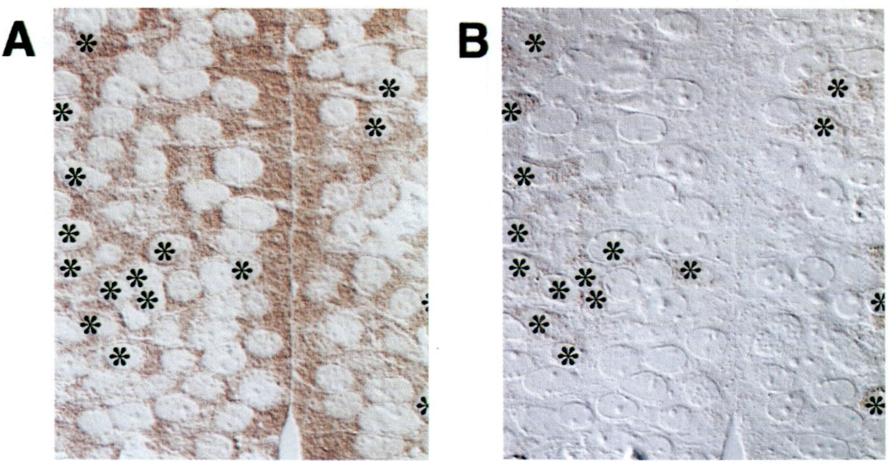

FIG. 7A,B. The presence of m-Msi-1 and Hu double-positive cells in the embryonic spinal cord. Spinal cords from E10 mouse embryos were embedded in resin and cut into 1000-nm sections. **A** and **B** show serial sections. **A** 14B8 (anti m-Musashi-1; Kaneko and Okano, unpublished results). **B** 16A11 (anti-Hu) [31]. *Asterisks* shows m-Msi-1 and Hu double-positive cells

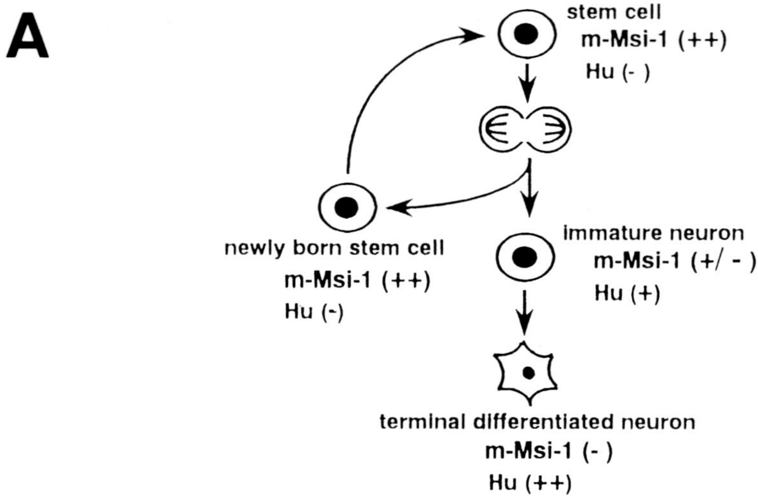
A
stem cell
m-Msi-1 (++)
Hu (-)
newly born stem cell
m-Msi-1 (++)
Hu (-)
immature neuron
m-Msi-1 (+/ -)
Hu (+)
terminal differentiated neuron
m-Msi-1 (-)
Hu (++)

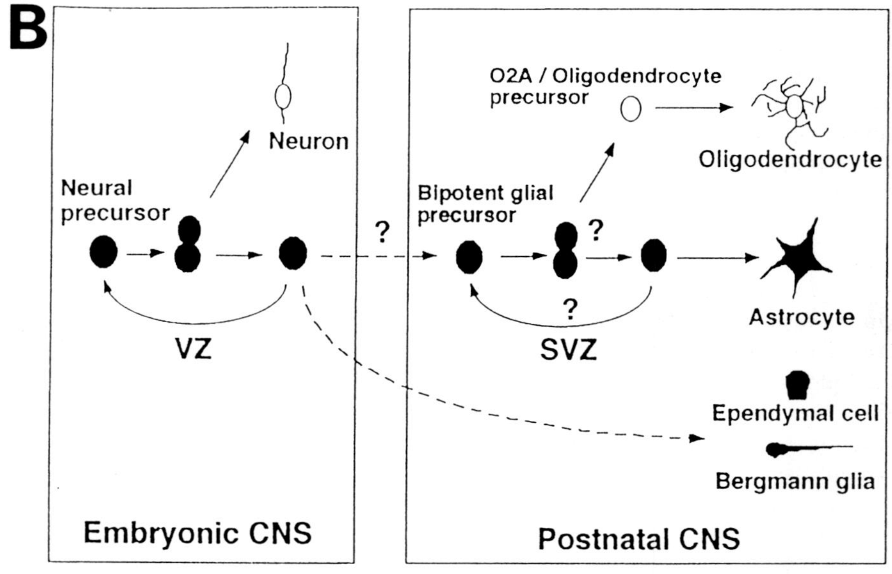
B
Neuron
Neural precursor
?
VZ
Embryonic CNS
O2A / Oligodendrocyte precursor
Oligodendrocyte
Bipotent glial precursor
?
?
Astrocyte
SVZ
Ependymal cell
Bergmann glia
Postnatal CNS

Fig. 8A,B. Summary of the m-Msi-1 expression pattern during CNS development. **A** The expression of m-Msi-1 and Hu and its relation to the putative asymmetric division of CNS stem cells during neurogenesis (taken from [23], with permission). During mammalian neural development, asymmetric cell division has been shown to occur at least during neurogenesis [16]. The expression patterns of m-Msi-1 and Hu were likely to be associated with self-renewing multipotent (or neuronal) precursor cells and postmitotic neurons, respectively. Because m-Msi-1 and Hu are transiently coexpressed within the same cell, the complementary expression of m-Msi-1 and Hu during neurogenesis may result from either the asymmetric sorting of protein molecules or their sequential expression in neurogenesis. As Hu expression is not observed in glial cells [19], the differential expression of m-Msi-1 and Hu is likely to be of biological significance only in neurogenesis. **B** Schematic representation of cells expressing m-Msi-1 during embryonic and postnatal CNS development (taken from [25], with permission). Solid patterns represent m-Msi-1-positive cells. In the embryonic CNS, m-Msi-1 is expressed in proliferating neural precursor cells in the ventricular zone (VZ). In the perinatal stage, the VZ shrinks and a second proliferative subventricular zone (SVZ) appears. Postnatally, m-Msi-1 expression is seen in proliferating glial precursor cells in the SVZ and in cells of astrocyte lineage, including ependymal cells and Bergmann glia. At present, it is not known whether neural precursors in the embryonic VZ generate glial precursors in the postnatal SVZ, or whether bipotential glial precursor cells in the postnatal SVZ generate the astrocyte and O-2A-oligodendrocyte lineage through an "asymmetric cell division." It will be interesting to determine whether m-Msi-1 is involved in this putative asymmetric cell divison

their expression patterns, number of RRMs, and sequence similarities. Such differences may be important in examining the functions of both subfamilies. It will be interesting to determine whether the sequential and cooperative actions of both subfamilies are indispensable for all stages of neural development from multipotent stem cells to mature neurons.

Acknowledgments. We thank Drs. Katsuhiko Mikoshiba (University of Tokyo, Japan), Craig Montell (Johns Hopkins University School of Medicine, USA), Noriko Osumi (National Institute of Neuroscience, Japan), Makoto Nakamura (National Institute of Basic Biology, Japan), Masaharu Ogawa (Kochi Medical School, Japan), Shigeo Hayashi (National Institute of Genetics, Japan), Tetsuo Noda (Tohoku University, Japan), and members of the Okano Laboratory (Osaka University Medical School, Japan) for their helpful cooperation and collaborations. We are also grateful to Drs. Susan McConnell (Stanford University, USA), and Yuh Nung Jan (University of California at San Francisco, USA) for invaluable discussions. We also thank Dr. Allan Spradling and members of his laboratory (Carnegie Institute, USA) for providing the P-element insertion lines, and the members of the Montell Laboratory for cooperation in the screening of the P-element insertion lines. This work was supported by grants to H.O. from the Ministry of Education, Science, Sports and Culture of Japan.

References

1. Okano H (1995) Two major mechanisms regulating cell-fate decisions in the developing nervous system. Dev Growth Differ 37:619–629
2. Okabe M, Sawamoto K, Imai T, et al (1997) Instinsic and extrinsic determinants regulating cell fact decision in developing nervous system. Dev Neurosci 19:9–16

3. Okano H, Okabe M, Taguchi A, et al (1997) Evolutionarily conserved mechanisms regulating neural development: lessons from the development of *Drosophila* peripheral nervous systems. Hum Cell 10:139–149
4. Sawamoto K, Okano H (1996) The cell-cell interactions during neural development: multiple types of lateral inhibitions regulating pattern formations in the developing *Drosophila* compound eye. Neurosci Res 26:205–214
5. Freeman M (1996) Reiterative use of the EGF receptor triggers differentiation of all cell types in the *Drosophila* eye. Cell 87:651–660
6. Okano H, Hayashi S, Tanimura T, et al (1992) Regulation of *Drosophila* neural development by a putative secreted protein. Differentiation 52:1–11
7. Freeman M, Klambt C, Goodman C, et al (1992) The *argos* gene encodes a diffusible factor that regulates cell fate decisions in the *Drosophila* eye. Cell 69:963–975
8. Kretzschmar D, Brunner A, Wiersdorff V, et al (1992) *giant lens*, a gene involved in cell determination and axonal guidance in the visual system of *Drosophila melanogaster*. EMBO J 11:2531–2539
9. Sawamoto K, Okano H, Kabayakawa Y, et al (1994) The function of *argos* in regulating cell fate decisions during the *Drosophila* eye and wing vein development. Dev Biol 164:267–276
10. Sawamoto K, Taguchi A, Jin M, et al (1998) Argos induces programmed cell death in the developing *Drosophila* eye by inhibition of the Ras pathway. Cell Death Differ 5:262–270
11. Hay BA, Wolff T, Rubin GM (1994) Expression of baculovirus P35 prevents cell death in *Drosophila*. Development (Camb) 120:2121–2129
12. Okabe M, Sawamoto K, Okano H (1996) The function of *Drosophila argos* gene product in the development of embryonic chordotonal organs. Dev Biol 175:37–49
13. Sawamoto K, Okabe M, Tanimura T, et al (1996) Argos, a diffusible inhibitor of signal transduction in the Ras/MAP kinase cascade for *Drosophila* eye and wing vein development. Dev Biol 178:13–22
14. Scweitzer R, Howes R, Smith R, et al (1995) Inhibition of *Drosophila* EGF receptor activation by the secreted protein Argos. Nature (Lond) 376:699–702
15. Flickinger TW, Maihle NJ, Kung HJ (1992) An alternatively processed mRNA from the avian *c-erbB* gene encodes a soluble, truncated form of the receptor that can block ligand-dependent transformation. Mol Cell Biol 12:883–893
16. Chenn A, McConell SK (1995) Cleavage orientation and the asymmetric inheritance of Notch 1 immunoreactivity in mammalian neurogenesis. Cell 82:631–641
17. Nakamura M, Okano H, Blendy J, et al (1994) MUSASHI, a neural RNA-binding protein required for *Drosophila* adult external sensory organ development. Neuron 13:67–81
18. Yao KM, Samson ML, Reeves R, White K (1993) Gene *elav* of *Drosophila melanogaster*: a prototype for neuronal-specific RNA binding protein gene family that is conserved in flies and humans. J Neurobiol 24:723–739
19. Szabo A, Dalmau J, Manley G, et al (1991) HuD, a paraneoplastic encephalomyelitis antigen, contains RNA-binding domains and is homologous to elav and sex-lethal. Cell 67:325–333
20. Okano HJ, Darnell R (1997) A hierarchy of Hu RNA binding proteins in developing and adult neurons. J Neurosci 17:3024–3037
21. Barami K, Iversen K, Furneaux H, Goldman S (1995) Early expression of Hu proteins by newly generated neurons of the adult avian forebrain. J Neurobiol 28:82–101
22. Wakamatsu Y, Weston JA (1997) Sequential expression and role of Hu RNA-binding proteins during neurogenesis. Development (Camb) 124:3449–3460
23. Sakakibara S, Imai T, Aruga J, et al (1996) Mouse-Musashi-1, a neural RNA binding protein highly enriched in the mammalian CNS stem cell. Dev Biol 176:230–242
24. Lendahl U, Zimmerman LB, McKay R (1990) CNS stem cells express a new class of intermediate filament protein. Cell 60:585–595

25. Sakakibara S, Okano H (1997) Expression of neural RNA-binding proteins in the postnatal CNS: implication of their roles in neural and glial cells development. J Neurosci 17:8300–8312
26. Levison SW, Goldman JE (1993) Both oligodendrocytes and astrocytes develop from progenitors in the subventricular zone of postnatal rat forebrain. Neuron 10:3494–3504
27. Richter K, Good PJ, Dawid IB (1990) A developmentally regulated, nervous system-specific gene in *Xenopus* encodes a putative RNA-binding protein. New Biol 2:556–565
28. Good P, Yoda A, Sakakibara S, et al (1998) Human Musashi homolog1 (MSI1) gene encoding the homologue of Musashi/Nrp-1, a neural RNA-binding protein putatively expressed in CNS stem cells and neural progenitor cells. Genomics (in press)
29. Pincus DW, Keyoung H, Restelli C, et al (1998) FGF2/BDNF-responsive neuronal progenitor cells in the adult human subependyma. Ann Neurol 43:576–585
30. Mlodzik M, Hiromi Y (1992) The enhancer trap method in *Drosophila*: its application to neurobiology. In: Cann PM (ed) Gene expression in neural tissues. Methods in neuroscience, vol 9. Academic Press, Orlando, pp 397–414
31. Marusich M, Furneaux H, Henion P, et al (1994) Hu neuronal proteins are expressed in proliferating neurogenic cells. J Neurobiol 25:143–155

Mechanism of Myogenic Differentiation Suppression by Notch and RBP-J

HISANORI KUROOKA, HISAMUNE KATO, YOSHIHITO TANIGUCHI, SHIGERU MINOGUCHI, TAKASHI SAKAI, KUMIKO TAMURA, and TASUKU HONJO

Summary. The intracellular region (RAMIC) of Notch directly interacts with the DNA-binding protein RBP-J and activates transcription. IC devoid of the RAM domain can also activate transcription, although less strongly than RAMIC. Three functional domains were identified in RAMIC; RAM and ankyrin regions for RBP-J binding, and an autonomous transactivation domain in the C-terminus. RAMIC or RBP-J-VP16 fusion protein has strong transactivation activity and markedly suppressed differentiation suppression of myogenic precursor C2C12 cells. IC showed suppression activity comparable to its transactivation activity. These results strongly suggest that suppression of myogenesis by Notch signaling results from transactivation of genes carrying the RBP-J-binding motifs.

Key words. Notch–RBP-J—Ankyrin repeats—Transactivation domain—Myogenic suppression

Introduction

The Notch receptor is involved in cell-fate determination of various lineages including the nerve, muscle, and germ cells in *Drosophila*. Most studies on biological function have been performed using the truncated intracellular forms of Notch, which show constitutively active phenotypes [1]. In general, Notch signaling brings cells into the proliferative phase and blocks cell differentiation. Expression of the mouse Notch 1 intracellular region suppresses myogenesis from myogenic precursor cells and neurogenesis from neuronal precursor cells [2,3].

The intracellular region of Notch (RAMIC) binds to a nuclear DNA-binding protein RBP-J (previously called RBP-Jk), a mammalian homolog of suppressor of hairless (*Su*(*H*)) at the RAM domain located immediately downstream of the transmembrane region, resulting in transactivation of genes carrying RBP-J-binding motifs (YGTGGGAA) in their promoter regions [4,5]. Here we show that transactivation of genes by Notch and RBP-J is responsible for suppression of myogenesis. In addition,

Department of Medical Chemistry, Kyoto University, Faculty of Medicine, Yoshida, Sakyo-ku, Kyoto 606, Japan

RAMIC is shown to contain three functional domains: RBP-J-interacting RAM and ankyrin regions, and the C-terminal transactivation domain.

Materials and Methods

All plasmids except GAL4-IC constructs for transactivation assay were described previously [6]. GAL4-IC-full (a.a. 1848–2531], -N (a.a. 1848–2193], and -C (a.a. 2194–2531] were derived from the yeast expression plasmid containing the full-length IC region [5]. The immunoprecipitation and differentiation suppression assay have been described previously [6]. The luciferase activity was measured 48 h after transfection and normalized according to the β-galactosidase activity. Autonomous transactivation function of IC was analyzed as follows. 0.35 μg of TK-MH 100 × 4 LUC [6] and 0.35 μg of each GAL4 construct were cotransfected into COS7 cells. Comparable expression of each fusion protein was checked by Western blot analysis with a polyclonal antibody against the GAL4 DNA-binding domain (data not shown).

Results and Discussion

IC devoid of the RAM domain from RAMIC (Fig. 1a) was reported to have weak but significant transactivation activity through HES-1 promoter in HeLa cells [7]. We have confirmed it using an Epstein–Barr virus promoter, ERE-TP1, as well as the HES-1 promoter in COS7 cells (Table 1). To examine the involvement of RBP-J in transactivation by IC, the physical interaction between the two was assessed by coimmunoprecipitation assay. IC was coimmunoprecipitated with RBP-J, although the precipitated amount of IC was much lower than that of RAMIC (Fig. 1b; lanes 1, 5). The M1 mutant of IC, which completely lost the transactivation activity (Table 1), was not coimmunoprecipitated with RBP-J (Fig. 1b; lane 3). We also investigated the interaction between IC and RBP-J using the mammalian two-hybrid assay. RBP-J-VP16 [8] together with GAL4-ankyrin augmented transcription in a dose-dependent

TABLE 1. Suppression activities of C2C12 cell differentiation by Notch and RBP-J derivatives

Experimental group	Transfected constructs	Differentiation (%) (nuclei counted)	Relative suppression (%)	Relative transactivation (%)	
				TP-1	HES-1
1	RAMIC	23 (333)	72	100	100
	IC	41 (242)	51	8	53
	IC-M1	83 (275)	0	0	0
2	RAM	60 (479)	6	0	0
3	RBP-J-VP16	9 (225)	90	100	100
	RBP-J	83 (638)	6	0	0

C2C12 cell monolayers were transfected with each indicated construct. The images of cells were incorporated and analyzed with Photoshop (Adobe Systems). Relative suppression activity was calculated by the ratio to the level of the nontransfected cells.
From [6], with permission.

manner, whereas GAL4 alone or the M1 mutation in the ankyrin repeats abolished the augmentation of transcription [6], in agreement with the coimmunoprecipitation assay. That the Notch intracellular region can bind to RBP-J at the ankyrin repeats as well as the RAM domain suggests that the transactivation activity of IC also involves RBP-J. To ensure it, transactivation activity of IC was tested using an RBP-J null mutant cell line. Both RAMIC and IC showed transactivation activity only in the presence of exogenous RBP-J [6]. Furthermore, transactivation activity of IC was blocked by competition with a mutant of RBP-J (R218H) that cannot bind to DNA [6]. These results convincingly demonstrate that the transactivation activity of IC is dependent on RBP-J.

Next, we examined an autonomous transactivation function of IC using heterologous GAL4 fusion proteins (Fig. 1a). GAL4-IC (full-length) activated transcription more than 60 fold as compared with GAL4 (data not shown). The N-terminal half of IC (GAL4-IC-N) displayed no activity, and the transactivation domain was found to be localized in the C-terminal half (GAL4-IC-C). Taken together, we concluded that RAMIC is composed of three separate domains, i.e., RAM and ankyrin regions for RBP-J binding, and a C-terminal autonomous transactivation domain (Fig. 1c).

We measured the relative myogenic suppression activities of RAMIC and its derivatives (Table 1] and compared them with their transactivation activities. RBP-J-VP16 and RAMIC markedly suppressed differentiation of myogenic precursor C2C12 cells. IC had an activity of myogenic suppression comparable to its transactivation activity. RBP-J and transactivation-incompetent derivatives of Notch 1 such as IC-M1 and the RAM domain alone had negligible levels of the differentiation suppression activity. Thus, transactivation activities of the Notch 1 intracellular region and its derivatives

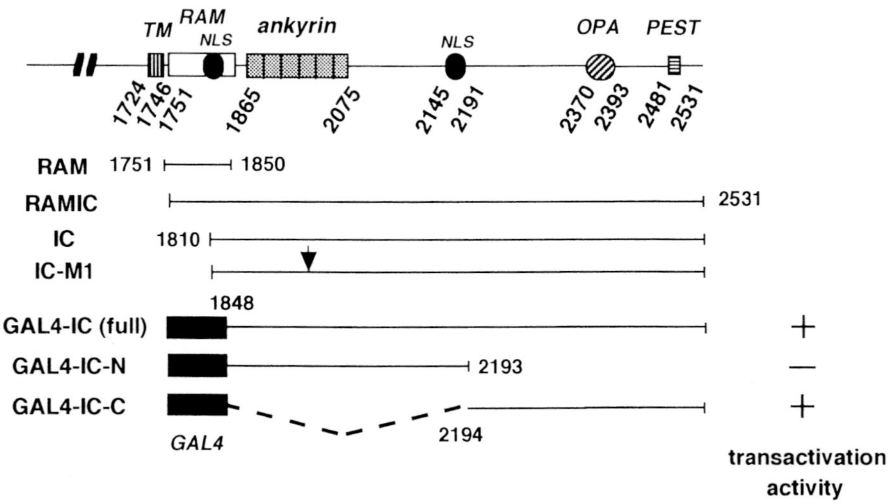

a

FIG. 1a. a Schematic representation of mouse Notch 1 intracellular region. The *top line* shows the amino acid positions with structural motifs. Truncated forms used in this study are shown *below*, and the mutation site is indicated by a *vertical arrow*. GAL4 fusion constructs are shown in the *lower part*

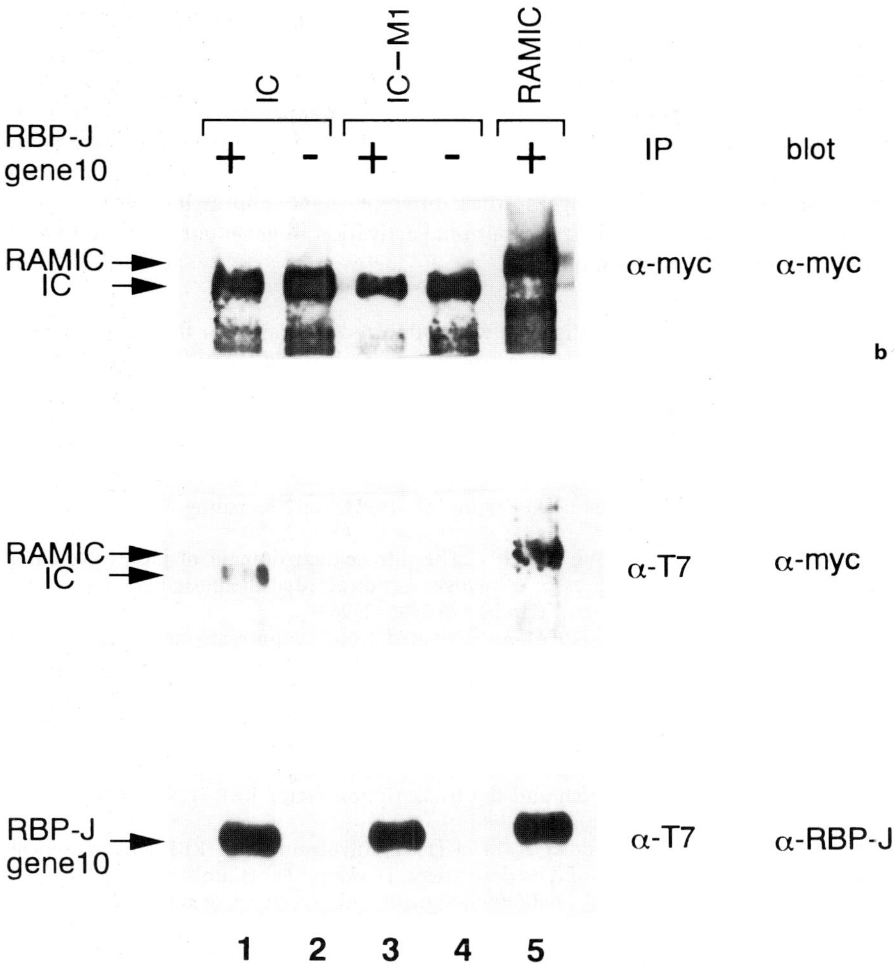

b

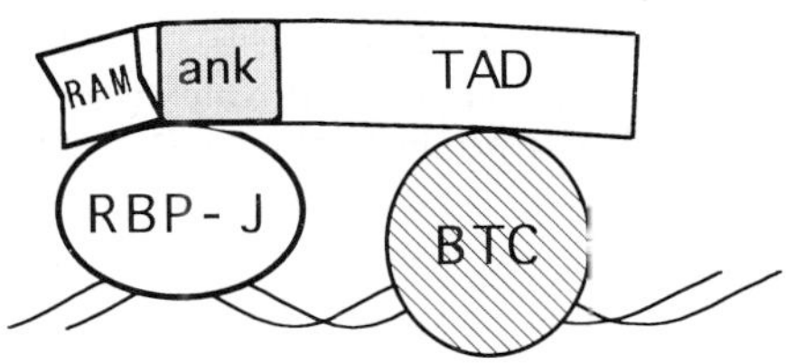

c

FIG. 1b,c. **b** Coimmunoprecipitation between IC and RBP-J. c-*myc* epitope-tagged IC, IC-M1, or RAMIC in pEF-BOSneo and gene 10 epitope-tagged RBP-J in pMKIT [6] were cotransfected into COS7 cells. The gene 10 epitope was detected by anti-T7 monoclonal antibody. The *upper and lower parts* confirm the same amounts of expressed proteins. From [6], with permission. **c** Transactivation by mouse Notch 1 intracellular region and RBP-J. RAMIC interacts with RBP-J at the two domains, the RAM domain and ankyrin repeats. The C-terminal transactivation domain (*TAD*) may interact with basal transcriptional complex (*BTC*) basically parallel their differentiation suppression activities. Furthermore, differentiation suppression of C2C12 cells by RBP-J-VP16 indicates that genes regulated by RBP-J/Notch are responsible for regulation of myogenic differentiation. That IC can not activate transcription in HeLa cells suggested the existence of RBP-J-independent pathway of Notch signaling [9]. However, the lack of transactivation activity by IC could result from inefficiency of the read-out construct used or the absence of cofactors. Our results strongly suggest that differentiation suppression of myogenic precusor cells is mediated by transcriptional activation of genes carrying the RBP-J-binding motifs in their promoters.

Acknowledgments. This investigation was supported by grants for the COE program from Ministry of Education, Science, Sports and Culture of Japan, Imori Industries Inc.

References

1. Artavanis-Tsakonas S, Matsuno K, Fortini ME (1995) Notch signaling. Science 268:225–232
2. Kopan R, Nye JS, Weintraub H (1994) The intracellular domain of mouse Notch: a constitutively activated repressor of myogenesis directed at the basic helix-loop-helix region of MyoD. Development (Camb) 120:2385–2396
3. Nye JS, Kopan R, Axel R (1994) An activated Notch suppresses neurogenesis and myogenesis but not gliogenesis in mammalian cells. Development (Camb) 120:2421–2430
4. Honjo T (1996] The shortest path from the surface to the nucleus: RBP-Jk/Su(H) transcription factor. Genes Cells 1:1–9
5. Tamura K, Taniguchi Y, Minoguchi S, et al (1995] Physical interaction between a novel domain of the receptor Notch and the transcription factor RBP-Jk/Su(H). Curr Biol 5:1416–1423
6. Kato H, Taniguchi Y, Kurooka H, et al (1997) Involvement of RBP-J in biological functions of mouse Notch1 and its derivatives. Development (Camb) 124:4133–4141
7. Jarriault S, Brou C, Logeat F, et al (1995) Signalling downstream of activated mammalian Notch. Nature (Lond) 377:355–358
8. Waltzer L, Bourillot PY, Sergeant A, et al (1995] RBP-J kappa repression activity is mediated by a co-repressor and antagonized by the Epstein-Barr virus transcription factor EBNA2. Nucleic Acids Res 23:4939–4945
9. Shawber C, Nofziger D, Hsieh JJD, et al (1996) Notch signaling inhibits muscle cell differentiation through a CBF1-independent pathway. Development (Camb) 122:3765–3773

Role of bHLH Proteins in Neuronal Cell-Fate Specification

Yuh-Nung Jan

In *Drosophila*, neural development is initiated by the action of proneural genes including genes of the *achaete-scute complex (AS-C)* and *atonal*, which encode bHLH proteins. There are four types of sensory organs in the *Drosophila* PNS, *AS-C* controls the formation of external sensory organs and most of the multiple dendritic neurons, whereas *atonal* controls the formation of chordotonal organs and photoreceptors. Between them, *atonal* and *AS-C* can account for the origin of almost the entire PNS I discuss the following issues:

1. As different proneural genes promote the formation of different types of neurons, an interesting question is where in the proneural genes does the neuronal type information reside?
2. The position at which the proneural gene is expressed determines the location of the neurons. We took *atonal* as a model system and analyzed the control of *atonal* expression pattern.

I also discuss the preliminary characterizations and the potential development roles of newly discovered bHLH proteins in the nervous system of *Drosophila* and mouse.

Howard Hughes Medical Institute and Department of Physiology & Biochemistry, University of California, San Francisco, CA, USA

Development of the Neural Crest

Marianne Bronner-Fraser

We are analyzing the cellular and molecular events underlying the formation, cell lineage decisions, and migration of neural crest cells. The neural crest is a population of migratory cells, arising from the ectoderm of vertebrate embryos, which gives rise to a diverse range of cell types. Following neurulation, neural crest cells emerge from the neural tube and undergo extensive movements along pathways that often are characteristic of their axial level of origin. After migration, these cells assume one of a number of possible fates, ranging from neurons and glia of the peripheral nervous system to pigment cells and cells of the facial skeleton. It has been classically assumed that the neural crest is a segregated population in the early ectoderm, lying between the neural plate and presumptive epidermis. However, our recent studies on avian embryos show that individual precursor cells within the "neural folds" can form neural tube (central nervous system), neural crest (peripheral nervous system), and epidermal derivatives. This led us to explore the interactions that impart the potential to form the neural crest. Interestingly, we found that neural crest cells are generated when epidermis and neural plate are juxtaposed, a classical type of embryonic induction. Our current goal is to characterize the inductive interactions that lead to formation of the neural crest and to examine how regional differences arise in the neural crest populations along the rostrodaudal axis.

Division of Biology, California Institute of Technology, Pasadena, CA 91125, USA

Asymmetric Divisions and Mammalian Neurogenesis

S. K. McConnell, A. Chenn, B. Chang, and A. Zhang

The cerebral cortex is patterned into layers, each of which contains neurons with similar morphologies and axonal connections. During early stages of development, neural progenitors in the cerebral cortex normally produce neurons of different layers during successive cell divisions. To address the question of how neurons are actually generated from progenitor cells, we followed dividing cells in living slices of ferret cortex using time-lapse confocal microscopy (Chenn and McConnell 1995; Cell 82:631–641). Our results provide evidence for the existence of asymmetric cell divisions within the ventricular zone, and suggest that the plane of cleavage of a dividing progenitor cell may correlate with the subsequent decision of its daughters to reenter the cell cycle or differentiate into a postmitotic neuron.

Notch 1 immunoreactivity is asymmetrically localized to the basal pole of mitotic progenitors, and appears to be inherited differentially in asymmetric divisions. We next examined the subcellular distribution of various proteins that colocalize with adherens junctions in epithelial cells. Published reports indicate that N-cadherin and the zonula occludins protein ZO-1 are highly enriched at the apical surface of the ventricular zone (Aaku-Saraste et al. 1996; Dev Biol 180:664–679). We found that several other proteins are localized apically, including β-catenin, γ-catenin, paxillin, focal adhesion kinase (FAK), and pp 120. Finally, phosphotyrosine immunoreactivity is concentrated apically, consistent with the suggestion that signaling molecules which localize to adherens junctions may be inherited differentially during asymmetric divisions.

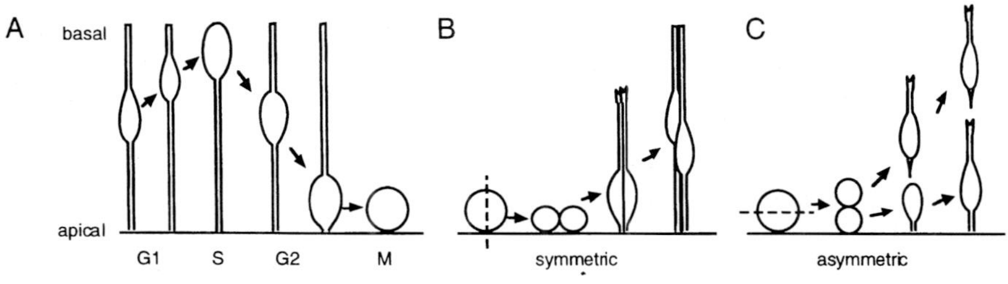

Department of Biological Sciences, Stanford University, Stanford, CA 94305, USA

Selected Publications

McConnell SK, Kaznowski CE (1991) Cell cycle dependence of laminar determination in developing neocortex. Science 254:282–285

O'Rourke NA, Dailey ME, Smith SJ, McConnell SK (1992) Diverse migratory pathways in the developing cerebral cortex. Science 258:299–302

Frantz GD, Weimann JM, Levin ME, McConnell SK (1994) Otx1 and Otx2 define layers and regions in developing cerebral cortex and cerebellum. J Neurosci 14:5725–5740

O'Rourke NA, Sullivan DP, Kaznowski CE, Jacobs AA, McConnell SK (1995) Tangential migration of neurons in the developing cerebral cortex. Development (Camb) 121:2165–2166

Chenn A, McConnell SK (1995) Cleavage orientation and the asymmetric inheritance of Notch 1 immunoreactivity in mammalian neurogenesis. Cell 82:631–641

Frantz GD, McConnell SK (1996) Restriction of late cerebral cortical progenitors to an upper-layer fate. Neuron 17:55–61

Bohner AP, Akers RM, McConnell SK (1997) Induction of deep layer cortical neurons in vitro. Development (Camb) 124:915–923

O'Rourke NA, Chenn A, McConnell SK (1997) Postmitotic neurons migrate tangentially in the cortical ventricular zone. Development (Camb) 124:997–1005

Caric D, Gooday D, Hill RE, McConnell SK, Price DJ (1997) Determination of the migratory capacity of embryonic cortical cells lacking the transcription factor Pax-6. Development (Camb) 124:5087–5096

Chenn A, Zhang YA, Chang B, McConnell SK (1998) Intrinsic polarity of mammalian neuroepithelial cells. Molec Cell Neurosci 11:183–193

Regulated Parameters of Cell Proliferation in Neocortical Histogenesis

T. Takahashi[1,2], S. Miyama[1], and V.S. Caviness, Jr.[2]

Summary. Production of neocortical neurons, neuronogenesis, occurs in a pseudostratified ventricular epithelium (PVE) and is regulated with great precision. Over the course of neuronogenesis in mice, each founder cell of the PVE gives rise to approximately 150 neurons in the course of 11 cell cycles. Although the onset, progression, and termination of neuronogenesis advance progressively with rostrolateral progression across the cerebral hemisphere, the set of values for the length of the G_1 phase (T_{G1}) and for Q (the fraction of cells that quits the cycle) for a specific cell cycle is invariable. Following neuronogenesis, the postmitotic neurons migrate to the cortex where they occupy laminar positions that are closely correlated with the cell cycle of origin. Despite the graded regional difference across the PVE in the progression of neuronogenesis, there is a further tight correlation between the cell cycle of origin and the layer destination of the neurons and the values of T_{G1} and Q. We suggest that specification of the narrow range of neuronal classes peculiar to each neocortical layer is coordinately regulated with those mechanisms regulatory to T_{G1} and Q.

Key words. Cell cycle—Cerebral histogenesis—Neuronal migration—Mouse—Q fraction

We review here a characterization of the overall series of events for cerebral histogenesis, built up from experiments in the developing murine cerebral wall. Although the experiments have been conducted in the mouse, we propose that the characterization of this process would be equally applicable to neocortical development in all species, if suitably calibrated across species.

The developmental sequence for cerebral histogenesis proceeds from neuronogenesis (the production of neocortical neurons), which occurs in a pseudostratified ventricular epithelium (PVE) at the margin of the ventricular system of the embryonic brain (Fig. 1). The full set of neurons in any sector of the neocortex are formed in an

[1] Department of Pediatrics, School of Medicine, Keio University, 35 Shinanomachi, Shinjuku-ku, Tokyo 160-8582, Japan
[2] Department of Neurology, Massachusetts General Hospital, Harvard Medical School, Fruit St., Boston, MA 02114, USA

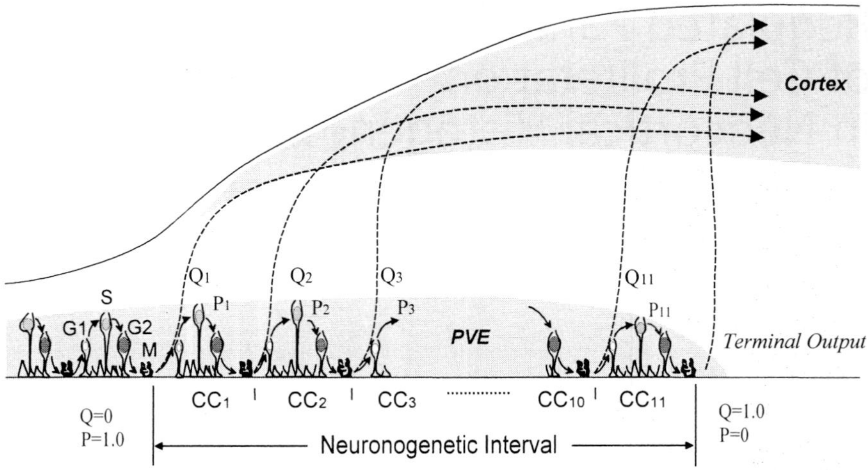

FIG. 1. Schematic representation of neocortical neuronogenesis and neuronal migration. The cells of the pseudostratified ventricular epithelium (PVE) undergo interkinetic nuclear migration (*curved arrows*) in the course of the cell cycle. Postmitotic cells either exit the cell cycle (Q fraction, Q_1–Q_{11}) or reenter S phase (P fraction, P_1–P_{11}). Before the onset of neuronogenesis, the P fraction is 1.0 and the Q fraction is 0. Q becomes greater than 0 with the onset of neuronogenesis ($Q_1 > 0$) and eventually reaches 1.0 at the end of the neuronogenetic interval, which corresponds to 11 integer cell cycles (CC_1–CC_{11}) in the mouse. P, reciprocally, follows a path from 1.0 to 0. At the completion of CC_{11} all postmitotic cells exit the PVE as Q (*terminal output*). Postmitotic neurons migrate across the intervening cerebral wall (*broken arrows*) to the developing cortex. Laminar positions in the cortex are closely correlated with the cell cycle of origin

inside-to-outside (i.e., layer 6 to 2) sequence with respect to the time of origin. The onset, progression, and termination of neuronogenesis advance progressively with rostrolateral progression across the PVE (Fig. 2). We have shown in the mouse that, independent of location in the PVE, the neuronogenetic process is divisible into 11 cell cycles occurring over a 6-day period [1,2] (Fig. 1).

Regulatory mechanisms governing the proliferative process in the course of the 11 cell cycles of mouse neocortical neuronogenesis involve changes both in the "time domain" and in the "output domain." With respect to regulation in the time domain, the duration of the G_1 phase of the cell cycle (T_{G1}), but no other proliferative parameters, accounts for a progressive elongation of the overall cell-cycle length in the course of the neuronogenetic interval [1,2] (Fig. 3A). With respect to regulation in the output domain, the fraction of cells that quit the cell cycle (Q fraction) advances with closely regulated precision with each successive cycle (Fig. 3B). It has been estimated from the fractional increase in Q after each cell cycle (Fig. 3B) that each founder cell will on average give rise to approximately 150 neurons in the course of 11 cell cycles during the interval of neuronogenesis [3]. Thus, only T_{G1} and Q fraction are regulated parameters of neuronogenesis and, at any given moment during neuronogenesis, the specific cell cycle of the 11-cycle series is operating with its cycle-specific values of T_{G1} and Q.

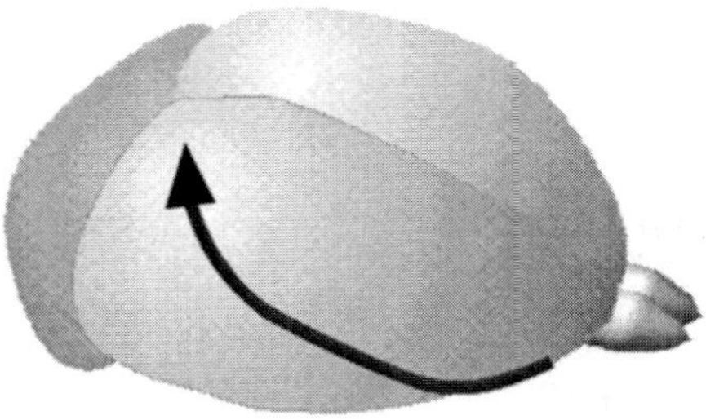

Fig. 2. The propagation of neocortical neuronogenesis. The process of neuronogenesis is initiated rostrolaterally in the PVE and then propagates along the rostral to caudal and lateral to medial axes of the epithelium. The rostrolateral to caudomedial resultant of the two gradient coordinates is represented as a *curved arrow* on the surface

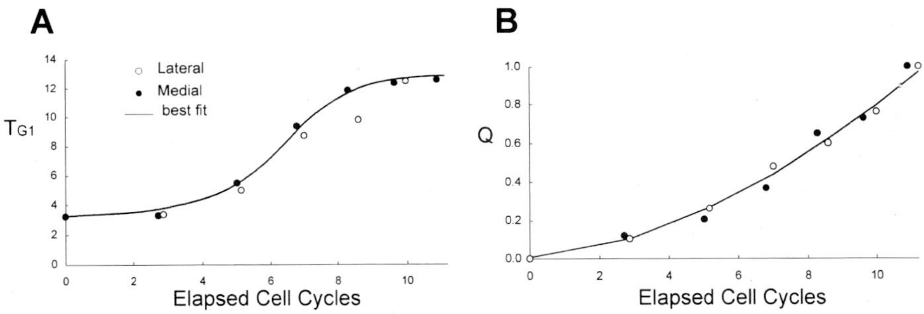

Fig. 3A,B. The progression in the length of the G_1 phase (T_{G1}) and Q fraction. With each successive cycle there is an advance in T_{G1} (**A**) and the Q fraction (**B**). The length of the overall cell cycle is approximately doubled during the interval of neuronogenesis, and this increase is essentially the result of the increase in T_{G1} [1,2]. Note a close correlation between cell-cycle number (abscissa) and cell-cycle parameters of neuronogenesis (i.e., T_{G1} and Q on the ordinate) despite the graded regional difference in the progression of neuronogenesis (medial region, *closed circles*, versus lateral region, *open circles*)

Neuronogenesis is followed by neuronal migration (Fig. 1). Young postmitotic neurons migrate from the PVE across the full width of the developing cerebral wall to gain their positions in the emerging cortex at the cerebral surface. Their laminar positions in the cortex are closely correlated with the cell cycle of origin [4]: the neurons produced from earlier cell cycles (cycles 1 through 7) are destined for layer 6 while progressively later produced cells are destined in sequence for layer 5 (cycle 8), layer 4 (cycle 9), and then layers 3 and 2 (cycles 10 and later). This second stage of development, like neuronogenesis, observes a rostrolateral-to-caudomedial gradient across the cerebral hemisphere (Fig. 2). Despite this graded regional difference in the progression of neuronogenesis, there is a tight correlation between the layer destination of the neurons and the cell cycle of origin [4].

The close correlation between proliferative parameters (i.e., T_{G1} and Q) of a given cell cycle of neuronogenesis and the specific set of neurons arising from the cycle implies that specification of the narrow range of projection neuronal classes peculiar to each neocortical layer is coordinately regulated with mechanisms regulatory to cell-cycle parameters.

References

1. Takahashi T, Nowakowski RS, Caviness VS Jr (1995) The cell cycle of the pseudostratified ventricular epithelium of the murine cerebral wall. J Neurosci 15:6046–6057
2. Miyama S, Takahashi T, Nowakowski RS, et al (1997) A gradient in the duration of the G_1 phase in the murine neocortical proliferative epithelium. Cereb Cortex 7:678–689
3. Takahashi T, Nowakowski RS, Caviness VS Jr (1996) The leaving or Q fraction of the murine cerebral proliferative epithelium: a general computational model of neocortical neuronogenesis. J Neurosci 16:6183–6196
4. Takahashi T, et al (1996) Intracortical distribution of a cohort of cells arising in the PVE. Soc Neurosci Abstr 22:284

The Roles of Wnt Signaling in the Developing Dorsal Neural Tube

Makoto Ikeya[1], Scott M. K. Lee[2], Andrew P. McMahon[2], and Shinji Takada[1]

Summary. Wnt-1 and Wnt-3a, which are members of the Wnt family of cysteine-rich secreted signaling molecules, are coexpressed at the dorsal midline of the developing neural tube. Although each signal is essential for several aspects of embryonic development, no dorsal neural tube phenotype was observed in each mutant, suggesting that Wnt signaling may be redundant. To address this possibility we generated compound mutant lacking both Wnt-1 and Wnt-3a signaling. The phenotype of the compound mutant indicated that these Wnt genes play essential roles for the formation of the neural crest and dorsolateral neural precursors in the neural tube.

Key words. Wnt—Cell signaling molecule—Roof plate—Neural crest—CNS

Introduction

Members of the Wnt family encode cysteine-rich signaling molecules, which act as autocrines and paracrines. Four of these, Wnt-1, Wnt-3, Wnt-3a, and Wnt-4, are expressed in overlapping regions within the dorsal central nervous system (CNS), predominantly at the roof plate from the forebrain to spinal cord [1]. However, Wnt-1, Wnt-3a, and Wnt-4 mutants, which have already been reported, show no obvious neural crest and CNS phenotype [2–5]. Because cell transformation studies indicate that Wnt-1 and Wnt-3a have similar activities [6] and expression of these two genes in the dorsal neural tube appears identical in both the temporal and spatial sense [1], functional redundancy between these two Wnt signals may obscure a possible role for Wnt signaling in neural crest formation and in dorso ventral patterning of the CNS. To address this possibility, we generated compound mutants lacking both Wnt-1 and Wnt-3a signaling.

[1] Center for Molecular and Developmental Biology, Faculty of Science, Kyoto University, Kitashirakawa, Sakyo-ku, Kyoto 606-8502, Japan
[2] Department of Molecular and Cellular Biology, The Biclabs, Harvard University, 16 Divinity Avenue, Cambridge, MA 02138, USA

Materials and Methods

Generation of mice heterozygous for null alleles of Wnt-1 and Wnt-3a has been described previously [2,4]. Compound heterozygotes were intercrossed to recover compound mutants. Genotypes were initially confirmed by southern blotting and polymerase chain reaction (PCR). Whole-mount in situ hybridization was performed as previously described [7].

Results and Discussion

Phenotype in Neural Crest Development

We first examined neural crest-derived structures. Neurofilament staining indicated that both neural crest-derived cranial and spinal ganglia formation were defective in compound mutants while it was unaltered in single mutants that were either wild type or heterozygous for mutations in the other Wnt member. For instance, in compound mutants, neurons derived from the proximal ganglion of cranial nerve IX (glossopharyngeal), which is formed by crest cells originating from rhombomere 6 within the hindbrain (r6) [8,9], were absent. In contrast, the distal ganglion, which is placodal in origin was present. In addition, there was a marked reduction in the proximal axons of cranial nerves V (trigeminal, r2-derived) and X (vagus, r7-derived). Similarly, in the trunk, there was a dramatic reduction in neurofilament staining in the cervical dorsal root ganglia.

Whole-mount in situ hybridization with probes recognizing Islet-1 and cadherin-6 [10], markers of neuronal and glial neural crest derivatives, respectively, within the cranial ganglia and dorsal root ganglia confirmed the reduction or absence of crest cells. In contrast, sympathetic ganglia, which express c-ret [11], were unaffected. In addition to the defect of neural and glial derivatives, we have also observed obvious defects in other neural crest derivatives, including part of the head skeleton and melanocyte precursors (Fig. 1).

The reduction of neural crest derivatives suggests that fewer neural crest cells may emerge in embryos lacking both Wnt-1 and Wnt-3a signaling. To address this possibility, we next examined expression of several molecular markers that are normally expressed in migrating neural crest cells, including AP-2. AP-2 is first expressed at 8.5 d.p.c. (days post coitus) in the dorsal neural plate, coincident with neural crest formation. By 9.5 d.p.c., cranial expression is absent in the neural tube but persists in migrating and maturing neural crest derivatives at cranial and spinal cord levels [12]. In compound mutants, a clear decrease was observed in migrating AP-2-positive cells in the hindbrain (Fig. 2a,b) and anterior trunk (Fig. 2c,d). Surprisingly, in contrast to their wildtype littermates, compound mutants also retained AP-2 expression within the dorsal CNS at 9.5 d.p.c. (Fig. 2c–f).

Thus, in the absence of Wnt-1 and Wnt-3a, there is both a reduction in neural crest cell formation and persistent expression of AP-2 at the dorsal midline. This prolonged expression of AP-2 within the dorsal CNS suggest that AP-2-expressing cells may still be differentiating at later stages. Alternatively, these cells may be retained at the midline because Wnt signaling promotes neural crest emigration. A correlation has

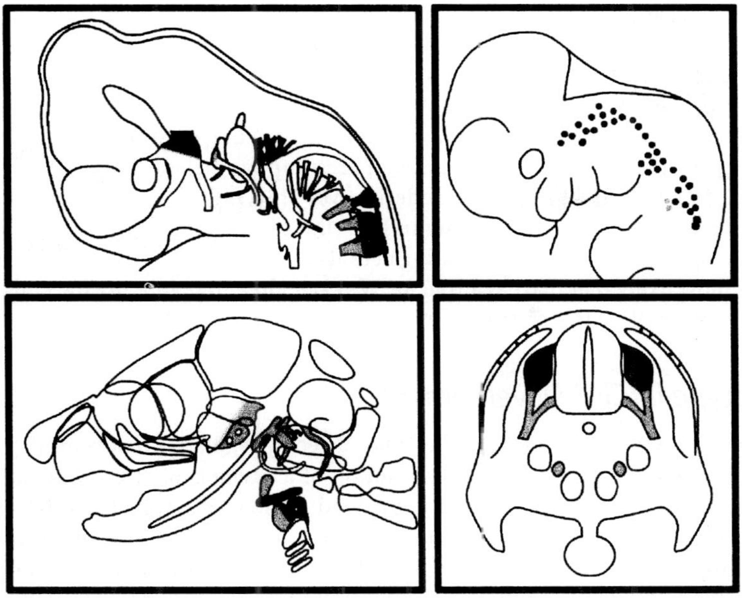

Fig. 1. Summary of defects in neural crest derivatives in Wnt-1/Wnt-3a compound mutants. Normally formed derivatives were marked with *gray* and defected derivatives with *black*. *Top left*: cranial ganglia; *top right*: melanocyte precursors; *bottom left*: head skeleton; *bottom right*: trunk ganglia

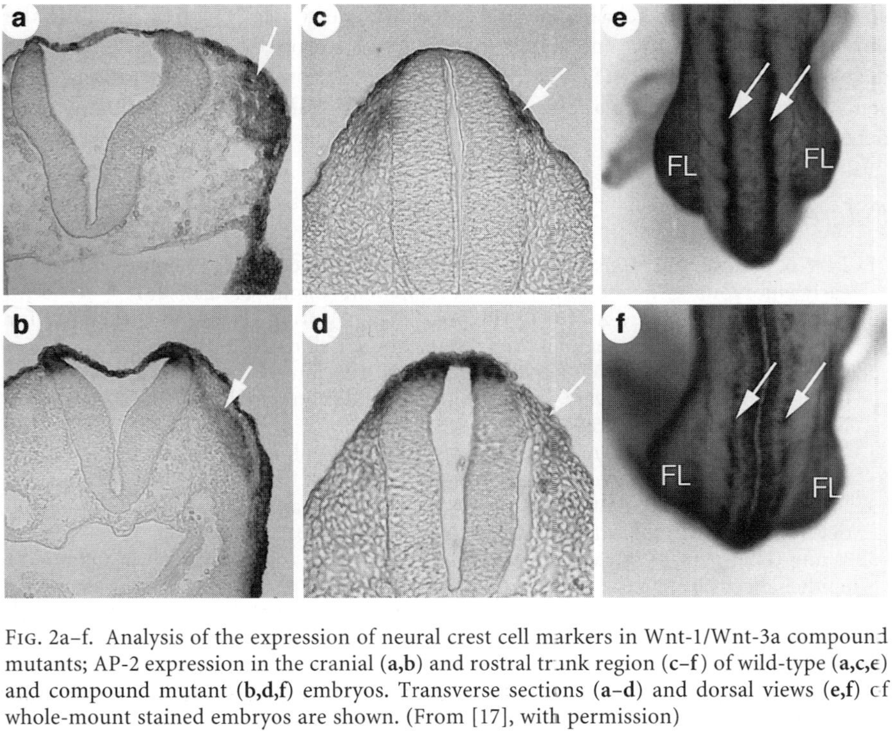

Fig. 2a–f. Analysis of the expression of neural crest cell markers in Wnt-1/Wnt-3a compound mutants; AP-2 expression in the cranial (**a,b**) and rostral trunk region (**c–f**) of wild-type (**a,c,e**) and compound mutant (**b,d,f**) embryos. Transverse sections (**a–d**) and dorsal views (**e,f**) of whole-mount stained embryos are shown. (From [17], with permission)

been observed in in vitro assays that examined neural crest induction by surface ectoderm between the appearance of Wnt-1 expression and emigration of neural crest cells [13]. We also observed reduction of late-formed neural crest cells at 11.5 d.p.c. by examining expression of TRP-2, a marker of presumptive melanocytes [14], which are dominant in late formed cranial crest derivatives [15].

In summary, loss of Wnt signaling in both the caudal hindbrain and spinal cord leads to a marked reduction in migrating neural crest cells. In the light of these results, it seems most likely that there is a role for Wnt signaling in regulating the emigration or proliferation of neural crest precursors within the dorsal CNS.

Phenotype in CNS Development

To determine whether Wnt signaling may directly regulate dorsoventral polarity within the CNS, we examined the distribution of a number of regionally expressed markers. Although spinal cord levels appeared normal, the hindbrain displayed a striking phenotype. Expression of Wnt-3a, Wnt-1, and Lmx-1b was slightly reduced but localized, as normal, to the roof plate. In contrast, expression of Pax-3, which occupies most of the dorsal half of the CNS, was dramatically reduced in the hindbrain of compound mutants. Dbx expression at the dorsalventral interface and Pax-6 in the ventrolateral hindbrain were normal.

A pronounced reduction in dorsolateral progenitors in the hindbrain suggests that Wnt signaling regulates the expansion of these cells, either as a survival factor or as a mitogen. The latter explanation is supported by transgenic analysis, which demonstrated that Wnt-1 can act as a potent mitogen when ectopically expressed within the dorsal CNS [16]. Many Wnt family members are regionally expressed within the developing mammalian CNS. Whether the local regulation of cell proliferation represents a general mechanism applicable to other Wnt signaling pathways remains to be determined.

References

1. Parr B, et al (1993) Mouse Wnt genes exhibit discrete domains of expression in the early embryonic CNS and limb buds. Development (Camb) 119:247–261
2. McMahon AP, Bradley A (1990) The Wnt-1 (int-1) proto-oncogene is required for development of a large region of the mouse brain. Cell 62:1073–1085
3. Thomas KR, Capecchi MR (1990) Targeted disruption of the murine int-1 proto-oncogene resulting in severe abnormalities in midbrain and cerebelar development. Nature (Lond) 346:847–850
4. Takada S, et al (1994) Wnt-3a regulates somite and tailbud formation in the mouse embryo. Genes Dev 8:174–189
5. Stark K, et al (1994) Epithelial transformation of metanephric mesenchyme in the developing kidney regulated by Wnt-4. Nature (Lond) 372:679–683
6. Wong GT, Gavin BJ, McMahon AP (1994) Differential transformation of mammary epithelial cells by Wnt genes. Mol Cell Biol 14:6278–6286
7. Yoshikawa Y, et al (1997) Evidence that absence of Wnt-3a signaling promotes neuralization instead of paraxial mesoderm development in the mouse. Dev Biol 183:234–242
8. D'Amico-Martel A, Noden DM (1983) Contributions of placodal and neural crest cells to avian cranial peripheral ganglia. Am J Anat 166:445–468

9. Lumsden A, Sprawson N, Graham A (1991) Segmental origin and migration of neural crest cells in the hindbrain region of the chick embryo. Development (Camb) 113:1281–1291
10. Inoue T, et al (1997) Cadherin-6 expression transiently delineates specific rhombomeres, other neural tube subdivisions and neural crest subpopulations in mouse embryos. Dev Biol 183:183–194
11. Pachnis V, Mankoo B, Costantini F (1993) Expression of the c-ret proto-oncogene during mouse embryogenesis. Development (Camb) 119:1005–1017
12. Chazaud C, et al (1995) AP-2.2, a novel gene related to AP-2, is expressed in the forebrain, limbs and face during mouse embryogenesis. Mech Dev 54:83–94
13. Dickinson ME, et al (1995) Dorsalization of the neural tube by non-neural ectoderm. Development (Camb) 121:2099–2106
14. Wehrle-Haller B, Weston JA (1995) Soluble and cell-bound forms of steel factor activity play distinct roles in melanocyte precursor dispersal and survival on the lateral neural crest migration pathway. Development (Camb) 121:737–742
15. Serbedzija GN, Bronner-Fraser M, Fraser SE (1989) A vital dye analysis of the timing and pathways of avian neural crest cell migration. Development (Camb) 106:809–816
16. Dickinson ME, Krumlauf R, McMahon AP (1994) Evidence for a mitogenic effect of Wnt-1 in the developing mammalian central nervous system. Development (Camb) 120:1453–1471
17. Ikeya M, et al (1997) Wnt signalling required for expansion of neural crest and CNS progenitors. Nature (Lond) 389:966–970

Physical and Functional Links Between Deltex and Notch Receptor in *Drosophila melanogaster*

Kenji Matsuno[1,2], Mikiko Ito[2], Noriyuki Kishi[1], Satoshi Suzuki[1], Spyros Artavanis-Tsakonas[3], and Hideyuki Okano[1,2]

Summary. Local cell–cell interactions are essential for the development of multicellular organisms [1]. Involvement of Notch signaling in cell–cell interactions regulating various cell-fate determinations has been documented [2–8]. The genetic components of Notch signaling have been identified and analyzed in *Drosophila* and *Caenorhabditis elegans* [1,3–6]. Recently, it has been found that the components of Notch signaling are conserved in vertebrates [6–8]. Although several biochemical models of Notch signaling have been proposed, the molecular picture of Notch signaling is still elusive [6]. An involvement of *deltex* in Notch signaling has been studied in *Drosophila* as well as in humans [5,6,9–14]. We describe here recent results supporting the idea that Deltex is a positive regulator of Notch signaling and may function through an physical interaction with the Ankyrin repeats of the Notch intracellular domain. An interplay involving Deltex, Suppressor of Hairless (Su(H)), and the intracellular domain of Notch is also discussed.

Key words. Notch—Deltex—Cell signaling—Cell–cell interaction—Lateral inhibition

Introduction

Development of multicellular organisms relies on cell–cell communication to achieve orchestrated pattern formation, cell-fate determinations, morphogenesis, etc. [1,15]. One mechanism adopted for cell–cell interactions is diffusion of secreted molecules

[1] Department of Neuroanatomy, Osaka University Medical School, 2-2 Yamadaoka, Suita, Osaka 565-0871, Japan
[2] CREST, Japan Science and Technology Corporation, 2-2 Yamadaoka, Suita, Osaka 565-0871, Japan
[3] Howard Hughes Medical Institute, Department of Cell Biology and Biology, Yale University, New Haven, CT 06536, USA

[15], including morphogens, for instance, Hedgehog and transforming growth factor-β (TGF-β) [15]. Another mechanism involves direct cell–cell contact to send the signal to neighboring cells [1]. The second mechanism is referred to as juxtacrine signaling. A signaling pathway mediated by the Notch receptor (Notch signaling) is classified as juxtacrine signaling. Notch is a transmembrane receptor, and its ligands also have a transmembrane domain [4–7].

Notch signaling has been studied extensively in *Drosophila* and in *Caenorhabditis elegans* [1–6]. These studies revealed the developmental roles of Notch signaling and led to the identification of genes that function in Notch signaling [2–6]. In *Drosophila*, Notch signaling is known to be involved in the binary cell-fate decisions of a very broad spectrum of cells [2–6]. Especially, an involvement of Notch signaling in "lateral inhibition" or "lateral specification" has been proposed [1–6]. Lateral specification is a mechanism to single out a cell or a certain rate of cells from a population of cells that have equal potential to make a choice between binary cell fates [1–6]. Several lines of evidence have been provided to support an involvement of Notch signaling in lateral inhibition, especially from the studies of insect neurogenesis and cell-fate determination in *C. elegans*. For example, it is known that neuroblast segregation in the insect neuroectoderm is controlled by cell–cell interaction [16]. Ablation of a neuroblast leads to recruitment of a neighboring epidermoblast to the neuroblast fate, suggesting that a neuroblast prevents the surrounding epidermoblast from taking the neuroblast fate via a cell–cell communication [16]. Because *Drosophila* embryos carrying mutations in the *Notch* locus show hyperplasia of neuroblasts at the expense of epidermoblasts, Notch is believed to be involved in the regulation of neuroblast segregation [3,5,6].

In *C. elegans*, existence of a feedback loop involving the transcription of *lin-12*, encoding a Notch-like receptor, and *lag-2*, a ligand for Lin-12, has been reported [17]. As a mechanism acting in lateral specification, the presence of such a feedback loop has been predicted theoretically [1,5]. However, it is also clear that Notch does not function exclusively in lateral specification. It has been shown that Notch signaling is involved in inductive signal events [18]. Furthermore, function of Notch signaling in pattern formation has been documented in *Drosophila* as well as in mammalian appendages [19].

Recently, it has been found that the components of Notch signaling are conserved in vertebrates [6–8,20]. Studies of vertebrate Notch signaling suggest that biochemical mechanisms of Notch signaling are also conserved from *Drosophila* to mammals [6–8,20,21]. In mammalian systems, links between the aberrant activation of Notch signaling and carcinogenesis have been discovered [22,23]. Moreover, it has been suggested that two human syndromes, CADASIL (for cerebral autosomal dominant arteriopathy with subcortical infarcts and leukoencephalopathy) and Alagille syndrome, are caused by mutations in human *Notch-3* and human *Jagged-1*, encoding a ligand for human Notch, respectively [24–26].

Although recent studies employing molecular genetics and biochemistry have provided several models of Notch signaling, the molecular picture of Notch signaling is still obscure. We describe here recent results regarding *Drosophila* Deltex, which is a positive regulator of Notch signaling, and discuss the biochemical function of Deltex in Notch signaling.

deltex Is Involved in Notch Signaling

In principle, it is probably true that the Notch receptor family is activated by the binding of their transmembrane ligands. At least some of the signal, although probably not all (see below), is transduced from the Notch receptor to the nucleus via Su(H)/RBP-Jκ/Lag-1, an effector protein family in Notch signaling [8,27,28]. It has been demonstrated that the target genes of Notch signaling, basic helix-loop-helix (bHLH) genes in *Enhancer of split* [*E(spl)*] complex and *vestigial* gene, are transcriptionally activated by the binding of Su(H) to the promoter regions of these genes [29–31]. However, the biochemical nature of receptor activation has not been understood.

At least three hypotheses have been proposed to explain Notch receptor activation: (i) a nuclear translocation of Su(H) is induced by the activation of the Notch receptor; (ii) the intracellular domain of Notch is cleaved proteolytically and translocates to the nucleus to form a transcriptional complex with Su(H); (iii) an activation of the Notch receptor induces a posttranslational modification of Su(H). In any case, understanding of the first biochemical event occurring after binding of the ligand should be important in understanding the molecular nature of Notch signaling. In this sense, Deltex might be a good candidate of a factor involved in activation of the Notch receptor, because Deltex binds to the intracellular domain of Notch irrespective of the ligand binding, and it has been suggested that Deltex functions upstream of Notch biochemically [12,13].

An involvement of Deltex in Notch signaling was first noticed by a genetic modifier screen [9,32]. This screen was designed to isolate the dominant suppressors of lethality associated with hyperactivation of the Notch receptor [32]. Four genes, *Notch*, *Delta*, *mastermind (mam)*, and *deltex* were isolated [9,32]. Among these genes, *Notch*, *Delta*, and *mam* are known as neurogenic genes [3,5,32]. Therefore, functional involvement of *Notch*, *Delta*, and *mam* in Notch signaling had already been established by independent research [33]. Identification of *deltex* by the genetic screen which also isolate these three genes strongly supports an idea that *deltex* is involved in Notch signaling.

The presence of genetic interactions among a group of genes has been considered as genetic evidence supporting the idea that the products of these genes are involved in a common biochemical pathway, although this may not be always true. It has been shown that *deltex* has genetic interactions with the genes involved in Notch signaling [9–11]. Phenotypic analyses of *deltex* also support the idea that the *deltex* gene is involved in Notch signaling [10]. For example, loss-of-function *deltex* has phenotypes in common with various loss-of-function Notch mutants, although it has not been known whether *deltex* has a neurogenic phenotype [10]. In summary, the genetic evidence provided from these studies suggests that *deltex* functions as a positive regulator in Notch signaling.

Deltex Physically Interacts with the Intracellular Domain of Notch

Genetic evidence suggesting that Deltex is a positive regulator or an effector in Notch signaling fascinated us to investigate the biochemical function of Deltex in Notch signaling. First, we analyzed the physical interaction between the intracellular domain of Notch and Deltex, because it is known that Deltex localizes to the cytoplasm in *Drosophila* tissue culture cells and in vivo [12,13]. In *Drosophila* tissue culture cells, it was found that a full-length Notch and Deltex are colocalized [12]. This colocalization was mediated by the Ankyrin repeats in the intracellular domain of Notch [12]. A yeast "interaction trap" analysis suggested that this colocalization results from the physical interaction between the Notch Ankyrin repeats and Deltex [12,13].

Physical Interaction Between the Notch Ankyrin Repeats and Deltex May Be Essential for the Deltex Function in Notch Signaling

It has been proposed that Deltex can be divided into three domains, referred as domains I, II, and III [13] (Fig. 1). Using the interaction trap system and the colocalization assay in *Drosophila* tissue culture cells, the domain I of Deltex, which is located in the N-terminal region of the protein, was found to be sufficient and necessary to bind to the Ankyrin repeats of Notch [13]. The Deltex domains II and III include a proline-rich motif and a RING-H2 finger domain, respectively, although in this review these two motifs are not discussed in detail [13] (Fig. 1). It has been suggested that these two motifs are also involved in protein–protein interactions [13].

To understand the biochemical function of the binding between Deltex and the intracellular domain of Notch, several mutant Deltex proteins were overexpressed in vivo [13]. Overexpression of a full-length Deltex or a Deltex polypeptide composed only of domain I resulted in generation of phenotypes similar to that induced by an activated form of Notch [13,34]. These phenotypes are similar to those observed in various gain-of-function mutants of *Notch* [13]. Therefore, overexpression of the full-length Deltex or the Deltex domain I probably activates Notch signaling ectopically in vivo. Furthermore, expression of these two proteins could rescue loss-of-function mutants of *deltex* [13]. On the other hand, overexpression of a Deltex polypeptide composed only of domains II and III did not produce any detectable phenotypes and failed to rescue the *deltex* mutants [13]. These results, therefore, suggest that Deltex functions through physical interaction with the Ankyrin repeats of the Notch intracellular domain.

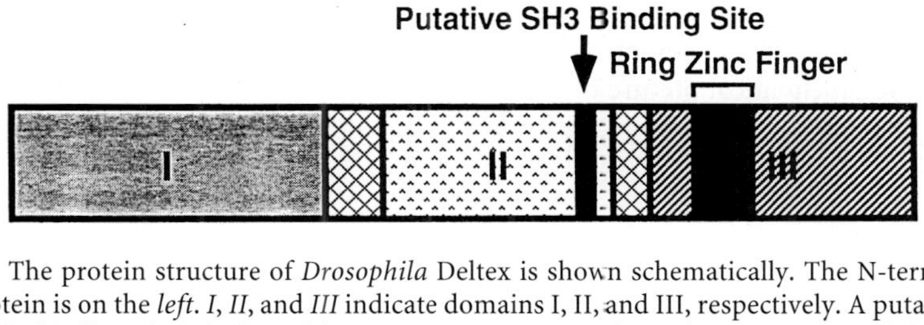

Fig. 1. The protein structure of *Drosophila* Deltex is shown schematically. The N-terminal of the protein is on the *left*. *I*, *II*, and *III* indicate domains I, II, and III, respectively. A putative SH-3 domain-binding site is indicated by an *arrow*; a RING-H2 finger (Ring Zinc Finger) domain is shown by a *filled rectangle*; OPA repeats are shown by *hatched rectangles*

Interplay Among Deltex, Su(H), and the Intracellular Domain of Notch

Deltex is not the only protein that binds to the intracellular domain of Notch. *Drosophila* Su(H), an effector protein in Notch signaling, also binds to at least two independent regions of the Notch intracellular domain [8,27,35]. Interplay among the protein–protein interactions involving the Notch intracellular domain, Deltex, and Su(H) has been studied [13].

Fortini and Artavanis-Tsakonas have shown that subcellular localization of Su(H) is regulated by Notch signaling in a *Drosophila* cell line [27]. The Su(H) protein overexpressed in the *Drosophila* cells localizes to the nucleus [27]. However, when Su(H) was coexpressed with a full-length Notch, Su(H) was detected in the cytoplasm in most of the cells [27]. The distribution of the two proteins in the cytoplasm and the membrane was analyzed, and colocalization of them was observed [27]. The retention of Su(H) in the cytoplasm was not induced if Su(H) was coexpressed with a mutant Notch receptor that lacks the Ankyrin repeats [27]. In the yeast interaction trap system, physical interaction between the Notch intracellular domain and Su(H) was demonstrated [8,27]. These results suggest that the cytoplasmic retention of Su(H) is caused by the interaction between Su(H) and the intracellular domain of Notch [27].

The effect induced by binding of the ligand to the Notch receptor on cytoplasmic sequestration of Su(H) was tested. If the *Drosophila* tissue-cultured cells overexpressing either Notch or Delta, a Notch ligand, were mixed, these cells form cell aggregates via extracellular interaction between Notch and Delta [36]. The subcellular localization of Su(H) in the Notch-expressing cells that formed cell aggregates with Delta-expressing cells was examined [27]. In these cells, cytoplasmic retention of Su(H) was inhibited, and Su(H) localized to the nucleus, suggesting that activation of the Notch receptor promotes the nuclear translocation of Su(H) [27] (Fig. 2).

Because a positive Notch ligand, Delta, apparently inhibits cytoplasmic sequestration of Su(H), an effect of Deltex that was known as a cytoplasmic positive regulator of signaling on subcellular localization of Su(H) was also tested [13]. Under the condition that Su(H) is sequestered in the cytoplasm by the co-expression of the full-length Notch, Deltex was overexpressed further [13]. In the cells expressing Notch, Su(H), and Deltex simultaneously, the cytoplasmic retention of Su(H) was inhibited, and Su(H) was detected in the nucleus in most of the cells [13] (Fig. 2). Therefore, Delta and Deltex, two factors that function positively in Notch signaling, apparently promote the nuclear translocation of Su(H) in *Drosophila* tissue culture cells [13,27] (Fig. 2). As mentioned, it has been demonstrated that Su(H) binds to the promoter region of target genes of Notch signaling and induces the transcription of these genes [29–31]. Taking these observations together, nuclear translocation of Su(H) may be a part of the biochemical picture of Notch signaling.

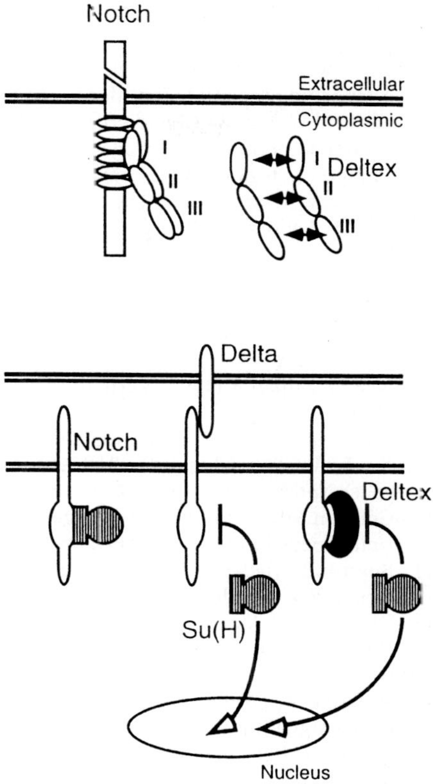

FIG. 2. Physical interactions involving Deltex and a model of Notch signaling mediated by Su(H). *Top panel*: physical interactions involving Deltex and the intracellular domain of Notch. Deltex binds to the Ankyrin repeats of the Notch intracellular domain. Deltex interacts homotypically, although these interactions are not discussed here. *Bottom panel*: nuclear translocation of Su(H) is promoted by Delta and Deltex. Su(H) is sequestered in the cytoplasm when co-expressed with Notch via its interaction with the intracellular domain of Notch. The Su(H) protein translocates into the nucleus when Notch binds to Delta or Deltex

Conclusion

As discussed, nuclear translocation of Su(H) may play a part in Notch signaling. However, it is also clear that this model is oversimplified. According to this model, when no Notch is present, for example, in null mutants of *Notch*, Su(H) should localize to the nucleus at the highest level and therefore Notch signaling should be maximized. Thus, this hypothesis requires at least an additional regulatory mechanism to avoid this contradiction. Subcellular localization of Su(H) has been studied extensively in vivo [37]. In most cells, the Su(H) protein was detected exclusively in the nucleus, although in a small number of cells, promotion of Su(H) nuclear translocation could be induced by an overexpression of Deltex [37]. Therefore, it is not clear that the nuclear translocation of Su(H) observed in *Drosophila* tissue culture cells represents the biochemical nature of Notch signaling in vivo. It may be possible that some physical interactions are exaggerated under the conditions involving the overexpressed proteins, although these bindings are important for the signaling. Further analysis should be done to construct a molecular model of Deltex function in Notch signaling.

References

1. Greenwald I, Gubin GM (1992) Making a difference: the role of cell–cell interactions in establishing separate identities for equivalent cells. Cell 68:271–281
2. Ghysen A, Dambly-Chaudière C, Jan LY, et al (1993) Cell interactions and gene interactions in peripheral neurogenesis. Genes Dev 7:723–733
3. Campos-Ortega JA (1993) Early neurogenesis in *Drosophila melanogaster*. In: Bate M, Martinez-Arias A (eds) The development of *Drosophila melanogaster*. Cold Spring Harbor Laboratory Press, NY, pp 1091–1129
4. Muskavitch MAT (1994) Delta-Notch signaling and *Drosophila* cell fate choice. Dev Biol 166:415–430
5. Simpson P (1994) The Notch receptor. Landes, Austin
6. Artavanis-Tsakonas S, Matsuno K, Fortini ME (1995) Notch signaling. Science 268:225–232
7. Nye JS, Kopan R (1995) Vertebrate ligands for Notch. Curr Biol 5:966–969
8. Honjo T (1996) The shortest path from the surface to the nucleus: RBP-Jκ/Su(H) transcription factor. Genes Cells 1:1–9
9. Xu T, Artavanis-Tsakonas S (1990) *deltex*, a locus interacting with the neurogenic genes *Notch*, *Delta* and *mastermind* in *Drosophila melanogaster*. Genetics 126:665–677
10. Gorman MJ, Girton RJ (1992) A genetic analysis of *deltex* and its interaction with the *Notch* locus in *Drosophila melanogaster*. Genetics 131:99–112
11. Busseau I, Diederich RJ, Xu T, et al (1994) A menber of the Notch group of interacting loci, *deltex* encodes a cytoplasmic basic protein. Genetics 136:585–596
12. Diederich RJ, Matsuno K, Hing H, et al (1994) Cytosolic interaction between deltex and Notch ankyrin repeats implicates deltex in the Notch signaling pathway. Development (Camb) 120:473–481
13. Matsuno K, Diederich RJ, Go MJ, et al (1995) Deltex acts as a positive regulator of Notch signaling through interactions with the Notch ankyrin repeats. Development (Camb) 121:2633–2644
14. Matsuno K, Eastman DS, Mitsiades T, et al (1998) Human Deltex: a conserved regulator of Notch signaling. Nat Genet 19:74–79
15. Lawrence PA, Struhl G (1996) Morphogens, compartments, and pattern: lessons from *Drosophila*? Cell 85:951–961
16. Doe CQ, Goodman CS (1985) Early events in insect neurogenesis. II. The role of cell interactions and cell lineage in the determination of neural precursor cells. Dev Biol 111:206–219
17. Wilkinson HA, Fitzgerald K, Greenwald I (1994) Reciprocal changes in expression of the receptor *lin-12* and its ligand *lag-2* prior to commitment in a *C. elegans* cell fate decision. Cell 79:1187–1198
18. Spana EP, Doe CQ (1996) Numb antagonizes Notch signaling to specify sibling neuron cell fates. Neuron 17:21–26
19. Shubin N, Tabin C, Carroll S (1997) Fossils, genes and the evolution of animal limbs. Nature (Lond) 388:639–648
20. Blaumueller CM, Artavanis-Tsakonas S (1997) Comparative aspects of Notch signaling in lower and higher eukaryotes. Perspect Dev Neurobiol 4:325–343
21. Kageyama R, Nakanishi S (1997) Helix-loop-helix factors in growth and differentiation of the vertebrate nervous system. Curr Opin Genet Dev 7:659–665
22. Ellisen LW, Bird J, West DC, et al (1991) *TAN-1*, the human homolog of the *Drosophila Notch* gene, is broken by chromosomal translocations in T-lymphoblastic neoplasms. Cell 66:649–661
23. Jhappan C, Gallahan D, Stahle C, et al (1992) Expression of an activated *Notch*-related *int-3* transgene interferes with cell differentiation and induces neoplastic transformation in mammary and salivary glands. Genes Dev 6:345–355

24. Joutel A, Corpechot C, Ducros A, et al (1996) Notch 3 mutations in CADASIL, a hereditary adult-onset condition causing stroke and dementia. Nature (Lond) 383:707–710
25. Oda T, Elkahloun AG, Pike BL, Okajima K, et al (1997) Mutations in the human *Jagged 1* gene are responsible for Alagille syndrome. Nat Genet 16:235–242
26. Li L, Krantz ID, Deng Y, et al (1997) Alagille syndrome is caused by mutations in human *Jagged 1*, which encodes a ligand for Notch 1. Nat Genet 16:243–251
27. Fortini ME, Artavanis-Tsakonas S (1994) The Suppressor of Hairless protein participates in Notch receptor signaling. Cell 79:273–282
28. Christensen S, Kodoyianni V, Bosenberg M, et al (1996) *lag-1*, a gene required for *lin-12* and *glp-1* signaling in *Caenorhabditis elegans*, is homologous to human CBF1 and *Drosophila* Su(H). Development (Camb) 122:1373–1383
29. Lecourtois M, Schweisguth F (1995) The neurogenic Suppressor of Hairless DNA-binding protein mediates the transcriptional activation of the Enhancer of split complex genes triggered by Notch signaling. Genes Dev 9:2598–2608
30. Bailey AM, Posakony JW (1995) Suppressor of Hairless directly activates transcription of Enhancer of split complex genes in response to Notch receptor activity. Genes Dev 9:2609–2622
31. Kim J, Sebring A, Esch JJ, et al (1996) Integration of positional signals and regulation of wing formation and identity by *Drosophila vestigial* gene. Nature (Lond) 382:133–138
32. Xu T, Rebay I, Fleming RJ, et al (1990) The *Notch* locus and the genetic circuitry involved in early *Drosophila* neurogenesis. Genes Dev 4:464–475
33. Compos-Ortega JA (1988) Cellular interaction during early neurogenesis of *Drosophila melanogaster*. Trends Neurosci 11:400–405
34. Rebay I, Fehon RG, Artavanis-Tsakonas S (1993) Specific truncations of *Drosophila* Notch define dominant activated and dominant negative forms of the receptor. Cell 74:319–329
35. Matsuno K, Go MJ, Sun X, et al (1997) Suppressor of Hairless-independent events in Notch signaling imply novel pathway elements. Development (Camb) 124:4265–4273
36. Fehon RG, Kooh PJ, Rebay I, et al (1990) Molecular interactions between the protein products of the neurogenic loci *Notch* and *Delta*, two EGF-homologous genes in *Drosophila*. Cell 61:523–534
37. Gho M, Lecourtois M, Géraud G, et al (1996) Subcellular localization of Suppressor of Hairless on *Drosophila* sense organ cells during Notch siganling. Development (Camb) 122:1673–1682

Perpendicular Contact Guidance of CNS Neurons: Is It Operative in Brain Cortex Development?

Katsuhiko Ono[1], Isao Nagata[2], Taro Hama[3], and Norio Nakatsuji[3]

Summary. We have reported that rodent CNS neuroblasts show perpendicular contact guidance on aligned neurite bundles in microexplant cultures of the cerebellar and cerebral cortices. Such perpendicular orientation of the CNS neurons occurs even on artificially fabricated parallel micro-structures on a quartz plate, indicating the importance of the physical surface structure. Perpendicularly oriented cells extended many side spikes from the leading neurite process, and they made close contact to parallel microstructures of the substrate. Here we present data that suggest operation of such a unique guidance system in the brain cortex development. First, detailed examination of the morphology of vertically migrating granule cells in the postnatal mouse cerebellum revealed the presence of many side spikes extending horizontally from the vertically oriented leading cell process, and these made direct contact to the horizontal parallel fibers. Second, ventricular cells were labeled with a fluorescent dye in organ culture of mouse cerebral cortex or by injection of DiI microcrystals into the fetal ventricle. Examination of the brain slices after a few days visualized many labeled neuroblasts after migration, and there were horizontally oriented cells in the cortical plate as well as in the intermediate zone. Perpendicular contact guidance can explain horizontal orientation in the cortical plate in which the cells are surrounded by vertical parallel structures.

Key words. Cell migration—Neuron—Brain cortex—Contact guidance—Cerebellum

---

[1] Department of Anatomy, Shimane Medical University, 89-1 Enyacho, Izumo, Shimane 693, Japan

[2] Department of Cell Biology, Tokyo Metropolitan Institute for Neuroscience, 2-6 Musashidai, Fuchu, Tokyo 183, Japan

[3] Mammalian Development Laboratory, National Institute of Genetics, and Department of Genetics, The Graduate University for Advanced Studies, 1111 Yata, Mishima, Shizuoka 411-8540, Japan

Perpendicular Contact Guidance Discovered and Examined in Culture of CNS Neuroblasts

Perpendicular contact guidance was discovered when we made a microexplant culture of neonatal mouse cerebellar cortex [1–4]. Granular cell neurons extended radial bundles of neurites from explants. These cells first showed typical contact guidance and migrated outward along the radial neurite bundle (Fig. 1). Then, these neurons changed their orientation by 90° and extended new cell processes and migrated perpendicular to the radial neurite bundle in the absence of glial elements. These perpendicular processes have microspikes that attach to the aligned substratum of neurites. In such cultures, neurite bundles were probably equivalent to the parallel fibers in vivo, and the cells oriented perpendicular to the neurites correspond to the vertically migrating cells in vivo during the later phase of cerebellar cortex development.

Not only the cerebellar granule cells but almost all neuroblasts isolated from various parts of the CNS exhibited perpendicular contact guidance in addition to parallel guidance [5]. Neurons from the peripheral nervous system, however, never showed such perpendicular orientation. In time-lapse video recording, single neurons frequently changed orientation by making right or left turns during their migration on the aligned neurite bundle.

Next, we examined whether such perpendicular guidance requires specific neurite bundles for manifestation. For such experiments, microstructures were fabricated on the quartz surface to create aligned structures similar to the neurite bundle [6]. CNS neuroblasts exhibited both perpendicular and parallel orientation on such artificial substratum, indicating the importance of the physical surface structure. Perpendicular cell processes are attached to the aligned ridges with many microspikes.

In a different line of experiments, we made chimeric mice using an embryonic stem cell line expressing the *E. coli lacZ* gene for easy histochemical staining. Analysis of such chimeric brains [7] revealed clear columnar patches of tightly packed labeled neurons in the cortical plate of the E18 fetus cerebral cortex, indicating that most neuroblasts migrated radially during early development of the cortex. Cerebral cortex of the adult chimeras, however, showed patches made of much more dispersed labeled neurons, thus suggesting horizontal migration of neurons during the later phase of the cortex development.

Perpendicular guidance may play important roles in development of the CNS tissue architecture. Therefore, we looked for presence of two prominent features of the perpendicular contact guidance, the formation of many side spikes from the leading cell process and their attachment to the parallel structures.

Vertical Migration of Cerebellar Granule Cells

Vertical migration of granule cells in the vertebrate cerebellar cortex is probably the most defined type of neuronal migration in the developing CNS [8–10]. Previous studies indicated that such granule cells migrate along the surface of radially oriented Bergmann glia [8,11,12]. However, the geometric arrangement of tissue structures

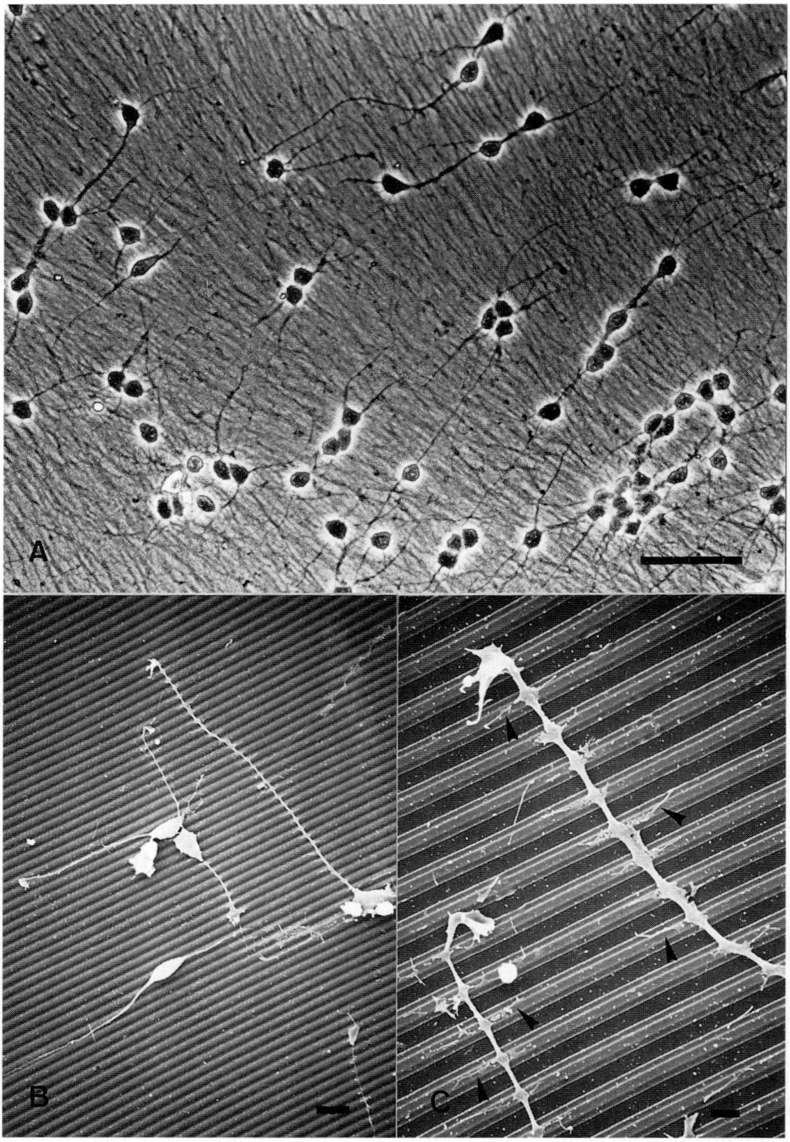

FIG. 1A–C. Perpendicular contact guidance in culture. **A** Phase-contrast photomicrographs of cerebellar granule cells exhibit perpendicular contact guidance on the parallel neurite bundle that grew radially from a microexplant of the cerebellar external granular layer [1] after 4 days in culture of a P2 cerebellum explant. *Bar* 50 μm. **B**, **C** Scanning electron photomicrographs of cerebral cortical neuroblasts on the fabricated microstructure [6]. They were dissociated from the E14 cerebral cortex and cultured for 2 days. *Arrowheads* in **C** indicate side spikes extending from a neurite. *Bars* 10 or 2 μm surrounding migrating granule cells suggests other possible mechanisms. For example, they are directly associated with parallel fibers, and also with Purkinje cell dendrites and inhibitory neurons in the deep layer of the external granular layer and the molecular layer. Most of these structures are orthogonally arranged around the vertically migrating granule cells. If the migratory cells have the ability to recognize such orthogonal geometry, the latter may be an important factor for guiding their migration in vivo. A hypothesis that fits this idea very well is perpendicular contact guidance.

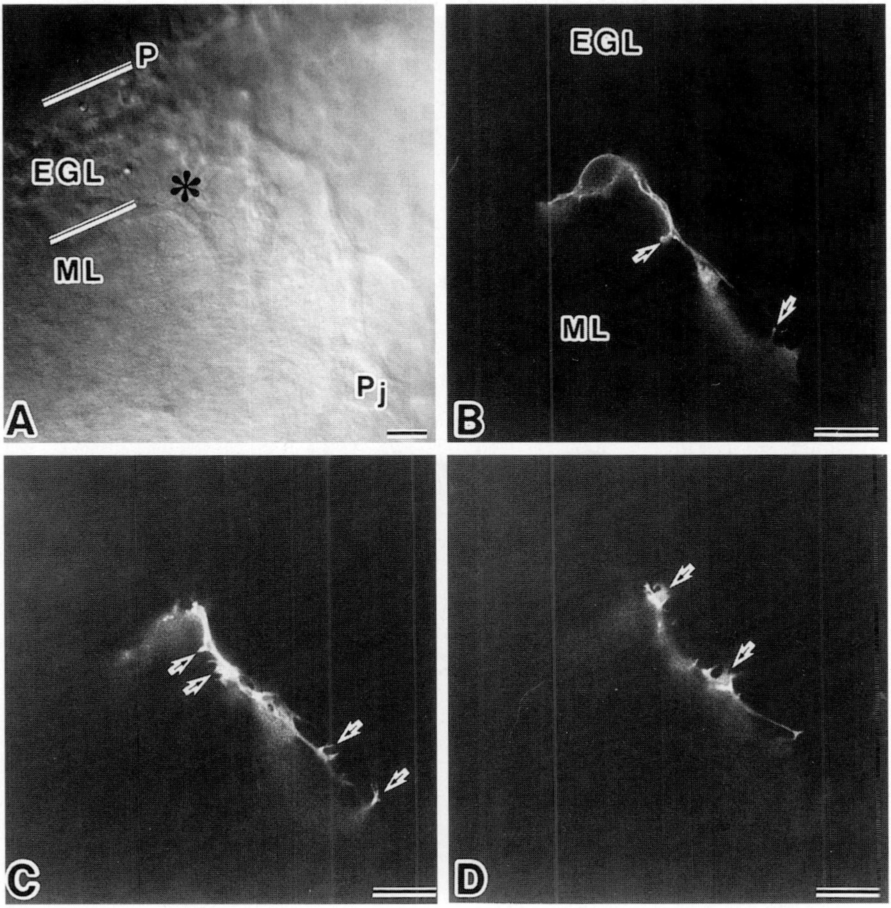

FIG. 2A–D. Laser scanning confocal microscopy of a DiI-labeled cerebellar granule cell [13] by confocal images at three different focal plane levels (**B–D**) and a differential interference optics image (**A**) of a cerebellar slice. *Asterisk* in **A** indicates the position of the DiI-labeled granule cell perikaryon shown in **B–D**. The granule cell has a horizontally oriented perikaryon with a thick vertical process extending from one side of the cell body and many side spikes (*arrows*) extending in the direction parallel to the parallel fibers. Location of the pial surface (*P*), external granular layer (*EGL*), molecular layer (*ML*), and Purkinje cell layer (*Pj*) are indicated in **A**. *Bars* 10 μm

We reexamined details of the morphology of vertically migrating granule cells in the postnatal mouse cerebellum [13]. The first sign of inward migration was extension of short vertical filopodium-like processes from the sides of the perikarya of tangentially oriented granule cells, followed by a change of orientation of cell bodies to the vertical axis. The T-shaped migratory cells formed many filopodia (side spikes) from both their vertical leading processes and perikarya at right angles to the vertical axis. The presence of such side spikes was confirmed with both a Golgi method and laser scanning confocal microscopy of DiI-labeled granule cells (Figs. 2,3). Further, transmission electron microscopy indicated that such migratory cells made direct contacts by means of their side spikes with parallel fibers [13].

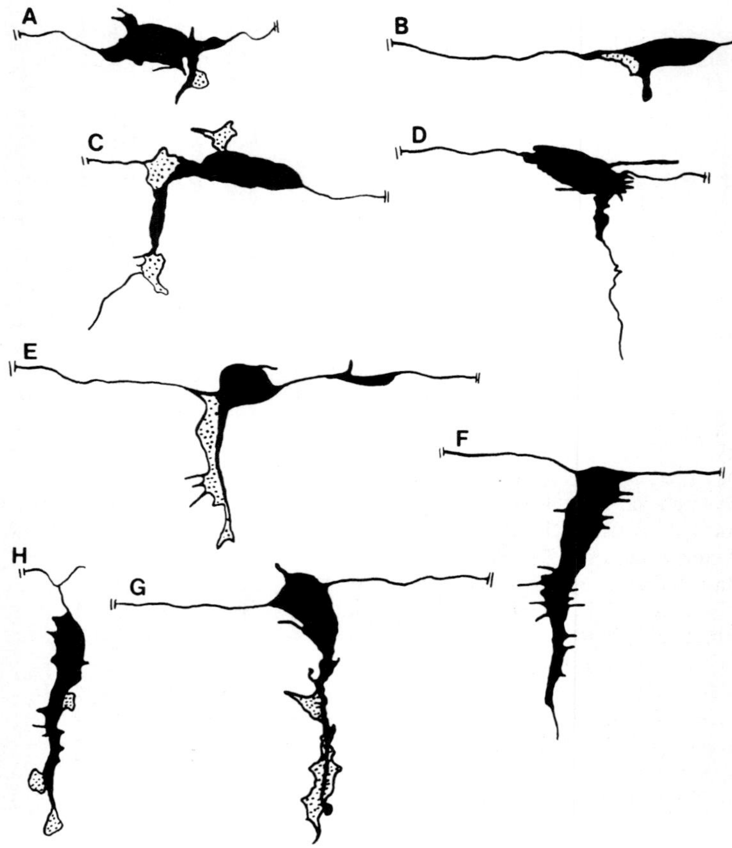

FIG. 3A–H. Camera lucida drawing of vertically migrating cerebellar granule cells examined by a Golgi method [13]. The cells have many horizontally oriented side spikes extending from the vertical leading cell process. *Dotted areas* in the drawings represent lamellipodia. **A–D** Tangential cells with vertical processes of various lengths. **E** A round cell with a vertical process has a lamellipodium along the entire length of the process. **F** A T-shaped, elongated perikaryon with many side spikes. **G** A cell with an inverted conical shape has weblike lamellipodia and a growth cone at the tip of the vertical process. **H** Vertical cells with trailing processes

These results suggested that during vertical migration growth cone-related structures of the leading processes of granule cells adhere to and probably recognize tangentially oriented parallel fibers. Therefore, the mechanisms of the vertical guidance and migration of granule cells in the cerebellar cortex seem to be multiple, involving not only contact guidance by the radial fibers of Bergmann glia but also perpendicular contact guidance by the tangential parallel fibers.

Horizontal Migration of Neuroblasts in the Cerebral Cortical Plate

Ventricular cells were labeled with fluorescent dye (PKH26) in organ or slice culture of the mouse cerebral cortex isolated from E17–E19 fetuses. After allowing cell migration for 1 day, horizontally oriented neuroblasts were frequently observed not only in the intermediate zone but also in the cortical plate, which is made of radially (vertically) aligned bundles of the radial glia and neuroblasts, and in which there is no horizontal structure. In addition, these cells had many side spikes extending vertically from the horizontal cell process.

Further, injection of DiI microcrystals into the fetal ventricle and examination of the brain slices after a few days visualized many DiI-labeled neuroblasts after migration from the ventricular layer. There were horizontally oriented cells in the cortical plate as well as in the intermediate zone. Therefore, we speculate that such horizontally oriented neurons in the cortical plate is exhibiting perpendicular contact guidance among the radially oriented bundle of cell processes of other neurons and radial glia, by being attached to them with side spikes.

References

1. Nakatsuji N, Nagata I (1989) Paradoxical perpendicular contact guidance displayed by mouse cerebellar granule cell neurons in vitro. Development (Camb) 106:441–447
2. Nagata I, Nakatsuji N (1990) Granule cell behavior on laminin in cerebellar microexplant cultures. Dev Brain Res 52:63–73
3. Nagata I, Nakatsuji N (1994) Migration behavior of granule cell neurons in cerebellar cultures. I. A PKH26 labeling study in microexplant and organotypic cultures. Dev Growth Differ 36:19–27
4. Ono K, Nakatsuji N, Nagata I (1994) Migration behavior of granule cell neurons in cerebellar cultures. II. An electron microscopic study. Dev Growth Differ 36:29–38
5. Nagata I, Nakatsuji N (1991) Rodent CNS neuroblasts exhibit both perpendicular and parallel contact guidance on the aligned parallel neurite bundle. Development (Camb) 112:581–590
6. Nagata I, Kawana A, Nakatsuji N (1993) Perpendicular contact guidance of CNS neuroblasts on artificial microstructures. Development (Camb) 117:401–408
7. Nakatsuji N, Kadokawa Y, Suemori H (1991) Radial columnar patches in the chimeric cerebral cortex visualized by use of mouse embryonic stem cells expressing β-galactosidase. Dev Growth Differ 33:571–578
8. Rakic P (1971) Neuron-glia relationship during granule cell migration in developing cerebellar cortex. A Golgi and electron microscopic study in *Macaca rhesus*. J Comp Neurol 141:283–312
9. Altman J (1972) Postnatal development of the cerebellar cortex in the rat. III. Maturation of the components of the granular layer. J Comp Neurol 145:353–398

10. Hatten ME, Heintz N (1995) Mechanisms of neural patterning and specification in the developing cerebellum. Annu Rev Neurosci 18:385–408
11. Das GD, Lammert GL, Mcallister JP (1974) Contact guidance and migratory cells in the developing cerebellum. Brain Res 69:13–29
12. Hatten ME (1990) Riding the glial monorail: a common mechanism for glial-guided neuronal migration in different regions of the developing mammalian brain. Trends Neurosci 13:179–184
13. Ono K, Shokunbi T, Nagata I, et al (1997) Filopodia and growth cones in the vertically migrating granule cells of the postnatal mouse cerebellum. Exp Brain Res 117:17–29

Part 2
Regional and Cellular Specification

Pathfinding and Cell-Fate Specification of Zebrafish Motoneurons

JUDITH S. EISEN

Summary. Normal animal behavior requires precise innervation of muscles by appropriate motoneurons. To learn how this precision arises, we have been investigating specification and pathfinding of individually identified motoneurons in embryonic zebrafish. I review our studies suggesting that motoneuron specification results from integration of responses to several signaling pathways and that myotome-derived cues participate in motor growth cone guidance.

Key words. Zebrafish—Motoneurons—Fate specification—Pathfinding—Identified neurons

Introduction

The ability of an animal to carry out its normal behavioral repertoire depends on exquisitely precise synaptic connections between motoneurons and muscles. Here I review our studies in zebrafish of motor axon pathfinding and motoneuron fate specification, designed to reveal mechanisms underlying motoneuron specificity.

Some Zebrafish Spinal Motoneurons Are Uniquely Identified

Zebrafish have two distinct classes of axial motoneurons [1]. Primaries (PMNs) are large, early-developing motoneurons present in small numbers. Secondaries (SMNs) are smaller, later-developing motoneurons that are far more numerous. Each of the three zebrafish PMNs, CaP, MiP, and RoP, can be recognized uniquely by its cell body location, its innervation territory, and its time of growth cone extension (Fig. 1a) [2,3]. During extension, each growth cone in turn elongates along a common pathway to a choice point at the nascent horizontal myoseptum where it pauses for a variable period of time before extending along its cell-specific pathway. A fourth PMN, VaP, is randomly present in only about half the spinal hemisegments (Fig. 1a) [4]. The VaP

Institute of Neuroscience, 1254 University of Oregon, Eugene, OR 97403, USA

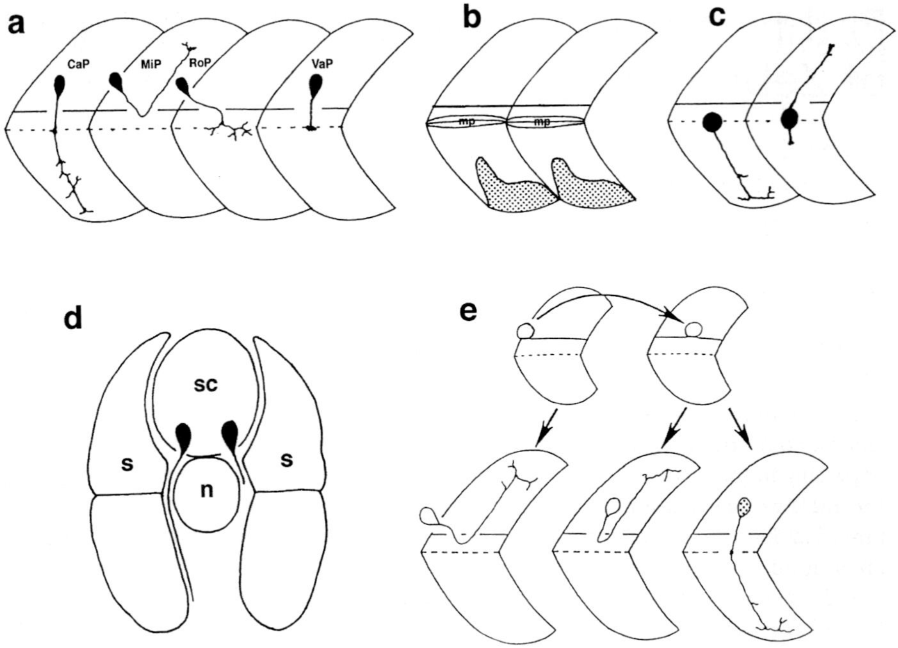

FIG. 1. **a** Primary motoneurons (PMNs) are individually identified. Side view of four segments of the trunk of **a** zebrafish embryo 24 hours post fertilization (24h), anterior to the left and dorsal to the top. **a** shows a single PMN in each segment. The *solid horizontal line* represents the ventral aspect of the spinal cord and the *dashed horizontal line* represents the nascent horizontal myosepta. **b** Somites generate the sclerotome and myotome. Side view of two segments of the trunk of a 24h zebrafish embryo showing muscle pioneer cells (*mp*) at the level of the nascent horizontal myoseptum and the sclerotome (*stippling*) just beginning to migrate dorsally from the ventral aspect of each somite. Essentially all the somite that does not form sclerotome forms myotome. **c** PMN growth cones recognize specific pathways. Side view of two segments shows normal pathway selection by *CaP* (*left*) and *MiP* (*right*) motoneurons transplanted to the level of the nascent horizontal myoseptum. **d** Permissiveness of dorsal myotome for CaP axons changes over time. Cross section of a 24h zebrafish embryo shows the spinal cord (*sc*), notochord (*n*), and somites (*s*). The CaP on the *left* shows the normal morphology in control embryos in which the CaP axon has extended on a permissive ventral myotome. The CaP axon has the same normal morphology when it extends on a young dorsal myotome. The CaP on the *right* illustrates abnormal morphology when CaP fails to extend an axon on an older dorsal myotome. **e** Positional cues may specify motoneurons. A cell in the MiP position (*upper left*) normally develops a MiP morphology (*lower left*). When transplanted from the MiP position to the CaP position (*upper right*) late, it develops a normal MiP axonal trajectory (*lower middle*). When the transplant is done early, the cell develops a normal CaP morphology (*lower right*) and expresses *islet2* (*stippling*), a gene normally expressed by CaP and not by MiP

cell body is located directly adjacent to the CaP cell body; most VaPs die. SMNs extend growth cones several hours later than PMNs [3,5]. At least some SMN growth cones appear to fasciculate with PMN axons, and the two cell types innervate the same targets [6].

Mesodermally Derived Signals Participate in Motor Axon Guidance

Guidance signals may be myotome-derived. Somite generates both myotomes and sclerotomes (Fig. 1b). The somite-derived sclerotome is implicated in pathfinding by chick limb motor axons. However, ablation of zebrafish sclerotome had essentially no effect on PMN pathfinding, suggesting that guidance cues are myotome derived [7].

An identified choice point is unnecessary for motor axon guidance. All PMN growth cones pause at a choice point [2,3,8], the muscle pioneer cells (Fig. 1b). It is appealing to speculate that the *netrin*-expressing [9,10] muscle pioneers are necessary for PMN pathway choice. However, motoneuron projections appear normal in mutants lacking muscle pioneers, such as *no tail* [11], and following muscle pioneer ablation [8], suggesting this is not the case.

PMN growth cones distinguish among cell-specific pathways. Individual PMNs transplanted to the choice point where they had access to all three cell-specific pathways made normal pathway choices (Fig. 1c) [12]. This result suggests that each pathway has distinct cues recognized by PMN growth cones, an idea supported by isolation of mutations affecting pathfinding only by CaP motoneurons [13,14].

Notochord-derived signals may establish differences between pathways. Localization of the CaP pathway to ventral myotome requires notochord-derived signals [15], because dorsal myotomes deprived of notochord exposure are permissive for CaP axons, while dorsal myotomes that have been exposed to notochord are nonpermissive (Fig. 1d). This distinction appears important for pathway choice, because CaP axons can distinguish permissive from nonpermissive myotomes in choice tests.

Myotome-derived and motoneuron-derived signals may act together during guidance. Although some zebrafish motor axon guidance cues appear to derive from myotome, experimental evidence suggests that interactions among PMNs affect more subtle aspects of pathfinding. For example, ablation of CaP does not alter the MiP growth cone pathway choice [16]; however, it alters retention of a transient MiP processes [17], an effect enhanced by ablation of both CaP and muscle pioneers [8].

Motoneuron Specification Results from Integration of Several Signaling Pathways

Midline-derived signals specify motoneuron fate. Spinal cord neurons develop characteristic fates at specific dorsoventral locations, and midline signals, in particular Sonic hedgehog (Shh), appear to pattern ventral spinal cord [18]. Mutant analysis suggests midline signals also specify zebrafish motoneurons and that PMNs and SMNs may be specified by temporally separated Shh signaling [19,20].

Local signals regulate motoneuron number and position. How can a single morphogen specify distinct cell types at the same dorsoventral spinal cord position? Studies in mice raise the possibility that motoneurons may influence neighboring cells [21]. Our studies suggest that lateral inhibition, mediated by homologs of the fly neurogenic genes, *Notch* and *Delta*, contributes to specification of identified spinal cord neurons in zebrafish [22]. Zebrafish have at least four *delta* genes [22–24] and at least four *notch* genes [25,26]. Expression of a dominant negative (dn) *Xenopus laevis* Delta [27] results in formation of supernumerary PMNs and concomitant loss of nearby SMNs and interneurons [22], suggesting that lateral inhibition normally regulates these cell fate choices in the zebrafish spinal cord.

Lateral inhibition may regulate reponse to inductive signals. The time at which cells are competent to respond to inductive signals appears to influence their specification for particular fates [28,29]. Regulation of zebrafish PMN and SMN specification fits well with this model [22]. We propose that lateral inhibition regulates the number of cells competent to respond to any particular instructive signal, thus diversifying neuronal fates in the zebrafish ventral spinal cord.

Positional Cues May Direct Expression of Homeobox Genes That Define Motoneuronal Fates

Combinations of LIM homeobox genes define motoneuron subpopulations. In developing chick embryos, different expression patterns of LIM homeobox genes correlate with spinal motoneuron identity as defined by cell body position and axonal trajectory [30]. Zebrafish PMNs can also be distinguished by their patterns of expression of the LIM homeobox genes *isl1*, *isl2*, and *lim3* [31–33].

Positional cues may direct axonal trajectories. Some PMNs can alter their axonal trajectories when transplanted to new spinal cord positions [34]. LIM homeobox genes may mediate positional regulation of PMN fate, since transplanted PMNs with altered axonal trajectories also altered LIM homeobox gene expression patterns (Fig. 1e) [33].

Competition may determine fates of specific primary motoneurons. In many vertebrate species motoneurons are overproduced by twofold and die during later development [35,36], probably from competition for limited target-derived trophic support [37]. In zebrafish, a single PMN, VaP, typically dies [4]. VaP is initially specified as CaP, and interactions with CaP are responsible for VaP's demise [38]. Thus, CaP and VaP form an equivalence pair in which CaP is the preferred fate and the VaP fate is determined by interactions between the two cells. It is currently unknown how these interactions relate to the competitive interactions described for motoneurons in amniotes.

Closing Remarks

We are now forming a picture of the processes regulating cell-fate specification and pathfinding by zebrafish motoneurons. Early signals such as Shh appear to induce competent cells in the ventral spinal cord to take on fates that depend on the timing of signal exposure. Interactions among these cells, such as Delta/Notch-mediated lateral inhibition, then sort out specific fates and regulate the number and position in which motoneurons develop. Cell-fate specification involves initiating expression of genes, such as LIM homeobox genes, that likely control axonal pathfinding, probably by regulating expression of receptors that recognize cell-specific axon pathways that lead toward particular targets. Later interactions among motoneurons then further refine motoneuron number. At the same time that motoneuron fates are being specified, the same notochord-derived signals are specifying the fates of cells within the myotome that are the likely source of cues for motor axon guidance. Of course, this is still only the broadest outline of these processes, and there is clearly much left to do to really understand how it all works.

Acknowledgments. I thank all my collaborators who contributed to the studies described here, the support services of the University of Oregon, and Bruce Appel and Bruce Bowerman for comments on earlier drafts of the manuscript. Research in my lab is supported by NIH grants NS23915 and HD22486.

References

1. Myers PZ (1985) Spinal motoneurons of the larval zebrafish. J Comp Neurol 263:555–561
2. Eisen JS, Myers PZ, Westerfield M (1986) Pathway selection by growth cones of identified motoneurons in live zebrafish embryos. Nature (Lond) 320:269–271
3. Myers PZ, Eisen JS, Westerfield M (1986) Development and axonal outgrowth of identified motoneurons in the zebrafish. J Neurosci 6:2278–2289
4. Eisen JS, Pike SH, Romancier B (1990) An identified neuron with variable fates in embryonic zebrafish. J Neurosci 10:34–43
5. Pike SH, Melançon EF, Eisen JS (1992) Pathfinding of zebrafish secondary motoneurons in the absence of normal pioneer axons. Development (Camb) 114:825–831
6. Westerfield M, McMurray J, Eisen JS (1986) Identified motoneurons and their innervation of axial muscles in the zebrafish. J Neurosci 6:2267–2277
7. Morin-Kensicki EM, Eisen JS (1997) Sclerotome development and peripheral nervous system segmentation in embryonic zebrafish. Development (Camb) 124:159–167
8. Melancon E, Liu DWC, Westerfield M, et al (1997) Pathfinding by identified zebrafish motoneurons in the absence of muscle pioneers. J Neurosci 17:7796–7804
9. Lauderdale JD, Davis NM, Kuwada JY (1997) Axon tracts correlate with netrin-1a expression in the zebrafish embryo. Mol Cell Neurosci 9:293–313
10. Blader P, Fischer N, Gradwohl G, et al (1997) The activity of Neurogenin1 is controlled by local cues in the zebrafish embryo. Development (Camb) 124:4557–4569
11. Halpern ME, Ho RK, Walker C, et al (1993) Induction of muscle pioneers and floor plate is distinguished by the zebrafish *no tail* mutation. Cell 75:99–111
12. Gatchalian CL, Eisen JS (1992) Pathway selection by growth cones of ectopic motoneurons in embryonic zebrafish. Neuron 9:105–112
13. Beattie CE, Eisen JS (1996) Mutations affecting motoneuronal pathfinding in embryonic zebrafish. Soc Neurosci Abstr 22:1716
14. Beattie CE, Stoesser KL, Eisen JS (1995) Motoneuronal mutants in embryonic zebrafish. Soc Neurosci Abstr 21:1511
15. Beattie CE, Eisen JS (1997) Notochord alters the permissiveness of myotome for pathfinding by an identified motoneuron in embryonic zebrafish. Development (Camb) 124:713–720
16. Eisen JS, Pike SH, Debu B (1989) The growth cones of identified motoneurons in embryonic zebrafish select appropriate pathways in the absence of specific cellular interactions. Neuron 2:1097–1104

17. Pike SH, Eisen JS (1990) Interactions between identified motoneurons in embryonic zebrafish are not required for normal motoneuron development. J Neurosci 10:44–49
18. Tanabe Y, Jessell TM (1996) Diversity and pattern in the developing spinal cord. Science 274:1115–1123
19. Beattie CE, Hatta K, Halpern ME, Liu H, Eisen JS, Kimmel CB (1997) Temporal separation in specification of primary and secondary motoneurons in zebrafish. Dev Biol 187:171–182
20. Halpern ME, Hatta K, Amacher SL, et al (1997) Genetic interactions in zebrafish midline development. Dev Biol 187:154–170
21. Pfaff SL, Mendelsohn M, Stewart CL, et al (1996) Requirement of LIM homeobox gene *Isl1* in motor neuron-dependent step in interneuron differentiation. Cell 84:309–320
22. Appel B, Eisen JS (1998) Regulation of neuronal specification in the zebrafish spinal cord by Delta function. Development (Camb) 125:371–380
23 Dornseifer P, Takke C, Campos-Ortega JA (1997) Overexpression of a zebrafish homologue of the *Drosophila* neurogenic gene *Delta* perturbs differentiation of primary neurons and somite development. Mech Dev 63:159–171
24. Haddon C, Smithers L, Schneider-Maunoury S, et al (1998) Multiple delta genes and lateral inhibition in zebrafish primary neurogenesis. Development (Camb) 125:359–370
25. Bierkamp C, Campos-Ortega JA (1993) A zebrafish homologue of the *Drosophila* neurogenic gene Notch and its pattern of transcription during early embryogenesis. Mech Dev 43:87–100
26. Westin J, Lardelli M (1997) Three novel *Notch* genes in zebrafish: implications for vertebrate *Notch* gene evolution and function. Dev Genes Evol 207:51–63
27. Chitnis A, Henrique D, Lewis J, et al (1995) Primary neurogenesis in *Xenopus* embryos regulated by a homologue of the *Drosophila* neurogenic gene Delta. Nature (Lond) 375:761–766
28. Fortini ME, Rebay I, Caron LA, et al (1993) An activated Notch receptor blocks cell-fate commitment in the developing *Drosophila* eye. Nature (Lond) 365:555–557
29. Dorsky RI, Rapaport DH, Harris WA (1995) Xotch inhibits cell differentiation in the *Xenopus* retina. Neuron 14:487–496
30. Tsuchida T, Ensini M, Morton SB, et al (1994) Topographic organization of embryonic motor neurons defined by expression of LIM homeobox genes. Cell 79:957–970
31. Inoue A, Takahashi M, Hatta K, et al (1994) Developmental regulation of *Islet-1* mRNA expression during neuronal differentiation in embryonic zebrafish. Dev Dyn 199:1–11
32. Tokumoto M, Gong Z, Tsubokawa T, et al (1995) Molecular heterogeneity among primary motoneurons and within myotomes revealed by the differential mRNA expression of novel *Islet-1* homologs in embryonic zebrafish. Dev Biol 171:578–589
33. Appel B, Korzh V, Glasgow E, et al (1995) Motoneuron fate specification revealed by patterned LIM homeobox gene expression in embryonic zebrafish. Development (Camb) 121:4114–4125
34. Eisen JS (1991) Determination of primary motoneuron identity in developing zebrafish embryos. Science 252:569–572
35. Hamburger V, Oppenheim RW (1982) Naturally occurring neuronal death in vertebrates. Neurosci Commun 1:39–55
36. O'Brien MK, Oppenheim RW (1989) Development and survival of thoracic motoneurons and hindlimb musculature following transplantation of the thoracic neural tube to the lumbar region in the chick embryo: anatomical aspects. J Neurobiol 21:313–340
37. Oppenheim RW (1989) The neurotrophic theory and naturally occurring motoneuron death. TINS (Trends Neurosci) 12:252–257
38. Eisen JS (1992) The role of interactions in determining cell fate of two identified motoneurons in the embryonic zebrafish. Neuron 8:231–240

Essential Role of Zic Genes in Neural Induction and Brain Pattern Formation

KATSUHIKO MIKOSHIBA[1,2], JUN ARUGA[2], TAKEHARU NAGAI[1,2], and KATSUNORI NAKATA[1,2]

Summary. Zic genes belong to a group of genes coding for zinc finger proteins expressed in the adult mouse cerebellum. These genes are the vertebrate homologs of *Drosophila* odd-paired, which plays an essential role in parasegmental subdivision and in visceral mesoderm development. The expression of the three Zic genes was first detected at gastrulation. At neurulation, the expression became restricted to the dorsal neural ectoderm and dorsal paraxial mesoderm. During organogenesis, the three genes were expressed in specific regions of several developing organs, including dorsal areas of the brain and spinal cord. Zic 1-deficient mice showed cerebellar abnormality with a missing lobule in the anterior lobe. Zic 1 may determine the cerebellar folial pattern principally via regulation of cell proliferation in the external germinal layer. It was found that the human X-linked situs abnormality (situs inversus, situs ambiguus) is caused by missense or nonsense mutation of the Zic 3 gene. To clarify the role of Zic in the early stage of neural development, we have cloned the *Xenopus* homolog of Zic, Xzic3. Xzic3 expression began at the prospective neuroectoderm as soon as neural induction occurred and was induced by blockage of bone morphogenetic protein (BMP-) mediated signaling in the *Xenopus* embryo. Overexpression of Xzic3 causes neural hyperplasia involved with induction of all the proneural genes reported so far. In addition to the change in the neural tissue, the generation of neural crest-derived tissue is promoted. These results suggest an essential role of Xzic3 in the induction of the neuroectoderm.

Key words. Zic—Zinc finger protein—Cerebellum—opa gene—BMP-4

Introduction

The coordinated process of body pattern formation in vertebrates is governed by a number of genes, the expression of which is regulated developmentally. Many of the genes that are involved in the formation of the vertebrate body are evolutionally

[1]Department of Molecular Neurobiology, Institute of Medical Science, The University of Tokyo, 4-6-1 Shirokanedai, Minato-ku, Tokyo 108-8639, Japan
[2]Developmental Neurobiology Laboratory, RIKEN Brain Science Institute, 2-7 Hirosawa, Wako, Saitama 351-0198, Japan

conserved in *Drosophila*. The genes are often grouped by similarity or a conserved structural motif.

The mouse Zic gene encodes a zinc finger protein and is expressed in the matured cerebellum, highly restricted to the granule cells, but is also expressed in the developing central nervous system [1]. Four Zic genes are identified, and they belong to the same gene family. They are highly similar each other in the common exon–intron boundaries.

A recently found *Drosophila* pair-rule gene, *odd-paired*, is highly homologous to the Zic gene family. The similarity is not only the zinc finger motif, but also the exon–intron boundary, which are the same as those of the mouse Zic gene family. These findings suggest that the Zic gene family and *Drosophila odd-paired* are derived from a common ancestral gene. The *opa* mutation causes severe pair-rule-type segmentation abnormalities and impairment in the timing of activation of the segment polarity genes *wingless* and *engrailed*. The vertebrate homologs of these genes, En1, En2, and Wnt-1, were shown to be involved in vertebrate neural pattern formation. Zic1, Zic2, Zic3, and Zic4 were determined to be localized to mouse chromosome 9, 14, X, and 9 using an interspecific backcross panel.

Expression of the Zic Genes at Gastrulation and Neurulation Stages

The expression of the three Zic genes was first detected at gastrulation in a spatially restricted manner. At neurulation, the expression became restricted to the dorsal neural ectoderm and dorsal paraxial mesoderm. During organogenesis, the three genes were expressed in specific regions of several developing organs, including dorsal areas of the brain, spinal cord, paraxial mesenchyme, and epidermis, the marginal zone of the neural retina, and distal regions of the developing limb. For all stages, significant differences in the spatial expression of Zic1, Zic2, and Zic3 were observed (Fig. 1). Furthermore, the expression of Zic genes in Pax 3, Wnt-1, and Wnt-3a mutant embryos suggested that Zic genes are not primarily regulated by the three genes that were expressed in dorsal areas similar to Zic genes. However, in open brain, a mutant with severe neural tube defects, and in Wnt-3a mutant mice, the expression of Zic genes was changed. The changed expression pattern in Wnt-3a mutant mice suggests that Zic genes in the neural tube are regulated by factors from the notochord.

We then compared the expression of the three mouse Zic genes. Generally, the three genes are expressed in the dorsal structure of the embryo. From 8.0 to 16.5 days postcoitum (dpc), all three genes were expressed in the dorsal one-third of the spinal cord and dorsal mesoderm-derived structure. Although expression along the dorsoventral and the mediolateral axes was quite similar among the three genes, the expression profile along the anteroposterior axis seemed to be varied. Zic1 is relatively evenly distributed along the anteroposterior axis, but no expression was detected in the tail region. In contrast, Zic2 and Zic3 were highly expressed in the head and tail bud, but weakly expressed in the trunk region. These findings suggested that the three genes could have complementary roles along the anteroposterior axis. Differences

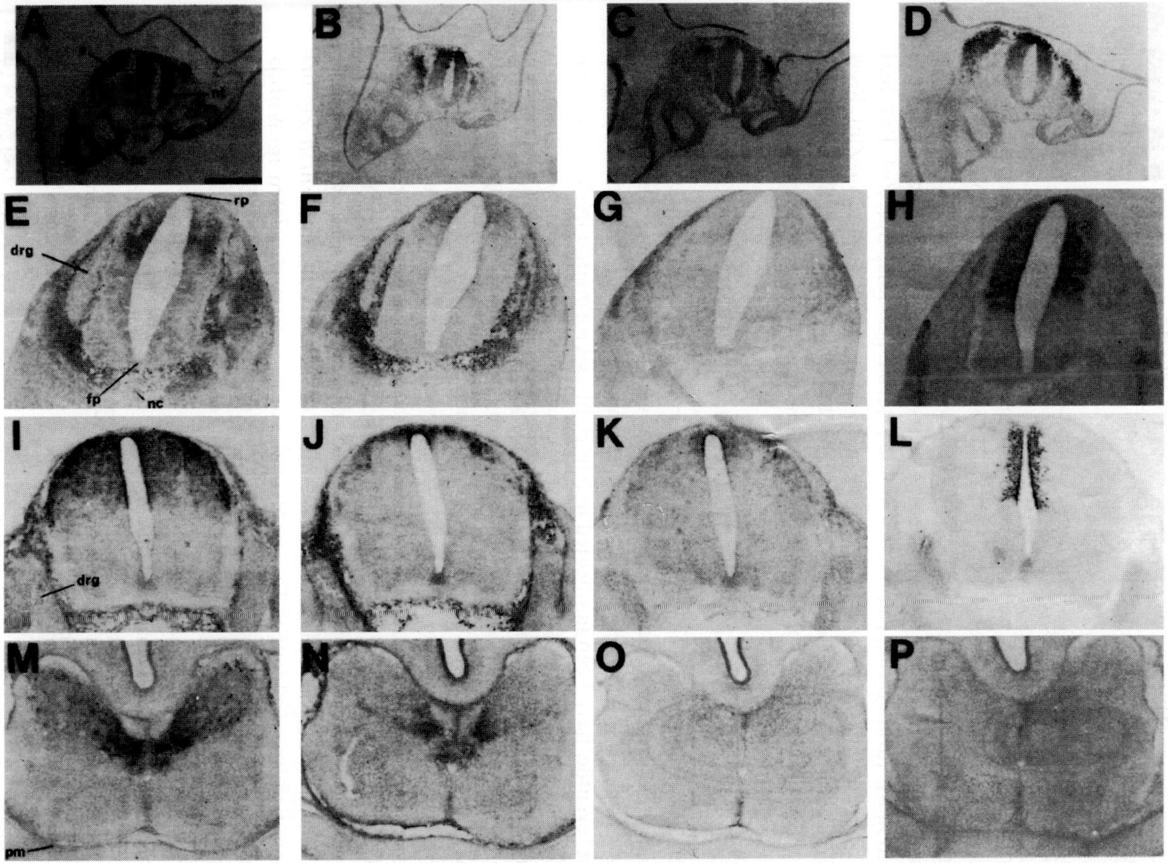

FIG. 1A–P. Dorsoventral expression of Zic genes in a series of transverse sections through the caudal region of the cervical spinal cord. Zic1 (**A, E, I, M**), Zic2 (**B, F, J, N**), Zic3 (**C, G, K, O**), and Pax3 (**D, H, L, P**) expression was examined at 8.0 (**A–D**), 10.5 (**E–H**), 12.5 (**I–L**), and 14.5 dpc (**M–P**). *drg*, dorsal root ganglion; *fp*, floor plate; *nc*, notochord; *nt*, neural tube; *pc*, pericardioperitoneal canal; *pm*, primitive meninx; *rp*, roof plate; *s*, somites. *Bar* 200 mm between Zic1, and Zic2 and Zic3, were also found in neuroectoderm formation at the primitive streak stage. Both Zic2 and Zic3 are expressed in the primitive ectoderm, whereas Zic1 was not (Fig. 1); it was detected after mesoderm and neuroectoderm formation. The induction of Zic1 expression seems to be closely associated with mesoderm and neuroectoderm formation. Our findings suggest that Zic genes are involved in many developmental processes.

Zic1 Is Involved in Cerebellar Development

To clarify the role of the Zic1 gene, we have generated mice deficient in Zic 1. Homozygous mice showed remarkable ataxia during postnatal development. Nearly all the mice died within 1 month. Their cerebella were hypoplastic, and a lobule was missing in the anterior lobe. A bromodeoxyuridine labeling study indicated a reduction both in the proliferating cell fraction in the external germinal layer, from 14 days postcoitum, and in forward movement of the external germinal layer. These findings suggest that Zic1 may determine the cerebellar foliar pattern principally via regulation of cell proliferation in the external germinal layer.

Zic3 Is a Causal Gene for X-Linked Situs Abnormalities

Aberrant left–right axis development can lead to randomization of individual organ position (situs ambiguus) or to mirror-image reversal of all lateralized structures (situs inversus). It was reported that sonic hedgehog and the vertebrate homolog of wingless, cWnt8c, have been implicated in left–right axis formation in the chick. The *Drosophila* homolog of Zic is required for the appropriate expression of wingless, a segment polarity gene that interacts with hedgehog to maintain parasegmental identity. No genes have been convincingly associated with human situs abnormalities. Initially, there were reports suggesting that mutations in the gap junction gene, connexin 43, may be related to situs ambiguus. However, connexin 43 mutations have not been detected in follow-up studies. The molecular genetic analysis of a group of familial and sporadic cases with possible X-linked inheritance was carried out to elucidate human situs abnormalities. By positional cloning of the Zic3 gene, one frameshift, two missense, and two nonsense mutations have been identified in familial and sporadic situs ambiguus. The frameshift allele is also associated with situs inversus among some heterozygous females. Zic3 may function in the earliest stages of left–right axis formation. Zic3 is the first gene unequivocally associated with human situs abnormalities.

Role of *Xenopus* Zic3, a Primary Regulator in Both Neural and Neural Crest Development

In *Xenopus laevis*, the neuroectoderm is induced on the dorsal side by signals such noggin, chordin, and follistatin emanating from the Spemann organizer during early gastrulation. BMP4 has an inhibitory effect on ectoderm to differentiate into neuroectoderm, and the blockade of the signal triggers the neuroectoderm formation. Noggin, chordin, and follistatin organizer signals may act as endogenous blockages for BMP4 by direct binding. Several basic helix-loop-helix transcription factors have been shown to participate in vertebrate neural development. Neurogenin, NeuroD, XASH-3, and ZATH-3 are shown to have neurogenic activity, as their overexpression in *Xenopus* embryos results in enlargement of neural tissue or ectopic neural cell induction. The link between the BMP4 signal and proneural genes is not well understood.

Xenopus *Zic3 Is Expressed in the Prospective Neural Plate in the Gastrula and Neural Folds in the Neurula*

To clarify the role of Zic in the early stage of neural development, we have cloned the *Xenopus* Zic3. Zic3 expression was first detected in the early gastrula in the dorsal lip and the prospective neural plate was detected by whole-mount in situ hybridization. Expression decreased in the dorsal lip and increased in the prospective neural plate as gastrulation proceeded. In the late gastrula, Zic3 expression diminished gradually in the central region. At the neural plate stage, Zic3 was expressed strongly in the prospective region of the mesencephalon and anterior rhombencephalon. Zic3 then became stronger in expression in the anterior neural folds, whereas that in the trunk neural folds remained weak. At early tailbud stages, Zic3 expression was gradually restricted to the dorsal region of forebrain, midbrain, and hindbrain and weakly to the dorsal region of the trunk. After the mid-tailbud stage, Zic3 expression diminished in the diencephalon and additional expression became clear in the lateral mesoderm of the tailbud region. Zic3 expression was restricted in the dorsal part of the neural tube.

Zic3 Is Induced by Blockade of BMP4-Mediated Signaling

To examine whether Zic3 is induced in animal cap explant under such a condition, we cultured animal caps in dispersal to the time point equivalent to midneurula, and the expression of Zic 3 and other genes was tested by RT-PCR (Fig. 2A,B). Zic3 was expressed in dispersed animal caps with neural marker NCAM, which was not detected in intact caps in which the epidermal marker, epidermal keratin, expressed. Because the neuralization that occurred in the dispersed animal caps is considered to result from the attenuation of BMP4-mediated signal, we tested if blockage of BMP4-mediated signal actually can induce Zic3 expression in vivo. A dominant-negative form of BMP receptor, or noggin, was overexpressed in the ventral region by injection into the two-cell-stage embryo. The embryos were subjected to in situ hybridization with Zic3 probe at the early gastrula stage. Both BMP4 mRNA and noggin mRNA were able to induce ectopic Zic3 mRNA in the ventromarginal zone in the gastrula. Blockage of BMP4 signaling is sufficient to induce Zic3 in vivo.

Overexpression of Zic3 in Xenopus *Embryos Leads to Hyperplasia of Neural and Neural Crest Tissue*

Zic3 overexpression in the developing embryos results in expansion of the neural plate and hyperplasia of the neural tissue (Fig. 3). Moreover, several neural markers are induced in the Zic3 overexpressed animal caps. These data indicate that Zic3 acts

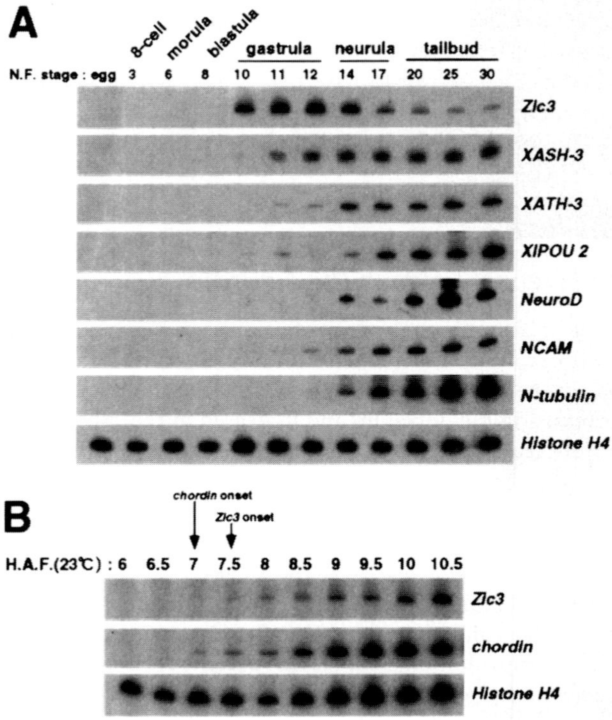

FIG. 2A,B. Temporal expression profiles of Zic3 and neural marker genes during development. **A** RNA was extracted from embryos at the indicated stage of development, and the content of Zic3 and neural marker genes (XASH-3, XATH-3, XIPOU 2, NeuroD, NCAM, N-tubulin) mRNA were measured by RT-PCR. **B** To compare the onset of the expression of Zic3 and the neural inducer, chordin, during development, RNA was extracted from embryos at the indicated time after artificial fertilization and the expression level was determined by RT-PCR. The ubiquitous marker, Histone H4, served as a control for both **A** and **B**. *H.A.F.*, hours after fertilization

in the earliest phase of neural development and that Zic3 belong to a novel classes of genes involved in neural development, based on the following three points. First, initiation of Zic3 expression is in the late blastula, earlier than that of most proneural genes. Although Neurogenin transcripts are detected earlier than Zic3 expression, as a maternal message in our assay, the function of Neurogenin as a proneural gene seems to be activated after neural induction occurs (Fig. 4). Second, Zic3 is first expressed in almost the entire prospective neural plate region. This spatial expression pattern is considerably different from those of proneural genes. Neurogenin, NeuroD, and XATH-3 are expressed in the three longitudinal regions where the primary neurons are generated. As for ZASH-3 and XlPOU2, restricted expression in these regions has not been reported. However, ZASH-3 expression is detected in a restricted region of the neural plate particularly from its onset of expression. The ZASH-3 expression pattern is in contrast to that of Zic3, which is first expressed in almost the entire

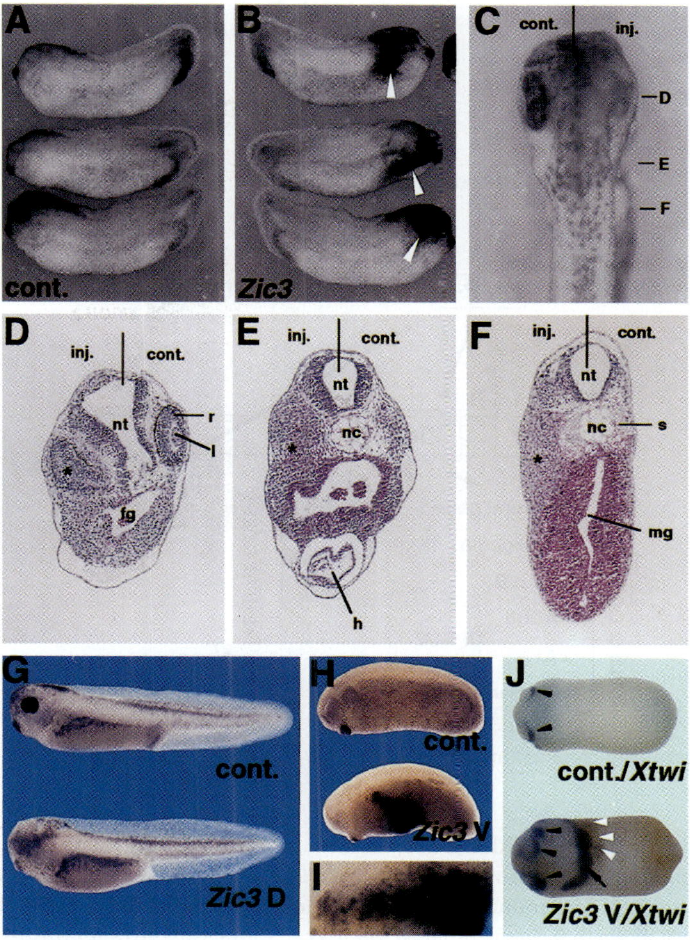

FIG. 3A–J. Zic3 overexpression induces neural tissue hyperplasia and ectopic pigment cell expression. Zic3 mRNA (**A–F**) was injected into one blastomere of two-cell-stage embryos. **A** Uninjected control side (stage 27). **B** Zic3 mRNA-injected side (stage 27). Clusters of pigment cells appeared in the anterior region (*arrowheads*). **C** Dorsal view of anterior region of the injected embryo (*inj.*) (stage 36). **D–F** Transverse sections of the embryo shown in C; the levels of the sections are indicated. The injected side is characterized by hyperplastic neural walls, distorted eye, and unidentified tissue, probably derived from the neural crest (*asterisk* in **D, E, F**) (*cont.*, control). Zic3 mRNA or control LacZ mRNA was injected into dorsoanimal (**G**) or ventroanimal (**H–J**) two blastomeres of the eight-cell-stage embryo and cultured to stage 36 (**G**), stage 25 (**H, I**), or stage 20 (**J**). Lateral view of control LacZ mRNA-injected embryos (**G** and **H**, *upper*) and Zic3 mRNA-injected embryos (**G** and **H**, *lower*). Eye abnormalities were observed in embryos injected with Zic3 mRNA into the dorsal blastomeres (**G**). Remarkable clusters of ectopic pigment cells appeared in embryos injected with Zic3 mRNA into the ventral blastomeres (**H**). **I** Higher magnification of the pigment cell clusters in **H**. **J** Xtwi expression in the embryos (*ventral view*) injected with LacZ mRNA (*upper*) or Zic3 mRNA (*lower*) into the ventroanimal two blastomeres at the eight-cell stage. Xtwi expression was observed in the head neural crest of the control embryo (*upper*; *arrowheads*). However, Zic3 mRNA injection induced ectopic Xtwi expression (*lower*; *arrow*) near the ectopic clusters of pigment cells (*white arrowheads*; the *brown* staining indicates the remnants of the pigments) in the ventral side of the embryo. Expansion of the Xtwi-expressing cephalic neural crest (*lower*; *black arrowheads*) was also observed in this embryo

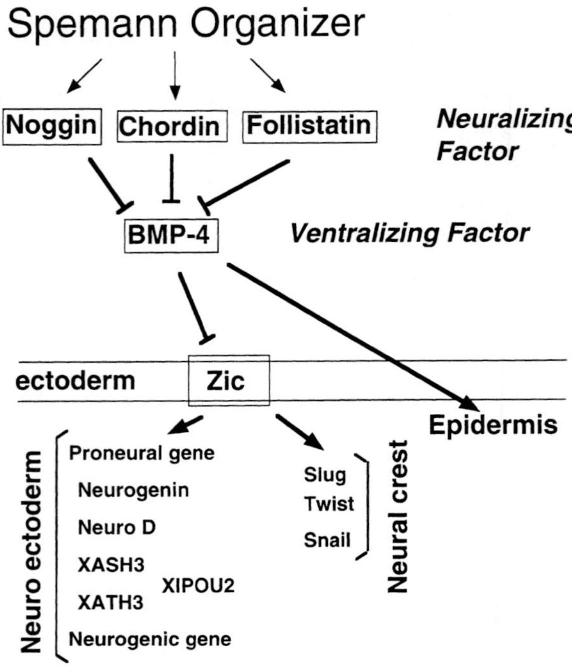

FIG. 4. Mechanism of neural induction

prospective neural plate. XlPOU2 is detected as early as stage 10 in the involuting mesoderm, but not in prospective neural plate where Zic3 is expressed. Finally, overexpression of Zic3 induces the expression of all the proneural genes in the animal cap. It is not likely that a gene other than Zic3 can induce all the other proneural genes based on their spatiotemporal expression patterns.

Zic3 Functions in Neural Crest Formation

Zic3 is expressed in the neural folds including the neural crest, and strongly in the cephalic region at stage 12–16. Furthermore, Zic3 overexpression induces ectopic pigment cells, which derive from the neural crest and neural crest markers; Xtwi and Wslu are induced in the Zic3 overexpressed animal cap explant. All these findings suggest that Zic3 has an essential role in neural crest induction.

References

1. Aruga J, Yokota N, Hashimoto M, Furuichi T, Fukuda M, Mikoshiba K (1994) A novel zinc finger protein, zic, is involved in neurogenesis, especially in the cell lineage of cerebellar granule cells. J Neurochem 63:1880–1890
2. Aruga J, Nagai T, Tokuyama T, Hayashizaki Y, Okazaki Y, Chapman VM, Mikoshiba K (1996) The mouse zic gene family: homologues of the *Drosophila* pair-rule gene odd paired. J Biol Chem 271:1043–1047

3. Aruga J, Yozu A, Hayashizaki Y, Okazaki Y, Chapman VM, Mikoshiba K (1996) Identification and characterization of Zic 4, a new member of the mouse Zic gene family. Gene 172:291–294
4. Nagai T, Aruga J, Takeda S, Günter,T, Spörle R, Schughart K, Mikoshiba K (1997) The expression of the mouse Zic1, Zic2, and Zic3 gene suggests an essential role for Zic genes in body pattern formation. Dev Biol 182:299–313
5. Nakata K, Nagai T, Aruga J, Mikoshiba K (1997) *Xenopus* Zic3, a primary regulator both in neural and neural crest development. Proc Natl Acad Sci USA 94:11980–11985
6. Aruga J, Minowa O, Yaginuma H, Kuno J, Nagai T, Noda T, Mikoshiba K (1998) Mouse Zic 1 is involved in cerebellar development. J Neurosci 18:284–293
7. Benedyk MJ, Mullen JR, Dinardo S (1994) odd-paired: a zinc finger pair-rule protein required for the timely activation of engrailed and wingless in *Drosophila* embryos. Genes Dev 8:105–117

Molecular Mechanisms for Regional Specification of the Developing Neuroepithelium

MASATO NAKAFUKU[1,$], YASUSHI NAKAGAWA[2,#], MASA-AKI TORII[1], MASAKI KATOH[1], EIJI MATSUNAGA[1], SHIN-ICHI FUKAMI[1], NORIKO OSUMI[2], and SHUN NAKAMURA[2]

The molecular basis for the highly organized structure of the brain has long been a central issue to understand in developmental neurobiology. The recent discoveries of genes with discrete expression domains in the developing brain has suggested that they play crucial roles in the early patterning of the neuroepithelium as well as in the subsequent specification of divergent neuronal phenotypes. In this study, we explored how expression of these region-specific genes are regulated in neuroepithelial cells, and how their expression correlates with regional specificity of the cells. First, by using in vitro culture of both primary and immortalized neuroepithelial cells from the rat brain at embryonic day 11.5, we demonstrate that differential expression patterns of *Pax-3*, *Pax-5*, *Pax-6*, *Dlx-1*, *Dlx-2*, *Emx2*, *Otx1*, and *Dbx* observed in vivo were strictly maintained even when the cells were isolated free from environment. Furthermore, distinct neuroepithelial cells that maintained distinct regional identities in vitro expressed different sets of genes including *Islet-1*, *Nkx-2.1*, and *Nkx-2.2* in response to Sonic hedgehog, which is a major inductive signal from the environment. These results suggest that certain cell-autonomous mechanisms operate in regulating both environmental signal-dependent and -independent expression of region-specific genes. Next, we asked whether expression of particular sets of genes observed in vitro actually correlates with particular regional phenotypes in vivo. When immortalized neuroepithelial cells were transplanted into the embryonic brain, they could be integrated into the host neuroepithelium at particular positions. Importantly, the preferential integration sites of the cells along the rostro-caudal and dorsoventral neuraxis in vivo appear to be consistent with their observed regional characters in vitro. Taken together, our in vitro and in vivo studies strongly support the idea that cell lineage plays an essential role in defining regional specificity of neuroepithelial cells during development of mammalian brain.

---

[1] Graduate School of Biological Sciences, Nara Advanced Institute of Science and Technology, Ikoma, Nara 630, Japan

[2] Division of Biochemistry and Cellular Biology, National Institute of Neuroscience, Kodaira, Tokyo 187-8502, Japan

[$] *Present address*: Division of Neurobiology, Department of Basic Neuroscience, The University of Tokyo, Graduate School of Medicine, 7-3-1 Hongo, Bunkyo-ku, Tokyo 113, Japan

[#] *Present address*: Molecular Neurobiology Laboratory, The Salk Institute for Biological Studies, La Jolla, CA, USA

Role of Eph Receptors in Patterning Avian Trunk Neural Crest Migration

Rusty Lansford[1], Catherine E. Krull[1], Nicholas W. Gale[2], Andres Collazo[1], Christophe Marcelle[1], George D. Yancopoulos[2], Marianne Bronner-Fraser[1], and Scott E. Fraser[1]

Summary. We show here that Eph-related receptor tyrosine kinases (RTKs) and their ligands function in the segmental migration of avian trunk neural crest cells through the somites. The EphB3 receptor of the EphB subfamily of RTKs localizes to the rostral half-sclerotome, including the neural crest. Ephrin-B1 exhibits a complementary pattern of expression in the caudal half-sclerotome. To test the functional significance of this striking asymmetry, ephrin-B1-Fc was added to intefere with receptor function in either whole trunk explants or neural crest cultures on alternating stripes of ephrin-B1-Fc. The resulting loss of the metameric migratory pattern and disorganization of neural crest cell movement demonstrate a functional role for Eph receptor–ligand interactions in patterning the neural crest.

Key words. Neural crest—Cell migration—Tyrosine kinase—Eph—Somite

Introduction

During embryonic development, neural crest cells migrate along stereotypic pathways to their target destinations where they differentiate into much of the peripheral nervous system, including sensory and sympathetic ganglia [1,2]. In the trunk of the avian embryo, neural crest migration through the somite is segmental, with neural crest cells entering the rostral half of each somitic sclerotome but avoiding the caudal half [3,4]. Previous studies have shown that cues intrinsic to the somites are responsible for this segmental migration and that similar mechanisms may influence the guidance of motor and sensory axons [5–7]. Inhibitory molecules localized to caudal sclerotomes or attractive molecules present in rostral sclerotomes could contribute to segmental patterning. A number of candidate molecules have the expected spatiotemporal localization; for example, chondroitin sulfate proteoglycans [8–10], T-cadherin [11], and peanut agglutinin (PNA)-binding molecules [8] are all expressed in the caudal half-sclerotome, whereas butyrylcholinesterase is expressed in the rostral half-

[1] Division of Biology and Beckman Institute, California Institute of Technology, Pasadena, CA 91125, USA

[2] Regeneron Pharmaceuticals, 777 Old Saw Mill River Road, Tarrytown, NY 10591, USA

sclerotome [12]. However, functional experiments are required to illuminate the role, if any, of these molecules in cell migration. We recently developed an explant preparation, which offers both direct visualization of neural crest migration in normal living tissue and accessibility to perturbing reagents [13,14]. Here, we report functional studies demonstrating a role for Eph receptor–ligand interactions in neural crest migration.

Several members of the Eph family of receptor tyrosine kinases (RTKs) and their ligands display intriguing patterns of expression in the developing nervous system, suggesting potential roles in early neural patterning [15–20]. Based on their binding specificities, the Eph-related RTKs can be divided into two major subclasses: EphB-related receptors, which interact primarily with a transmembrane subclass of ligands, and EphA-related receptors, which interact primarily with a GPI-linked subclass of ligands [15,18]. Eph receptor–ligand interactions have been implicated in axonal patterning events, such as the topographic organization of the retinotectal projection and axonal fasciculation [23–27]. Members of this family also are expressed in the developing forebrain and hindbrain [16,20,28–30] and appear to play a role in their regional organization [30–32]. The potential role of these molecules in the patterning of neural precursors in the developing peripheral nervous system remains largely unexplored.

Results

Eph Family Receptor EphB3 is Expressed in the Neural Crest and Rostral Sclerotome

To detect receptors of the EphB-related subclass of RTKs during trunk neural crest migration, a representative transmembrane ligand-Fc fusion protein (ephrin-B1-Fc) was used to stain whole chick embryos [18,21]. Because ligand–receptor interactions among members of a subclass are known to be promiscuous, the ligand ephrin-B1-Fc has the potential to interact with receptors EphB1, EphB2, and EphB3, but not EphB4 [18,19]. This cross-reactivity offers unique advantages both in surveying the patterns of expression and in perturbing their function.

Receptors recognized by ephrin-B1-Fc first appeared in the rostral portion of mature somites at the 15-somite stage (stage 12), coincident with the onset of neural crest migration (data not shown); this pattern persisted and became more distinct by the 25-somite stage (stage 15) (Fig. 1a) through the 35-somite stage (stage 19). Longitudinal sections revealed that the receptors were associated with the surface of cells in the rostral half-sclerotome (Fig. 1b). Identical results were observed in chick (Fig. 1) and quail (data not shown) embryos.

To identify the specific RTKs that may account for the staining associated with neural crest migratory pathways, we performed in situ hybridizations with quail-specific probes for two RTKs of the transmembrane ligand subclass: EphB3 [33], and EphB2 [26,34]. At the forelimb level of the 25-somite embryo, EphB3 transcripts localized within cells in the rostral half-sclerotome during neural crest migration (Fig. 1c). In contrast, no transcripts were apparent in the four or five most recently formed epithelial somites, before both the time of sclerotome formation and the entry of

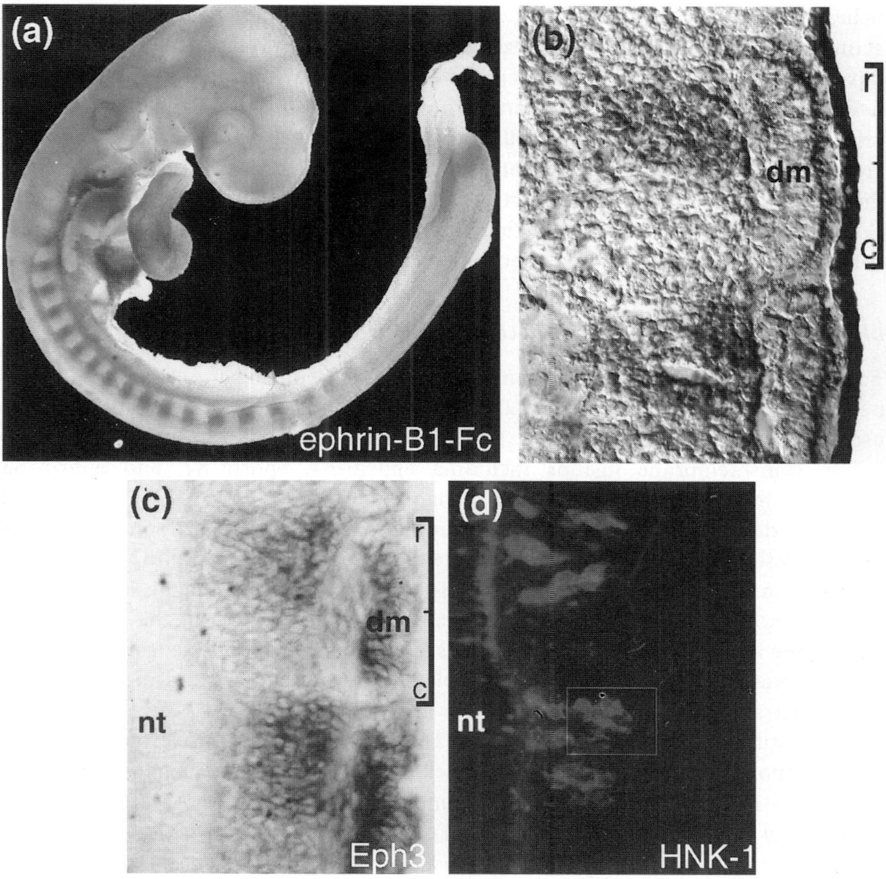

FIG. 1a–d. Staining of 25-somite chick embryos with ephrin-B1-Fc and hybridization with an Eph3 cDNA probe reveals that Eph-related receptor tyrosine kinases (RTKs) are distributed on cells in the rostral half of the somitic sclerotome, including neural crest. **a** In a whole-mount preparation, ephrin-B1-Fc staining, visualized with antihuman Fc antibody coupled to alkaline phosphatase, is distributed in a segmental fashion in the developing somites. **b** A longitudinal section shows ephrin-B1-Fc staining localized primarily to rostral sclerotome; the brackets indicate the rostrocaudal extent of the somite, with the tick mark representing the intrasomitic border. **c** A longitudinal section of another 25-somite embryo after whole-mount in situ hybridization shows that EphB3 transcripts are present throughout the rostral sclerotome. **d** The section in **c** was stained with the HNK-1 antibody to mark migrating neural crest cells within the rostral sclerotome. *r*, rostral; *c*, caudal; *nt*, neural tube; *dm*, dermomyotome. (Adapted from [54])

neural crest cells. Double-labeling with HNK-1 antibody, a neural crest marker, showed that all neural crest cells expressed EphB3 mRNA (Fig. 1d). Other EphB3-positive cells in the rostral half-sclerotome and dermomyotome were HNK-1 negative, and even extended into the perinotochordal space, which is devoid of neural crest cells. These patterns indicate that cells within the rostral half-somite other than neural crest cells also express EphB3. In contrast to EphB3, EphB2 was weakly expressed in the rostral sclerotome only at later stages of neural crest migration (data not shown). Combined, these results indicate that migrating neural crest cells express EphB3, consistent with their ability to be recognized by ephrin-B1-Fc.

Ligand Ephrin-B1 Is Distributed in the Caudal Sclerotome

To assay the distribution of transmembrane ligands during the process of trunk neural crest migration, a representative receptor-Fc fusion protein (EphB2-Fc) was used to label whole chick embryos. EphB2-Fc has the potential to interact with Eph family transmembrane ligands including ephrin-B1, ephrin-B2, and ephrin-B3 [18,35]. At the 10-somite stage, ligands recognized by EphB2-Fc were highly expressed in the caudal portion of all somites, including immature epithelial somites (before neural crest migration). During active neural crest migration, EphB2-Fc staining persisted in the caudal half of each somite (Fig. 2a; 25-somite stage). Longitudinal sections showed that ligands were associated with the surfaces of the caudal half-sclerotome and some dermomyotome cells (Fig. 2b). Throughout neurulation, staining was present within the open neural plate (data not shown) and in the recently closed neural tube (Fig. 2a).

To identify the ligands bound by EphB2-Fc in the caudal half-sclerotome, in situ hybridization was performed in whole chick embryos with probes against ephrin-B1 and ephrin-B2 mRNA [36]. In 25-somite embryos, ephrin-B1 mRNA was observed in a segmental pattern in the caudal halves of somites along the body axis. Longitudinal sections confirmed that ephrin-B1 transcripts were expressed by caudal half-sclerotome cells (Fig. 2c), and were absent from the rostral half-sclerotome through which neural crest cells migrate (Fig. 2d). Ephrin-B2 was not expressed in the somite, but was localized to endothelial cells positioned at the intersomitic boundary (Fig. 2e). The chick homolog of ephrin-B3 has not been isolated; therefore, we did not analyze its pattern of expression during trunk neural crest migration.

Soluble Ephrin-B1 Disrupts the Segmental Pattern of Neural Crest Migration in Whole Trunk Explants

The reciprocal distribution of receptors and ligands suggests that they may play a potential role in conferring the segmental pattern of neural crest migration. Specifically, neural crest cells expressing the RTK EphB3 may interact with ephrin-B1-expressing caudal sclerotome cells, mediating an inhibitory event that prohibits entry of neural crest into this caudal domain. To test this hypothesis, soluble ephrin-B1-Fc was added to whole trunk explants [13] to disrupt receptor–ligand interactions. The subsequent distribution of migrating neural crest cells was assayed by HNK-1 antibody labeling or by prelabeling the neural tube with DiI [37]. In the presence of ephrin-B1-Fc (1 mg/ml and 10 mg/ml), the normally segmental migratory pattern of

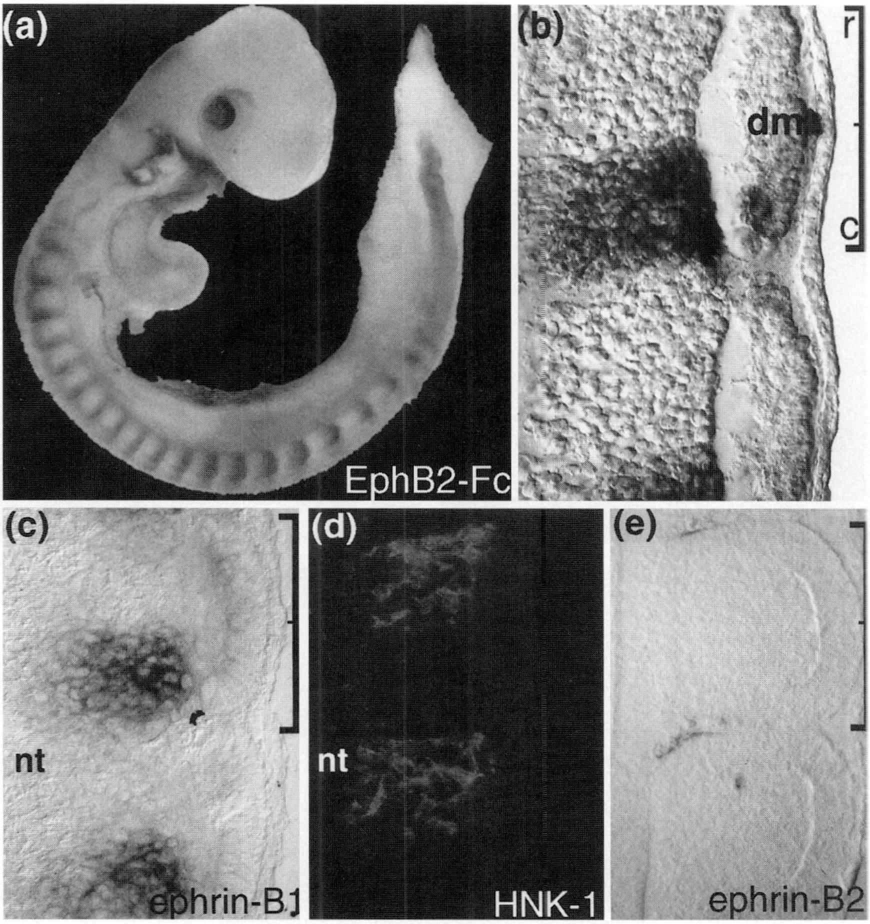

FIG. 2a–e. Staining of 25-somite chick embryos with EphB2-Fc and hybridization with ephrin-B1 cDNA probe reveals that transmembrane ligands are distributed on cells in the caudal sclerotome. **a** In a whole-mount preparation, EphB2-Fc staining, visualized with antihuman Fc antibody coupled to alkaline phosphatase, is found in caudal halves of somites. **b** A longitudinal section through the trunk reveals EphB2-Fc staining in caudal sclerotome as well as the dermomyotome. **c** A longitudinal section through an embryo after in situ hybridization with ephrin-B1 cDNA probe reveals that ephrin-B1 transcripts are present in caudal sclerotomal cells. **d** The same section as **c** after staining with the HNK-1 antibody to recognize migrating neural crest cells reveals that ephrin-B1 is expressed in a pattern complementary to the distribution of neural crest cells. **e** A longitudinal section of an embryo after hybridization with an ephrin-B2 probe reveals that these transcripts localize to endothelial cells at intersomitic borders, but not the sclerotome. *Brackets* indicate the rostrocaudal extent of a single somite, with the midline representing the intrasomitic border. *r*, rostral; *c*, caudal; *nt*, neural tube; *dm*, dermomyotome. (Adapted from [54])

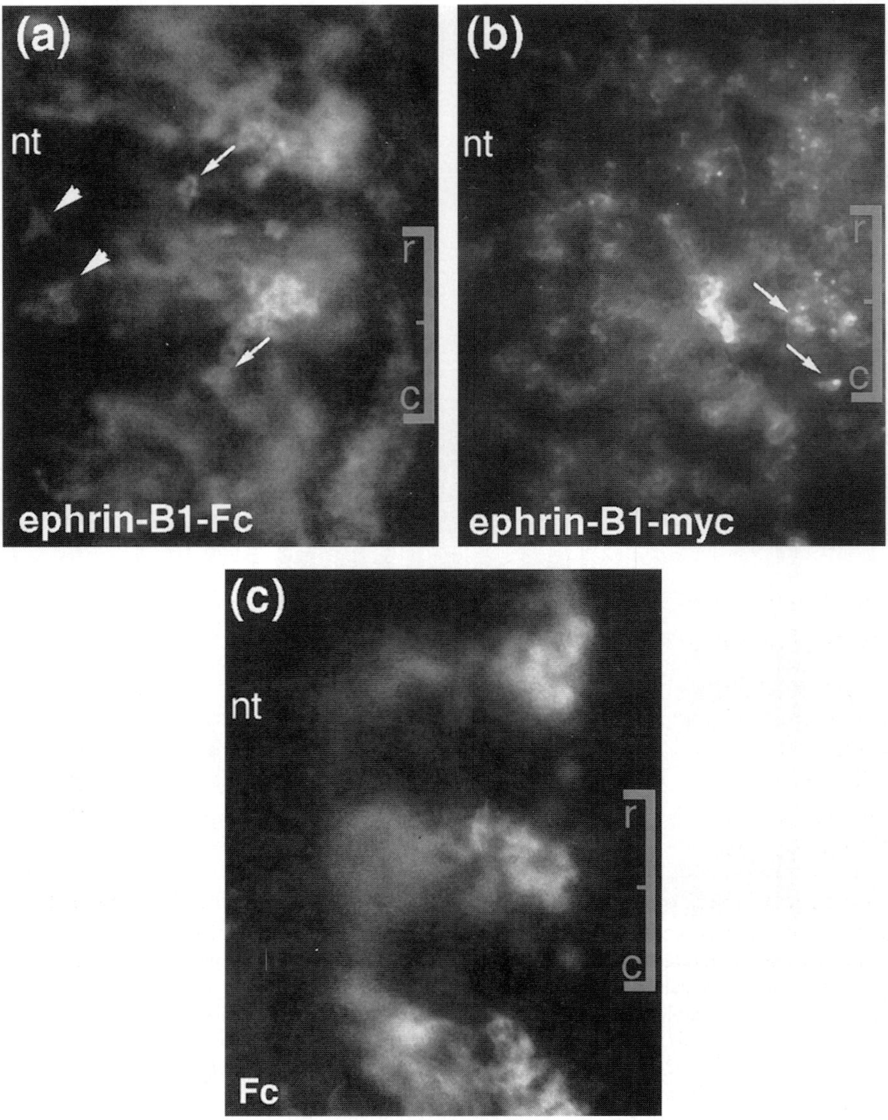

FIG. 3a–c. Soluble ephrin-B1 disrupts the segmental pattern of neural crest migration. Ephrin-B1-Fc **a**, Fc alone **b**, or ephrin-B1-myc **c** were added to trunk explants and static patterns of neural crest migration were visualized via HNK-1 antibody staining. *Arrowheads* indicate cells over the dorsal aspect of the neural tube; *small arrows* mark a few examples of neural crest cells migrating within the caudal portion of the somite; *brackets* indicate the rostrocaudal extent of a single somite, with the midline representing the intrasomitic border. *r*, rostral; *c*, caudal; *nt*, neural tube. (Adapted from [54])

neural crest cells was disrupted. Instead, labeled cells were observed in both rostral and caudal portions of the sclerotome (Fig. 3a, and data not shown), as confirmed by sectioning the explants. To rule out that the disruption resulted from the the dimeric nature of ephrin-B1-Fc, we used monomeric ephrin-B1-myc; in the presence of ephrin-B1-myc (10 mg/ml), neural crest cells were observed in both halves of the somite (Fig. 3c; $n = 16$). In contrast, treatment with soluble Fc (Fig. 3b; 10 mg/ml; $n = 20$), COS cell supernatant (concentrated similarly to ephrin-B1-myc; $n = 12$), or soluble EphB2-Fc (10–50 mg/ml; $n = 25$; data not shown) did not alter the segmental pattern of neural crest migration.

To quantitate these effects on neural crest patterning, we counted the numbers of cells in the rostral and caudal halves in ephrin-B1-Fc-treated somites, ephrin-B1-myc-treated somites, and control Fc-treated somites ($n = 16$ somites in 8 explants per condition, data not shown; also see [54]). In Fc-treated somites, 2% (8/488) of the neural crest cells entered the caudal sclerotome. In contrast, 33% (296/896) of neural crest in ephrin-B1-Fc-treated somites and 47% (486/1035) of neural crest in ephrin-B1-myc-treated somites entered the caudal sclerotome (data not shown). These results indicate that binding the EphB3 receptor with excess, soluble ephrin-B1 protein allows neural crest cells to enter the caudal sclerotome domain.

Time-Lapse Analysis of Neural Crest Migration in the Presence of Ephrin-B1

To better define the cell migratory defects in the presence of ephrin-B1-Fc, DiI-labeled neural crest cells were followed in explants using time-lapse video microscopy. In control explants [13], the first neural crest cells entered only the rostral half-sclerotome. Exposure to soluble ephrin-B1-Fc resulted in neural crest cells emigrating directly from the neural tube into both rostral and caudal somitic halves (data not shown).

The trajectories and rates of migration for individual cells in trunk explants treated with ephrin-B1-Fc, EphB2-Fc, or Fc were acquired from the time-lapse recordings with an automated cell-tracking program (S. Speicher and G. Solomon, unpublished data). In ephrin-B1-Fc-treated explants, the trajectories of cells in both the rostral and caudal somite halves appeared more erratic and disorganized than those in control cultures (Fig. 4). Surprisingly, the ephrin-B1-Fc-treated cells often moved backward from somite to the neural tube. Quantitative measurements of cell migration show that the mean rates of migration were not significantly different between ephrin-B1-Fc-treated cells (10.8 μm/h ± 3.1 SD; $n = 53$ cells), untreated (14.9 μm/h ± 3.0 SD; $n = 12$ cells), or EphB2-Fc- (13.7 μm/h ± 3.8 SD; $n = 10$) or Fc-treated cells (17.6 μm/h ± 2.8 SD; $n = 14$ cells), and were similar to those reported previously [13]. However, significant differences were noted between ephrin-B1-Fc-treated and Fc- or EphB2-Fc-treated cells with respect to the continuance of cell migration, defined as the displacement distance of a cell divided by its total trajectory length (data not shown) [54]. This is manifest in the disordered migratory trajectories of neural crest cells in ephrin-B1-Fc-treated explants, compared with the linear migratory trajectories observed in controls (compare Fig. 4). Analysis of the cell trajectories of ephrin-B1-Fc-treated cultures showed some of them crossing the intrasomitic border (7/40 ephrin-B1-Fc-treated cells). In contrast, such crossings rarely were observed under

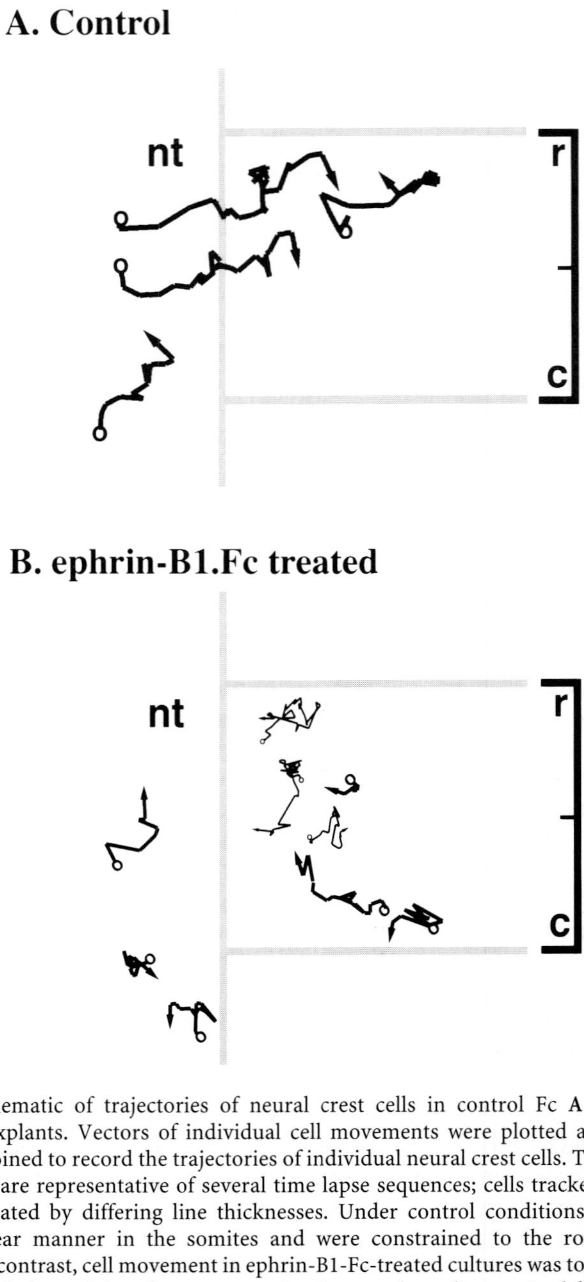

FIG. 4A,B. Schematic of trajectories of neural crest cells in control Fc **A** and ephrin-B1-Fc-treated **B** explants. Vectors of individual cell movements were plotted at half-hour time intervals and joined to record the trajectories of individual neural crest cells. The cell migration vectors shown are representative of several time lapse sequences; cells tracked from different films are indicated by differing line thicknesses. Under control conditions, cells moved in an overall linear manner in the somites and were constrained to the rostral half of the sclerotome. In contrast, cell movement in ephrin-B1-Fc-treated cultures was tortuous, with cells located in both halves of the sclerotome moving in a random, disorganized fashion. (Adapted from [54])

control conditions (1/14 Fc-treated cells and 0/12 untreated cells). Thus, the bearing and continuance of neural crest migration, but not its rate, was altered by exogenous ephrin-B1-Fc.

The disruptive effect of dimeric ephrin-B1-Fc does not appear to result from an alteration of the polarity of the somitic sclerotome. The distribution pattern of molecular markers characteristic of either the rostral or caudal half-sclerotome were normal in ephrin-B1-Fc-treated explants: heparan sulfate proteoglycans were present in their typical position in the rostral half-sclerotome, and PNA-binding glycoproteins and chondroitin sulfate proteoglycans were found in their characteristic location in the caudal half-sclerotome (data not shown).

Ephrin-B1 Inhibits Neural Crest Migration In Vitro

The trunk explant system cannot distinguish whether exogenous ephrin-B1 acts directly on neural crest cells or indirectly by binding to rostral sclerotome. To distinguish between these possibilities, we have examined the direct effects of exogenous ephrin-B1 protein on neural crest migration in vitro. Neural crest cells were plated on a patterned two-dimensional substrate composed of alternating stripes [38,39] of multimeric ephrin-B1-Fc plus fibronectin versus fibronectin alone. The behavior of these cells was compared with cells on control substrates composed of (1) alternating stripes of Fc (plus fibronectin) versus fibronectin (2) alternating stripes of fibronectin, and (3) alternating stripes of Fc antibody (plus fibronectin) versus fibronectin alone.

Cultured neural crest cells bind ephrin-B1-Fc, indicating that they maintained receptor expression (Fig. 5A,B). Neural crest cells migrated uniformly on all control substrates (Fig. 5C). In contrast, individual neural crest cells avoided migrating on stripes containing ephrin-B1-Fc ($n = 31$), and instead migrated on stripes composed of fibronectin alone (Fig. 5D). Time-lapse video microscopy of migrating the neural crest on these striped substrates indicates that the neural crest displays a typical avoidance response consisting of lamellipodial collapse and process retraction in the presence of ephrin-B1. The typical broad lamellipod on the leading process of migrating neural crest cells underwent collapse and retraction within minutes of crossing into an ephrin-B1 stripe (data not shown) [54]. This avoidance behavior did not occur immediately on contact with the border of ephrin-B1 but after neural crest entry and process extension on the ephrin-B1 stripe. In contrast, cells on control substrates did not demonstrate collapse behavior upon crossing stripes (data not shown).

To examine further the specificity of these effects and to verify that soluble ephrin-B1 interferes with the normal receptor–ligand interactions, we studied the effects of adding soluble ephrin-B1 on the behavior of neural crest cells on patterned ephrin-B1/fibronectin substrates. After addition of soluble dimeric ephrin-B1-Fc (10 μg/ml; 1.7×10^{-7} M; $n = 10$), neural crest cells migrated equally well on all lanes (Fig. 5E). Analogous results were obtained with monomeric ephrin-B1-myc (10 μg/ml; 3.3×10^{-7} M; $n = 3$). The similarity of the two effects argues that a major mechanism of action of soluble ligand is via competitive inhibition: ephrin-B1-Fc causes a small but significant receptor autophosphorylation (data not shown) [54], whereas ephrin-B1-myc causes no signal transduction [20]. In contrast to ligand fusion proteins, soluble EphB2-Fc had no obvious effect on the patterned migration on ephrin-B1/fibronectin

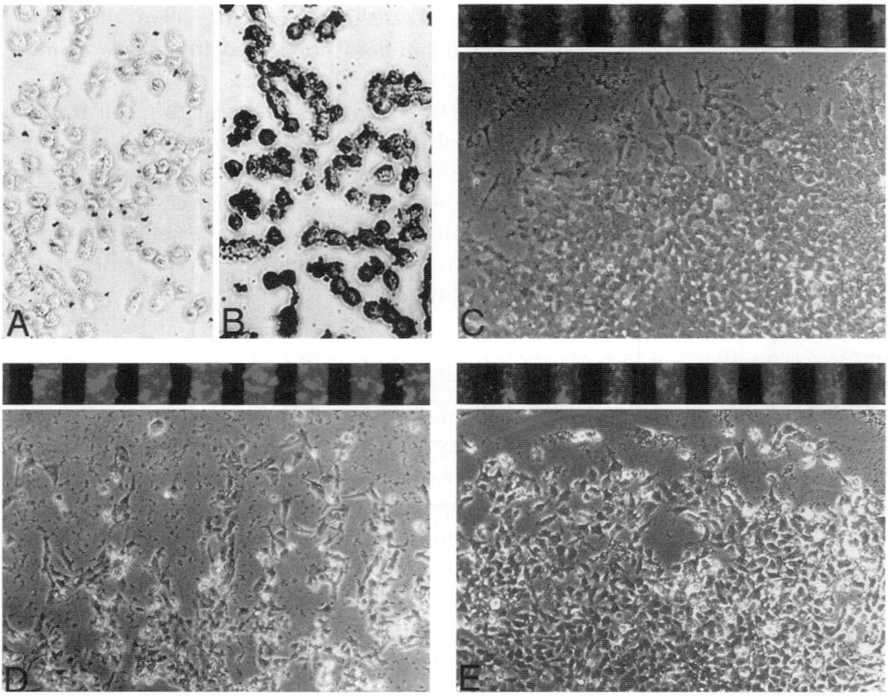

FIG. 5A–E. Alternating lanes of ephrin-B1-Fc in vitro inhibit neural crest migration in a manner that is mitigated by addition of soluble ephrin-B1-Fc. Avoidance by neural crest of ligand-Fc lanes involves process collapse/retraction. **A,B** Neural crest cells grown in vitro were labeled with Fc alone (**A**) or ephrin-B1-Fc (**B**) visualized with antihuman Fc antibody coupled to alkaline phosphatase. **A** Control cultures displayed no staining. **B** Ephrin-B1-Fc bound specifically to neural crest cells, indicating that receptors were maintained in vitro. **C** When migrating neural crest cells were exposed to a substrate composed of alternating lanes of fibronectin versus Fc/fibronectin, they migrated uniformly on both substrates. **D** When migrating neural crest cells were exposed to a substrate composed of alternating lanes of fibronectin versus antibody-aggregated ephrin-B1-Fc/fibronectin, they migrated selectively on the fibronectin lanes, avoiding the lanes containing multimeric ephrin-B1-Fc. **E** When soluble ephrin-B1-Fc was added to the culture medium of neural crest migrating on alternating lanes of fibronectin versus antibody-aggregated ephrin-B1-Fc/fibronectin, they migrated equally well on all lanes. Analogous results were observed with monomer ephrin-B1-myc (data not shown), demonstrating that soluble ephrin-B1 functions as a competitive inhibitor of multimeric ephrin-B1-Fc. (Adapted from [54])

(data not shown). Our results demonstrate that lanes of ephrin-B1-Fc can directly inhibit neural crest migration; furthermore, they verify that soluble ephrin-B1 can function as a competitive antagonist of this inhibition.

Treatment with Ephrin-B1 Induces Receptor Phosphorylation

The phosphorylation status of endogenous EphB2 receptor expressed by COS7 cells was examined after addition of 1 or 10 mg/ml of clustered or dimeric ephrin-B1-Fc.

Ephrin-B1-Fc, even at 1 mg/ml concentration, induced saturating levels of tyrosine phosphorylation of EphB2 (data not shown). In contrast, substantially lower levels of tyrosine phosphorylation were observed with 1 or 10 mg/ml of unclustered dimeric ephrin-B1-Fc. EphB2 tyrosine phosphorylation was not observed in untreated COS7 cells (data not shown) [54] or in ephrin-B1-myc-treated COS7 cells [21].

Disussion

In the present study, we have examined whether members of the Eph family of receptors and ligands participate in the segmental patterning of the neural crest migration. Ligand-Fcs and receptor-Fcs were used to identify the distribution of their respective subclass of receptors and ligands [18]. Staining with ephrin-B1-Fc and in situ hybridization with receptor-specific probes showed the strong expression of EphB3 (see Fig. 1) and weak expression of EphB2 in the rostral half-sclerotome. The strong binding of ephrin-B1-Fc and expression of EphB3 by neural crest cells as they migrate is consistent with the previous observation that ligand ephrin-B1 binds with high affinity to EphB3 [18,40]. The cognate ligands are present in the caudal half of each sclerotome, from which neural crest cells are normally absent, first appearing before neural crest migration (see Figure 2). Using in situ hybridization with ligand-specific probes, we found that the ligand ephrin-B1, but not ephrin-B2, was expressed in the caudal half-sclerotome in a pattern consistent with EphB2-Fc labeling (Fig. 2). Based on studies of axon guidance and fasciculation [41,42] an emerging theme is that interactions between Eph family of receptor tyrosine kinases and their ligands mediate repulsive signals that constrain growing axons to particular regions in the developing nervous system. The complementary expression patterns of the EphB3 receptor and the ephrin-B1 ligand in avian somites raises the intriguing possibility that Eph interactions constrain the migration of neural crest cells to the rostral sclerotome.

To test the functional significance of this complementary distribution of receptors and ligands, we conducted both in vivo and in vitro experiments of neural crest migration. In a whole trunk explant system [13], addition of soluble ephrin-B1 disrupted the patterned migration of neural crest cells in their native environment (see Fig. 3). Neural crest cells were observed within both rostral and caudal halves of each somitic sclerotome after exposure to ephrin-B1-myc or ephrin-B1-Fc, suggesting that the normal inhibitory action of the caudal half-sclerotome on neural crest cells had been blocked. A significant increase in the numbers of neural crest cells within whole somites in the presence of ephrin-B1-myc or ephrin-B1-Fc compared to controls was noted. One possibility to account for this increase in cell number is that neural crest entered the somite prematurely in treated explants. Using an in vitro assay (alternating lanes of ephrin-B1/fibronectin), we determined that ephrin-B1 ligand can directly inhibit neural crest cells (see Fig. 5), causing a collapse and retraction of cell processes, analogous to growth cone collapse behavior displayed by neurons in culture [23].

Of note, the repulsive activity of ephrin-B1 on migrating neural crest in vitro does not appear as an immediate consequence of cell contact with ligand but instead develops over 30 min. Addition of soluble ephrin-B1 (either monomeric or dimeric) blocked this inhibition, resulting in neural crest cells migrating uniformly on all lanes. Because monomeric ephrin-B1-myc fails to activate signal transduction [21], it functions solely as an antagonist; dimeric ephrin-B1-Fc activates low levels of receptor phosphorylation (Fig. 5f) and may function as both an antagonist and a weak agonist. These results clearly show that (1) substrate-bound ephrin-B1 directly inhibits neural crest cell migration; and (2) soluble ephrin-B1 disrupts this inhibition. Thus, soluble ephrin-B1 acts as a competitive inhibitor of ligand binding. The normal rostrocaudal patterning of the somites (as assayed by the metameric pattern of heparan sulfate proteoglycan in the rostral half-sclerotome and peanut lectin and chondroitin sulfate proteoglycan in the caudal half-sclerotome), combined with our in vitro data, suggests that the primary effect of soluble ephrin-B1 is on the neural crest.

Time-lapse cinematography revealed major defects and decreased continuance in the trajectories of the ephrin-B1-Fc-treated neural crest cells (Fig. 4; and data not shown) [54]. Although their overall rates of migration were similar to control and EphB2-Fc-treated explants, neural crest cells moved in a circular and disoriented fashion in the presence of ephrin-B1-Fc, sometimes moving across the intrasomitic border and backward into the neural tube. This disorientation is likely caused by activation of signal transduction pathways in neural crest cells surrounded by dimeric ephrin-B1-Fc, which may function as a weak agonist. In support of this possibility, preliminary data suggest that neural crest cells display normal continuance in the presence of monomeric ephrin-B1-myc, which does not activate signaling [21]. These data, in combination, provide interesting mechanistic information about how Eph receptors might exert their function in neural crest cells, suggesting a potential link between receptors and cytoskeletal elements.

Our previous studies demonstrated that treatment of trunk explants with PNA, which binds to cells in the caudal sclerotome [8,43] resulted in neural crest cells migrating through both rostral and caudal halves of the somite [13]. However, time-lapse recordings reveals different migratory behaviors in the presence of PNA compared to ephrin-B1-Fc. In soluble ephrin-B1-Fc-treated explants, both rostrally and caudally-located neural crest cells traveled in a nonlinear, disoriented manner (see Fig. 4). In PNA-treated explants, by contrast, rostrally located neural crest cells moved normally, while caudally located neural crest cells traveled in a highly disorganized manner [13], resembling neural crest cells in the presence of dimeric ephrin-B1-Fc (Fig. 4). This similarity suggests that neural crest cells may be receiving repulsive signals in both cases, either from the exogenous ephrin-B1-Fc or from the ephrin-B1-expressing caudal sclerotome cells. However, induction of an abnormal collapse response cannot be the sole explanation for the disruption of the segmental pattern. Ephrin-B1-myc, which does not stimulate receptor phosphorylation and signaling [21], alters segmental migration without eliciting abnormal cell behaviors. The differential effects observed with PNA and ephrin-B1-Fc treatment demonstrate there are likely to be multiple inhibitory mechanisms in the caudal sclerotome.

Addition of EphB2-Fc had no effect on the segmental pattern, rates, or continuance of neural crest migration either in vitro or in vivo. This may result from the highly adaptive nature of Eph family receptor–ligand interactions, as observed in previous studies on the growth of optic axons [23,38]. In stripe assays, optic axons can adapt to high basal levels of ligand and grow normally on concentrations dramatically larger than normally required for growth cone collapse and repulsion. Similarly, the axons become exquisitely sensitive to very low levels of ligand in conditions with no ligand. Thus, if even a fraction of the ligand remained free to interact in the presence of

EphB2-Fc, adaptation would result in more sensitive cells that should generate a near-normal response. Alternatively, the failure of the inhibition may suggest a higher order interaction between receptors and ligands, or an asymmetry in Eph receptor–ligand interactions, that are not abolished by EphB2-Fc.

Our functional studies, performed on neural crest cells migrating in situ, are in agreement with complementary studies using different assay systems. Both ephrin-A5 and ephrin-A2 act to inhibit axon outgrowth in the retinotectal system [22,23,27]; ephrin-A5 has also been implicated in axon fasciculation by cortical neurons [25]. Eph-related RTKs and ligands act in earlier events of neuronal patterning: expression of a truncated EphA4 receptor disrupts the normal segmental pattern of *Krox-20* expression in the hindbrain of frogs and zebrafish [31] and the RTKs EphA4, EphB1, and ligand ephrin-B2 influence the segmental organization of branchial arch neural crest migration (Smith et al., in man uscript).

Chick and rat both exhibit metameric expression patterns of EphB receptors and ephrin-B ligands within the developing trunk somites, but different members are expressed in each species. Recently, Wang and Anderson demonstrated in rat an inhibitory influence in vitro of the RTK ligands ephrin-B2 and ephrin-B1 on trunk neural crest cells and motor axons [36]. The authors reported that rat neural crest cells primarily express EphB2, whereas chick neural crest cells predominantly express EphB3 (Fig. 1). In addition, EphB2 and ephrin-B2 are expressed by rat rostral and caudal sclerotome [36], respectively, whereas EphB3 and ephrin-B1 are present in chick rostral and caudal sclerotome (Figs. 1 and 2). In rat, ephrin-B1 transcripts are expressed by cells at the dorsal lip of the dermomyotome [36] and not by caudal sclerotome cells as observed here in chick and quail (Fig. 2). These findings suggest that there is evolutionary conservation within this Eph subclass but not necessarily of individual receptors and their ligands.

Acknowledgments. Parts of this manuscript have been reproduced from [54] with permission. This work was supported by USPHS HD-15527 and a grant from the Muscular Dystrophy Association to MBF and MH-49176 to S.E.F.

References

1. Le Douarin NM (1982) The neural crest. Cambridge University Press, Cambridge
2. Weston J (1970) The migration and differentiation of neural crest cells. Adv Morphol 8:41–114
3. Rickmann M, Fawcett JW, Keynes RJ (1985) The migration of neural crest cells and the growth of motor axons through the rostral half of the chick somite. J Embryol Exp Morphol 90:437–455
4. Bronner-Fraser M (1986) Analysis of the early stages of trunk neural crest migration in avian embryos using monoclonal antibody HNK-1. Dev Biol 115:44–55
5. Keynes RJ, Stern CD (1984) Segmentation in the vertebrate nervous system. Nature (Lond) 310:413–429
6. Teillet M, Kalcheim C, LeDouarin NM (1987) Formation of the dorsal root ganglion in the avian embryo: segmental origin and migratory behavior of neural crest progenitor cells. Dev Biol 120:329–347
7. Bronner-Fraser M, Stern C (1991) Effects of mesodermal tissues on avian neural crest cell migration. Dev Biol 143:213–217

8. Oakley RA, Tosney KW (1991) Peanut agglutinin and chondroitin-6-sulfate are molecular markers for tissues that act as barriers to axon advance in the avian embryo. Dev Biol 147:187–206
9. Landolt RM, Vaughan L, Winterhalter KH, Zimmerman DR (1995) Versican is selectively expressed in embryonic tissues that act as barriers to neural crest cell migration and axon outgrowth. Development (Camb) 121:2303–2312
10. Perris R, Krotoski D, Lallier T, Domingo C, Sorrel JM, Bronner-Fraser M (1991) Spatial and temporal changes in the distribution of proteoglycans during avian neural crest development. Development (Camb) 111:583–599
11. Ranscht B, Bronner-Fraser M (1991) T-cadherin expression alternates with migrating neural crest cells in the trunk of the avian embryo. Development (Camb) 111:15–22
12. Layer PG, Alber R, Rathjen FG (1988) Sequential activation of butyrylcholinesterase in rostral half somites and acetylcholinesterase in motoneurons and myotomes preceding growth of motor axons. Development (Camb) 102:387–396
13. Krull CE, Collazo A, Fraser SE, Bronner-Fraser M (1995) Segmental migration of trunk neural crest: time-lapse analysis reveals a role for PNA-binding molecules. Development (Camb) 121:3733–3743
14. Krull CE, Kulesa P (1998) Embryonic explant and slice preparations for studies of cell migration and axon guidance. In: dePablo F, Ferrus A, Stern CD (eds) Current topics in developmental biology, vol 36. Academic Press, San Drego (in press)
15. Brambilla R, Klein R (1995) Telling axons where to grow: a role for Eph receptor tyrosine kinases in guidance. Mol Cell Neurosci 6:487–495
16. Becker N, Seitanidou T, Murphy P, Mattei M-G, Tipilko P, Nieto MA, Wilkinson DG, Charnay P, Gilardi-Hebenstreit P (1994) Several receptor tyrosine kinase genes of the Eph family are segmentally expressed in the developing hindbrain. Mech Dev 47:3–17
17. Flenniken AM, Gale NW, Yancopoulos GD, Wilkinson DG (1996) Distinct and overlapping expression patterns of ligands for Eph-related receptor tyrosine kinases during mouse embryogenesis. Dev Biol 179:382–401
18. Gale NW, Holland S, Valenzuela D, Flenniken A, Pan L, Henkemeyer M, Strebhardt K, Hirai H, Wilkinson DG, Pawson T, Davis S, Yancopoulos GD (1996) Eph receptors and ligands comprise two major specificity subclasses, and are reciprocally compartmentalized during embryogenesis. Neuron 17:9–19
19. Brambilla R, Brucker K, Orioli D, Bergemann AD, Flanagan JG, Kelin R (1996) Similarities and differences in the way transmembrane-type ligands interact with the Elk subclass of Eph receptors. Mol Cell Neurosci 8:199–209
20. Henkemeyer M, Marengere LEM, McGlade J, Oliver JP, Conlon RA, Holmyard DP, Lewtin K, Pawson T (1994) Immunolocalization of the Nuk receptor tyrosine kinase suggests roles in segmental patterning of the brain and axonogenesis. Oncogene 9:1001–1014
21. Davis S, Gale NW, Aldrich TH, Maisonpierre PC, Lhotak V, Pawson T, Goldfarb M, Yancopoulos GD (1994) Ligands for the Eph-related receptor tyrosine kinases that require membrane attachment or clustering for activity. Science 266:816–819
22. Cheng H-J, Nakamoto N, Bergemann AD, Flanagan JG (1995) Complementary gradients in expression and binding of ELF-1 and Mek4 in development of the topographic retinotectal projection map. Cell 82:371–381
23. Drescher U, Kremoser C, Handwerker C, Loschinger J, Noda M, Bonhoeffer F (1995) In vitro guidance of retinal ganglion cell axons by RAGS, a 25-kDa tectal protein related to ligands for Eph receptor tyrosine kinases. Cell 82:359–370
24. Holash JA, Pasquale EB (1995) Polarized expression of the receptor protein-tyrosine kinase CEK5 in the developing avian visual system. Dev Biol 172:683–693
25. Winslow JW, Moran P, Valverde J, Shih A, Yuan JQ, Wong SC, Tsai SP, Goddard A, Henzel WJ, Hefti F, Beck KD, Caras IW (1995) Cloning of Al-1, a ligand for an Eph-related tyrosine kinase receptor involved in axon bundle formation. Neuron 14:973–981

26. Kenny D, Bronner-Fraser M, Marcelle C (1995) The receptor tyrosine kinase QEK5 m-RNA is expressed in a gradient within the neural retina and the tectum. Dev Biol 172:708–716
27. Nakamoto M, Cheng H-J, Friedman GC, McLaughlin T, Hansen MJ, Yoon CH, O'Leary DDM, Flanagan JG (1996) Topographically specific effects of ELF-1 on retinal axon guidance in vitro and retinal axon mapping in vivo. Cell 86:755–766
28. Macdonald R, Xu Q, Barth KA, Mikkola I, Holder N, Fjose A, Krauss S, Wilson SW (1994) Regulatory gene expression boundaries demarcate sites of neuronal differentiation in the embryonic zebrafish forebrain. Neuron 13:1039–1053
29. Bergemann AD, Cheng HJ, Brambilla R, Klein R, Flanagan JG (1995) ELF-2, a new member of the Eph ligand family, is segmentally expressed in mouse embryos in the regions of the hindbrain and newly forming somites. Mol Cell Biol 15:4921–4929
30. Henkemeyer M, Orioli D, Henderson JT, Saxton TM, Roder J, Pawson T, Klein R (1996) Nuk controls pathfinding of commissural axons in the mammalian central nervous system. Cell 86:35–46
31. Xu Q, Alldus G, Holder N, Wilkinson DG (1995) Expression of truncated Sek-1 receptor tyrosine kinase disrupts the segmental restriction of gene expression in the *Xenopus* and zebrafish hindbrain. Development (Camb) 121:4005–4016
32. Xu Q, Alldus G, MacDonald R, Wilkinson DG, Holder N (1996) Function of the Eph-related kinase RTK1 in patterning of the zebrafish forebrain. Nature (Lond) 381:319–322
33. Sajjadi FG, Pasquale EB (1993) Five novel avian Eph-related tyrosine kinases are differentially expressed. Oncogene 8:1807–1813
34. Pasquale EB (1991) Identification of chicken-embryo kinase-5, a developmentally regulated receptor-type tyrosine kinase of the Eph family. Cell Regulation 2:523–534
35. Gale NW, Flenniken A, Compton DC, Jenkins N, Copeland NG, Gilbert DJ, Davis S, Wilkinson DG, Yancopoulos GD (1996) Elk-L3, a novel transemembrane ligand for the Eph family of receptor tyrosine kinases, expressed in embryonic floor plate, roof plate and hindbrain segments. Oncogene 13:1343–1352
36. Wang HU, Anderson DJ (1996) Roles of Eph family transmembrane ligands in repulsive guidance of trunk neural crest migration and motor axon outgrowth. Neuron 18:383–396
37. Serbedzija GN, Bronner-Fraser M, Fraser SE (1989) A vital dye analysis of the timing and pathways of neural crest cell migration. Development (Camb) 86:809–819
38. Walter J, Kern-Veits B, Huf J, Stolze B, Bonhoeffer F (1987) Recognition of position-specific properties of tectal cell membranes by retinal axons in vitro. Development (Camb) 101:685–696
39. Vielmetter J, Stolze B, Bonhoeffer F, Stuermer CAO (1990) In vitro assay to test differential substrate affinities of growing axons and migratory cells. Exp Brain Res 81:283–287
40. Brambilla R, Schnapp A, Casagranda F, Labrador JP, Bergemann AD, Flanagan JG, Pasquale EB, Klein R (1995) Membrane-bound LERK2 ligand can signal through three different Eph-related receptor tyrosine kinases. EMBO J 14:3116–3126
41. Tessier-Lavigne M (1995) Eph receptor tyrosine kinases, axon repulsion and the development of topographic maps. Cell 82:345–348
42. Friedman GC, O'Leary DDM (1996) Eph receptor tyrosine kinases and their ligands in neural development. Curr Opin Neurobiol 6:127–133
43. Davis JA, Cook GMW, Stern CD, Keynes RJ (1990) Isolation from chick somites of a glycoprotein fraction that causes collapse of dorsal root ganglion growth cones. Neuron 2:11–20
44. Irving C, Nieto MA, DasGupta R, Charnay P, Wilkinson DG (1996) Progressive spatial restriction of Sek-1 and Krox-20 gene expression during hindbrain segmentation. Dev Biol 173:26–38

45. Pandey A, Shao H, Marks RM, Polverini PJ, Dixit VM (1995) Role of B61, the ligand for the Eck receptor tyrosine kinase, in TNF-a-induced angiogenesis. Science 268:567–569
46. Marcelle C, Eichmann A (1992) Molecular cloning of a family of protein kinase genes expressed in the avian embryo. Oncogene 7:2479–2487
47. Bohme B, Holtrich U, Wolf G, Luzius H, Grzeschik KH, Strebhardt K, Rubsamen-Waigmann H (1993) PCR-mediated detection of a new human receptor tyrosine kinase, HEK2. Oncogene 8:2857–2862
48. Ruiz JC, Conlon FL, Robertson EJ (1994) Identification of novel protein kinases expressed in the myocardium of the developing mouse heart. Mech Dev 48:153–164
49. Henrique D, Adam J, Myat A, Chitnis A, Lewis J, Ish-Horowicz D (1995) Expression of a Delta homologue in prospective neurons in chick. Nature (Lond) 375:787–790
50. Tucker GC, Aoyama H, Lipinski M, Tursz T, Thiery JP (1984) Identical reactivity of monoclonal antibodies HNK-1 and NC-1: conservation in vertebrates on cells derived from the neural primordium and on some leukocytes. Cell Differ 14:223–230
51. Lallier T, Bronner-Fraser M (1991) Avian neural crest cell attachment to laminin: involvement of divalent cation dependent and independent integrins. Development (Camb) 113:1069–1084
52. Hamburger V, Hamilton HL (1951) A series of normal stages in the development of the chick embryo. J Morphol 88:49–92
53. Eph Nomenclature Committee (1997) Unified nomenclature for Eph family receptors and their ligands, the ephrins. Cell 90:403–404
54. Krull CE, Lansford R, Gale NW, Collazo A, Marcelle C, Yancopoulos GD, Fraser SE, Bronner-Fraser M (1997) Interactions of Eph-related receptors and ligands confer rostrocaudal pattern to trunk neural crest. Curr Biol 7:571–580

Induction and Maintenance of *Hoxb4* Expression in the Developing Hindbrain

ALEX GOULD[1], NOBUE ITASAKI[2], and ROBB KRUMLAUF[2]

Summary. The developing vertebrate hindbrain is organized into a series of lineage-restricted units called rhombomeres, each of which expresses a different array of *Hox* genes. These genes encode transcription factors that are implicated in controlling morphological differences between rhombomere (r) derivatives. It has recently been shown in chick embryos that the *Hox* expression status of a rhombomere is not determined and can be reprogrammed by heterotopic grafting to a more posterior location. Transposition experiments in the chick have also been used to investigate the properties of the signals that can reprogram *Hoxb4* expression, and these demonstrate that somites have a graded ability to induce neural expression that depends on both their developmental stage and axial level of origin. To identify response elements mediating neural *Hoxb4* induction by paraxial mesoderm, we have used chick-transgenic mouse graftings to show that a 9-kb murine transgene can be appropriately induced by chick posteriorizing signals. Functional dissection of the transgene identified a neural enhancer (HRE) that is conserved between mouse, fish, and chick and is implicated in stabilizing the normal r6/7 boundary of *Hoxb4* expression after the time of rhombomere formation. We show that the HRE is an autoregulatory element that can also respond to activation by other *Hox* genes, including *Hoxd4* and *Hoxb5*. This suggests that the HRE may be required for sustaining the initial response to induction by paraxial mesoderm and could also be a component of the ratchet mechanism involved in maintaining *Hoxb4* expression in anterior rhombomere transpositions.

Key words. Induction—Somites—Autoregulation—*Hoxb4*—Transgenic mice

Introduction

Rhombomeres form during hindbrain development as a consequence of an underlying molecular segmentation process. Fate-mapping studies in the chick show that this results in a subdivision of the hindbrain into six to eight lineage-restricted

Laboratories of [1] Mammalian Development and [2] Developmental Neurobiology, MRC National Institute for Medical Research, The Ridgeway, Mill Hill, London, NW7 1AA, UK cellular units that each give rise to distinct neuronal populations (reviewed in [1,2]). The rhombomere-restricted expression of *Hox* genes is tightly coupled to the segmentation process per se and has been widely implicated in the specification of rhombomere (r) identity (reviewed in [3]). However, little is known about the timing of events or signals that establish and maintain the segmental domains of *Hox* expression.

Reciprocal transpositions between presumptive r2 and r4 in the chick embryo have suggested that *Hoxb1* expression in rhombomere 4 is neural plate autonomous from as early as the 6s stage [4–7]. This led to the idea that rhombomeric *Hox* expression, once established by early prepatterning cues, is determined and independent of environmental influences.

We discuss here the results of recent heterotopic grafting experiments that challenge this view [8–10] and demonstrate that hindbrain *Hox* expression in chick and mouse can be reprogrammed by signals from the paraxial mesoderm. We focus on the *Hoxb4* gene, normally expressed in the postotic hindbrain up to the r6/7 boundary, and show that transgenic approaches in the mouse [11] may be used to identify distinct *Hoxb4* enhancers mediating different aspects of the reprogramming response.

Hoxb4 Induction Following Posterior Transposition of Rhombomeres in the Chick

Heterotopic grafting experiments have demonstrated that ectopic *Hoxb4* gene expression can be induced in r2–r6 following posterior transpositions to the rostral spinal cord in 5s–10s chick embryos. This induction demonstrates that the hindbrain is plastic and that *Hoxb4* expression can be reprogrammed at these stages (Fig. 1A–E). In contrast, anterior transpositions of r7–r8 into the rostral hindbrain do not cause this gene to be switched off. Thus, *Hoxb4* expression in caudal rhombomeres can be maintained in a preotic environment in an autonomous manner [8–10]. The appearance of hindbrain plasticity only in A-to-P transpositions is also revealed using other *Hox* markers and suggests the presence of a ratchet-like mechanism where changes in cell fate can occur in one direction only, anterior to posterior.

Axial and Temporal Differences in the *Hoxb4*-Inducing Properties of Chick Somites

What is the source of the posteriorizing signals that induce the changes in *Hoxb4* expression detailed here? One possibility is that planar signals in the neural ectoderm might be primarily responsible for the induction [9]. However, in anterior transpositions where *Hoxb4* expression is maintained, little or no expression is induced in surrounding host neural plate, suggesting that the posterior neural tube itself is not the primary source of the inductive signal [10]. In contrast, anterior transpositions of chick paraxial mesoderm indicate that caudal somites are capable of posteriorizing adjacent preotic rhombomeres and inducing ectopic *Hoxb4* expression (Fig. 2A–F).

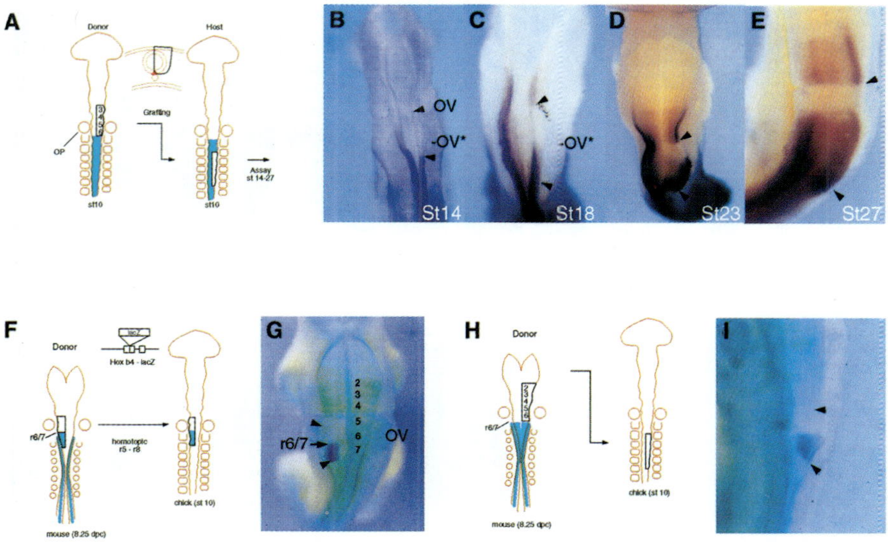

FIG. 1A–I. Induction of *Hoxb4* expression in posteriorly transposed rhombomeres of chick and chick/mouse grafts. **A** Scheme of chick A–P transposition of r3–r6 at stage 10. **B–E** Time course of induction, analyzed by *Hoxb4* in situ hybridization. The graft is negative for *Hoxb4* at *st 14* (**B**), but is stably induced in the posterior region between stages 18 and 27. In **B**, there is an ectopic otic vesicle (*OV**) lateral to the graft, often seen with r5–r6 transpositions. Carbon particles in **C** and **E** indicate precise borders of graft. **F** Scheme of homotopic r5–r8 grafting from donor 8.25 dpc *Hoxb4/lacZ* transgenic mouse line into chick host. **G** Transgene expression in interspecies chick/mouse homotopic graft shows the boundary maps to proper r6/7 junction. The reporter signal is in the posterior half of the graft, with a sharp boundary at r6/7 (arrow). **H** Scheme of r2–r6 posterior transposition using *Hoxb4/lacZ* donor tissue. **I** Transgene expression in mouse/chick grafted embryo. Note *lacZ* expression is only induced in the posterior part of the graft, as seen in analogous chick/chick graftings (**C–E**). *Arrowheads*, borders of graft; *OV*, otic vesicle; *OP*, otic placode; *r*, rhombomere. (From [10], with permission)

To investigate if somite-inducing ability varies with axial level of origin, individual somites from s1 to s10 were placed adjacent to r4 and assayed for *Hoxb4* induction (Fig. 2G–J). At stage 10, only somites posterior to the level of s4 or s5 display inducing potential (Fig. 2G). However, s1–s4 from earlier embryos are able to induce expression, showing that the boundary in inducing ability is not fixed but progressively sweeps caudally to the s5/6 junction with increasing developmental age (Fig. 2K,L). Thus, paraxial mesoderm has a graded ability to induce *Hoxb4* that is influenced by both age and axial level of origin.

Conservation of *Hoxb4* Induction Between Chick and Mouse

We grafted hindbrain tissue from *Hoxb4/lacZ* lines of transgenic mice [12] into host chick embryos to test whether murine *Hoxb4* could be induced in a similar way to

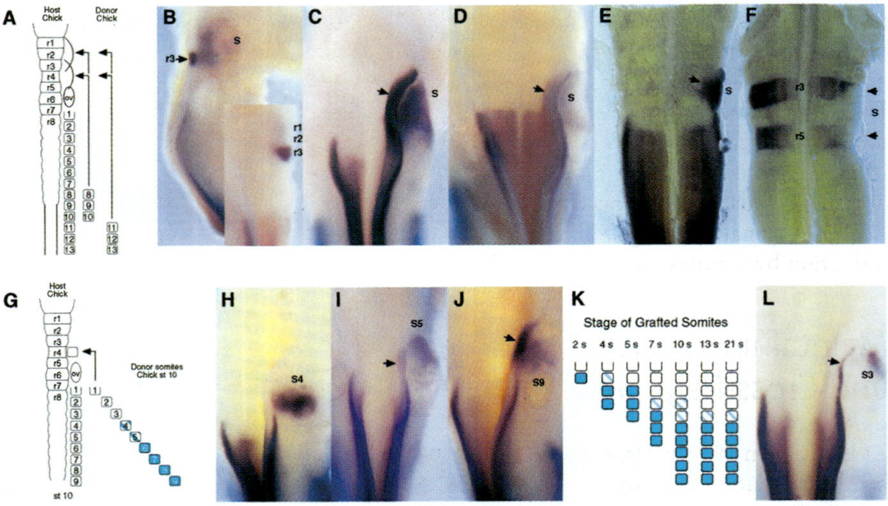

FIG. 2A–L. Differential ability of somites to induce *Hox* expression. **A** Scheme for unilateral anterior transposition of somite blocks adjacent to hindbrain. **B** Embryo with s8–s10 at the r1–r3 level; *Hoxb4* induction is only seen in *r3*. *Inset* shows dorsal view of embryo. **C** Embryo with s8–s10 at the r3–r5 level probed for *Hoxb4*. **D** Embryo with s8–s10 at the level of r3–r5 probed for *Hoxa3*. **E** Hindbrain flat mount of s8–s10 at the level of r3–r5 probed for *Hoxb4*. Induction is often limited to the region directly adjacent to ectopic somites. **F** Hindbrain flat mount of s8–s10 at r3–r5 level probed for chick *Eph A2* (*cek8*). Downregulation in r3 and r5 shows somites influence other segmental markers. **G** Scheme for single somite transpositions from st10 donors. **H–J** Embryos with single somites grafted at the r4 level and probed for *Hoxb4*. **H** s4, no induction. **I** s5, weak induction. **J** s9, full induction. **K** Summary of single somite-inducing potential from different stages of donor. *Solid blue color* in **G** and **K** indicates full inducing capability; *blue stripes* indicate variable potential. **L** Embryo with s3 from a 5-somite-stage donor at the r4 level, probed for *Hoxb4*. *Arrows*, sites of induction; *S*, position of the grafted somite. (From [10], with permission)

chick *Hoxb4*. Homotopic grafts indicated that reporter gene expression is properly activated in these interspecies chimeras (Fig. 1F,G). Posterior heterotopic grafts of mouse r1–r6 or r2–r6 to the chick postotic hindbrain/anterior spinal cord region resulted in *Hoxb4/lacZ* inductions with a similar spatial and temporal profile to that seen for chick *Hoxb4* (Fig 1H,I). As chick somites are also able to induce murine *Hoxb4/lacZ* expression (data not shown), the combined interspecies results indicate that the mechanism of plasticity and its underlying signals are conserved. This finding suggests that a transgenic analysis of mouse *Hoxb4* regulation would be a feasible approach for dissecting the response elements downstream of the hindbrain induction event.

A Candidate Late-Response Element to *Hoxb4* Induction in the Mouse

The *Hoxb4/lacZ* transgene used contains 9 kb of murine DNA that recapitulates the full neural pattern and so must include all the sequences necessary for a response to somite induction. Sequence comparisons have identified a conserved region (CR3), present in both chick and mouse *Hoxb4* genes, as one of several neural enhancer activities residing within this genomic region. Transgenic analysis has shown that this is the only enhancer capable of specifying a r6/7 expression boundary [11]. However, although murine *Hoxb4* protein is expressed in the pre-rhombomeric hindbrain, CR3 is first activated only after morphologically visible rhombomere boundaries are first apparent. Thus, this enhancer may be involved in the maintenance but not establishment of normal *Hoxb4* gene expression and could mediate a late response to *Hoxb4* induction by somites.

Maintaining *Hoxb4* Expression by Auto-, Para-, and Cross-Regulation

As it is known that the *Drosophila* ortholog of *Hoxb4* maintains its expression, in part, by autoregulation [13], the ability of CR3 to function as a *Hox* response element was tested (Fig. 3C–H). Driving *Hoxb4* expression throughout the mouse embryo under the control of a β-actin promoter results in the ectopic activation of CR3 within the anterior CNS, suggesting that CR3 responds to autoregulation (Fig. 3C–H). However, in complementary loss-of-function experiments, we found that the functions of *Hoxb4* and its paralog, *Hoxd4*, both need to be removed to abolish CR3 activity in r7, thus indicating that CR3 responds similarly to autoregulation or pararegulation (Fig. 3A,B).

While the group 4 genes (*Hoxb4* and *Hoxd4*) are involved in setting the r6/7 boundary of CR3, more 5′ genes, such as *Hoxb5*, are involved in the cross-regulatory activation of CR3 in more caudal regions of the CNS (Fig. 3F; and data not shown). The results of the loss and gain of *Hox* function experiments described here, together with other studies in *Drosophila* [11], indicate that CR3 is a target for the action of multiple *Hox* genes. For these reasons we refer to CR3 as the *H*ox *R*esponsive *E*nhancer (HRE). The response of the HRE is likely to involve direct binding of Hox proteins because it contains two in vitro *Hox* binding sites, HS1 and HS2, and transgenic analysis shows that these sites are required for CR3 activity in vivo (Fig. 4; and data not shown).

The Caudalizing Signal and Ratchet Model

In a variant of the activation-transformation model, proposed for neural induction during normal *Xenopus* development [14–17], plasticity of *Hox* expression in the hindbrain might be explained by an A–P gradient in posteriorizing influence that emanates from paraxial mesoderm. Progressively more posterior regions of the CNS would be exposed to progressively higher levels of the mesodermal signal. In A-to-P transpositions, preotic hindbrain tissue would come under the influence of higher signal levels, switch on more 5′ *Hox* genes such as *Hoxb4*, and thus adopt a more posterior cell fate.

Exposure to lower levels of the posteriorizing signal during P-to-A transpositions of postotic hindbrain might then be expected to result in switching off *Hoxb4* expression. However, this is clearly not the case. Instead, it appears that the initial exposure

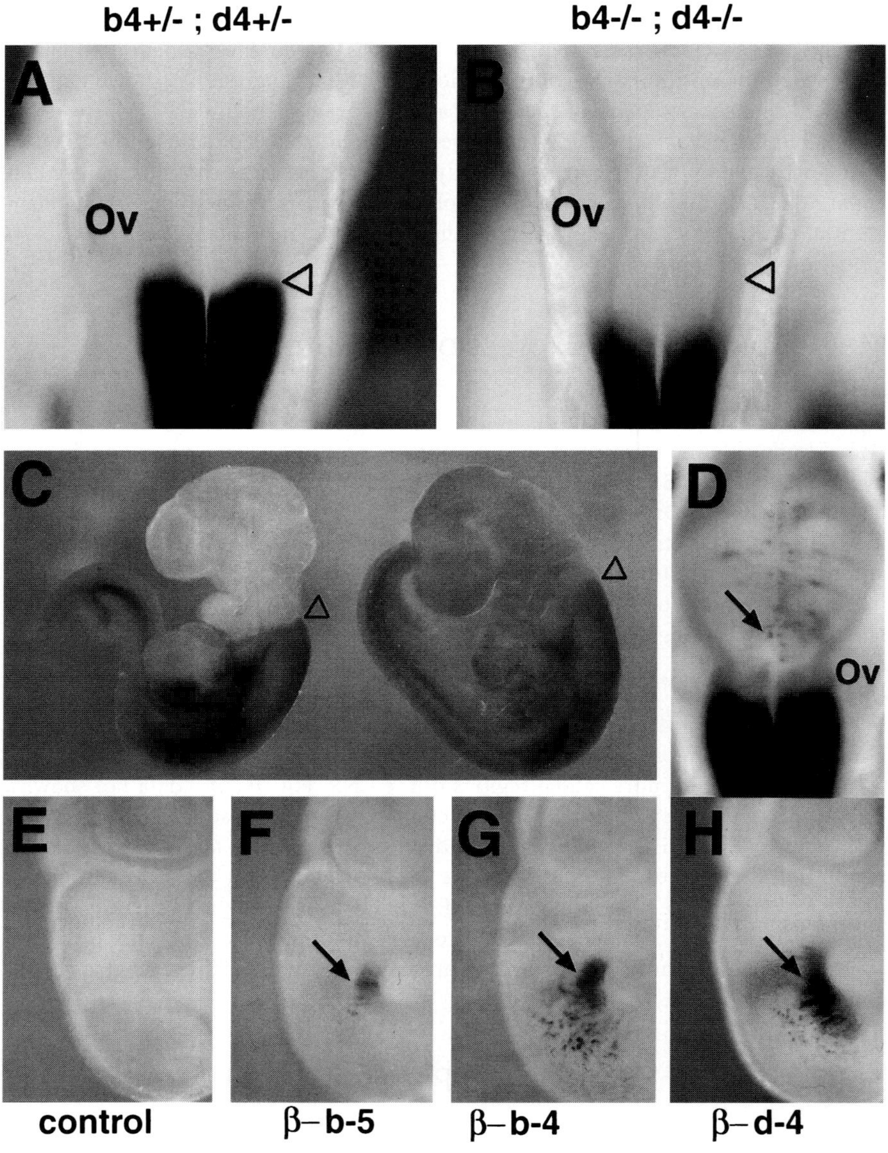

FIG. 3A–H. The HRE (Hox responsive enhancer) is regulated by group 4 and 5 genes. **A,B** Dorsal views of 10.5 dpc embryos transgenic for an HRE/*lacZ* construct and heterozygous (**A**) or homozygous (**B**) for mutant alleles of both *Hoxb4* and *Hoxd4* show that absence of both genes leads to loss of *lacZ* expression from the vicinity of r6/7. **C** Comparison between wild-type (*left*) and β-actin *Hoxb4* transgenic (*right*) 9.75 dpc embryos immunostained for *HOXB4*. **D** Dorsal view of hindbrain of 10.5 dpc embryo carrying a β-actin *Hoxd4* construct and showing ectopic sites of expression (*arrows*). **E–H** Regions of control embryos transgenic for CR3/*lacZ* alone (**E**) or embryos also transgenic for β-actin *Hoxb5* (**F**), β-actin *Hoxb4* (**G**), and β-actin *Hoxd4* (**H**), all showing ectopic midbrain expression (*arrows*). (From [11], with permission)

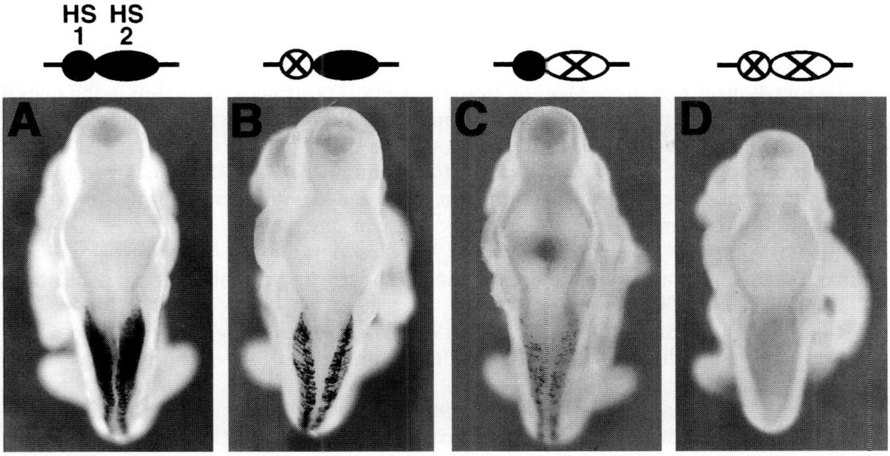

FIG. 4A–D. *Hox* binding sites *HS1* and *HS2* are essential components of the HRE. Dorsal views of 10.5 dpc embryos transgenic for a wild-type HRE/*lacZ* reporter construct (**A**) or variant constructs containing mutations in HS1 (**B**), HS2 (**C**), or both *Hox* binding sites (**D**). Mutations are indicated above each panel. (From [11], with permission)

to high levels of caudalizing signal can be retained or remembered following anterior transposition. For the *Hoxb4* gene, at least one component of this ratchet-like mechanism might be provided by the auto-, para-, and cross-regulatory maintenance enhancer, HRE. Once a *Hox* feedback loop is initiated, it could be sustained in a neural plate-autonomous manner, even in the absence of mesodermal influences. In this regard, it is interesting that *Hoxb1* is also known to be regulated by a positive feedback loop that could contribute to sustaining expression following r4 to r2 transpositions [18].

Finally, we note that further experiments are required to determine whether the mechanisms controlling *Hoxb4* reprogramming, and hindbrain plasticity in general, actually play a role in the establishment of gene expression during normal hindbrain development.

References

1. Keynes R, Krumlauf R (1994) *Hox* genes and regionalization of the nervous system. Annu Rev Neurosci 17:109–132
2. Lumsden A, Krumlauf R (1996) Patterning the vertebrate neuraxis. Science 274:1109–1115
3. Krumlauf R (1994) *Hox* genes in vertebrate development. Cell 78:191–201
4. Guthrie S, Muchamore I, Kuroiwa A, et al (1992) Neuroectodermal autonomy of *Hox-2.9* expression revealed by rhombomere transpositions. Nature (Lond) 356:157–159
5. Kuratani SC, Eichele G (1993) Rhombomere transposition repatterns the segmental organization of cranial nerves and reveals cell-autonomous expression of a homeodomain protein. Development (Camb) 117:105–117
6. Prince V, Lumsden A (1994) *Hoxa-2* expression in normal and transposed rhombomeres: independent regulation in the neural tube and neural crest. Development (Camb) 120:911–923

7. Simon H, Hornbruch A, Lumsden A (1995) Independent assignment of antero-posterior and dorso-ventral positional values in the developing chick hindbrain. Curr Biol 5:205–214
8. Grapin-Botton A, Bonnin M-A, Le Douarin N (1997) *Hox* gene induction in the neural tube depends on three parameters: competence, signal supply and paralogue group. Development (Camb) 124:849–859
9. Grapin-Botton A, Bonnin M-A, Ariza-McNaughton L, et al (1995) Plasticity of transposed rhombomeres: *Hox* gene induction is correlated with phenotypic modifications. Development (Camb) 121:2707–2721
10. Itasaki N, Sharpe J, Morrison A, et al (1996) Reprogramming *Hox* expression in the vertebrate hindbrain: influence of paraxial mesoderm and rhombomere transposition. Neuron 16:487–500
11. Gould A, Morrison A, Sproat G, et al (1997) Positive cross-regulation and enhancer sharing: two mechanisms for specifying overlapping *Hox* expression patterns. Genes Dev 11:900–913
12. Whiting J, Marshall H, Cook M, et al (1991) Multiple spatially specific enhancers are required to reconstruct the pattern of *Hox-2.6* gene expression. Genes Dev 5:2048–2059
13. Zeng C, Pinsonneault J, Gellon G, et al (1994) Deformed protein binding sites and cofactor binding sites are required for the function of a small segment-specific regulatory element in *Drosophila* embryos. EMBO J 13:2362–2377
14. Nieuwkoop P (1952) Activation and organisation of the central nervous system in amphibians. J Exp Zool 120:1–108
15. Nieuwkoop P (1985) Inductive interactions in early amphibian development and their general nature. J Embryol Exp Morphol 89(suppl):333–347
16. Toivonen S, Saxen L (1968) Morphogenetic interaction of presumptive neural and mesodermal cells mixed in different ratios. Science 158:539–540
17. Slack JMW, Tannahill D (1992) Mechanism of anteroposterior axis specification in vertebrates. Lessons from the amphibians. Development (Camb) 114:285–302
18. Pöpperl H, Bienz M, Studer M, et al (1995) Segmental expression of *Hoxb1* is controlled by a highly conserved autoregulatory loop dependent upon *exd/Pbx*. Cell 81:1031–1042

The Induction and Patterning of Spinal Motor Neurons

YASUTO TANABE and THOMAS M. JESSELL

Summary. In the developing spinal cord, the distinct neuronal cell types are generated along its dorsoventral axis. A combination of in vitro assays of neural cell differentiation and studies of ventral patterning in mutant mice has resulted in two main findings: (i) the generation of different cell types at distinct positions in the ventral half of the neural tube is established by a gradient of Shh signaling activity, and (ii) Pax6 functions as a key intermediate in the regulation of progenitor cell identity and neuronal fate in response to Shh signaling.

Key words. Induction—Patterning—Sonic hedgehog—Cell-fate specification—Motor neurons

Introduction

The patterning of cell types during vertebrate embryogenesis is controlled by organizing centers, specialized cell groups that direct the fate of adjacent cells through the secretion of inductive factors. During the early development of the nervous system, cell types generated in the ventral half of the neural tube acquire their identity and the pattern of differentiation in response to signals provided by an axial mesodermal organizing center, the notochord [1].

How does notochord signaling control ventral cell pattern? Assays of neural cell differentiation in vitro and analyses of mutant mouse and zebrafish embryos [2] first provided evidence that the notochord is the source of two separable inductive signals: a signal which acts locally to induce the differentiation of floor plate cells at the ventral midline of the neural tube and a signal which acts at longer range to induce motor neurons (MNs) and ventral interneurons in more dorsal positions. Floor plate differentiation requires induction by the notochord, but once induced floor plate cells express the same short- and long-range signaling activities as the notochord [2–4].

Howard Hughes Medical Institute, Department of Biochemistry and Molecular Biophysics, Columbia University, 701 West 168th Street, New York, NY 10032, USA

Induction of Motor Neurons by Sonic Hedgehog

The strongest candidate as a mediator of both the short- and long-range inductive activities of the notochord is a secreted glycoprotein, sonic hedgehog (Shh). Shh is expressed by the notochord and later by the floor plate over the period that these two midline cell groups exhibit their inductive activity (Fig. 1). In addition, studies in several vertebrate species have shown that ectopic expression of *Shh* can induce floor plate differentiation in vivo [5–7].

Studies of cell differentiation in neural plate explants in vitro have begun to provide a more detailed description of the mechanisms by which Shh signaling controls the differentiation of MNs and other ventral cell types [8]. The differentiation of both

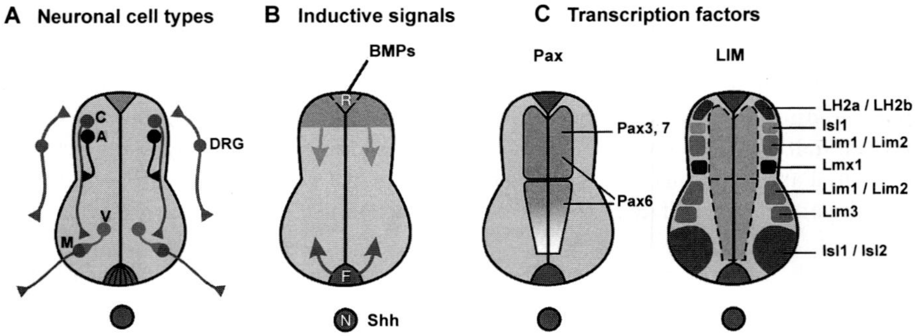

FIG. 1. **A** The distinct neuronal cell types generated in the developing spinal cord. Neuroepithelial cells proliferate and differentiate into neurons located at different dorsoventral positions. Subclasses of commissural (*C*) and association (*A*) neurons differentiate dorsally, close to the roof plate, whereas motor neurons (*M*) and ventral interneurons (*V*) differentiate ventrally near the floor plate. Dorsal root ganglion (*DRG*) neurons are generated from postmigratory neural crest cells. **B** Diagram of the source of ventralizing (sonic hedgehog, *Shh*) and dorsalizing (*BMPs*) inductive signals in the developing spinal cord. Shh is initially expressed in the axial mesoderm, and BMPs originate in the epidermal ectoderm flanking the lateral edges of the neural plate. At neural fold stages, Shh begins to be expressed by floor plate (*F*) cells at the midline of the neural plate and BMPs by cells in the dorsal tips of the neural folds. After neural tube closure, BMP expression is lost from the epidermal ectoderm except at the dorsal midline, but BMPs are now expressed in the roof plate (*R*) and adjacent dorsal neural tube. At the onset of neuronal differentiation, BMP expression persists in the dorsal neural tube, and Shh expression is maintained in the floor plate. **C** The *left-hand* diagram shows *Pax* genes expression during spinal cord development. During the transformation of the neural plate into the neural tube in the chick embryo, the regulated expression of three *Pax* genes—*Pax3*, *Pax6*, and *Pax7*—subdivides the neural tube into distinct domains. After neural tube closure, *Pax3* and *Pax7* expression is restricted to the dorsal half of the neural tube, whereas *Pax6* is expressed by more ventral cells in a gradient. *Pax6* is also expressed by cells in the dorsal half of the neural tube. The *right-hand* diagram shows that subsets of neurons derived from different domains of the ventricular zone can be distinguished by the expression of *LIM* homeodomain proteins. Motor neurons express Isl1/Isl2; certain dorsal commissural neurons express LH2a/LH2b; and dorsal ipsilateral interneurons express Isl1. The axonal projection patterns of the remaining interneuron classes have not been established. Motor neurons can be further subdivided into columnar subsets on the basis of a more complex LIM homeodomain protein code

floor plate cells and MNs can be assessed in such explants using a series of molecular markers [9]. Floor plate differentiation can be assessed by expression of the winged helix transcription factor, HNF3β [10,11]. The transcription of this gene in neural plate cells in vitro appears to be a direct response to notochord-derived signals [11]. Moreover, misexpression of HNF3β in the neural tube is sufficient to trigger ectopic floor plate differentiation [12,13]. Expression of the chemoattractant *Netrin-1* provides an independent marker of later stages of floor plate differentiation. The differentiation of MNs can be assessed by expression of the LIM homeodomain proteins Isl-1 and Isl-2 [14], and by the later markers *choline acetyltransferase (ChAT)* and the cell surface immunoglobulin-like protein SC1 [15].

Neural plate explants grown alone in vitro do not express floor plate and MN markers but when grown in contact with notochord are induced to express *HNF3β* and *Netrin-1*, indicating the differentiation of floor plate cells. The same explants contain cells that express Isl1/2 and SC1, and also *ChAT*, indicative of the differentiation of MNs. The floor plate- and MN-inducing activities of the notochord can be separated experimentally by their dependence on contact between the inducing and responsive tissue. In the absence of contact, the notochord still induces MN differentiation, albeit less effectively, but does not induce floor plate differentiation. Thus, a notochord-derived diffusible factor can induce MNs in the absence of floor plate differentiation.

To what extent does Shh mimic the contact-dependent and diffusible activities of the notochord? Neural plate explants grown in contact with COS cells transfected with *Shh* generate both floor plate cells, assessed by expression of HNF3β and *Netrin-1*, and MNs, assessed by Isl1/2$^+$/SC1$^+$ neurons and *ChAT*. In contrast, neural plate explants grown in the absence of contact with COS cells transfected with *Shh* do not express *HNF3β* or *Netrin-1*, indicating that floor plate differentiation has not occurred. Nevertheless, MN differentiation is still induced in the absence of contact. Thus, the diffusible activity derived from *Shh*-transfected COS cells can induce the entire complement of MN markers detected in response to induction by the notochord. Expression of *Shh*, therefore, confers COS cells with a local floor plate-inducing activity and a long-range MN-inducing activity that does not elicit floor plate differentiation.

The possibility that COS cells transfected with *Shh* are induced to secrete a distinct factor that mediates the floor plate- or MN-inducing activities can be excluded by transfecting *Shh* directly into cells within neural plate explants. Such neural plate explants expressed *HNF3β*, *Netrin-1*, *Isl1*, and *Isl2*, confirming that Shh is sufficient to induce floor plate and MN differentiation. The expression of *Isl1* in neural plate explants transfected with *Shh* either precedes or occurs coincidentally with that of *HNF3β*. Thus, MN differentiation in neural plate explants transfected with *Shh* occurs before or synchronously with floor plate differentiation.

The evidence that Shh accounts for both the floor plate- and MN-inducing activity of the notochord and floor plate is strengthened by the finding that the recombinant amino-terminal autoproteolytic cleavage product of Shh also has floor plate- and MN-inducing activity [16]. In addition, antibodies that inhibit Shh signaling in vitro have been shown to block the ability of the notochord and floor plate to induce floor plate cells and MNs, supporting the idea that Shh mediates notochord signaling [17,18]. Furthermore, mice in which the *Shh* gene has been inactivated by gene targeting fail to generate ventral cell types in the CNS [19]. Taken together, these studies show that Shh is necessary and sufficient for the induction of ventral cell types.

Progressive Steps in Ventral Cell Differentiation

The demonstration that Shh signaling is required for MN differentiation, however, leaves unresolved many details of the progression of neural progenitor cell differentiation. In this section, we discuss evidence that addresses when and how Shh controls the identity and pattern of MNs and other cell types generated in the ventral neural tube.

An early step in this process appears to be the regulation of transcription factor expression in cells at different mediolateral positions of the neural plate. When the caudal neural plate is formed, cells at all mediolateral positions express the homeodomain-containing transcription factors Pax3, Pax7, Msx1, and Msx2 [20–22]. The expression of these genes is rapidly repressed in medial neural plate cells by a Shh-mediated signal from the notochord [18,20,22], and after neural tube closure, their expression is restricted to proliferating cells in the dorsal neural tube (Fig. 1) [18,20,22]. The repression of *Pax3* and *Pax7* expression in neural plate cells appears to be a prerequisite for the generation of ventral cell types. Lateral neural plate cells that have not been exposed to Shh maintain Pax7 expression and rapidly lose the competence to generate floor plate cells and MNs in response to Shh [18]. Furthermore, misexpression of *Pax3* in ventral regions of the spinal cord in transgenic mice blocks floor plate differentiation [23]. The rapid time course of the repression of *Pax3*, *Pax7*, and *Msx1/2* by Shh [18,20,22] raises the possibility that these genes may be direct targets for the intracellular transduction events triggered in neural plate cells by Shh.

On exposure to Shh, cells in the medial region of the neural plate are converted to a ventralized state (defined operationally as the repression of *Pax3* and *Pax7* expression), and they also acquire the capacity to give rise to floor plate cells, MNs, or ventral interneurons (Fig. 2). The selection of one of these three cell fates appears to be regulated by a second phase of Shh signaling. Cells at the midline of the neural plate appear to be exposed to Shh generated locally by the notochord, thus directing floor plate cell fate through the expression of HNF3β [12,13] At later stages, Shh activity in adjacent regions of the ventral neural tube appears to define whether ventralized progenitors give rise to MNs or to ventral interneurons [18].

Threshold Responses to Shh Signaling

How are the identity and position of differentiation of ventral cell types controlled by Shh signaling? Studies of neuronal cell differentiation in vitro have provided evidence that the exposure of ventral progenitor cells to small differences in Shh concentration determines their eventual fate. The Shh concentration threshold required to generate specific cell types in vitro is inversely related to the distance from the ventral midline in vivo at which each neuronal type is generated. Low concentrations of Shh induce the differentiation of ventral interneurons. Increasing the Shh concentration two- to threefold induces MNs. A further two- to threefold increase in Shh concentration

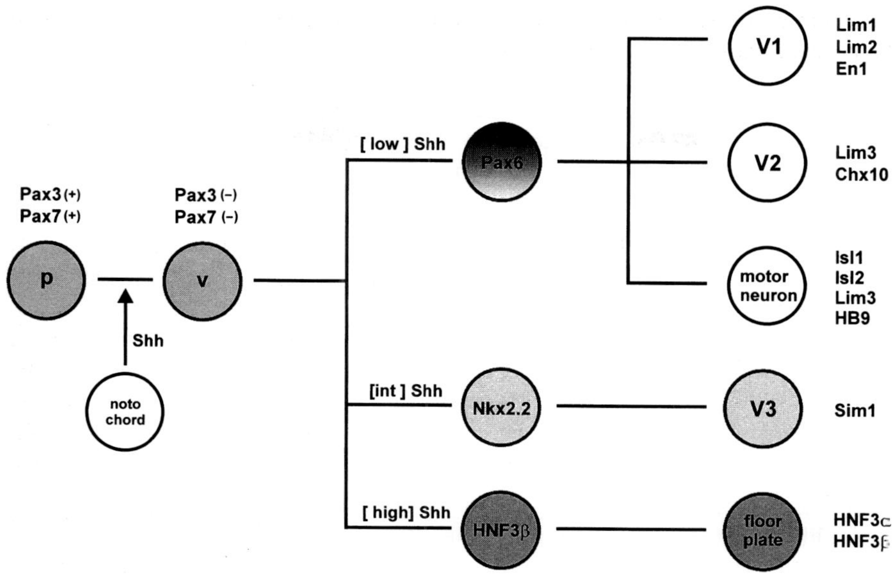

FIG. 2. Ventral inductive signaling. The graded activity of Shh and a proposed progenitor lineage of ventral neurons. Shh derived from the notochord acts on naive progenitor cells (*p*) within the neural plate to specify them as ventral progenitors (*v*). In a second phase of Shh signaling, three ventral progenitor subdomains are established by the expression of HNF3β, Nkx2.2, and Pax6 according to their Shh concentration requirements. Distinct neuronal cell types are generated from these individual progenitor domains. In the Pax6 domain there is a gradient of dorsal-high to ventral-low Pax6 expression. The subsequent expression of transcription factors defines *V1*, *V2*, and *V3* cell groups, distinct populations of ventral interneurons (for details, see [24])

induces floor plate cells at the expense of MNs and ventral interneurons (Fig. 2) [24].

Additional in vitro studies using function-blocking anti-Shh antibodies have shown that ventralized progenitors require Shh signaling until late in their final progenitor cell division cycle to generate MNs. This finding is reminiscent of studies showing that the laminar identity of cortical neurons is determined late in their final progenitor division cycle [25]. At the onset of MN differentiation in higher vertebrates, the notochord has been displaced ventrally and is no longer close to the neural tube. Thus, the Shh required to convert ventralized progenitors into MNs is likely to derive from the floor plate.

It has been difficult, however, to visualize the low levels of Shh protein that are presumed to diffuse from the notochord and floor plate. Nevertheless, several lines of evidence indicate that Shh does act at long range and in addition suggest that the range of Shh signaling initially extends throughout the ventral neural tube. First, the ability of anti-Shh antibodies to block MN differentiation in ventral neural tissue devoid of the floor plate indicates that the conversion of ventral progenitors into MNs depends on the diffusion of Shh from the floor plate [18]. Second, during the early period of Shh signaling, the domain of the neural tube over which Pax7 expression is repressed extends about 10 cell diameters from the notochord [18]. Third, an elevated level of *patched* mRNA, diagnostic of the exposure of cells to Hh signaling in flies and vertebrates [26,27], is detected in the medial neural plate and ventral neural tube over a domain similar to that in which Pax7 is repressed [27,28]. Finally, because the progenitors of MNs and ventral interneurons remain sensitive to the ambient Shh concentration until shortly before they leave the cell cycle [18], long-range Shh signaling is likely to involve direct action on ventral progenitor cells and not the activation of secondary inductive signals.

Regulation of Pax6 and Nkx2.2 and the Control of Ventral Neuronal Fates

What are the genes expressed by ventral progenitor cells that respond to graded Shh signals and determine neuronal fate? One critical output of Shh signaling appears to be the regulation of *Pax6* gene expression [24,29]. Pax6 expression is extinguished at high Shh concentrations, resulting in the generation of a distinct progenitor cell domain, defined by expression of Nkx2.2.

The regulation of Pax7, Pax6, and Nkx2.2 expression to Shh signaling may underlie the progression of progenitor cell specification within the ventral neural tube. The expression of Pax7 in vitro is repressed by Shh at a concentration eightfold lower than that required to repress Pax6. Thus, the early exposure of neural cells to low concentrations of Shh provided by the notochord appears to be sufficient to extinguish expression of Pax7 from ventral cells without repressing Pax6. The stable inheritance of the early state of Pax7 expression [18] may contribute to the establishment of a sharp boundary of Pax7 expression in the neural tube. Pax6 expression, in contrast to that of Pax7, remains sensitive to the ambient Shh concentration during the entire early period of ventral neurogenesis. This feature of Pax6 regulation may underlie the formation of the ventral low to dorsal high gradient of Pax6 in the ventral neural tube. The prolonged sensitivity of Pax6 to the local Shh concentration may also explain the late dependence of MN generation on Shh signaling.

Taken together, these studies suggest that the identity and pattern of progenitor cell types generated in the ventral half of the neural tube are controlled in large part by Shh signaling through actions at multiple concentration thresholds. Ventral progenitor cells respond to graded Shh signaling with the establishment of an inverse gradient of Pax6 protein expression. The gradient of Pax6 that is observed in vivo can be recapitulated in vitro by exposure of neural cells to a tenfold range in Shh concentration [24]. Thus, it is likely that a similar Shh concentration range is sufficient to specify the distinct cell types that are generated in the ventral neural tube in vivo. The conversion of Pax6 progenitors into Nkx2.2 progenitors in response to a twofold change in Shh-N concentration indicates, in addition, that abrupt changes in gene expression and progenitor cell itydentity occur in response to graded Shh signals.

Is Pax6 itself involved in the control of ventral neuronal pattern? It has been possible to address this question by analyzing ventral patterning in small eye (*Sey*) mutant mice that lack Pax6 function [24]. Elimination of Pax6 results in a dorsal-to-ventral transformation in the identity of ventrally located progenitor cells and in a consequent change in MN fate. The lack of Pax6 function in dorsally located ventral progenitor cells results in the loss of certain ventral interneurons. Pax6 is therefore a key intermediate in the Shh-dependent control of neuronal subtype identity in the ventral spinal cord and hindbrain. There is a marked dorsal expansion of Nkx2.2 expression in *Sey/Sey* mice, indicating that the dorsal extent of the Nkx2.2 progenitor cell domain is defined indirectly by the Shh concentration required for extinction of Pax6. In *Sey/Sey* embryos, the dorsal limit of the expanded Nkx2.2 domain could reflect the position at which Shh activity falls below a threshold necessary to activate Nkx2.2 expression. Alternatively, Shh signaling could define additional domains of gene expression in a Pax6-independent manner, and Nkx2.2 expression may be inhibited dorsally by such genes.

Although the expression of Pax6 is graded within the ventral neural tube, it remains unclear whether this gradient is itself instructive in the control of neuronal fate or is an incidental consequence of the requirement for graded Shh signaling. In the spinal cord and hindbrain, it remains possible that Pax6 functions at a single threshold level, defining solely Pax6-off and Pax6-on progenitor cell states. If this is the case, Shh signaling may, independently of Pax6, establish additional distinctions in progenitor cell identity that contribute to the specification of ventral interneuron fate.

Conclusions

Future studies on the role of Shh in MN generation need to investigate in greater detail how small differences in extracellular Shh concentration generate MN subtypes. Defining the intracellular transduction events elicited by Shh in neural cells and the downstream targets of the conserved Shh signaling pathway involved in MN differentiation will be essential steps in resolving this issue.

In addition, studies of *Sey* mice have indicated that Pax6 does not function directly in the pathway of MN differentiation. Rather, the main function of Pax6 appears to be to repress the expression of Nkx2.2, a factor that blocks somatic MN differentiation [24]. These observations imply the existence of other transcription factors that are expressed within Pax6 progenitor cells which function in a more active manner to promote somatic MN differentiation. Identification of these factors should provide an important step forward in defining the molecular pathway by which Shh signaling controls MN differentiation.

References

1. Tanabe Y, Jessell TM (1996) Diversity and pattern in the developing spinal cord. Science 274:1115–1123
2. Placzek M (1995) The role of the notochord and floor plate in inductive interactions. Curr Opin Genet Dev 5:499–506
3. Yamada T, Placzek M, Tanaka H, et al (1991) Control of cell pattern in the developing nervous system: polarizing activity of the floor plate and notochord. Cell 64:635–647
4. Placzek M, Jessell TM, Dodd J (1993) Induction of floor plate differentiation by contact-dependent, homeogenetic signals. Development (Camb) 117:205–218

5. Echelard Y, Epstein DJ, St Jacques B, et al (1993) Sonic hedgehog, a member of a family of putative signaling molecules, is implicated in the regulation of CNS polarity. Cell 75:1417–1430
6. Krauss S, Concordet JP, Ingham PW (1993) A functionally conserved homolog of the *Drosophila* segment polarity gene *hh* is expressed in tissues with polarizing activity in zebrafish embryos. Cell 75:1431–1444
7. Roelink H, Augsberger A, Heemskerk J, et al (1994) Floor plate and motor neuron induction by Shh, a vertebrate homolog of hedgehog expressed by the notochord. Cell 76:761–775
8. Tanabe Y, Roelink H, Jessell TM (1995) Induction of motor neurons by sonic hedgehog is independent of floor plate differentiation. Curr Biol 5:651–658
9. Yamada T, Pfaff SL, Edlund T, et al (1993) Control of cell pattern in the neural tube: motor neuron induction by diffusible factors from notochord and floor plate. Cell 73:673–686
10. Ruiz i Altaba A, Prezioso VR, Darnell JE, et al (1993) Sequential expression of HNF-3α and HNF-3β by embryonic organizing centres: the dorsal lip/node, notochord, and floor plate. Mech Dev 44:91–108
11. Ruiz i Altaba A, Placzek M, Baldassare M, et al (1995) Early stages of notochord and floor plate development in the chick embryo defined by normal and induced expression of HNF3β. Dev Biol 170:299–313
12. Sasaki H, Hogan BLM (1994) HNF-3β as a regulator of floor plate development. Cell 76:103–115
13. Ruiz i Altaba A, Cox C, Jessell TM, et al (1993) Ectopic neural expression of a floor plate marker in frog embryos injected with the midline transcription factor Pintallavis. Proc Natl Acad Sci USA 90:8268–8272
14. Tsuchida T, Ensini M, Morton SB, et al (1994) Topographic organization of embryonic motor neurons defined by expression of LIM homeobox genes. Cell 79:957–970
15. Tanaka H, Obata K (1984) Developmental changes in unique cell surface antigens of chick embryo spinal motor neurons and ganglion cells. Dev Biol 106:26–37
16. Roelink H, Porter J, Chiang C, et al (1995) Floor plate and motor neuron induction by different concentrations of amino-terminal cleavage product of sonic hedghog autoproteolysis. Cell 81:445–455
17. Marti E, Bumcrot DA, Takada R, McMahon AP (1995) Requirement of 19K form of sonic hedgehog for induction of distinct ventral cell types. Nature (Lond) 375:322–325
18. Ericon J, Morton S, Kawakami A, et al (1996) Two critical periods of sonic hedgehog signaling required for the specification of motor neuron identity. Cell 87:661–673
19. Chiang C, Litingtung Y, Lee E, et al (1996) Cyclopia and defective axial patterning in mice lacking sonic hedgehog gene function. Nature (Lond) 383:407–413
20. Goulding MD, Lumsden A, Gruss P (1993) Signals from the notochord and floor plate regulate the region-specific expression of two *Pax* genes in the developing spinal cord. Development (Camb) 117:1001–1016
21. Stuart ET, Kioussi C, Gruss P (1994) Annu Rev Gen 28:219
22. Liem KF, Tremml G, Roelink H, et al (1995) Dorsal differentiation of neural plate cells induced by BMP-mediated signals from epidermal ectoderm. Cell 82:969–979
23. Tremblay P, Pituello F, Gruss P (1996) Inhibition of floor plate differentiation by Pax3: evidence from ectopic expression in transgenic mice. Development (Camb) 122:2555–2567
24. Ericson J, Rashbass P, Schedl A, et al (1997) Pax6 controls progenitor cell identity and neuronal fate in response to graded Shh signalling. Cell 90:169–180
25. McConnell SK, Kaznowski CE (1991) Cell cycle dependence of laminar determination in developing neocortex. Science 254:282–285
26. Perrimon N (1995) Hedgehog and beyond. Cell 80:517–520

27. Goodrich LV, Johnson RL, Milenkovic L, et al (1996) Conservation of the hedgehog/patched signaling pathway from flies to mice: induction of a mouse patched gene by hedgehog. Genes Dev 10:301–312
28. Marigo V, Tabin CJ (1996) Regulation of patched by sonic hedgehog in the developing neural tube. Proc Natl Acad Sci USA 93:9346–9351
29. Osumi N, Hirota A, Ohuchi H, et al (1997) Pax-6 is involved in the specification of hindbrain motor neuron subtype. Development (Camb) 124: 2961–2972

Islet-3 in the Regional Specification of the Brain

Yutaka Kikuchi[1], Hiroshi Segawa[2], Mika Tokumoto[3], Yoshiki Hotta[4], Keiichi Uyemura[1], and Hitoshi Okamoto[1]

Summary. Islet-3 is an LIM/homeodomain protein that is expressed specifically in the eyes and the presumptive tectum in the central nervous system of zebrafish (*Danio rerio*) embryos. Overexpression of the protein (LIM^{Isl-3}) consisting only of the Islet-3 LIM domains in embryos specifically prevented formation of the optic vesicles, caused abnormal termination of the expression of *wnt1*, *engrailed2*, and *pax2* in the mesencephalic and metencephalic region between 14 and 20 h post fertilization and severely impaired morphogenetic movement in this region between 20 and 26 hs, which should normally lead to formation of the cerebellar primordium. Such defects were all rescued by simultaneous overexpression of Islet-3, suggesting that LIM^{Isl-3} acted as a specific dominant-negative variant of Islet-3. These data suggest that Islet-3 is activated by putative LIM-binding cofactors and functions to promote evagination of the optic vesicles and to maintain reciprocal interaction between the mesencephalon and the mesencephalic-metencephalic boundary essential for normal development of this region.

Key words. Zebrafish—Islet-3—LIM/homeodomain protein—Eye—Cerebellum

Introduction

Zebrafish (*Danio rerio*) embryos are transparent and have a relatively simple central nervous system with many of their early neurons identifiable by their stereotyped positions, shapes, and axonal trajectories [1,2]. These features have made zebrafish embryos a suitable material for analyzing the roles of transcription factors in neural differentiation. We and others have identified three homologs of Islet-1, a LIM/homeodomain-type transcription factor, and termed them zebrafish Islet-1, Islet-2, and Islet-3 [3–5]. By about 24 h, the expression of Islet-3 mRNA is almost completely

[1] Department of Physiology and [2] Department of Rehabilitation Medicine, School of Medicine, Keio University, 35 Shinanomachi, Shinjuku-ku, Tokyo 160-8582, Japan
[3] Division of Cellular Communication, Department of Cell Biology, National Institute for Basic Biology, Okazaki, Aichi 444-0867, Japan
[4] National Institute of Genetics, 1111 Yata, Mishima, Shizuoka 411-8540, Japan

restricted to the eyes and the presumptive tectal region of the mesencephalon [5]. This expression pattern suggests that Islet-3 is involved in the regional specification of the eyes and the midbrain. Accumulated evidence has supported a model in which the LIM domains of the LIM/homeodomain proteins negatively regulate the homeodomain function, and such an inhibition is removed in vivo by binding of activating protein to the LIM domain [6]. Accordingly, we overexpressed the protein ($LIM^{Isl\text{-}3}$) consisting only of the LIM domains of Islet-3 in zebrafish embryos, expecting that it could block Islet-3 by titrating out putative activating molecules [7].

Overexpression of $LIM^{Isl\text{-}3}$ Caused Defects in Optic Vesicle Formation

At 15.5 h, the zebrafish embryos injected at the one-cell stage with mRNA encoding $LIM^{Isl\text{-}3}$ were completely devoid of the optic vesicles (Fig. 1A,C), which were present in the normal embryos (Fig. 1B,D). The remaining structures of the body such as the array of chevron-shaped somites were indistinguishable between the injected and the normal embryos.

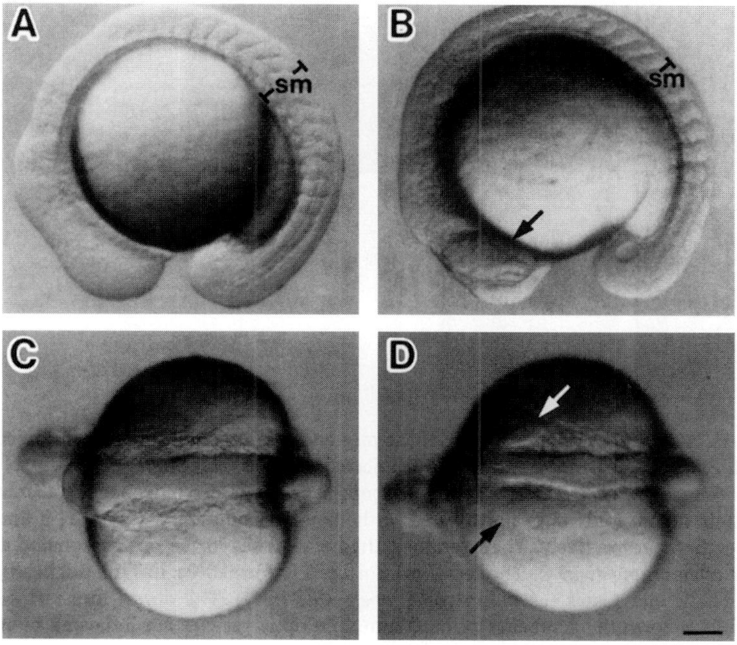

Fig. 1A–D. Eyeless zebrafish embryos induced by overexpression of $LIM^{Isl\text{-}3}$. **A,C** 15.5-h embryo injected with mRNA encoding $LIM^{Isl\text{-}3}$ at the one-cell stage. **B,D** 15.5-h normal embryo. The optic vesicles indicated by *arrows* in the normal embryos (**B,D**) are absent in the injected embryos (**A,C**). Lateral views with anterior to the left and dorsal to the top (**A,B**); dorsal views with anterior to the left (**C,D**). *sm*, somites. *Bar* 100 mm

The mes and met Region of the Injected Embryo Show Severely Impaired Morphogenetic Movement

In the normal embryo, the tissues in the mes and met region undergo dynamic morphogenetic movement between 20 and 26h. The mesencephalon widens by bilateral shifts of the dorsal tissues (alar plates), opening a transient ventricular space medially. The alar plates then start to thicken inwardly to occupy the dorsomedial ventricular space again, giving rise to the tectal primordium (Fig. 2D). This movement of the mesencephalic tissues is accompanied by transformation of the dorsal tissues in the mes-met boundary into a constriction, the isthmus, consisting of folded neural tissues (Fig. 2C,D). These folds constitute the cerebellar primordium.

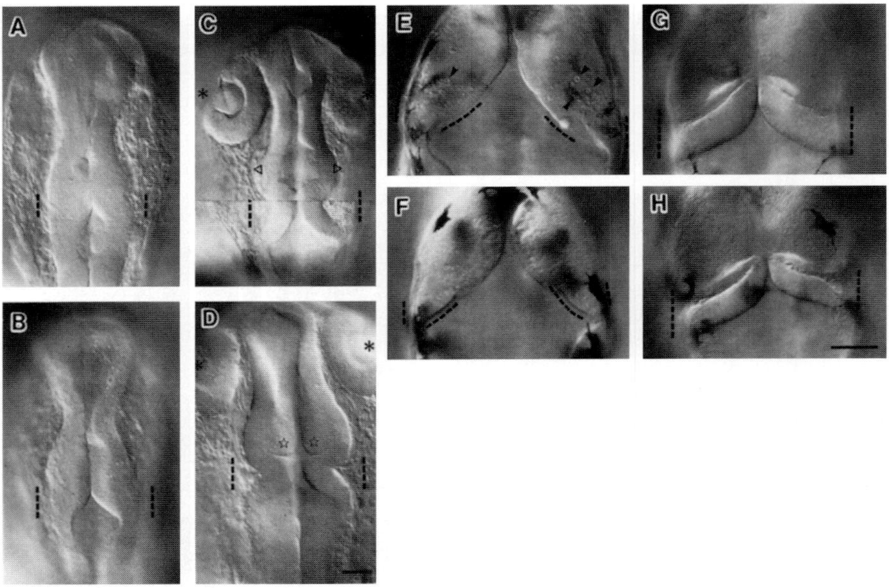

FIG. 2A–H. Morphogenetic movement of the mes and met region in normal and injected live zebrafish embryos. **A,B** A 24-h and a 26-h injected embryo, respectively. The dorsal mesencephalic tissues of the injected embryos did not show dynamic morphogenetic movement. The isthmic neural folds were not formed in the mes-met boundary. **C,D** A 24-h and a 26-h normal embryo, respectively. The laterally shifted alar plates in the mesencephalon are indicated by *open triangles* in **C**. The tectal primordium is formed by inward thickening of the mesencephalic alar plates (as indicated by *open stars* in **D**). The dorsal mes-met boundary tissues fold to form the cerebellar primordium. The optic vesicles are indicated by *asterisks*. **E,F** 32-h and 48-h injected embryos, respectively. The dorsal tissues (outlined by *broken lines*) in the mes-met boundary contained pyknotic dead cells (some of which are indicated by *arrowheads*) in E, and remained as rudimentary swellings attached to the caudal ends of the mesencephalon in **F.G,H** 32-h and 48-h normal embryos. The mes-met boundaries in **A–D,G,H** are also indicated by *broken lines*. Dorsal views with anterior to the top (**A–H**). *Bars* 50 mm (A–D), 50 mm (E–H)

The dorsal mesencephalic tissue did not show such morphogenetic movement as observed in the normal embryo, and the isthmic fold was not formed in the mes-met boundary (Fig. 2A, B). In the live injected embryos, the dorsal tissue in the mes-met boundary contained pyknotic dead cells around 32h (Fig. 2E), and remained only as a rudimentary swelling attached to the caudal edge of the mesencephalon at 48h (Fig. 2F). As a result, the injected embryo almost completely lacked the cerebellar primordium.

Gene Expression Specific to the mes and met Region Terminates Prematurely in the Injected Embryos

To examine whether anomaly at the gene expression level precedes various morphological abnormalities in the mes and met region of the injected embryo, we examined the expression pattern of the *wnt1* gene as well as the *pax2* and *engrailed2* (*eng2*) genes, all of which are expressed in the mes and met region and are essential for normal development of this region.

Expression of all three genes appeared normal in the mes and met region of the injected embryos at 10h. By 20h, however, expression of *wnt1*, *eng2*, and *pax2* in the mes and met region was completely lost except for expression of *wnt1* along the dorsal midline (Fig. 3A,C,E). In contrast, their expression in the rest of the body remained intact, e.g., *wnt1* expression along the dorsal midline of the hindbrain and the spinal cord (Fig. 3A,B), *eng2* expression in the muscle pioneers (Fig. 3C,D), and *pax2* expression in the presumptive optic stalks, the otic vesicles, the dorsal commissural neurons in the spinal cord, and the pronephros (Fig. 3E,F).

Islet-3 Function in the mes and met Region

Initiation and maintenance of expression of the genes homologous to *wnt1*, *pax2*, and *eng2* have been shown to be essential for normal development of the mes and met region in various vertebrates [8,9]. For maintaining expression of these genes later in development, reciprocal interaction between the mesencephalon and the mes-met boundary plays an important role. The mes-met boundary tissue transplanted in the caudal forebrain induced *Wnt1* expression in the neighboring host tissue and transformed it into an ectopic midbrain [10,11], suggesting that the mes-met boundary induces mesencephalic differentiation. Recent data have suggested that FGF8 intrinsically expressed in this boundary may act as a signaling molecule mediating this induction [12]. Presence of inductive signaling in the opposite direction, i.e., from the mesencephalon to the mes-met boundary, has also been suggested. Although *Wnt1* is not expressed in the mes-met boundary in normal mouse development, *Wnt1* mutant mice lack, both of the midbrain and of the cerebellum, a derivative of this boundary [13–15]. In such mice, *En1* expression is initiated normally but is lost later, demonstrating that *Wnt1* expression in the mesencephalon is necessary for the maintenance of *En1* expression. Recent data have further suggested that the signaling mediated by Wnt1 regulates development of the mes and met region mainly by activating *En*

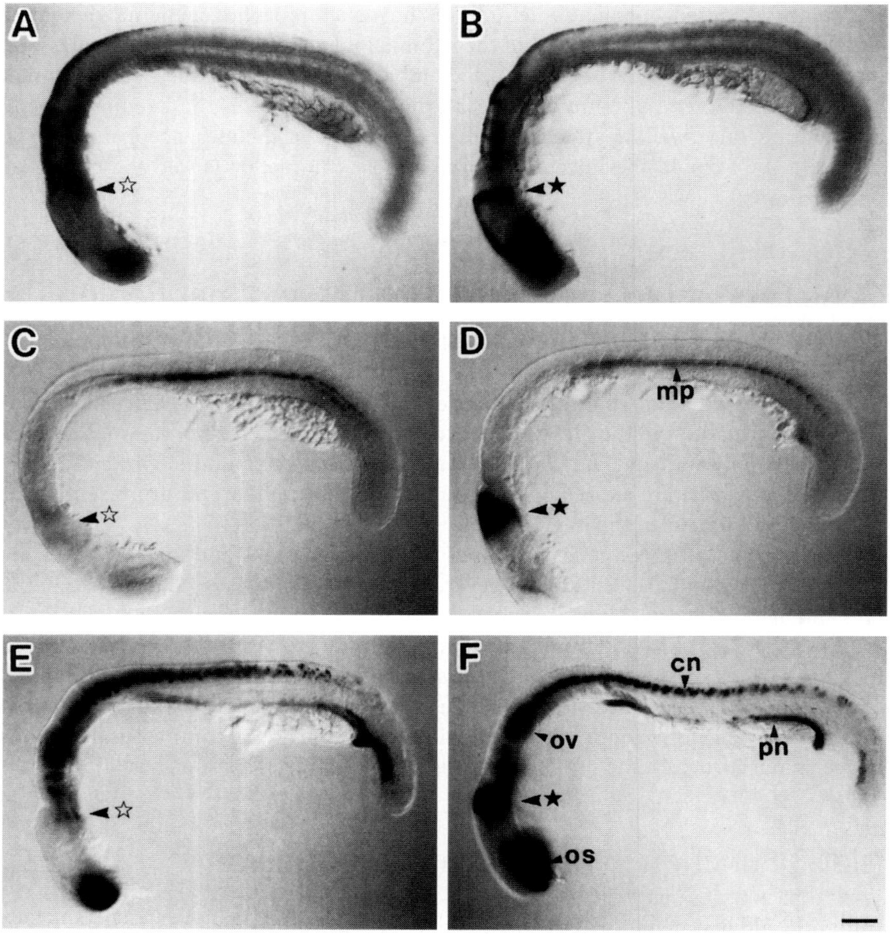

FIG. 3A–F. Expression patterns of *wnt1*, *eng2*, and *pax2* mRNA in 20-h zebrafish embryos. **A,B** *wnt1* expression in the 20-h injected and normal embryos, respectively. **C,D** *eng2* expression in the 20-h injected and normal embryos, respectively. **E,F** *pax2* expression in the 20-h injected and normal embryos, respectively. The expression of these genes in the mes and met region, indicated by *closed stars* in the normal embryos (**B,D,F**), was completely lost in the injected embryos, as indicated by *open stars* in **A,C,E**. In contrast, *wnt1* expression at the dorsal midline of the hindbrain and the spinal cord, *eng2* expression in the muscle pioneers (*mp*), and *pax2* expression in the presumptive optic stalk (*os*), the otic vesicles (*ov*), the dorsal commissural neurons in the spinal cord (*cn*), and the pronephros (*pn*) were normal in the injected embryos, compared to the expression patterns in the normal embryos. The yolks have been removed. Lateral views with anterior to the left and dorsal to the top. *Bar* 100 mm

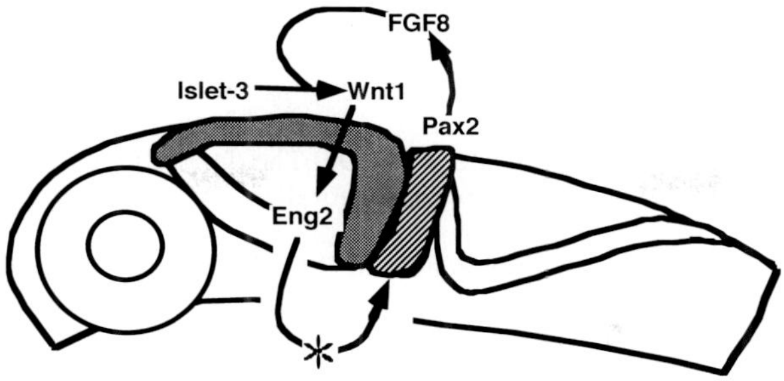

FIG. 4. A possible role for Islet-3 in the reciprocal interaction between the mesencephalon and the mes-met boundary. Induction of one molecule by the other is symbolized by an *arrow* connecting the two involved molecules. *Wnt1* expression in the caudal mesencephalon and *pax2* expression in the mes-met boundary are indicated by *shading* and *hatching*, respectively. A molecule (*asterisk*), which is released from the mesencephalon and mediates inductive signaling for the maintenance of *pax2* expression in the mes-met boundary, has not been identified

transcription, and that En protein then activates a putative mesencephalic signal that promotes normal development of the mesencephalon and the mes-met boundary [16].

Islet-3 expression becomes restricted to the presumptive tectal region of the mesencephalon as the mes and met region transforms into the tectal and cerebellar primordia. Taking this characteristic expression pattern into consideration, we putatively fit Islet-3 in the perspective of the aforementioned reciprocal interaction as the following (also as illustrated in Fig. 4), i.e., Islet-3 may function either in conditioning the mesencephalic tissue to be capable of expressing *wnt1* in response to the FGF8 signal emanating from the mes-met boundary or more directly in transduction of the FGF8 signal within the mesencephalic cells to maintain the *wnt1* gene expression. We further speculate that the severe impairment of morphogenetic movement in the mes and met region of the injected embryo was caused by disruption of this reciprocal interaction.

Acknowledgments. We are grateful to Drs. S. Krauss, R.T. Moon, M. Westerfield, and D.L. Turner for providing us with the *pax2* and *pax6* probes, a *wnt1* probe, an *eng2* probe, and a plasmid vector (pCS2), respectively; and to Drs. R. Ho and N. Hopkins for instruction in blastodermal cell transplantation. We also thank Drs. I.B. Dawid, H. Westphal, and their colleagues for discussion; S.W. Wilson, R. Toyama, M. Taira, A. Chiba, and A. Thomson for critical comments; and M. Hayashi, T. Akiyama, H. Endo, and K. Nakamura for excellent assistance. This research was supported by grants-in-aid and the new program system from the Ministry of Education, Science, Sports and Culture of Japan, special coordination funds for promoting science and technology from the Agency of Science and Technology of Japan, and the Uehara Foundation. Y.K. received a Fellowship for Junior Scientists from the Japan Society for the Promotion of Science.

References

1. Chitnis AB, Kuwada JY (1990) Axonogenesis in the brain of zebrafish embryos. J Neurosci 10:1892–1905
2. Wilson SW, Ross LS, Parret T, Easter SS (1990) The development of a simple scaffold of axon tracts in the brain of the embryonic zebrafish, *Brachydanio rerio*. Development (Camb) 108:121–145
3. Appel B, Korzh V, Glasgow E, Thor S, Edlund T, Dawid IB, Eisen JS (1995) Motoneuron fate specification revealed by patterned LIM homeobox gene expression in embryonic zebrafish. Development (Camb) 121:4117–4125
4. Inoue A, Takahashi M, Hatta K, Hotta Y, Okamoto H (1994) Developmental regulation of Islet-1 mRNA expression during neuronal differentiation in embryonic zebrafish. Dev Dyn 199:1–11
5. Tokumoto M, Gong Z, Tsubokawa T, Hew CL, Uyemura K, Hotta Y, Okamoto H (1995) Molecular heterogeneity among primary motoneurons and within myotomes revealed by the differential mRNA expression of novel Islet-1 homologs in embryonic zebrafish. Dev Biol 171:578–589
6. Dawid IB, Toyama R, Taira M (1995) LIM domain proteins. CR Acad Sci Paris 318:295–306
7. Kikuchi Y, Segawa H, Tokumoto M, Tsubokawa T, Hotta Y, Uyemura K, Okamoto H (1997) Ocular and cerebellar defects in zebrafish induced by overexpression of the LIM domains of the Islet-3 LIM/homeodomain protein. Neuron 18:369–382
8. Bally-Cuif L, Wassef M (1995) Determination events in the nervous system of the vertebrate embryo. Curr Opin Genet Dev 5:450–458
9. Joyner AL (1996) Engrailed, Wnt and Pax genes regulate mid-brain-hindbrain development. Trends Genet 12:15–20
10. Bally-Cuif L, Wassef M (1994) Ectopic induction and reorganization of Wnt-1 expression in quail/chick chimeras. Development (Camb) 120:3379–3394
11. Marin F, Puelles L (1994) Patterning of the embryonic avian midbrain after experimental inversions:a polarizing activity from the isthmus. Dev Biol 163:19–37
12. Crossley PH, Martinez S, Martin GR (1996) Midbrain development induced by FGF8 in the chick embryo. Nature (Lond) 380:66–68
13. McMahon AP, Bradley A (1990) The Wnt-1(int-1) proto-oncogene is required for development of a large region of the mouse brain. Cell 62:1073–1085
14. McMahon AP, Joyner AL, Bradley A, McMahon JA (1992) The midbrain-hindbrain phenotype of Wnt-1^-/Wnt-1^- mice results from stepwise deletion of engrailed-expressing cells by 9.5 days postcoitum. Cell 69:581–595
15. Mikkola I, Fjose A, Kuwada JY, Wilson S, Guddal PH, Krauss S (1992) The paired domain-containing nuclear factor pax[b] is expressed in specific commissural interneurons in zebrafish embryos. J Neurobiol 23:933–946
16. Danielian PS, McMahon AP (1996) Engrailed-1 as a target of the Wnt-1 signaling pathway in vertebrate midbrain development. Nature (Lond) 383:332–334

Reelin/CR-50 and the Formation of Cortical Layers

Masaharu Ogawa[1], Takaki Miyata[1], Kazunori Nakajima[2], and Katsuhiko Mikoshiba[2,3]

After closing of the neural tube and formation of the telencephalic vesicle, newly generated neuronal cells at the ventricular zone are addressed to their proper positions by neuronal migration and form the specific laminar patterns of neurons in cerebral cortex. The *reelin* gene, encoding a protein Reelin, was recently identified; it is responsible for the *reeler* mutant mouse phenotype that displays abnormal laminar organization of several brain structures as a consequence of a defect in neuronal migration.

We have generated a monoclonal alloantibody CR-50, which recognizes the Reelin protein, by immunizing *reeler* mice with homogenates of normal embryonic brains. Using this antibody we have characterized spatial and temporal relationships between the expression patterns of Reelin and the formation of cortical layers in cerebral, cerebellar, and hippocampal cortices. Our investigations showed that the Reelin is a secreted ECM-like glycoprotein and localized extracellularly in the marginal zone before and during the formation of cortical layers in each cortices. In cerebral and hippocampal cortices, Reelin is produced by Cajal-Retzius cells, that are among the first neurons to be generated in the cerebral cortex, and remains closely associated with the surface of their cell bodies and processes. Administration of CR-50 antibody to normal dissociated cortical neurons recapitulates the abnormal reaggregation that occurs with reeler neurons. CR-50 also altered the formation of hippocampal layering in vivo. In other *reeler* mutant mice Orleans, a 220-bp deletion near the 3′-end, the truncated Reelin exists in Cajal-Retzius cells; however, it was not found on the surfaces of living Orleans cells. These observations indicate that the extracellulary localized Reelin, in particular, the CR-50 epitope region, is responsible for the reeler phenotype in vitro and in vivo. It is generally accepted that newly generated neurons

[1] Department of Physiology, Kochi Medical School, Kochi 783, Japan
[2] Molecualr Neurobiology Laboratory, Tsukuba Life Science Center, The Institute of Physical and Chemical Research (RIKEN), Ibaraki 305, Japan
[3] Department of Molecular Neurobiology, Institute of Medical Science, The University of Tokyo, 4-6-1 Shirokanedai, Minato-ku, Tokyo 108-8639, Japan

at the ventricular zone migrate out radially along the radial glial fibers into the cortical plate, which develops beneath the Cajal-Retzius cells. Our observations indicate that the localized expression of Reelin underlies the molecular mechanism by which layer formation is initiated. In conjunction with the development of cortical layers, the spatial and temporal patterns of expression of Reelin in cerebellum, hippocampal proper and dentate gyrus are presented.

Pax-6 Is Involved in Specification of Ventral Cell Types in the Hindbrain

Noriko Osumi[1] and Masato Nakafuku[2]

Summary. *Pax-6* encodes a transcription factor and is expressed in several discrete domains of the developing CNS. To explore the roles of *Pax-6* in neural development, we have investigated expression of several region-specific genes in the hindbrain and spinal cord of homozygous *Small eye* rat (*rSey*[2]) embryos. We found that the expression domain of *Nkx-2.2*, which is normally observed adjacent to the floor plate, expanded dorsally in the mutant. This is suggested to be consistent with the transformation of dorsal somatic motor type neurons into the ventral branchiomotor type. In addition, the mutant hindbrain and spinal cord lacked *En-1* expression in the presumable ventral interneurons. In contrast, both dorsal and ventral markers that are expressed outside the *Pax-6* domain did not show any alteration. Taking these results together, it is suggested that *Pax-6* is involved in patterning the ventral neural tube, through which is regulated specification of ventral cell types. Furthermore, loss of *En-1* expression and altered expression of *Wnt-7b* in the mutant imply common genetic interactions among *Pax-6-paired*, *Wnt-7b-wingless*, and *En-1-engrailed* both in vertebrates and in vertebrates.

Key words. *Pax-6*—Hindbrain—Motor neuron—Interneuron—Ventral patterning

Introduction

The control of movement is regulated by coordinated activity of distinct classes of motor neurons and interneurons located within the spinal cord and brainstem. These neurons differentiate from progenitor cells in the distinct regions of the neural tube along the dorsoventral (D-V) axis. Especially, motor neurons and ventral interneurons are generated in the ventral half of the neural tube in response to the inductive action of sonic hedgehog (shh), a protein secreted from the notochord and floor plate

[1] National Institute of Neuroscience, National Center of Neurology and Psychiatry, 4-1-1 Ogawa-higashi, Kodaira, Tokyo 187-8502, Japan
[2] Division of Neurobiology, Department of Neuroscience, The University of Tokyo Graduate School of Medicine, 7-3-1 Hongo, Bunkyo-ku, Tokyo 113, Japan

[1]. It appears, however, that the shh signaling systems alone cannot account for the generation of distinct ventral cell types. The precise molecular mechanisms underlying complex D-V patterning and subsequent specification of the diverse cell types of the neural tube remains unclear.

Members of the mammalian *Pax* genes exhibit remarkable patterns of expression in the developing CNS [2,3]. For example, *Pax-3* and *Pax-7* are expressed in the dorsal part of the neural tube, while *Pax-6* expression is confined to the ventral domain excluding the floor plate and motor column [4,5]. Because region-specific expression of the *Pax* genes generally precedes the onset of neuronal differentiation, it is speculated that these genes are involved in the establishment of positional identities of neural precursors along the dorsoventral (D-V) axis. In chick embryos, removal or grafting of the notochord alters the expression patterns of *Pax-3* and *Pax-6* genes and modifies neuronal cell types [6]. Furthermore, overexpression of the *Pax-3* gene in the entire neural tube perturbs floor plate differentiation as judged by the loss of expression of *shh* and *HNF3β* [7]. Its ectopic expression in the ventral side of the neural tube also causes bilateral fusion of *Pax-6* domains at the ventral midline, yet motor neurons are differentiated in the ventral spinal cord. These observations strongly support the idea that *Pax* genes function as regulators of D-V patterning in the neural tube. Therefore, it is interesting to know whether loss of *Pax-6* function affects regional identity within the neural tube through altered expression of other region-specific genes.

Functions of *Pax-6* during neural development have been intensively studied by means of its mutant mice and rat [8–10]. In the mouse mutant defects of cranial motor neurons and ventral interneurons, marked by *En-1*, have been reported recently [11,12]. We have also demonstrated the phenotype of cranial motor neurons in our rat mutant [10]. Based on this information, we examined the expression patterns of various region-specific genes in the mutant rat hindbrain in detail. Here we show altered expression of *Nkx-2.2* and *En-1* in the mutant rat neural tube, suggesting that Pax-6 is essential to patterning the ventral neural tube and consequently in controlling the fate of motor neurons and ventral interneurons. Furthermore, the fact that expression of *Wnt-7b* and *En-1* is specifically missing in *Pax-6* mutant embryos strongly suggests the presence of regulatory interactions among particular members of the *Pax*, *Wnt*, and *En* families during development of the vertebrate hindbrain, which may be analogous to the molecular mechanisms that govern pattern formation in the fly.

Materials and Methods

Small eye rats (*rSey*^{2}) were found and maintained in the Sprague-Dawley (SD) rat colony of Safety Research Laboratories (Yamanouchi Pharmaceutical, Tokyo, Japan) [10]. Embryos were obtained from intercrosses of heterozygous rats or of wild-type SD rats, with the day of the vaginal plug designated as 0 days. Homozygotes were distinguishable by their external morphology from embryonic day 11 (E11) onward.

Methods for in situ hybridization on cryosections and in whole mounts have been described previously [13]. The antisense RNA probe for the rat *Pax-6* gene was synthesized from the template plasmid described previously and recognized both the wild-type and mutant *Pax-6* transcripts [9]. All other rat cDNAs were obtained by RT-PCR in which cDNA templates were synthesized from the total RNA taken from E11.5 rat embryo brain [13]. Oligonucleotides used to amplify the cDNAs are as follows: for *Pax-3*, 5′-AAGCTTACCGAGGCCCGAGTG-3′ and 5′-CTGCAGCTGGCTGACACC-GTG-3′; *Wnt-1*, 5′-CTCTTTGGCCGAGAGTTCGTG-3′ and 5′-AGACGAGCTGT-TGCAAGCTCG-3′; *Wnt-3*, 5′-GAAGGCTGGAAGTGGGGCGGC-3′ and 5′-ACGC-AATGGCATTTCTCCTTCCG-3′; *Wnt-7b*, 5′-GGCATCGACTTTTCTCGTCGC-3′ and 5′-CGAGGTGCGGTTGCACAGAC-3′; *En-1*, 5′-GACAGTGGCGGTGGTAGT-3′ and 5′-GAGGAGCCTGGAGGTGGC-3′, and for *Nkx-2.2*, 5′-GGGGGACGCAGGCAAG-AAG-3′ and 5′-CATCCCTTTTCCGCCTACA-3′. The cDNA fragments were cloned into pBluescript SK II (Stratagene), and antisense RNA probes were generated with T3 or T7 RNA polymerase. Corresponding sense probes were used to check the specificity of hybridization signals. Samples were photographed under a Nomarski or dissection microscope.

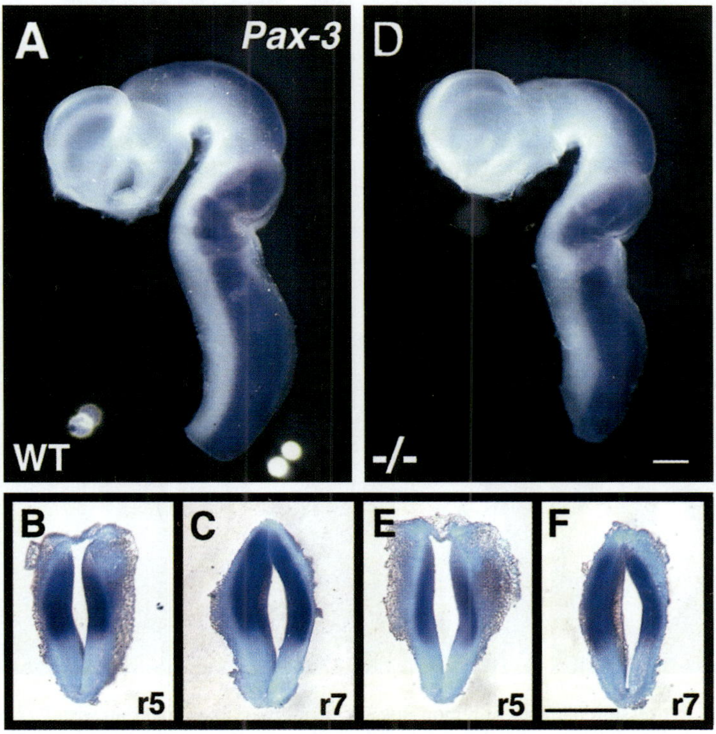

FIG. 1A–F. Expression patterns of *Pax-3* in the neural tube of E12 wild type (**A–C**) and homozygous mutant (**D–F**) embryos. Whole-mount brain (*upper*) and fragments at the level of r5 (*middle*) and r7 (*lower*). *Pax-3* was expressed in the dorsal region of the neural tube throughout the hindbrain and spinal cord in both wild-type and homozygous embryos in a similar manner. *Bar* 500 μm

Results and Discussion

Here we have examined the expression patterns of various region-specific genes in detail. In E12 wild-type embryos, *Pax-3* was expressed in the dorsal region of the neural tube (Fig. 1A–C) where its ventral margin overlaps with the dorsal boundary of the *Pax-6* positive domain (data not shown). No change of expression was observed in homozygous embryos (Fig. 1D–F). We also examined the expression patterns of *Wnt-1* and *Wnt-3*. These genes were expressed in the dorsalmost region in the neural tube in E12 wild-type embryos (Fig. 2A,B,E,F). Homozygous embryos did not show any

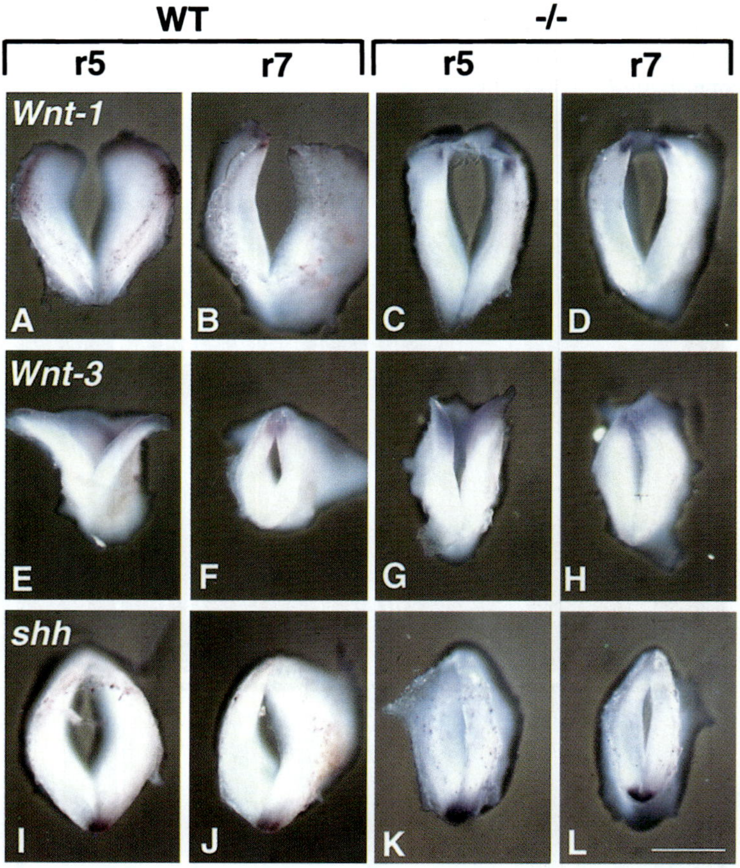

FIG. 2A–L. Expression patterns of region-specific genes that do not overlap with *Pax-6*. Whole-mount brain fragments were hybridized with probes for *Wnt-1* (**A–D**), *Wnt-3* (**E–H**), and *shh*. Shown are the r5 (**A,C,E,G,I,K**) and r7 (**B,D,F,H,J,L**) levels in E12 wild-type (*left two columns*) and homozygous (*right two columns*) embryos. Expression of *Wnt-1* and *Wnt-3* genes was dorsally restricted both in the wild-type and homozygous embryos (**A–H**). *Shh* was expressed in the floor plate both in wild-type (**I,J**) and homozygous (**K,L**) embryos. *Bar* 500 μm difference in expression of these dorsal *Wnt* genes (Fig. 2C, D, G, H). Thus loss of function of *Pax-6* did not affect expression of the dorsal markers, suggesting that *Pax-6* is not involved in the dorsal patterning of the neural tube.

We next observed several ventral markers of the neural tube. Transcripts of *shh* were normally observed In the ventral midline structures, i.e., the floor plate (Fig. 2I,J) and notochord (data not shown), which is outside of the *Pax-6* domain. No alteration of *shh* expression was detected in the homozygous mutants (Fig. 2K,L). In contrast, a dramatic change was observed in the expression pattern of *Nkx-2.2*. This gene was expressed in thin bilateral longitudinal stripes in the wild-type neural tube (Fig. 3A–C), whereas in the homozygous rat brain the expression expanded dorsally to make the stripes wider (Fig. 3D–F). In normal development the Nkx-2.2 domain does not overlap with Pax-6 ([11] and unpublished data). Therefore, it is likely that Pax-6 inhibits expression of *Nkx-2.2* in a direct or nondirect manner.

It was recently shown that shh signaling controls neuronal fate by establishing different progenitor cell populations in the ventral neural tube that are defined by the expression of Pax-6 and another transcription factor, Nkx-2.2, together with dorsal expansion of *Nkx-2.2* in the mouse *Pax-6* mutant [11]. Because somatic motor (SM)

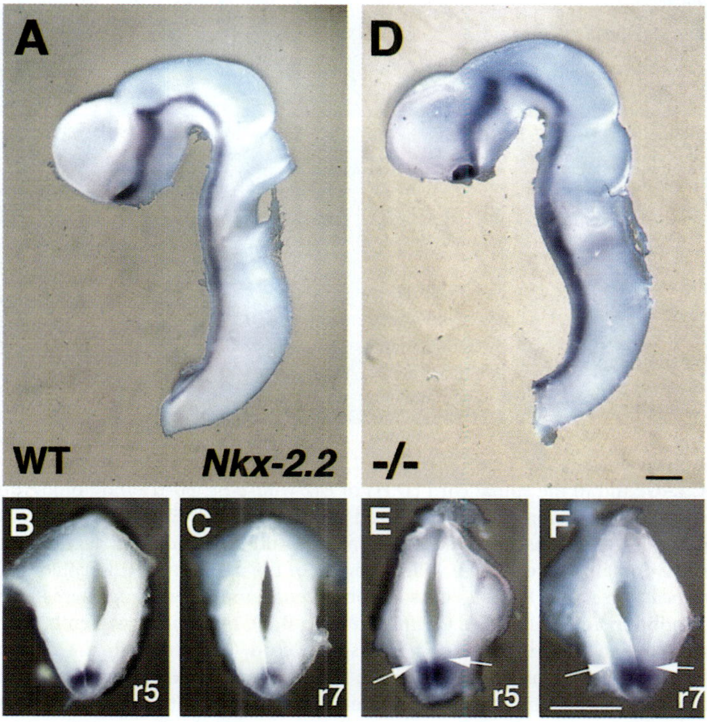

Fig. 3A–F. Expression patterns of *Nkx-2.2* in the neural tube of E12 wild-type (**A–C**) and homozygous mutant (**D–F**) embryos. Whole-mount brain (*upper*) and fragments at the level of r5 (*middle*) and r7 (*lower*). Expression of *Nkx-2.2* expands dorsally (*arrows*) to make the longitudinal stripe wider in the mutant hindbrain and spinal cord. *Bar* 500 μm neurons seem to be transformed into the branchiomotor (BM) neurons in the *Pax-6* mutant [10,11], it is supposed that BM neurons are generated from the Nkx-2.2-positive neuroepithelial precursors in the hindbrain. However, this should be carefully considered because such transformation is not seen in spinal cord levels where altered expression of *Nkx-2.2* is similarly observed.

As mentioned, loss of the ventral interneurons is affected in the *Pax-6* mutant [11,12]. In the wild-type E12 rat embryo, a cluster of *En-1*-positive cells was located in the ventrolateral region within the neural tube (Fig. 4B). At this stage, these *En-1*-positive cells were still negative for MAP2 and neurofilament (data not shown), suggesting that the cells are precursors or immature types of ventral interneurons. We

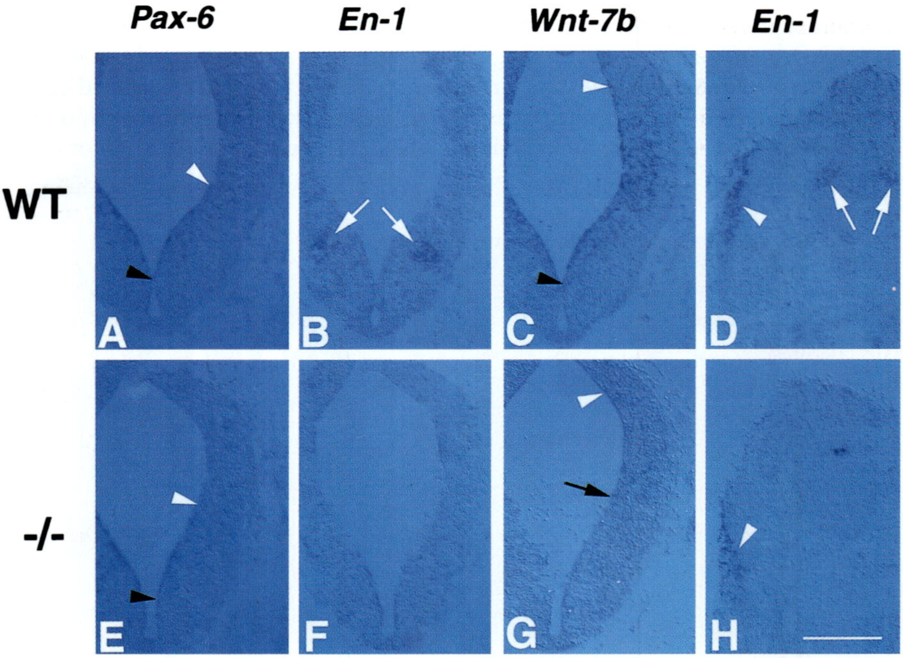

FIG. 4A–H. Expression patterns of *Pax-6*, *En-1*, and *Wnt-7b* genes at the r7 level in the hindbrain (**A–C**,**E–G**) and in the spinal cord (**D**,**H**) of E12 wild-type (**A–D**) and homozygous (**E–H**) embryos. Horizontal sections were hybridized for *Pax-6* (**A**,**E**), *En-1* (**B**,**F**,**D**,**H**), and *Wnt-7b* (**C**,**G**). **A–C** and **E–G** are consecutive sections. In both wild-type and homozygous embryos, the *Pax-6*-positive domain is seen in the ventral region of the neural tube, indicated by *black* and *white arrowheads* in **A** and **E**. *En-1*$^+$ cells are observed in the ventral intermediate zone of the wild-type hindbrain and spinal cord (*white arrows* in **B**,**D**). In homozygous embryos, no *En-1*$^+$ cells are detected in the hindbrain (**F**) and spinal cord (**H**), although the dermonyotome expresses the *En-1* gene similarly with that in the wild-type (*white arrowhead* in **D** and **H**). Comparing consecutive sections, the *Pax-6* domain overlapped with the *Wnt-7b* domain in its ventral region in normal development (**A**,**C**). In the homozygous embryo, expression of *Wnt-7b* disappears in the ventral region, but is observed in the dorsal region, indicated by a *black arrow* and *white arrowheads*. Note that the position of the ventral limit of *Wnt-7b* expression (*black arrow*) is comparable with dorsal limit of *Pax-6* expression (compare **E** and **G**) in the homozygote. *Bar* 200 μm could confirm the previous data in mice by showing no *En-1*-positive cells in either the hindbrain or the spinal cord of the homozygote (Fig. 4F,H). *En-1* expression in other regions such as the midbrain/hindbrain boundary (data not shown) and the somites (arrowhead in Fig. 4H) remained unaffected, suggesting that this defect is specific for the interneuron lineage.

We have previously shown that ventral expression of *Wnt-7b*, another member of the *Wnt* family, is specifically missing in the *Pax-6* mutant rat neural tube [10]. To be emphasized here is that the domain where *Wnt-7b* expression is lost corresponds to the *Pax-6* expression domain (compare Fig. 4C and G). Moreover, *En-1*-positive cells adjoined the ventral margin of the *Pax-6*/*Wnt-7b* domain in the wild-type embryo (Fig. 4B,C). Expression of *Pax-6* precedes that of *Wnt* genes and *En-1* in the hindbrain [5,14,15]. Therefore, our results suggest that expression of *Wnt-7b* and *En-1* in the hindbrain is regulated by *Pax-6*. In *Drosophila*, several lines of evidence suggest regulatory cascades among *prd*, *wg*, and *en*, the fly homologs of *Pax*, *Wnt*, and *En*, respectively [16]. Similar mechanisms have also been implied in the midbrain–hindbrain boundary in zebrafish and mouse embryos. Our results demonstrating that *Wnt-7b* and *En-1* are regulated by *Pax-6* in the mammalian hindbrain provide another example of such a genetic pathway conserved during evolution.

Acknowledgments. Sincere gratitude is extended to Prof. Kazuhiro Eto for his support and encouragement during the present study, and to Ms. Arisa Hirota for her technical assistance. This work was supported by grants-in aid for scientific research from the Japanese Ministry of Education, Science, Sports and Culture (No. 09260237, 09275234) and Core Research for Evolution Science and Technology, sponsored by Japan Science and Technology.

References

1. Tanabe Y, Jessell TM (1996) Diversity and pattern in the developing spinal cord. Science 274:1115–1123
2. Stoykova A, Gruss P (1994) Roles of Pax-genes in developing and adult brain as suggested by expression patterns. J Neurosci 14:1395–1412
3. Rubenstein JL, Martinez S, Shimamura K, et al (1994) The embryonic vertebrate forebrain: the prosomeric model. Science 266:578–580
4. Goulding MD, Chalepakis G, Deutsch U, et al (1991) *Pax-3*, a novel murine DNA binding protein expressed during early neurogenesis. EMBO J 10:1135–1147
5. Walther C, Gruss P (1991) *Pax-6*, a murine paired box gene, is expressed in the developing CNS. Development (Camb) 113:1435–1449
6. Goulding MD, Lumsden A, Gruss P (1993) Signals from the notochord and floor plate regulate the region-specific expression of two *Pax* genes in the developing spinal cord. Development (Camb) 117:1001–1016
7. Tremblay P, Pitello F, Gruss P (1996) Inhibition of floor plate differentiation by *Pax-3*: evidence from ectopic expression in transgenic mice. Development (Camb) 122:2555–2567
8. Hill RE, Favor J, Hogan BLM, et al (1991) Mouse *Small eye* results from mutation in a paired-like homeobox-containing gene. Nature (Lond) 354:522–525
9. Matsuo T, Osumi-Yamashita N, Noji S, et al (1993) A mutation in the *Pax-6* gene in rat small eye is associated with impaired migration of midbrain crest cells. Nature Genetics 3:299–304

10. Osumi N, Hirota A, Ohuchi H, et al (1997) Pax-6 is involved in specification of the hindbrain motor neuron subtype. Development (Camb) 124:2961–2967
11. Ericson J, Rashbass P, Schedl A, et al (1997) Pax6 controls progenitor cell identity and neuronal fate in response to graded Shh signaling. Cell 90:169–180
12. Burrill JD, Moran L, Goulding MD, et al (1997) PAX2 is expressed in multiple spinal cord interneurons, including a population of EN 1+interneurons that require PAX6 for their development. Development (Camb) 124:4493–4503
13. Nakagawa Y, Kaneko T, Ogura T, et al (1996) Roles of cell-autonomous mechanisms for differential expression of region-specific transcription factors in neuroepithelial cells. Development (Camb) 122:2449–2464
14. Osumi-Yamashita N, Kuratani S, Ninomiya Y, et al (1997) Cranial anomaly of homozygous *rSey* rat is associated with a defect in the migration pathway of midbrain crest cells. Dev Growth Differ 39:53–67
15. Parr BA, Shea MJ, Vassileva G, et al (1993) Mouse *Wnt* genes exhibit discrete domains of expression in the early embryonic CNS and limb buds. Development (Camb) 119:247–261
16. Nusse R, Varmus HE (1992) *Wnt* genes. Cell 69:1073–1083

Spatiotemporally Regulated Assortment of Activin and Bone Morphogenetic Protein Receptors in the Developing Anterior Pituitary of Mice

Kyoji Ohyama[1], Haruo Nogami[1], Masao Horie[1], Hitoshi Kawano[1], Mitsuyasu Kato[2], Kohei Miyazono[2], and Koki Kawamura[1]

Summary. We used immunohistochemistry to examine the localization of activin and bone morphogenetic protein receptors (ActRs, BMPRs) in the developing anterior pituitary of mice. Subtypes of the receptors (ActR-IB, -IIB, BMPR-IA, -II) were localized in a spatiotemporally regulated manner throughout embryonic stages. Act type IB receptor (ActR-IB) and BMP type II receptor (BMPR-II), in particular, showed characteristic localizations in the pituitary anlage. They were specifically localized at the rostral growth of the anlage, and the patterns of distribution were closely similar to that of α-glycoprotein subunit (α-GSU) positive cells. The present results suggest that these two types of receptors may play important roles in the development of α-GSU-positive cell lineages, the thyrotropes and gonadotropes, occurring in the anterior pituitary gland.

Key words. Activin receptor—BMP receptor—Anterior pituitary—Mouse—Development

Introduction

The anterior pituitary gland plays an essential role in homeostasis of mammals, especially in the secretion of hormones through an auto-regulatory feedback machinery [1]. However, the developmental processes and the underlying molecular mechanism are not fully understood. Rathke's pouch (RP), which originates from the oral ectoderm, is the anterior pituitary anlage [2]. Growth and differentiation of RP are thought to be influenced by diffusible signals originated from the basal diencephalon [3]. Because activins (Acts) and bone morphogenetic protein (BMPs) are known to be expressed in the diencephalic floor plate region by in situ hybridization analysis [4,5], these molecules are considered as candidate molecules that affect the growth and differentiation of RP [4,5]. However, it was not clear whether their receptors are

[1] Department of Anatomy, School of Medicine, Keio University, 35 Shinanomachi, Shinjuku-ku, Tokyo 160-8582, Japan
[2] Department of Biochemistry, The Cancer Institute, Japanese Foundation for Cancer Research, 1-37-1 Kami-ikebukuro, Toshima-ku, Tokyo 170, Japan

actually expressed at RP and accordingly on what types of anterior pituitary cells that are present in the embryonic stages. We examined immunohistochemically the localization of several members of Act receptors (ActRs) and BMP receptors (BMPRs) in the mouse embryo. We further examined their localizations by comparing with those of α-glycoprotein subunit (α-GSU), which encodes the α-subunit common to the gonadotropins (follicle-stimulating hormone, FSH, and luteinizing hormone, LH) and thyroid-stimulating hormone (TSH). We also discuss possible biological functions of ActRs and BMPRs in the development of the anterior pituitary gland of mice.

Results

The pituitary gland is composed of three lobes, the anterior, intermediate, and posterior (Fig. 1A). The anlage originated from the oral ectoderm at embryonic day 10 (E10) of mice. Then, the ectoderm formed a tubelike structure, RP, and grew toward the rostral direction forming the rostral growth and the principal lobe of the anterior pituitary gland (Fig. 1A). ActR-IB was first observed at E10 (data not shown) and was found, at E13, mainly in the rostral growth, with some also seen in the principal lobe

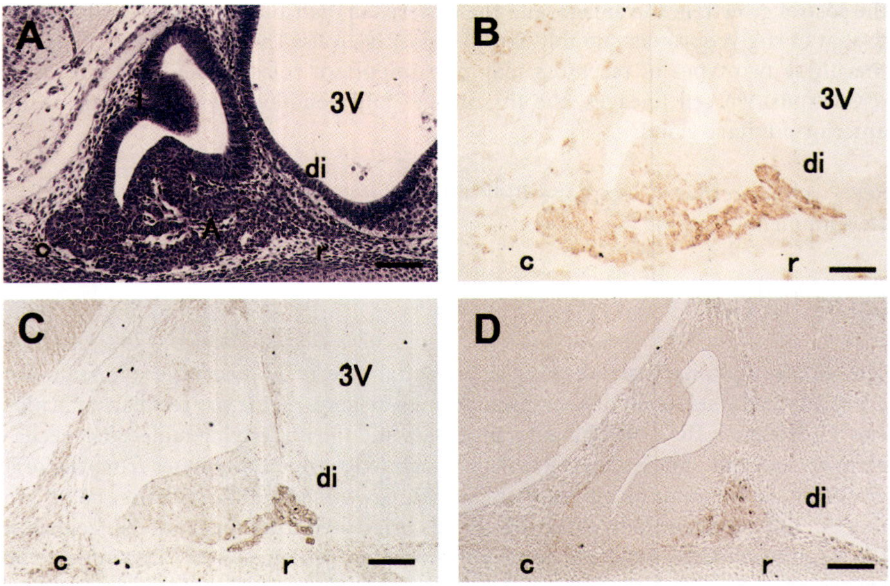

FIG. 1A–D. Cresyl violet staining (**A**) and immunohistochemistry of ActR-IB (**B**), BMPR-II (**C**), and α-glycoprotein subunit (α-GSU) (**D**) from sagittal sections through the primitive diencephalon and pituitary gland of E13 mice. The anterior pituitary anlage closely underlies the diencephalic floor (**A**). ActR-IB-positive cells are observed mainly in the rostral growth and a small number in the principal lobe of the anterior pituitary (**B**). BMPR-II-positive cells are located only in the rostral growth (**C**). α-GSU-positive cells are also located only in the rostral growth. Note the similar pattern of distribution in **B–D** (**D**). *A*, anterior lobe; *I*, intermediate lobe; *r*, rostral; *c*, caudal; *di*, diencephalon; *3V*, third ventricle. *Bars* 80 μm

(Fig. 1B). BMPR-II was expressed in the rostral growth more restrictedly than ActR-IB (Fig. 1C). α-GSU-positive cells were also confined to the rostral growth of the anterior pituitary at E13 (Fig. 1D).

We also examined the localization of other subtypes of ActRs and BMPRs. Both ActR-IIB and BMPR-IA appeared at E12 and peaked at E17, and their expressions were drastically reduced at E18 (data not shown). These features of their expressions were apparently different from those of ActR-IB or BMPR-II.

Discussion

In this study, we first demonstrated that ActR-IB and BMPR-II was specifically localized in the rostral growth of the anterior pituitary at E13 (Fig. 1B,C). Localization of the receptors was closely similar to that of α-GSU-positive cells (Fig. 1B–D). Some transcription factors such as Rpx and Prop-1 that contain homeodomain have been shown to be specifically expressed on the Pit-1-dependent cell lineages, somatotropes and lactotropes [6,7]. These factors have been shown to be expressed in the principal lobe but not in the rostral growth of the anterior pituitary [6,7]. Therefore, the localization of ActR-IB- and BMPR-II-positive cells was clearly different from that of Pit-1-dependent somatotropes and lactotropes in the developing anterior pituitary (Fig. 1B,C). From this study, these receptors were shown to be localized on α-GSU-positive cells that differentiate into thyrotropes and gonadotropes (Fig. 1B–D). Using double-staining techniques we also found that these receptors were not expressed on corticotropes immunoreactive for adrenocorticotropin (ACTH) (data not shown). On

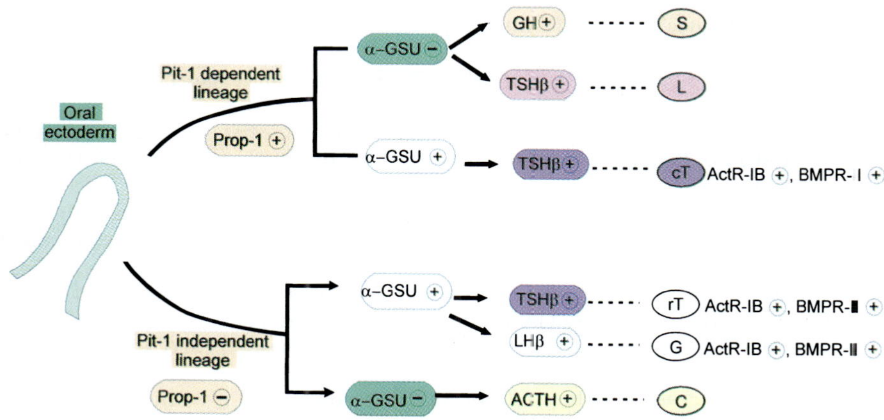

FIG. 2. Schematic diagram shows the cell-type specification in the developing anterior pituitary of mice as deduced from this study. Oral ectoderm invaginates at E10, and *Prop-1* specifies the progenitor cells into Pit-1-dependent cell lineages, somatotropes and lactotropes, in the ectoderm [6]. At E13, α-GSU-positive cells appear in the rostral growth (see Fig. 1). During mid- and late gestation of the embryo, the anterior pituitary cells that secrete growth hormone (*GH*), thyroid-stimulating hormone-β (*TSH-β*), luteinizing hormone-β (*LH-β*), and adrenocorticotropin (*ACTH*) become visible. Both *ActR-IB* and *BMPR-II* were apparently expressed on both thyrotropes and gonadotropes, which are descendants of *α-GSU*-positive cells

the basis of these data, we propose that both ActR-IB and BMPR-II are useful markers for both thyrotropes and gonadotropes in the developing anterior pituitary gland (Fig. 2). This notion is consistent with a recent report [8] suggesting that α-GSU-positive cells differentiate into thyrotropes and gonadotropes but not into corticotropes.

Biological functions of ActRs and BMPRs in the anterior pituitary anlage are, at present, still unknown. ActR-IB and BMPR-II were apparently expressed on the nonproliferative cells that were not immunoreactive for proliferating cell nuclear antigen (PCNA), as we observed at E11 and E13 (data not shown). These receptors were also still present on the anterior pituitary in later embryonic stages as well as in the adult (data not shown) when the differentiation into specific types of hormone-secreting cells has been completed.

One possible function of the ActRs and BMPRs is, therefore, the cell-type specification of the progenitor cells into α-GSU-positive immature thyrotropes and gonadotropes, as well as the maintenance of these types of cells. Future studies, for example, using transgenic mice expressing dominant-negative types of receptors, will elucidate the biological function of ActR-IB and BMPR-II in the developing anterior pituitary of mice.

Acknowledgments. We are grateful to National Hormone and Pituitary Program, NIH, Bethesda (USA), for the kind gift of antiserum to α-GSU. We also thank Mr. Funato and Mr. Sone for their technical assistance. This work was supported by grant from Keio University Medical Science Fund (Sakaguchi Memorial Foundation).

References

1. Voss JW, Rosenfeld MG (1992) Anterior pituitary development: short tales from dwarf mice. Cell 70:527–530
2. Jacobson AG, Miyamoto DM, Mai SH (1979) Rathke's pouch morphogenesis in the chick embryo. J Exp Zool 207:351–366
3. Daikoku S, Chikamori M, Adachi T, et al (1982) Effect of the basal diencephalon on the development of Rathke's pouch in rats: a study in combined organ cultures. Dev Biol 90:198–202
4. Jones CM, Lyons KM, Hogan BLM (1991) Involvement of bone morphogenetic protein-4 (BMP-4) and Vgr-1 in morphogenesis and neurogenesis in the mouse. Development (Camb) 111:531–542
5. Feijen A, Goumans MJ, van den Eijinden-van Raaij AJM (1994) Expression of activin subunits, activin receptors and follistatin in postimplantation mouse embryos suggests specific developmental functions for different activins. Development (Camb) 120:3621–3637
6. Hermesz E, Mackem S, Mahon KA (1996) Rpx: a novel anterior-restricted homeobox gene progressively activated in the prechordal plate, anterior neural plate and Rathke's pouch of the mouse embryo. Development (Camb) 122:41–52
7. Sornson MW, Wu W, Dasen JS, et al (1996) Pituitary lineage determination by the Prophet of Pit-1 homeodomain factor defective in Ames dwarfism. Nature (Lond) 384:327–333
8. Burrows HL, Birkmeier TS, Seasholtz AF, et al (1996) Targeted ablation of cells in the pituitary primordia of transgenic mice. Mol Endocrinol 10:1467–1477

Region-Specific Gene Expression During the Early Formation of Axonal Tracts in the Rat Forebrain

Yumiko Hatanaka

Summary. For clarification of elemental structures and their development, I have tried to isolate genes expressed in a region-specific manner in the developing forebrain. By use of a modified differential display method, two types of cDNAs, M1 and M2, were isolated. In situ hybridization analyses showed that M1 was expressed in the most ventral portion of the hippocampal rudiment from embryonic day (E) 12 onward. Its location corresponded to the fimbrial anlage. As the fimbria developed, M1 expression was restricted to the neuroepithelium flanked by fimbrial fibers. M2 expression in the E12 telencephalon was confined to a part of the medial cerebral wall, including the presumptive preoptic region and hippocampus. At the same stage, M2 expression also occurred in a part of the dorsal diencephalon adjacent to the M2-positive telencephalic region, with a caudal extension along the dorsal midline, and in the zona limitans intrathalamica. During later stages, M2 expression was found in association with axonal tracts, including the corpus callosum, hippocampal commissure, fimbria, optic nerve, stria medullaris, tract of the zona limitans, and habenulopeduncular tract. These findings indicate the presence of specialized cell populations that express M1 or M2 genes or both in paramedian regions of the forebrain, suggesting their involvement in the early specification of tract location.

Key words. Differential display—Gene expression—Development—In situ hybridization—Pathway formation

Introduction

The forebrain is among the most complex regions of the mammalian central nervous system. Its large number of functionally and anatomically discrete regions are connected by specialized pathways. Because of its structural complexity, it is not an easy matter to understand the mechanisms of forebrain development. One effective approach seems to be identification of structural elements by region-specific gene expressions.

Laboratory for Neuronal Recognition Molecules, Brain Science Institute, The Institute of Physical and Chemical Research, RIKEN, 2-1 Hirosawa, Wako, Saitama 351-0198, Japan

Recently, many putative transcription factors and diffusible factors have been isolated and shown to be expressed in segment-like patterns in the developing forebrain. Considering their expression patterns together with distinct cytoarchitectures, a "prosomeric model" has been proposed [1]. In this model, the forebrain is divided into six transverse domains, and each domain is further subdivided into several longitudinal compartments. However, precise subdivision in each domain, especially in the telencephalon, is not yet fully understood. To examine the precise subdivisions of the forebrain and their development in respect to future functionally and anatomically discrete regions, it is necessary to isolate additional genes that are expressed in a region-specific manner during early developmental stages.

Search for Genes Expressed in a Region-Specific Manner in the Forebrain

I focused on a search for differentially expressed genes in the dorsal telencephalon [2]. Dorsolateral and dorsomedial portions of embryonic day 13 (E13) rat telencephalon was used as a source for preparation of RNAs (Fig. 1A). At this stage, the proliferating neuroepithelial cell is the principal component of the cerebral wall. Total RNAs from each portion were compared by a modified differential display method with 80 combinations of primer sets. The patterns of amplified cDNA species of the two portions of the telencephalon were largely identical, but four bands, M1, M1′, M2, and M3, were displayed as region-specific expressed genes (Fig. 1B). They could be amplified only from the cDNA of dorsomedial portions but not from the cDNA of dorsolateral portions. Nucleotide sequence analysis showed that M1 and M1′ were identical and that M2 and M3 were related to each other, and that they did not exhibit significant homology to any known genes.

Expression Profiles of M1 and M2 Genes in the Embryonic Forebrain

In situ hybridization analysis was employed to determine the spatio temporal distribution of these transcripts [3]. To increase sensitivity, digoxigenin-labeled cRNA probes were synthesized from longer cDNA fragments of M1 and M2, which were obtained by screening of a cDNA library and the 5′-RACE method, respectively.

M1 expression was analyzed in frontal sections of E12 rat forebrain (Fig. 2a–h). The M1 hybridization signal was uniformly distributed in the medial telencephalon with obvious expression boundaries in about the middle of the telencephalon anteriorly (Fig. 2c) and at the posterior end of the telencephalon in the vicinity of the hippocampal rudiment posteriorly (Fig. 2e). In later stages (E16 onward, see Fig. 3), it became clear that the M1 expression area was flanked by the fimbria, suggesting that the area of M1 expression in the hippocampal rudiment corresponds to the location of the fimbrial anlage. Furthermore, these M1 expression areas lacked neuron-specific class III β-tubulin immunoreactivity, indicating that M1 is expressed in nonneuronal cells, probably glial cells.

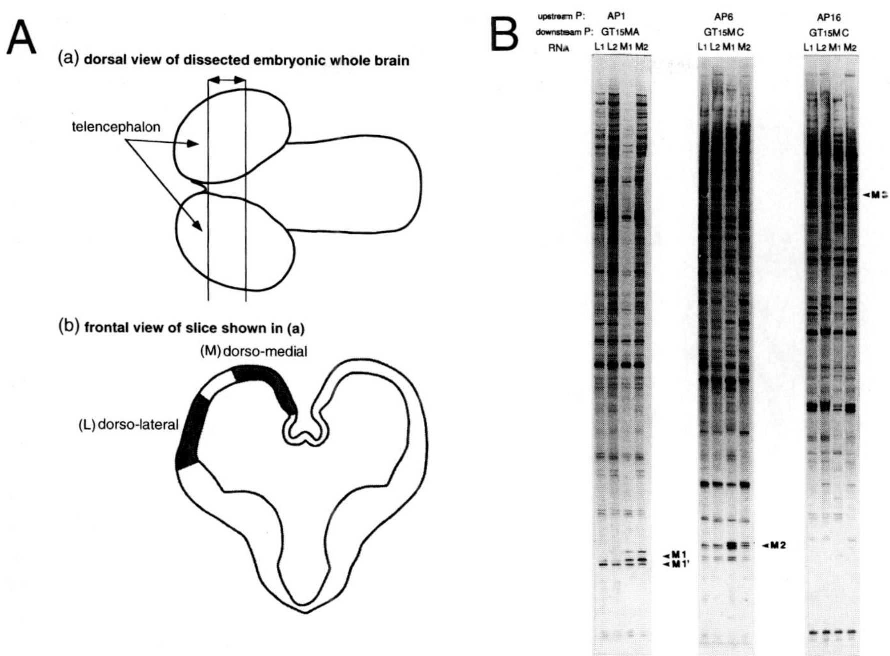

FIG. 1A,B. Genes expressed in dorsolateral (*L*) and dorsomedial (*M*) portions of the telencephalon were compared by a modified differential display method. **A(a)** Coronal slices were obtained from the middle portion of the telencephalon of E13 rats (*arrowheads*), and *L* and *M* portions were dissected out from the slices as indicated by shading in (**b**). **B** Result of differential display. Of more than 4000 cDNA fragments, 4 fragments termed M1, M1′, M2, and M3 were found only after amplification of cDNA derived from the *M* portion of the telencephalon. For confirmation of reproducibility, each experiment was carried out in duplicate [2]. (Reproduced with permission from Elsevier Science)

M2 expression was found in the medial telencephalon and adjacent dorsal diencephalon, extending across the junctional region sometimes called the eminentia thalami, and in the caudal diencephalon (Fig. 2a′–h′). M2 was also expressed in the zona limitans intrathalamica, which is the border between the developing dorsal and ventral thalami, and in the preoptic area and the optic epithelia. Anteriorly, the M2 expression area extended to the most rostral end of the medial telencephalon (Fig. 2a′), and likely included the commissural plate, where the hippocampal commissure and corpus callosum cross the midline at later developmental stages. The area of M2 expression in the epithalamic rudiment (Fig. 2e′) extended ventrally along the zona limitans intrathalamica and reached the presumptive mammillary region (Fig. 2f′,g′). M2-positive areas probably correspond to the location of future fiber tracts, and most of the areas lacked neuron-specific class III β-tubulin immunoreactivity at this stage. It was concluded that at least two identifiable areas of molecular specification occurred in the paramedian region of the telencephalon.

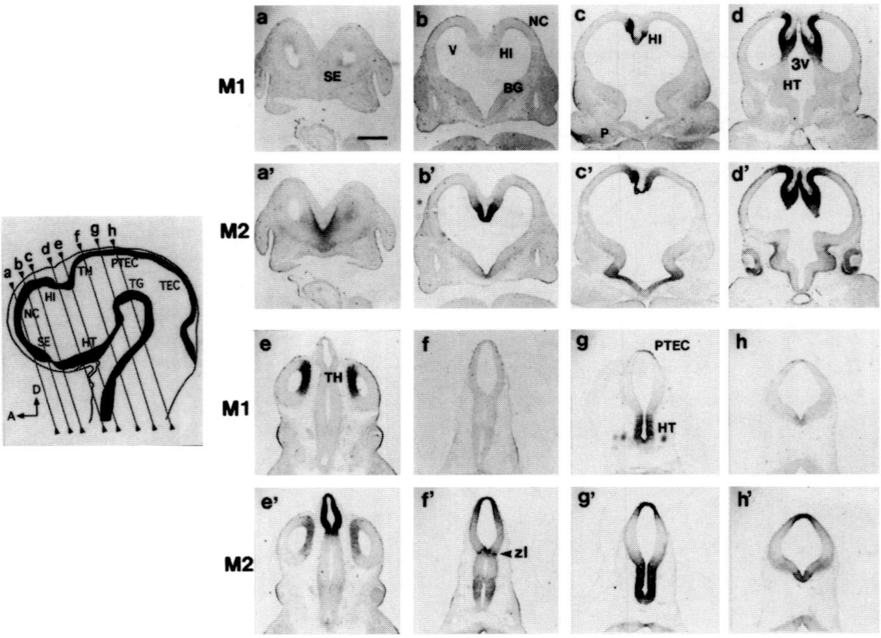

FIG. 2. M1 and M2 expression in the forebrain of an E12 rat. Serial frontal sections of E12 rat forebrain (locations are indicated on the *left*) were treated with DIG-labeled antisense riboprobes derived from M1 (**a–h**) or M2 (**a′–h′**) DNA fragments. Hybridized probes were detected with AP-conjugated anti-DIG antibody, followed by visualization with BCIP and NBT, which produces a blue precipitate. M1 signals were seen in a portion of hippocampal rudiment and in the caudal diencephalon. M2 signals were detected in the medial cerebral wall from the most anterior telencephalic level through the adjacent portion of the dorsal diencephalon, in the zona limitans intrathalamica (*zl*), in the developing eye, and in the caudal diencephalon. *BG*, basal ganglia; *HI*, hippocampus; *HT*, hypothalamus; *NC*, neocortex; *P*, preoptic area; *PTEC*, pretectum; *SE*, septum; *TEC*, tectum; *TG*, tegmentum; *TH*, thalamus; *3V*, third ventricle; *V*, lateral ventricle [3]. (Reproduced with permission from John Wiley & Sons, Inc.)

Expression of M1 and M2 with Respect to Axonal Tracts

Highly restricted expression of M1 in the most ventral portion of the hippocampal rudiment, and M2 expression along presumptive axonal tracts in early stages of their formation, raised the possibility that these genes are expressed in cells, possibly of neuroglial lineage, that could serve as region-defining markers of the tracts, as a scaffold for the tracts, and even potentially as guides for pioneer axons. To examine this, sections adjacent or near adjacent to those hybridized with M1 or M2 probes were processed for immunohistochemical detection of axons with antineuron-specific β-tubulin, GAP-43, or TAG-1 antibodies [3].

Strong GAP-43-immunopositive fibers were observed in the fimbria from E16 onward (Fig. 3), 4 days after the appearance of M1 expression in the fimbrial anlage. Timing of M2 expression in the presumptive optic tract (from E11), zona limitans

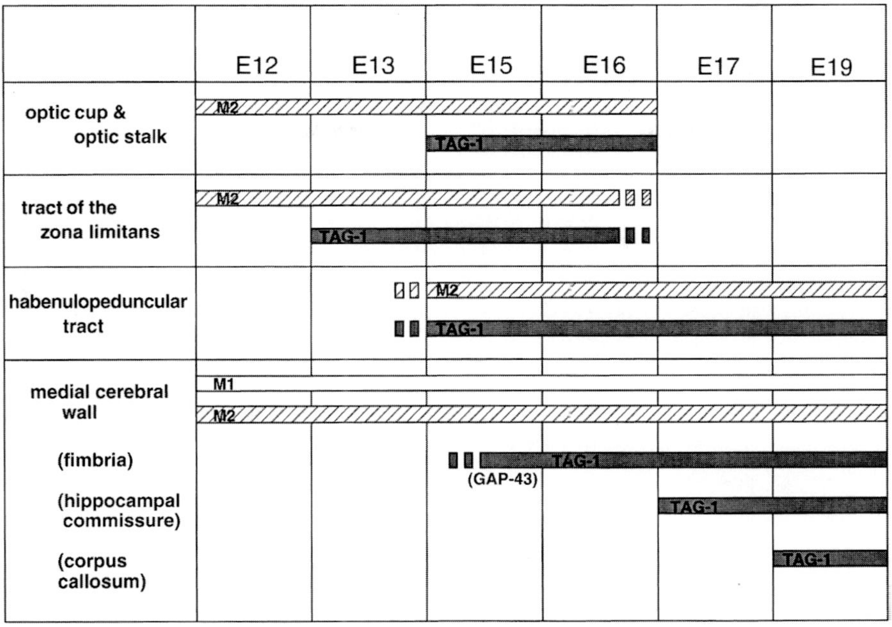

FIG. 3. Summary of spatiotemporal appearance of M1 and M2 signals and of neuron-specific β-tubulin-, GAP-43-, or TAG-1-immunopositive fibers in location for axonal tracts, including optic stalk, tract of the zona limitans, habenulopeduncular tract, fimbria, hippocampal commissure, and corpus callosum. *Lightly shaded bar,* M1 expression; *hatched bars,* M2 expression; *darkly shaded bars,* neuron-specific β-tubulin, GAP-43, or TAG-1 immunoreactivity. *Broken bars* indicate weak expression. Note that M1 and M2 expression occurred earlier than or at least contemporaneous with the appearance of immunopositive fibers in these tracts

intrathalamica (from E12), habenulopeduncular tract (from E13), and medial cerebral tracts including fimbria, hippocampal commissure, and corpus callosum (from E12) was earlier than or at least contemporaneous with the appearance of β-tubulin, GAP-43, or TAG-1-immunopositive fibers in these tracts. Thus, it is likely that M1- or M2-expressing cells are involved in the early regional specification of tract location in the forebrain.

Acknowledgments. This work was supported by the Frontier Research Program, RIKEN, Japan, and the Human Frontier Science Program.

References

1. Rubenstein JLR, Martinez S, Shimamura K, et al (1994) The embryonic vertebrate forebrain: the prosomeric model. Science 266:578–580
2. Hatanaka Y (1997) Early molecular specification of the hippocampal rudiment: isolation of genes expressed in a region-specific manner in the embryonic telencephalon. Dev Brain Res 98:65–73
3. Hatanaka Y, Jones EG (1998) Early region-specific gene expression during tract formation in the embryonic rat forebrain. J Comp Neurol 395:296–309

Part 3
Trophic Factors

Nerve Growth Factor-Induced Cell Death During Neural Development

José María Frade and Yves-Alain Barde

Summary. Nerve growth factor (NGF) is best known for its trophic role, including the prevention of programmed cell death in specific populations of neurones. This property can be accounted for by the activation of tyrosine kinase receptors. In addition, NGF can induce the death of neurones expressing the neurotrophin receptor p75 ($p75^{NTR}$) in the developing retina and other regions of the central nervous system (CNS). Microglial cells (the resident macrophages of the CNS) that invade the developing retina from the retroocular mesenchyme have been identified as the source of NGF in this tissue. In vitro, NGF adsorbed onto beads can reproduce the death caused by NGF, an effect that requires transcription and translation.

Key words. NGF—Receptors—Cell death—Retina—Microglia

Introduction

Nerve growth factor (NGF) was the founding member of the neurotrophin family of growth factors and has well-established trophic functions on selective neuronal populations [1]. The trophic role exerted by NGF can be accounted for by its binding to the receptor tyrosine kinase TrkA. Surprisingly, recent results indicate that this neurotrophin can also kill neurons during normal development (for review, see [2]). This results from the apparently unique property of NGF to activate the neurotrophin receptor p75 ($p75^{NTR}$), a member of the tumor necrosis factor (TNF) receptor and Fas (APO-1/CD95) family.

Materials and Methods

Fertilized eggs from White Leghorn hens were incubated at 38.5°C in 70% humidity and were staged according to Hamburger and Hamilton [3]. Whole eyes from HH18 embryos were carefully dissected out of the surrounding mesenchyme and lens tissue,

Max Planck Institute of Neurobiology, Department of Neurobiochemistry, Am Klopferspitz 18a, D-82152 Planegg-Martinsried, Germany and cultured embedded within 10-μl three-dimensional collagen gels (Boehringer Mannheim, Mannheim, Germany) with the vitreal surface of their neural retinae completely covered with NGF- or cytochrome *c*-adsorbed beads as described [4]. Cycloheximide (Sigma, St. Louis, MO, USA) was used at 2 μg/ml and actinomycin D (Sigma) at 2.5 μg/ml.

Cell death was quantified in neural retinae from the eye explants using an enzyme immunoassay based on a combination of antibodies recognizing histones and DNA as described previously [5]. Absorbance values were normalized with respect to the values obtained with control retinas.

Results and Discussion

NGF as a Cell Death-Inducing Factor in the Avian Retina

Programmed cell death is observed during early stages of development in the central retina. This process seems to be related to an intrinsic morphogenetic process creating space in the central retina for the axons converging to form the optic nerve [6,7]. During this time, $p75^{NTR}$ is expressed by essentially all dying cells and NGF is present in the retina, but TrkA is not expressed [5]. When monoclonal antibodies to NGF or polyclonal antibodies to $p75^{NTR}$ are applied during such early stages, cell death is dramatically reduced [5]. Microglial cells that invade the retinal tissue have been recently shown to be the source of NGF killing retinal cells early in development [4]. In particular, when eyes were isolated from the surrounding mesenchyme at stages preceding the period of microglial invasion, no NGF was detected in the neural retinas and cell death was much reduced compared to the control situation. Active participation of microglial cells in cell death was further demonstrated by adding purified retinal microglial cells to microglia-free optic cups: cell death was observed in these reconstitution experiments [4] and could be blocked with anti-NGF antibodies, as well as by a cyclic peptide previously shown to disturb the interaction between NGF and $p75^{NTR}$ [8].

Cell Death Induced in Eye Explants by NGF Requires Ongoing Gene Expression

Although soluble NGF used at physiological concentration was not able to induce apoptosis in retinas of HH18 eyes, as previously described in vivo [5,9], NGF adsorbed onto glass beads lying on the vitreal surface of the retinae for 16 h did induce apoptosis. Thus, this treatment triggered an increase of 2.9 fold in the levels of apoptosis with respect to the contralateral eye (treated with cytochrome *c*-absorbed beads [4]). This result thus indicates that NGF is sufficient for the induction of apoptosis when it is locally applied. This effect may require high local concentrations of NGF to recruit $p75^{NTR}$ efficiently and cause oligomerization of ligand–receptor complexes. The way in which NGF is presented by microglial cells to cells destined to be eliminated is not known, but one striking fact is that other ligands causing cell death, such as FAS ligand or TNF, typically exist in a membrane-bound form (for review, see [10]). No transmembrane form of NGF has been described so far, but neurotrophins are very basic proteins, and it is conceivable that NGF secreted by microglial cells remains bound to the cell surface.

When retinae treated with NGF-adsorbed beads were compared with control retinae, an increase of 1.6 fold in the levels of free nucleosomes was observed after 2.5 h (Fig. 1). To test if NGF-mediated cell death depends on gene expression, we used blockers of transcription and translation. Both actinomycin D (2.5 μg/ml) and cycloheximide (2 μg/ml) were able to reverse the induction of cell death induced by solid-phase-adsorbed NGF (Fig. 1). This requirement for gene transcription differs from the action of other cell death-inducing ligands, including Fas ligand in particular, which has an extremely fast action [11] and is independent on transcription and protein synthesis [12].

NGF-Mediated Cell Death in Other Systems

There is increasing evidence for a role of NGF in inducing cell death through $p75^{NTR}$ in development and after lesion. Thus, $p75^{NTR}$-positive, TrkA-negative cholinergic neurons have been noted in the brain of $p75^{NTR}$ knockout animals. This population, which is normally eliminated within 2 weeks after birth in wild-type animals, can be rescued

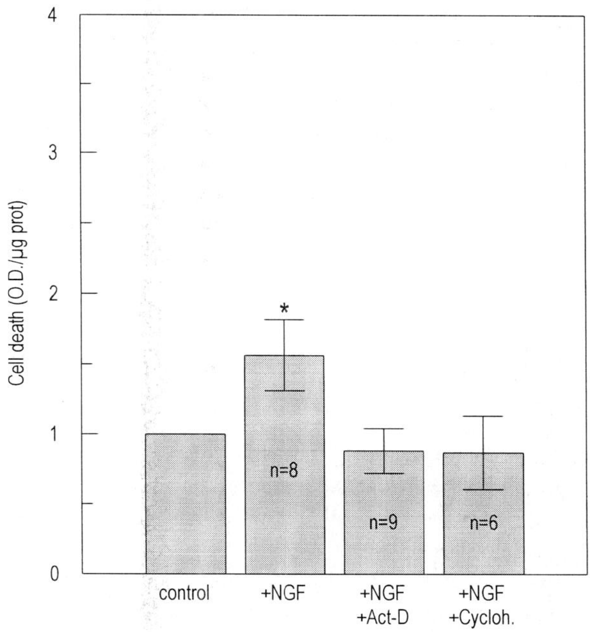

FIG. 1. Bead-adsorbed nerve growth factor (NGF) (+*NGF*) induces cell death in cultured chick retinae (HH18; see Methods). Cell death was quantified by measuring nucleosomes in the supernatant of homogenates of neural retinae. The addition of actinomycin D (2.5 μg/ml, +*Act-D*) or of cycloheximide (2 μg/ml, +*Cycloh.*) completely blocked the effect of bead-adsorbed NGF on the retinae. *Control*, bead-adsorbed cytochrome *c*-treated retinae. Means ± SEM; *asterisk*, $P < 0.05$ (Student's *t*-test)

following subcutaneous injections of a cyclic peptide preventing the binding of NGF to $p75^{NTR}$ [8].

Scattered observations can also be found in the literature indicating that the application of NGF can cause cell death in vivo. Thus, increased motoneuron death (over that caused by axotomy alone) was observed after NGF application in the spinal cord and in the facial nucleus [13,14]. Interestingly, a recent report showed that the absence of $p75^{NTR}$ in knock out mice increases somewhat the survival of axotomized motoneurons in the facial nucleus and, most impressively, elongation of their axons after section of the facial nerve [15]. One possible explanation of these results is that the absence of $p75^{NTR}$ in Schwann cells may increase their survival following axotomy, thus providing better conditions for axonal regeneration [15]. Finally, the intraocular injection of NGF into the eye of chick embryos has been reported to increase cell death in the isthmooptic nucleus, a collection of neurons that, in birds, project to the retina [16]. As $p75^{NTR}$ is expressed by these neurons, either during development or after axotomy, the possibility exists that NGF kills these cells by a $p75^{NTR}$-dependent mechanism.

It appears then that NGF, best known for its trophic functions, might also cause cell death through its apparently unique ability to activate $p75^{NTR}$ [17]. This effect may be not only of relevance during normal development, but also in the adult nervous system either during neurodegenerative diseases or after lesions, conditions in which $p75^{NTR}$ is often reexpressed on neurons and neurons are surrounded by microglial cells.

References

1. Levi-Montalcini R (1987) The nerve growth factor: thirty-five years later. EMBO J 6:1145–1154
2. Frade JM, Barde Y-A (1998) Nerve growth factor: two receptors, multiple functions. BioEssays 20:137–145
3. Hamburger V, Hamilton H (1951) A series of normal stages in the development of chick embryo. J Morphol 88:49–92
4. Frade JM, Barde Y-A (1998) Microglia-derived nerve growth factor causes cell death in the developing retina. Neuron 20:35–41
5. Frade JM, Rodríguez-Tébar A, Barde Y-A (1996) Induction of cell death by endogenous nerve growth factor through its p75 receptor. Nature (Lond) 383:166–168
6. Cuadros MA, Ríos A (1988) Spatial and temporal correlation between early nerve fiber growth and neuroepithelial cell death in the chick embryo retina. Anat Embryol 178:543–551
7. Martín-Partido G, Rodríguez-Gallardo L, Alvarez I, et al (1988) Cell death in the ventral region of the neural retina during the early development of the chick embryo eye. Anat Rec 222:272–281
8. Van der Zee CEEM, Ross GM, Riopelle RJ, et al (1996) Survival of cholinergic forebrain neurons in developing $p75^{NGF}$-deficient mice. Science 274:1729–1732
9. Frade JM, Bovolenta P, Martínez-Morales JR, et al (1997) Control of early cell death by BDNF in the chick retina. Development (Camb) 124:3313–3320
10. Nagata S (1997) Apoptosis by dead factor. Cell 88:355–365
11. Weis M, Schlegel J, Kass GEN, et al (1995) Cellular events in Fas/APO-1-mediated apoptosis in JURKAT T lymphocytes. Exp Cell Res 219:699–708

12. Luciani MF, Golstein P (1994) Fas-based d10S-mediated cytotoxicity requires macromolecular synthesis for effector cell activation but not for target cell death. Philos Trans R Soc Lond B Biol Sci 345:303–309
13. Miyata Y, Kashihara Y, Homma S, et al (1986) Effects of nerve growth factor on the survival and synaptic function of Ia sensory neurons axotomized in neonatal rats. J Neurosci 6:2012–2018
14. Sendtner M, Holtmann B, Kolbeck R, Thoenen H, et al (1992) Brain-derived neurotrophic factor prevents the death of motoneurons in newborn rats after nerve section. Nature (Lond) 360:757–759
15. Ferri CC, Moore FA, Bisby MA (1998) Effects of facial nerve injury on mouse motoneurons lacking the p75 low-affinity neurotrophin receptor. J Neurobiol 34:1–9
16. Von Bartheld CS, Kinoshita Y, Prevette D, et al (1994) Positive and negative effects of neurotrophins on the isthmo-optic nucleus in chick embryos. Neuron 12:639–654
17. Carter D, Kaltschmidt C, Kaltschmidt B, et al (1996) Selective activation of NF-κB by nerve growth factor through the neurotrophin receptor p75. Science 272:542–545

Changes in the Neurotrophic Factor Survival Requirements of Peripheral Nervous System Neurons During Development

Alun M. Davies

Summary. Several populations of neurons switch their survival requirements from one set of neurotrophic factors to another during development. The most comprehensive evidence for neurotrophic factor switching has come from studying embryonic mouse trigeminal ganglion neurons. In vitro studies of the survival responses of these neurons to neurotrophins and in vivo studies of the timing of neuronal death in embryos that have null mutations in neurotrophin and neurotrophin receptor genes have shown that most trigeminal neurons switch responsiveness from BDNF and NT3 to NGF early in development. This switch is triggered by signals that act on the neurons during the switchover period. The timing of neurotrophin synthesis in the periphery accords with the changing neurotrophin requirements of the neurons. Late in fetal development, trigeminal neurons acquire the ability to survive in response to additional neurotrophic factors such as the neurotrophic cytokines and GDNF. Contrary to the widely held view that NT3 is required for sympathetic neuroblast survival, our studies of the timing of sympathetic neuron death in $NT3^{-/-}$ mice have shown that sympathetic neurons have a late, not early, survival requirement for endogenous NT3 after the neurons have become dependent on NGF for survival.

Key words. Neurotrophin—Sensory neuron—Sympathetic neuron—Trk—Development

Introduction

Neurons are generated in excess in the developing vertebrate nervous system, and those that are superfluous to requirements die shortly after their axons reach their targets. To survive, each neuron must procure an adequate supply of one or more neurotrophic factors, and the limited supply of these factors governs the number of neurons that survive [1,2]. An increasing number of proteins have been shown to promote the survival of specific kinds of neurons. The most extensively characterized group of neurotrophic factors are the neurotrophins, which comprise nerve growth

School of Biomedical Sciences, Bute Medical Building, University of St. Andrews, St. Andrews, Fife KY16 9AJ, Scotland factor (NGF), brain-derived neurotrophic factor (BDNF), neurotrophin-3 (NT3), NT4/5, and NT6 [3]. Another family of neurotrophic factors includes glia cell-derived neurotrophic factor (GDNF) and neurturin [4,5]. A third group of more distantly related proteins that promote neuronal survival includes ciliary neurotrophic factor (CNTF), leukemia inhibitory factor (LIF), oncostatin-M (OSM), and cardiotrophin-1 (CT-1) [6].

Neurotrophic factors have several other functions in the nervous system in addition to promoting neuronal survival. These include regulating neuroblast proliferation, enhancing neuronal differentiation, promoting axonal branching, and influencing synaptic function. It is becoming apparent that several neurotrophic factors have a diversity of effects on a variety of different cell types outside the nervous system [1,2,7]. These actions are beyond the scope of this chapter, which focuses on the changing neurotrophic factor requirements of sensory and sympathetic neurons during development.

Trigeminal Ganglion Neurons

Comprehensive evidence for a switch in the neurotrophin survival requirements of developing neurons from one set of neurotrophins to another has come from studying the sensory neurons of the embryonic mouse trigeminal ganglion. When cultured in defined medium at E10, the stage when the earliest trigeminal axons are starting to grow to their peripheral targets, these neurons initially survive independently of neurotrophins. Between 24 and 48 h in vitro, corresponding to the time when the earliest trigeminal axons approach their peripheral targets in vivo, the neurons die unless BDNF or NT3 is present in the culture medium [8,9]. At this stage, NGF has a negligible effect on survival, but in cultures established at successively later stages NGF supports an increasing proportion of the neurons. Concomitant with the acquisition of NGF responsiveness, the neurons lose responsiveness to BDNF and NT3. During the switchover period, there is negligible additional neuronal survival in cultures containing NGF plus BDNF or NT3, indicating that the neurons pass through a transitional phase when they are capable of responding to each of these neurotrophins [8]. The acquisition and loss of responsiveness to different neurotrophins is not a simple on–off phenomenon occurring at a defined stage in development. As neurons mature in vivo, the length of time they are able to survive with BDNF or NT3 in vitro decreases with age, whereas the length of time they are able to survive with NGF increases with age [9].

Studies of the timing of neuronal death in the trigeminal ganglia of embryos that have null mutations in the neurotrophin genes and in the *trkA*, *trkB*, and *trkC* genes, which encode receptor tyrosine kinases for NGF, BDNF, and NT3, respectively, have provided further support for the occurrence of neurotrophin switching during development. In *trkB*$^{-/-}$ embryos, the number of trigeminal neurons undergoing apoptosis is markedly increased during the early developmental stages when neurons are responsive to BDNF in vitro. In *trkA*$^{-/-}$ embryos, neuronal apoptosis is markedly elevated later in development when neurons are responsive to NGF in vitro [10]. Thus, many neurons depend on TrkB signaling for survival in the early trigeminal ganglion before becoming dependent on TrkA signaling. Although early trigeminal neurons survive in culture equally well with either of the two preferred TrkB ligands, BDNF and NT4 [11], BDNF appears to be the physiologically relevant TrkB ligand for these neurons in vivo because there is a significant reduction in the neuronal complement of the neonatal trigeminal ganglia of *BDNF*$^{-/-}$ mice [12,13] but not *NT4*$^{-/-}$ mice [14,15].

A large decrease in the neuronal complement of the trigeminal ganglion of *NT3*$^{-/-}$ embryos takes place before the peak of naturally occurring neuronal death [16,17], implying that endogenous NT3 is also required at an early stage in trigeminal ganglion development. The precise role of NT3 in the ganglion during this stage is, however, controversial. The finding that the majority of the dying cells in the early trigeminal ganglia of *NT3*$^{-/-}$ embryos had incorporated bromodeoxyuridine (BrdU) administered 5 h earlier and expressed nestin, a marker for precursor cells, together with the finding that there is a reduction in the number of proliferating cells in early trigeminal ganglia of *NT3*$^{-/-}$ embryos, led to the conclusion that NT3 promotes the survival of proliferating precursor cells [16]. In contrast, the finding that the complement of neurons and nonneuronal cells (mostly precursor cells) is unchanged in the trigeminal ganglia of *NT3*$^{-/-}$ embryos during the earliest stages of gangliogenesis, and that the subsequent rapid depletion of neurons initially occurs without significant change in the number of nonneuronal cells and proliferating cells in the ganglion, suggested that NT3 acts as a survival factor for at least a proportion of early neurons [17].

The changing neurotrophin responses of trigeminal neurons are accompanied by changes in the expression of neurotrophins and their receptors. BDNF and NT3 mRNAs are expressed in the peripheral trigeminal territory before the arrival of the earliest sensory axons [8,18], and nerve growth factor (NGF) mRNA and protein are expressed later with the arrival of sensory axons [19]. The levels of BDNF and NT3 mRNAs are initially highest in the mesenchyme through which the axons grow to the periphery [8,17,18], whereas NGF mRNA is expressed predominantly in the target field epithelium [19]. TrkB mRNA is expressed at very low levels in the trigeminal ganglion during the period of neurotrophin independence when the sensory earliest axons are growing toward their targets [20]. The onset of BDNF dependence is correlated with an increase in the expression of trkB mRNA, whereas loss of BDNF responsiveness is correlated with an increasing expression of trkB transcripts coding for variants that lack the catalytic tyrosine kinase domain and function as negative modulators of BDNF signaling [20]. Although TrkA and p75 (a neurotrophin receptor that enhances the survival response of embryonic trigeminal neurons to NGF [21]) are expressed at very low levels in the trigeminal ganglion before the axons reach their targets [22,23], the acquisition of a sustained NGF survival response is associated with a marked increased expression of these receptors in trigeminal neurons [22].

Recent studies suggest that the switch from BDNF/NT3 to NGF dependence results from signals that act on the neurons during the switchover period in vivo. When trigeminal neurons are cultured before they respond to NGF, they survive with BDNF well beyond the switchover period from BDNF to NGF dependence, and when these early neurons are switched from BDNF to NGF after various times in culture, they die as rapidly as neurotrophin-deprived neurons. However, neurons switched from BDNF or NT3 to NGF in cultures set up at stages throughout the switchover period exhibit an NGF survival response that improves with age [9].

Additional changes occur in the neurotrophic factor responses of trigeminal neurons after their axons have reached and innervated their targets and become dependent on NGF for survival. The majority of these neurons acquire a late survival response to the neurotrophic cytokines CNTF, LIF, OSM, and CT-1 and to GDNF several days after they have become dependent on NGF for survival [24]. Likewise, embryonic chicken trigeminal neurons also acquire a late survival response to GDNF several days after they start responding to neurotrophins [25]. However, whereas the physiological relevance of the in vitro responses of developing trigeminal neurons to neurotrophins has been substantiated by detailed developmental studies of the timing of neuronal loss in mouse embryos with mutated neurotrophin and *trk* genes, similar detailed studies have yet to be carried out on the timing of sensory neuron death in embryos with null mutations in genes encoding other neurotrophic factors and their receptors.

Neurotrophin Switching in Other Populations of Sensory Neurons

Several studies have suggested that a proportion of neurons in embryonic dorsal root ganglia (DRG) change their neurotrophin survival requirements during development. The in vitro survival of most of early DRG neurons from the chicken embryo is promoted by either NGF or BDNF, whereas later in development DRG contain two distinct subpopulations of neurons that are dependent on either NGF or BDNF for survival [26,27]. Likewise, NGF, BDNF, NT3, and NT4 increase the number of surviving neurons in cultures of early DRG cells from rat and chicken embryos [28,29], whereas only NGF among the neurotrophins has a major survival-promoting effect on neonatal DRG neurons in culture [28]. These results suggest that many DRG neurons are responsive to multiple neurotrophins early in development before becoming more restricted in the survival requirements later on. Furthermore, the finding that the loss of DRG neurons in *trkA*$^{-/-}$; *trkB*$^{-/-}$ double knockout mice is not significantly greater than that occurring in *trkA*$^{-/-}$ mice [30] suggests that at least some DRG neurons are dependent on both TrkA and TrkB signaling for survival.

The number of cells undergoing apoptosis is markedly increased in the DRG of *NT3*$^{-/-}$ and *trkC*$^{-/-}$ embryos during the early stages of ganglion formation [31–34]. This results, at least in part, from the death of postmitotic neurons, because in the DRG of E11 *NT3*$^{-/-}$ embryos there is a significant reduction in the number of neurofilament-positive cells in the absence of any change in the precursor population [34], suggesting the endogenous NT3 is required for the survival of a proportion of newly differentiated neurons. In addition to a direct effect on the survival of recently differentiated neurons, there is growing evidence that NT3 also acts on sensory neuron precursor cells. *TrkC* mRNA [35–37] and TrkC protein [29] are expressed by the majority of cells in DRG during the early stages of gangliogenesis and become restricted to a subset of large neurons later in development. Function-blocking anti-NT3 [38] and anti-TrkC [29] antibodies cause a substantial reduction in the number of neurons in DRG when administered to avian embryos during gangliogenesis. A transient elevation in neurogenesis between E11 and E12 in the DRG of *NT3*$^{-/-}$ embryos, concomitant with a reduction in precursor cell numbers that is not apparently caused by the death of these cells or a reduction in their rate of proliferation [34], suggests that in the absence of NT3 precursor cells differentiate into neurons prematurely. The consequent depletion of the precursor cell pool results in the failure of neurons to accumulate in the DRG of NT3-deficient embryos between E12 and E13 when more than half the neurons are normally generated [34]. This finding suggests that endogenous NT3 plays a role in keeping precursor cells in the proliferative state and prevents their differentiation. Paradoxically, NT3 administration to quail embryos during gangliogenesis causes a substantial reduction in the neuronal complement of DRG [39]. This is associated with a reduction in cell division in early DRG, suggesting that exogenous NT3 inhibits precursor cell proliferation in this experimental paradigm. However, the levels of exogenous NT3 may have been high enough to activate nonpreferred Trk receptors, which may not be physiologically relevant at this stage of development. In addition to regulating precursor cell proliferation, it has been suggested that endogenous NT3 is required for precursor cell survival because the majority of cells undergoing apoptosis in the early DRG of *NT3*$^{-/-}$ embryos (recognized by the TUNEL technique for DNA fragmentation) had also incorporated BrdU administered 5h earlier [33]. However, in another study, BrdU-positive/neurofilament-positive cells, but not BrdU-positive/TUNEL-positive cells, were observed in the early DRG of *NT3*$^{-/-}$ embryos, suggesting that the survival of postmitotic neurons, not precursor cells, is compromised in the absence of endogenous NT3 [34].

In addition to a subset of DRG neurons, there is evidence that another population of neural crest-derived neurons also changes its neurotrophin survival requirements during development. The neurons of the jugular ganglion are supported by BDNF, NT3, and NGF early in their development in vitro. With increasing embryonic age, these neurons lose responsiveness to BDNF and NT3 but retain responsiveness to NGF [40].

Neurotrophin Switching Does Not Occur in All Sensory Neurons

In contrast to populations of neural crest-derived sensory neurons, several populations of placode-derived sensory neurons do not appear to switch responsiveness between neurotrophins during the early stages of their development. For example, the BDNF-dependent neurons of the ventrolateral part of the embryonic chicken trigeminal ganglion do not show early survival responses to NGF or NT3, and the BDNF-responsive and NT3-responsive nodose neurons do not show an early response to NGF [40]. Genetic evidence for the existence of populations of sensory neurons that depend on just one neurotrophin for survival during development has come from studying the spiral ganglion of mice with targeted null mutations in the neurotrophin and *trk* genes. The number of spiral ganglion neurons is unchanged in *trkA*$^{-/-}$ mice, but is reduced markedly in *NT3*$^{-/-}$ or *trkC*$^{-/-}$ mice and reduced to a lesser extent in *BDNF*$^{-/-}$ or *trkB*$^{-/-}$ mice. However, in *NT3*$^{-/-}$; *BDNF*$^{-/-}$ or *trkC*$^{-/-}$; *trkB*$^{-/-}$ double knockout mice, the neuronal loss is the sum of the losses in the single neurotrophin or Trk knockouts [30,41]. These observations suggest that the spiral ganglion contains separate subsets of neurons that depend on either NT3 or BDNF for survival, but not both. Thus, neurotrophin switching does not appear to be a universal feature of all populations of developing sensory neurons. It is possible, however, that other neurotrophic factors may play a role in regulating the survival of these populations of neurons during development.

Why some sensory neurons switch neurotrophin and others do not is unclear. In this respect it may be significant that sensory neurons that switch dependence from BDNF/NT3 to NGF survive only briefly without neurotrophins in culture during the earliest stages of their development [8,26,42], whereas sensory neurons that do not switch have a regulated period of neurotrophin independence that is correlated with target distance [43]. Because the peripheral axons of sensory neurons that undergo a switch in neurotrophin dependence appear to be exposed to BDNF and NT3 en route to their peripheral targets [8,18,32,34], it is likely that these neurotrophins play a role in sustaining neurons before their axons reach the peripheral tissues where NGF is produced in proportion to innervation density [19,44]. Sensory neurons that do not exhibit a switch in neurotrophin dependence survive in vitro for different lengths of time without neurotrophins during the early stages of their development; neurons with more distant targets survive longer than neurons with nearby targets [43]. Thus, it is possible that these populations of neurons are not dependent on intermediate support before encountering their targets.

Sympathetic Neurons

Paravertebral sympathetic ganglia initially contain many dividing cells that possess a variety of neuronal characteristics. Although these neuroblasts survive for up to several days in culture in the absence of added neurotrophins [45], a number of in vitro studies have shown that NT3, but not NGF, is capable of enhancing their survival during this early stage of development [46–48]. In cultures set up later in development when the neurons have begun to innervate their targets, most of the neurons are supported by NGF and only high levels of NT3 are effective in promoting survival [46,47]. Increased apoptosis of sympathetic neuroblasts leading to a failure to generate the normal number of neurons in the superior cervical sympathetic ganglion (SCG) has been reported in *NT3*$^{-/-}$ mouse embryos [49]. Early sympathetic ganglia express high levels of *trkC* mRNA [47,48,50] and are surrounded by non-neuronal cells expressing NT3 mRNA [51]. At later embryonic stages, *trkC* mRNA expression declines [47,48,50,51], *trkA* mRNA expression increases [47,52,53] and NGF is synthesized in the tissues innervated by sympathetic neurons [54]. At these later embryonic stages and during the postnatal period, the dependence of the majority of sympathetic neurons on target-derived NGF has been established in classic studies showing that anti-NGF antisera eliminate these neurons [55] and in studies of *NGF*$^{-/-}$ and *trkA*$^{-/-}$ mice in which almost all sympathetic neurons are lost by the first postnatal week [50,56,57].

The generally accepted interpretation of these studies is that sympathetic neuroblasts are initially dependent on locally produced NT3 for survival and later in development lose responsiveness to NT3 and become dependent on target-derived NGF for survival. However, several recent observations cast some doubt on this interpretation. Administration of a specific anti-NT3 antiserum to postnatal rats results in the death of up to 80% of SCG neurons [58], suggesting that the majority of SCG neurons are dependent on endogenous NT3 as well as NGF during this very late stage of their development. Although *trkC* mRNA is prominently expressed in early sympathetic ganglia, detailed comparison of the number of neurons in the SCG of *trkC*$^{-/-}$ and wild-type mice at stages throughout development has revealed no statistically significant differences [50], suggesting that NT3 signaling via TrkC receptors is not required for the genesis or survival of sympathetic neurons in vivo. Because NT3 is also able to promote the survival of sympathetic neurons by signaling via TrkA [59], it is possible that NT3 could be acting via TrkA to promote the survival of sympathetic neuroblasts. However, the number of neurons in the SCG of *trkA*$^{-/-}$ mice is very similar to that of wild-type mice up to E15.5 and decreases markedly at E17.5 and later stages because of increased apoptosis during the period of naturally occurring neuronal death [50]. Thus, NT3 cannot be acting exclusively through TrkA to promote the survival of SCG sympathetic neuroblasts in the early SCG.

Finally, doubt has been cast on the proposed role of NT3 in promoting the survival of sympathetic neuroblasts in vivo from the analysis of a large number of SCG of *NT3*$^{-/-}$ and wild-type embryos that has revealed no significant differences in the total number of neurons or in the levels of GAPDH, L27, trkA, or p75 mRNAs at E16, although significant reductions are evident at E18 and later ages [60]. Likewise, no differences have been observed in the number of SCG neurons of wild-type and *NT3*$^{-/-}$ mice at E15.5, although at E17.5 and later stages *NT3*$^{-/-}$ mice have smaller numbers of SCG neurons than wild-type mice (N. Francis, S. Landis, I. Fariñas, and L. F. Reichardt, personal communication). Thus, in at least the rostral part of the developing sympathetic chain, sympathetic neurons have a late, not early, survival requirement for NT3 that becomes manifest after they have become dependent on NGF for survival.

Acknowledgments. My thanks to Isabel Fariñas, N. Francis, Story Landis, and Louis Reichardt for allowing me to mention their unpublished data. Research in my laboratory is supported by grants from the Wellcome Trust, Cancer Research Campaign and Action Research.

References

1. Davies AM (1994) Role of neurotrophins in the developing nervous system. J Neurobiol 25:1334–1348
2. Lewin GR, Barde YA (1996) The physiology of neurotrophins. Annu Rev Neurosci 19:289–317
3. Bothwell M (1995) Functional interactions of neurotrophins and neurotrophin receptors. Annu Rev Neurosci 18:223–253
4. Lin LH, Doherty DH, Lile JD, et al (1993) GDNF: a glial cell-derived neurotrophic factor for midbrain dopaminergic neurons. Science 260:1130–1132
5. Kotzbauer PT, Lampe PA, Heuckeroth RO, et al (1996) Neurturin, a relative of glial-cell-line-derived neurotrophic factor. Nature (Lond) 384:467–470
6. Stahl N, Yancopoulos GD (1994) The tripartite CNTF receptor complex: activation and signalling involves components shared with other cytokines. J Neurobiol 25:1454–1466
7. Thoenen H (1995) Neurotrophins and neuronal plasticity. Science 270:593–598

8. Buchman VL, Davies AM (1993) Different neurotrophins are expressed and act in a developmental sequence to promote the survival of embryonic sensory neurons. Development (Camb) 118:989–1001
9. Paul G, Davies AM (1995) Trigeminal sensory neurons require extrinsic signals to switch neurotrophin dependence during the early stages of target field innervation. Dev Biol 171:590–605
10. Piñón LGP, Minichiello L, Klein R, et al (1996) Timing of neuronal death in *trkA*, *trkB* and *trkC* mutant embryos reveals developmental changes in sensory neuron dependence on Trk signalling. Development (Camb) 122:3255–3261
11. Davies AM, Horton A, Burton LE, et al (1993) Neurotrophin-4/5 is a mammalian-specific survival factor for distinct populations of sensory neurons. J Neurosci 13:4961–4967
12. Ernfors P, Lee KF, Jaenisch R (1994) Mice lacking brain-derived neurotrophic factor develop with sensory deficits. Nature (Lond) 368:147–150
13. Jones KR, Farinas I, Backus C, et al (1994) Targeted disruption of the BDNF gene perturbs brain and sensory neuron development but not motor neuron development. Cell 76:989–999
14. Conover JC, Erickson JT, Katz DM, et al (1995) Neuronal deficits, not involving motor neurons, in mice lacking BDNF and NT4. Nature (Lond) 375:235–238
15. Liu X, Ernfors P, Wu H, et al (1995) Sensory but not motor neuron deficits in mice lacking NT4 and BDNF. Nature (Lond) 375:238–241
16. ElShamy WM, Ernfors P (1996) Requirement of neurotrophin-3 for the survival of proliferating trigeminal ganglion progenitor cells. Development (Camb) 122:2405–2414
17. Wilkinson GA, Fariñas I, Backus C, et al (1996) Neurotrophin-3 is a survival factor in vivo for early mouse trigeminal neurons. J Neurosci 16:7661–7669
18. Arumae U, Pirvola U, Palgi J, et al (1993) Neurotrophins and their receptors in rat peripheral trigeminal system during maxillary nerve growth. J Cell Biol 122:1053–1265
19. Davies AM, Bandtlow C, Heumann R, et al (1987) Timing and site of nerve growth factor synthesis in developing skin in relation to innervation and expression of the receptor. Nature (Lond) 326:353–358
20. Ninkina N, Adu J, Fischer A, et al (1996) Expression and function of TrkB variants in developing sensory neurons. EMBO J 15:6385–6393
21. Davies AM, Lee KF, Jaenisch R (1993) p75-deficient trigeminal sensory neurons have an altered response to NGF but not to other neurotrophins. Neuron 11:565–574
22. Wyatt S, Davies AM (1993) Regulation of expression of mRNAs encoding the nerve growth factor receptors p75 and trkA in developing sensory neurons. Development (Camb) 119:635–648
23. Schropel A, von Shack D, Dechant G, et al (1995) Early expression of the nerve growth factor receptor trkA in chick sympathetic and sensory ganglia. Mol Cell Neurosci 6:544–556
24. Horton AR, Bartlett PF, Pennica D, et al (1998) Cytokines promote the survival of cranial sensory neurons at different developmental stages. Eur J Neurosci 10:673–679
25. Buj-Bello A, Buchman VL, Horton A, et al (1995) GDNF is an age-specific survival factor for sensory and autonomic neurons. Neuron 15:821–828
26. Ernsberger U, Rohrer H (1988) Neuronal precursor cells in chick dorsal root ganglia: differentiation and survival in vitro. Dev Biol 126:420–432
27. Lindsay RM, Thoenen H, Barde YA (1985) Placode and neural crest-derived sensory neurons are responsive at early developmental stages to brain-derived neurotrophic factor. Dev Biol 112:319–328
28. Memberg SP, Hall AK (1996) Proliferation, differentiation, and survival of rat sensory neuron precursors in vitro require specific trophic factors. Mol Cell Neurosci 6:323–335

29. Lefcort F, Clary DO, Rusoff AC, et al (1996) Inhibition of the NT3 receptor TrkC, early in chick embryogenesis, results in severe reductions in multiple neuronal subpopulations in the dorsal root ganglia. J Neurosci 16:3704–3713
30. Minichiello LFP, Vazquez E, et al (1995) Differential effects of combined trk receptor mutations on dorsal root ganglion and inner ear sensory neurons. Development (Camb) 121:4067–4075
31. Tessarollo L, Vogel KS, Palko ME, et al (1994) Targeted mutation in the neurotrophin-3 gene results in loss of muscle sensory neurons. Proc Natl Acad Sci USA 91:11844–11848
32. White FA, Silos-Santiago I, Molliver DC, et al (1996) Synchronous onset of NGF and TrkA survival dependence in developing dorsal root ganglia. J Neurosci 16:4662–4672
33. ElShamy WM, Ernfors P (1996) A local action of neurotrophin-3 prevents the death of proliferating sensory neuron precursor cells. Neuron 16:963–972
34. Farinas I, Yoshida CK, Backus C, et al (1996) Lack of neurotrophin-3 results in death of spinal sensory neurons and premature differentiation of their precursors. Neuron 17:1065–1078
35. Lamballe F, Smeyne RJ, Barbacid M (1994) Developmental expression of trkC, the neurotrophin-3 receptor, in the mammalian nervous system. J Neurosci 14:14–28
36. Kahane N, Kalcheim C (1994) Expression of trkC receptor mRNA during development of the avian nervous system. J Neurobiol 25:571–584
37. Zhang D, Yao L, Bernd P (1994) Expression of *trk* and neurotrophin mRNA in dorsal root and sympathetic ganglia of the quail during development. J Neurobiol 25:1517–1532
38. Gaese F, Kolbeck R, Barde YA (1994) Sensory ganglia require neurotrophin-3 early in development. Development (Camb) 120:1613–1619
39. Ockel M, Lewin GR, Barde YA (1996) In vivo effects of neurotrophin-3 during sensory neurogenesis. Development (Camb) 122:301–307
40. Buj-Bello A, Pinon LG, Davies AM (1994) The survival of NGF-dependent but not BDNF-dependent cranial sensory neurons is promoted by several different neurotrophins early in their development. Development (Camb) 120:1573–1580
41. Ernfors P, Van Der Water T, Loring J, et al (1995) Complementary roles of BDNF and NT3 in vestibular and auditory development. Neuron 14:1153–1164
42. Wright EM, Vogel KS, Davies AM (1992) Neurotrophic factors promote the maturation of developing sensory neurons before they become dependent on these factors for survival. Neuron 9:139–150
43. Vogel KS, Davies AM (1991) The duration of neurotrophic factor independence in early sensory neurons is matched to the time course of target field innervation. Neuron 7:819–830
44. Harper S, Davies AM (1990) NGF mRNA expression in developing cutaneous epithelium related to innervation density. Development (Camb) 110:515–519
45. Ernsberger U, Edgar D, Rohrer H (1989) The survival of early chick sympathetic neurons in vitro is dependent on a suitable substrate but independent of NGF. Dev Biol 135:250–262
46. Dechant G, Rodriguez TA, Kolbeck R, et al (1993) Specific high-affinity receptors for neurotrophin-3 on sympathetic neurons. J Neurosci 13:2610–2616
47. Birren SJ, Lo L, Anderson DJ (1993) Sympathetic neuroblasts undergo a developmental switch in trophic dependence. Development (Camb) 119:597–610
48. DiCicco-Bloom E, Friedman WJ, Black IB (1993) NT-3 stimulates sympathetic neuroblast proliferation by promoting precursor cell survival. Neuron 11:1101–1111
49. El Shamy WM, Linnarsson S, Lee KF, et al (1996) Prenatal and postnatal requirements of NT-3 for sympathetic neuroblast survival and innervation of specific targets. Development (Camb) 122:491–500
50. Fagan AM, Zhang H, Landis S, et al (1996) TrkA, but not trkC, receptors are essential for survival of sympathetic neurons in vivo. J Neurosci 16:6208–6218

51. Verdi JM, Groves AK, Farinas I, et al (1996) A reciprocal cell-cell interaction mediated by NT-3 and neuregulins controls the early survival and development of sympathetic neuroblasts. Neuron 16:515–527
52. Verdi JM, Anderson DJ (1994) Neurotrophins regulate sequential changes in neurotrophin receptor expression by sympathetic neuroblasts. Neuron 13:1359–1372
53. Wyatt S, Davies AM (1995) Regulation of nerve growth factor receptor gene expression in sympathetic neurons during development. J Cell Biol 130:1435–1446
54. Korsching S, Thoenen H (1988) Developmental changes of nerve growth factor levels in sympathetic ganglia and their target organs. Dev Biol 126:40–46
55. Levi-Montalcini R (1987) The nerve growth factor 35 years later. Science 237:1154–1162
56. Crowley C, Spencer SD, Nishimura MC, et al (1994) Mice lacking nerve growth factor display perinatal loss of sensory and sympathetic neurons yet develop basal forebrain cholinergic neurons. Cell 76:1001–1011
57. Smeyne RJ, Klein R, Schnapp A, et al (1994) Severe sensory and sympathetic neuropathies in mice carrying a disrupted Trk/NGF receptor gene. Nature (Lond) 368:246–249
58. Zhou X, Rush RA (1995) Sympathetic neurons in neonatal rats require endogenous neurotrophin-3 for survival. J Neurosci 15:6521–6530
59. Davies AM, Minichiello L, Klein R (1995) Developmental changes in NT3 signalling via TrkA and TrkB in embryonic neurons. EMBO J 14:4482–4489
60. Wyatt S, Piñón LGP, Ernfors P, et al (1997) Sympathetic neuron survival and TrkA expression in NT3-deficient mouse embryos. EMBO J 16:3115–3123

BDNF Exerts a Preventive Effect on Low K^+-Induced Apoptosis of Cerebellar Granule Neurons via Activation of PI3-Kinase

Masashi Yamada, Koji Shimoke*, Takekazu Kubo**, and Hiroshi Hatanaka

Summary. Cerebellar granule neurons obtained from 9-day-old rats die in an apoptotic manner when the neurons are cultured in serum-free medium containing a low concentration of potassium (5 mM). Brain-derived neurotrophic factor (BDNF) and a high concentration of potassium (26 mM) in the culture medium can effectively prevent this apoptosis. To examine which molecules are involved in the survival-promoting effect of BDNF on the apoptosis of cerebellar granule neurons, we used wortmannin, a specific inhibitor of phosphatidylinositol 3-kinase (PI3-K). Wortmannin inhibited the survival effect of BDNF and BDNF-induced increase of PI3-K activity in the cultured granule neurons. These results indicate that the preventive effect of BDNF on the apoptosis of cerebellar granule neurons is mediated by BDNF-stimulated activation of PI3-K.

Key words. Cell death—Wortmannin—BDNF—PI3-K—Neurotrophic factor

Introduction

During normal development, undesirable cells of animals are killed. This cell death also occurs in the nervous system, where about half of all generated neurons die during early development [1]. This phenomenon is usually referred to as naturally occurring, or programmed, cell death. Generally, this death occurs by apoptosis, which shows cell volume loss, chromatin condensation, and DNA fragmentation. Apoptosis is observed not only in the neurons during developmental stage but also in neurons affected by various neurodegenerative diseases. Neuronal death is often

Division of Protein Biosynthesis, Institute for Protein Research, Osaka University, 3-2 Yamadaoka, Suita, Osaka 565-0871, Japan

* *Present address*: Discovery Research Laboratory, Tanabe Seiyaku Co. Ltd., 3-16-89 Kashima, Yodogawa-ku, Osaka 532-8505, Japan

** *Present address*: Neuroscience Research Laboratory, Sankyo Co. Ltd., 1-2-58 Hiromachi, Shinagawa-ku, Tokyo 140-0005, Japan protected by specific neurotrophic factors [2,3]. Among these factors are neurotrophins, such as nerve growth factor (NGF), brain-derived neurotrophic factor (BDNF), neurotrophin-3 (NT-3), and NT-4/5.

Neurotrophins bind to and activate the Trk family of receptor tyrosine kinases to exert their effects [4]. NGF primarily binds to TrkA, BDNF and NT-4/5 bind to TrkB, and NT-3 binds to TrkC. Trks form homodimers following ligand binding. Then, they are activated and autophosphorylated on tyrosine residues. The activated Trks stimulate intracellular signaling by activating a variety of enzymes and effectors, including phospholipase C-γ, phosphatidylinositol 3-kinase (PI3-K), SHC, GRB-2, Ras, and mitogen-activated protein kinases.

PI3-K is activated by the neurotrophins as well as by various growth factors including epidermal growth factor, platelet-derived growth factor, and hepatocyte growth factor. In addition, several studies using PC12 cells and cultured cerebellar granule neurons have indicated that the activation of PI3-K is important for exertion of the survival-promoting effects of neurotrophic factors [5,6].

Cerebellar granule neurons obtained from 9-day-old rats are dependent for their survival on the potassium concentration in the medium [7–9]. A large number of cerebellar granule neurons are killed within a few days by lowering the K^+ concentration in the medium from 26 mM to 5 mM. This low-K^+-induced death of cerebellar granule neurons is characterized as apoptosis. In addition to high K^+, BDNF and NT-4/5 can prevent low-K^+-induced apoptosis in cerebellar granule neurons [8,9]. NT-3 has a small preventive effect on apoptotic cell death, but NGF has none [8]. These lines of evidence are consistent with the observation that TrkB and TrkC are expressed in the cerebellum, whereas TrkA is not [10].

In this report, we found that wortmannin, an inhibitor of PI3-K, attenuated the preventive effect of BDNF on the low-K^+-induced death of cultured cerebellar granule neurons. This finding indicates that the survival-promoting effect of BDNF is the result of PI3-K activation stimulated by BDNF.

Materials and Methods

Cerebellar Granule Neuron Culture

Primary culture of cerebellar granule neurons from dissociated cerebella of 9-day-old rats was performed as previously reported [8–10]. The neurons were cultured in high-K^+ minimum essential medium (MEM), supplemented with 5% heat-inactivated horse serum (HS), 2.2 mg/ml glucose, 2.2 mg/ml sodium bicarbonate, and 1 μM cytosine arabinoside (AraC), for 4 days. High-K^+ MEM was prepared by increasing the $KHCO_3$ concentration from the normal value of 5.4 to 26 mM, with the omission of the corresponding concentration of $NaHCO_3$. Neuronal survival was determined by the MTT assay as previously reported [8–10]. Hoechst 33 258 staining was done as follows; 24 h after change to serum-free MEM with or without factors, granule neurons were fixed in 4% paraformaldehyde, then were stained with Hoechst 33 258 as previously reported [9]. BDNF was provided by Regeneron (New York, NY, USA); Hoechst 33 258 and wortmannin were purchased from Wako (Osaka, Japan) and from Kyowa Hakko (Tokyo, Japan), respectively.

In Vitro PI-3 Kinase Assay

Cerebellar granule neurons were cultured in high-K^+ MEM containing 5% HS and 1 μM AraC for 4 days, and then the medium was replaced with serum-free MEM. After overnight culture, cells were exposed to 50 ng/ml BDNF and were washed twice with ice-cold buffer A (137 mM NaCl, 20 mM Tris-HCl pH 7.5, 1 mM $MgCl_2$, 1 mM $CaCl_2$, and 0.1 mM sodium orthovanadate). Then, cells were lysed with buffer B (buffer A containing 10% glycerol, 1% nonidet P-40, and 1 mM phenylmethylsulfonyl fluoride, PMSF). PI3-K activities in the lysates were assayed as previously reported [9].

Results

BDNF Prevents Low-K^+-Induced Apoptosis of Cultured Cerebellar Granule Neurons

After maturing cerebellar granule neurons in high-K^+ (26 mM) medium containing serum, changing to low K^+ (5 mM) serum-free medium caused a large number of

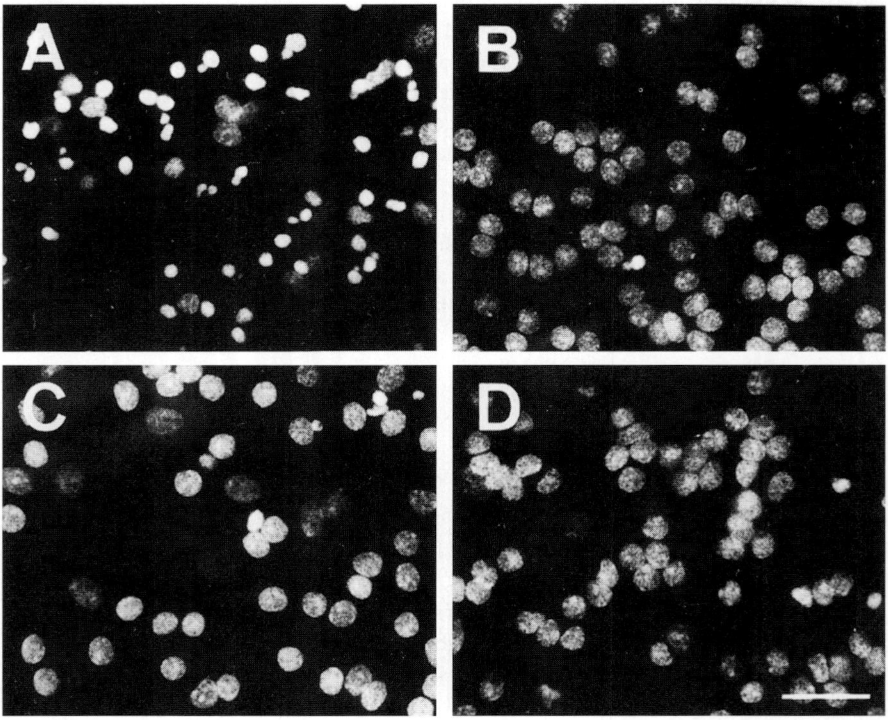

FIG. 1A–D. Prevention of the chromatin condensation of cultured cerebellar granule neurons in low-K^+ medium by brain-derived neurotrophic factor (BDNF) and cycloheximide. Cultured cerebellar granule neurons were maintained in high-K^+ MEM containing 5% HS (horse serum) and 1 μM AraC (cytosme arabmoside) for 4 days, then the neurons were cultured in low-K^+ (5 mM) medium for 1 day with no addition (**A**), BDNF (50 ng/ml) (**C**), or cycloheximide (0.3 μM) (**D**). Control neurons (**B**) were maintained in high-K^+ (26 mM) medium. Neurons were fixed and stained with Hoechst 33 258. *Bar*, 30 μm granule neurons to die [7]. Using this system, we have reported that BDNF, a neurotrophin, supports the survival of cerebellar granule neurons [8]. As shown in Fig. 1, we observed the morphology of the nuclei of granule neurons using Hoechst 33258 dye. The chromatin of the neurons in the low-K^+ medium was condensed, fragmented, and brightly stained. In contrast, the chromatin of the control cells in the high-K^+ medium was stained dispersedly. When BDNF was added to the cultures of low-K^+ medium, the chromatin was stained in a dispersed manner like that in high-K^+ medium. As the chromatin condensation is a hallmark of apoptosis, the granule neurons in low K^+ are thought to die in an apoptotic manner. This result is consistent with previous observations [7]. Therefore, these results suggested that the apoptosis of the cultured cerebellar granule neurons is prevented by BDNF.

BDNF Stimulates Activation of PI3-K in Cultured Cerebellar Granule Neurons

Recently, several reports have suggested that PI3-K is involved in promoting survival of cells, including PC12 cells and cerebellar granule neurons [5,6]. To examine whether BDNF activates PI3-K in the cerebellar granule neurons cultured in high-K^+ serum-containing medium for 4 days, we measured the activity of PI-3 kinase. Anti-phosphotyrosine (pTyr) immunoprecipitates from lysates of the cultured granule neurons treated with BDNF for various periods were subjected to in vitro PI-3 kinase assay. BDNF-induced increases in PI3-K activities were evident (Fig. 2). The increases in PI3-K activities were detected 30 min after the addition of BDNF. These results indicate that BDNF activates PI3-K in cultured granule neurons.

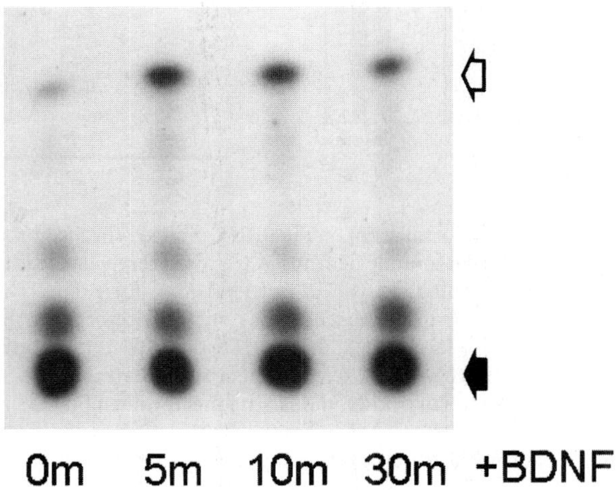

FIG. 2. BDNF-stimulated activation of PI3-K in cultured cerebellar granule neurons. Cultured cerebellar granule neurons were maintained in high-K^+ MEM containing 5% HS and 1 μM AraC for 4 days; the neurons were cultured in serum-free MEM overnight. Lysates were prepared from the neurons incubated without (*0 m*) or with 50 ng/ml BDNF for 5 min (*5 m*), 10 min (*10 m*), and 30 min (*30 m*). The activity of PI3-kinase was measured as described in Materials and Methods. *Closed arrow* indicates an origin; *open arrow* indicates the position of phosphatidylinositol 3-monophosphate, the product of PI3-kinase, estimated from the position of phosphatidylinositol 4-monophosphate as a standard

Wortmannin Inhibits the Survival-Promoting Effects of BDNF and BDNF-Induced PI3-K Activation on Cultured Cerebellar Granule Neurons

To examine whether PI3-K activation is involved in the survival-promoting effect of BDNF on the cultured cerebellar granule neurons, we used wortmannin, a specific inhibitor of PI-3 kinase. Wortmannin effectively blocked the BDNF-enhanced survival of cultured granule neurons (Fig. 3A). In addition, to confirm the inhibitory effect of wortmannin on the BDNF-stimulated increase in PI3-K activity, we measured PI3-K activity in the cultured granule neurons in the presence of wortmannin, which attenuated the BDNF-induced increase of PI3-K activity (Fig. 3B). These results suggested that activation of PI3-K is required for BDNF to promote the survival of cerebellar granule neurons.

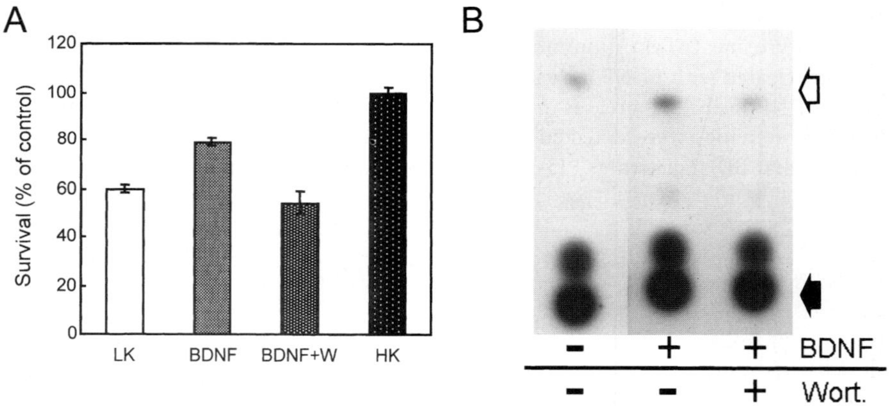

Fig. 3A,B. Inhibitory effects of wortmannin on the BDNF-induced survival promotion and PI3-K activation in cultured cerebellar granule neurons. **A** Cultured cerebellar granule neurons were maintained in high-K^+ MEM containing 5% HS and 1 μM AraC for 4 days. Then, the medium was changed to serum-free MEM in the absence (*LK*) or presence of 50 ng/ml BDNF (*BDNF*, *BDNF + W*) or serum-free high-K^+ (26 mM) MEM (*HK*); then 100 nM wortmannin was added to the cultures (*BDNF + W*) every 5 h. After 24 h, neuronal survival was determined by the MTT assay as described in Materials and Methods. The mean value obtained from cells maintained in serum-free high-K^+ MEM was defined as the control. Values are means ± SD ($n = 4$) [9]. **B** The medium was changed to serum-free MEM. The neurons were cultured overnight, then were or were not exposed to 50 ng/ml BDNF for 5 min. We added 100 nM wortmannin (*Wort.*) to the medium 30 min before exposure to BDNF, when we examined the effect of wortmannin. The activity of PI3-kinase was measured as described in Materials and Methods. *Closed arrow* indicates an origin; *open arrow* indicates the position of phosphatidylinositol 3-monophsphate, the product of PI3-kinase, estimated from the position of phosphatidylinositol 4-monophosphate as a standard

Discussion

We have shown here that BDNF activated PI3-K and that this activation is involved in exertion of the survival-promoting effect of BDNF in the cultured cerebellar granule neurons. PI3-K causes the elevation of intracellular concentrations of phosphatidylinositol 3,4-bisphosphate and phosphatidylinositol 3,4,5-trisphosphate. In addition, serine/threonine protein kinases, such as Akt [11] and $p70^{S6K}$ [12], are activated downstream of PI3-K. We observed that BDNF stimulated increases of Akt and $p70^{S6K}$ activities as well as that of PI3-K activity (data not shown). Dudek et al. [5] reported that activation of Akt is very important for survival promotion of cultured cerebellar granule neurons by insulin and insulin-like growth factor 1. Therefore, the BDNF-promoting survival of the cultured granule neurons may be also mediated by activation of Akt.

PI3-K is a heterodimer composed of an 85-kDa regulatory subunit (p85) and a 110-kDa catalytic subunit (p110). PI3-K is activated by association of p85 with phosphotyrosine residues on target proteins via SH2 domains of p85 [13]. PI3-K is activated by platelet-derived and hepatocyte growth factors through its binding to specific phosphotyrosine residues on autophosphorylated their receptors, respectively. However, the mechanism by which neurotrophins activate PI3-K remains unclear. There have been contradictory reports regarding whether NGF activates PI3-K via a direct association of PI3-K with the autophosphorylated TrkA or via an indirect interaction of PI3-K with the activated TrkA [14,15].

We have recently, reported that PI3-K is not activated through direct association with TrkB in response to BDNF in cultured cerebral cortical neurons [16]. In addition, we demonstrated that PI3-K activation may be mediated by association of PI3-K with tyrosine-phosphorylated adaptor proteins, such as insulin receptor substrate-1 and -2, in cultured cortical neurons. In the cultured cerebellar granule neurons, we could not also detect direct interaction of PI3-K with TrkB, but we could detect tyrosine-phosphorylated proteins associated with PI3-K in response to BDNF (data not shown). We suggest that BDNF-induced activation of PI3-K in the cultured granule neurons is mediated by the association with tyrosine-phosphorylated adaptor proteins but not by direct association with TrkB, as well as that in the cultured cortical neurons.

References

1. Oppenheim RW (1991) Cell death during development of the nervous system. Annu Rev Neurosci 14:453–501
2. Barde Y-A (1989) Trophic factors and neuronal survival. Neuron 2:1525–1534
3. Lindsay RM, Wiegand SJ, Altar CA, et al (1994) Neurotrophic factors: from molecule to man. Trends Neurosci 17:182–190
4. Glass DJ, Yancopoulos GD (1993) The neurotrophins and their receptors. Trends Cell Biol 3:262–268
5. Dudek H, Datta SR, Franke TF, et al (1997) Regulation of neuronal survival by the serine-threonine kinase Akt. Science 275:661–665
6. Yao R, Cooper GM (1995) Requirement for phosphatidylinositol-3 kinase in the prevention of apoptosis by nerve growth factor. Science 267:2003–2006

7. D'Mello SR, Galli C, Ciotti T, et al (1993) Induction of apoptosis in cerebellar granule neurons by low potassium: inhibition of death by insulin-like growth factor I and cAMP. Proc Natl Acad Sci USA 90:10989–10993
8. Kubo T, Nonomura T, Enokido Y, et al (1995) Brain-derived neurotrophic factor (BDNF) can prevent apoptosis of rat cerebellar granule neurons in culture. Dev Brain Res 85:249–258
9. Shimoke K, Kubo T, Numakawa T, et al (1997) Involvement of phosphatidylinositol-3 kinase in prevention of low K^+-induced apoptosis of cerebellar granule neurons. Dev Brain Res 101:197–206
10. Nonomura T, Kubo T, Oka T, et al (1996) Signaling pathway and survival effects of BDNF and NT-3 on cultured cerebellar granule cells. Dev Brain Res 97:43–50
11. Franke TF, Yang S, Chan TO, et al (1995) The protein kinase encoded by the *Akt* proto-oncogene is a target of the PDGF-activated phosphatidylinositol 3-kinase. Cell 81: 727–736
12. Chung J, Grammar TC, Lemon KP, et al (1994) PDGF- and insulin-dependent $pp70^{S6k}$ activation mediated by phosphatidylinositol-3-OH kinase. Nature (Lond) 370:71–75
13. Carpenter CL, Auger KR, Chanudhuri M, et al (1993) Phosphoinositide 3-kinase is activated by phosphopeptides that bind to the SH2 domains of the 85-kDa subunit. J Biol Chem 268:9478–9483
14. Ohmichi M, Decker SJ, Saltiel AR (1992) Activation of phosphatidylinositol-3 kinase by nerve growth factor involves indirect coupling of the trk proto-oncogene with *src* homology 2 domain. Neuron 9:769–777
15. Soltoff SP, Rabin SL, Cantley LC, et al (1992) Nerve growth factor promotes the activation of phosphatidylinositol 3-kinase and its association with the trk tyrosine kinase. J Biol Chem 267:17472–17477
16. Yamada M, Ohnishi H, Sano S, et al (1997) Insulin receptor substrate (IRS) -1 and IRS-2 are tyrosine-phosphorylated and associated with phosphatidylinositol 3-kinase in response to brain-derived neurotrophic factor in cultured cerebral cortical neurons. J Biol Chem 272:30334–30339

Brain-Derived Neurotrophic Factor Prevents Neuronal Cell Death Induced by Corticosterone

SHOEI FURUKAWA and ATSUMI NITTA

Summary. Glucocorticoids cause neuronal damage in the rat hippocampus, but the mechanism(s) of action underlying the neurodegenerative effects remain unknown. Brain-derived neurotrophic factor (BDNF), one of the neurotrophins, affects survival or differentiation of various types of neurons of the central nervous system (CNS) in culture and is known to prevent cell death of some of these neurons induced by various CNS insults. We investigated the effects of corticosterone on both neuronal cell death and BDNF synthesis using cultured hippocampal neurons. Most neurons survived 3 days after removal of serum, while corticosterone induced neuronal death as early as 1 day after the addition. The level of BDNF mRNA was decreased in a dose-dependent manner in the presence of corticosterone, while the β-actin mRNA level was unchanged. Intracellular BDNF content was also markedly reduced in the presence of corticosterone. These observations demonstrated that corticosterone causes a decrease in BDNF mRNA and protein in cultured hippocampal neurons. Finally, we found that corticosterone-induced neuronal death was significantly protected by the addition of BDNF. These observations suggest that the neurotoxic effect of corticosterone is mediated by suppression of synthesis of BDNF, which has a role in support of the survival of cultured neurons.

Key words. Corticosterone—Hippocampus—Neurons—Neuronal death—Brain-derived neurotrophic factor (BDNF)—Neurotrophins

Introduction

Brain-derived neurotrophic factor (BDNF) is one of the neurotrophin (NT) family proteins, which include nerve growth factor (NGF), NT-3, NT-4/5, and NT-6 [1,2]. Widespread expression of mRNAs of BDNF [3,4] and its receptor, TrkB [5], in the neurons of the central nervous system (CNS) suggests important roles for BDNF. BDNF affects the survival or differentiation of various types of CNS neurons in culture

Laboratory of Molecular Biology, Gifu Pharmaceutical University, 5-6-1 Mitahora-higashi, Gifu 502-8585, Japan

[6–9], and is known to prevent cell death of some of these neurons induced by various CNS insults [10,11].

Glucocorticoids are secreted at much higher concentrations as part of the stress response [12]. Recently, they have been reported to cause neuronal damage in the rat hippocampus; i.e., peripheral administration of corticosterone decreases neuronal cell number [13], and enhances both ischemia-induced neuronal death [14] and neurotoxicity of kainic acid [15]. However, the mechanism(s) of action underlying the neurodegenerative effects remains unknown, although one possibility is that glucocorticoids regulate the expression of neurotrophic factors. We have observed an effect of corticosterone on intracellular localization of BDNF-like immunoreactivity (BDNF-LI) in neurons of rat brain [16], which suggests that corticosterone influences BDNF synthesis and processing. Therefore, we investigated the effects of corticosterone on both neuronal cell death and BDNF synthesis using cultured hippocampal neurons.

Materials and Methods

Neurons were cultured from rat hippocampus of embryonic day 18 as described previously [17]. Cultures were started at 10^5 cells/cm^2 and maintained for 1 day in Dulbecco's modified Eagle's medium (Gibco BRL, Grand Island, NY, USA) containing 5% horse serum and fetal calf serum (Irvine Scientific, Irvme, CA, USA). Medium was then changed to a defined one containing insulin, transferrin, and progesterone. Corticosterone was added immediately or 4 days after the medium change, and neurons were cultured for another 1 day. Neuronal viability was estimated by counting living neuronal cells with a light microscope. Revere transcriptase-polymerase chain reaction (RT-PCR) was used to determine BDNF mRNA level as described [18,19]. Intracellular content of BDNF was measured by a sensitive enzyme immunoassay (EIA) for BDNF, which is based on the EIA system for NGF [20]. Corticosterone-binding sites were estimated using [^{3}H]corticosterone as described [21]. All experimental values were statistically analyzed with Turkey's test.

Results

Most neurons survived for 3 days, while corticosterone induced neuronal death even at 1 day after the addition (Fig. 1). The most effective concentration of corticosterone was 0.01 μM. Much higher concentrations were inversely less toxic. The amount of RT-PCR products from BDNF mRNA decreased with the increase in concentration of corticosterone added in a dose-dependent manner, while beta-actin mRNA level was unchanged at any dosage (Fig. 2). Intracellular BDNF content was calculated by dividing the total BDNF content for number of living neurons. BDNF content was markedly reduced in the presence of 0.01–1.0 μM corticosterone (Fig. 2). The most effective dosage was 0.01 μM as well as that observed in neuronal toxicity. These observations demonstrated that corticosterone causes a decrease in BDNF mRNA and protein in cultured hippocampal neurons.

We next tested responsiveness to corticosterone of neurons cultured for a longer time (4 days) in the defined medium. Interestingly, corticosterone induced not only

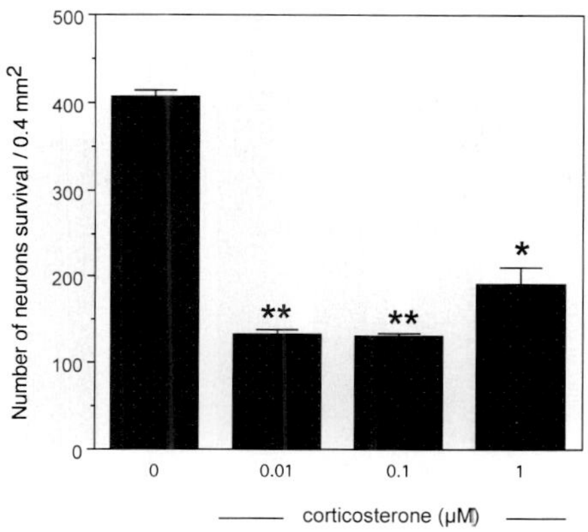

FIG. 1. Effects of corticosterone on survival of hippocampal neurons (10^5 cells/cm^2) cultured in the presence of various concentrations of corticosterone for 24h. Surviving neurons were counted for eight arbitrarily selected fields (total, 0.3 mm^2). Significant difference from values in absence of corticosterone: *, $P < 0.05$; **, $P < 0.01$ (Turkey's test)

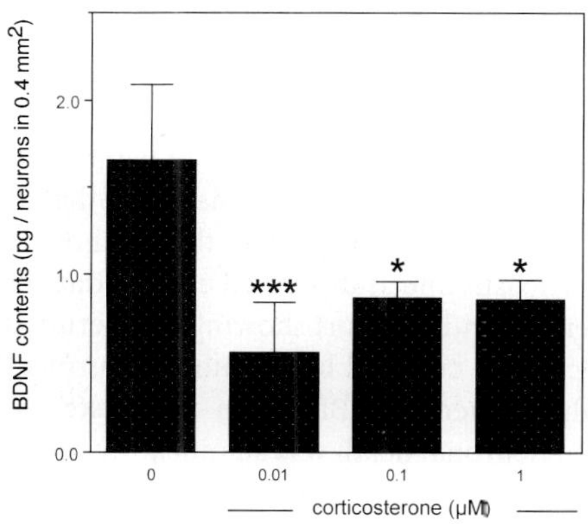

FIG. 2. Effects of corticosterone on brain-derived neurotrophic factor (BDNF) content in hippocampal neurons (10^5 cells/cm^2) cultured in the presence of various concentrations of corticosterone for 24h. Surviving neurons were counted at arbitrarily selected fields (total, 0.4 mm^2). Neurons were suspended in 0.1 M Tris-HCl buffer, pH 7.4, containing 1 M NaCl, 2% bovine serum albumin, 2 mM EDTA, 80 trypsin inhibitor units/l aprotinin, and 0.02% NaN_3, and pulse-sonicated for 30 s. The homogenates were centrifuged at 10,000 × *g* for 30 min; supernatants were provided for EIA after treatment with chloroform. The values were calculated as BDNF content/neuronal numbers in 0.4 mm^{-2}. Significant difference from values in absence of corticosterone: *, $P < 0.05$; ***, $P < 0.005$ (Turkey's test)

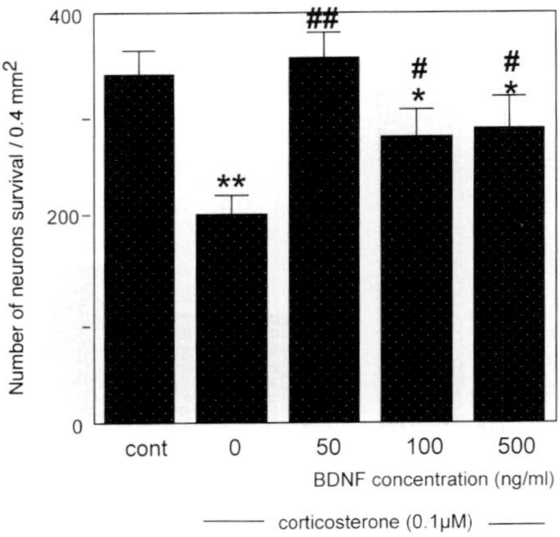

FIG. 3. Effects of addition of BDNF on corticosterone-induced death of hippocampal neurons (10^5 cells/cm^2) cultured for 24h with various concentrations of BDNF (0, 50, 100, or 500 ng/ml) in the presence of 0.1 μM of corticosterone or without both corticosterone and BDNF (*cont*). Numbers of surviving neurons were counted at arbitrarily selected fields (total, 0.4 mm^2). Significant difference from value obtained in absence of corticosterone: *, $P < 0.05$; **, $P < 0.005$ (Turkey's test). Significant difference from the value obtained without BDNF in the presence of corticosterone: #, $P < 0.05$; ##, $P < 0.001$ (Turkey's test)

neuronal death but also expression of BDNF mRNA and protein in the long-term-cultured neurons (data not shown). Quantification experiments of [^{3}H]corticosterone-binding sites clearly demonstrated that the binding sites were much fewer in the 4-day-cultured neurons than that in short-term-cultured neurons, which responded to corticosterone as described earlier (data not shown). This observation is in good agreement with a report showing that serum deprivation causes loss of corticosterone receptors in cultured hippocampal neurons [15].

Finally, we tested the influence of BDNF on corticosterone-responsive neurons. Corticosterone-induced neuronal death was significantly protected in the presence of BDNF (Fig. 3), suggesting that the neurotoxic effect of corticosterone is mediated by suppression of synthesis of BDNF, which has a supporting role in survival of cultured neurons.

Discussion

Exposure to stress decreases BDNF mRNA expression in the rat hippocampus, and administration of corticosterone reduces the level of BDNF mRNA in the hippocampal dentate gyrus in vivo [22]. Similar reduction by corticosterone is also observed in cultured hippocampal neurons [23]. In this study, we showed that corticosterone, the only glucocorticoid in rats, induces neuronal cell death from cultured hippocampal neurons at physiological concentrations (Fig. 1), accompanied by reduction of BDNF mRNA and intracellular BDNF content (Fig. 2). Previous reports demonstrated that corticosterone administration enhances hypoxic and hypoglycemic damage [12], accelerates ischemia-induced neuronal cell death [14], and amplifies neurotoxicity of kainic acid in the hippocampus [15]. In chronic administration, decrease in neuronal cell number is reported in the aged rat hippocampus [13]. These observations suggest a close relationship between reduction of BDNF synthesis and neuronal cell death.

However, there are no reports showing directly the toxicity of corticosterone on hippocampal neurons in vitro, probably because of inconsistency of the cultures. In fact, we found that expression of the corticosterone receptor is varied depending on time after serum removal. Neurons maintained for 4 days in a defined medium lost most of the binding sites for corticosterone, but neurons cultured for only 1 day did not (data not shown). Furthermore, the density of neurons in culture was also critical for detection of corticosterone toxicity; that is, corticosterone did not affect neuronal survival at much higher or lower cell density (data not shown), which may imply that the amount of BDNF secreted into the culture medium determines the influence of corticosterone. Recent findings [24] showing that autocrine production of BDNF is critical for survival of cultured hippocampal neurons support our consideration. BDNF is also known to be protective against some neuronal damage in vivo; that is, exogenously added BDNF prevents cell death of the hippocampal pyramidal neurons following ischemia [24] and of dopaminergic neurons induced by 6-hydroxydopamine [10] or axotomy [11]. In the case of corticosterone-induced cell damage, addition of BDNF is effective in preventing neuronal death (Fig. 3).

Our results, taken together with previous observations, suggest that corticosterone downregulates BDNF synthesis and results in neuronal damage of cultured hippocampal neurons.

Acknowledgment. The authors are grateful to Sumitomo Pharmaceutical Co. Ltd. for supply of recombinant human BDNF. This work was supported, in part, by a grant-in-aid for science research from the Ministry of Education, Science, Sports and Culture of Japan.

References

1. Barde YA (1990) The nerve growth factor family. Prog Growth Factor Res 2:237–248
2. Lindsay RM (1994) Neurotrophins and receptors. Prog Brain Res 103:3–14
3. Ernfors P, Wetmore C, Olson L, et al (1990) Identification of cells in rat brain and peripheral tissues expressing mRNA for members of the nerve growth factor family. Neuron 5:511–526
4. Maisonpierre PC, Belluscio L, Friedman B, et al (1990) NT-3, BDNF, and NGF in the developing rat nervous system: parallel as well as reciprocal patterns of expression. Neuron 5:501–509
5. Klein R, Martin-Zanca D, Barbacid M, et al (1990) Expression of the tyrosine kinase receptor gene trkB is confined to the murine embryonic and adult nervous system. Development (Camb) 109:845–850
6. Henderson CE, Camu W, Mettling C, et al (1993) Neurotrophins promote motor neuron survival and are present in embryonic limb bud. Nature (Lond) 363:266–270
7. Hyman C, Hofer M, Barde Y-A, et al (1991) BDNF is a neurotrophic factor for dopaminergic neurons of the substantia nigra. Nature (Lond) 350:230–232

8. Knusel B, Winslow JW, Rosenthal A, et al (1991) Promotion of central cholinergic and dopaminergic neuron differentiation by brain-derived neurotrophic factor but not neurotrophin-3. Proc Natl Acad Sci USA 88:961–965
9. Lindholm D, Dcchant G, Heisenberg CP, et al (1993) Brain-derived neurotrophic factor is a survival factor for cultured rat cerebellar granule neurons and protects them against glutamate-induced neurotoxicity. Eur J Neurosci 5:1455–1464
10. Spina MB, Squinto SP, Miller J, et al (1992) Brain-derived neurotrophic factor protects dopamine neurons against 6-hydroxydopamine and *N*-methyl-4-phenylpyridinium ion toxicity: involvement of the glutathione system. J Neurochem 59:99–106
11. Yan Q, Eliott J, Snider WD (1992) Brain-derived neurotrophic factor rescues spinal motor neurons from axotomy-induced cell death. Nature (Lond) 360:753–755
12. Scully JL, Otten U (1995) Glucocorticoids, neurotrophins and neurodegeneration. J Steroid Biochem Mol Biol 52:391–401
13. Mizoguchi K, Kunishita T, Chui DH, et al (1992) Stress induces neuronal death in the hippocampus of castrated rats. Neurosci Lett 138:157–160
14. Sapolsky RM, Pulsinelli WA (1985) Glucocorticoids potentiate ischemic injury to neurons: therapeutic implications. Science 229:397–400
15. Sapolsky RM (1990) Glucocorticoids, hippocampal damage and the glutamatergic synapse. Prog Brain Res 86:13–23
16. Nitta A, Fukumitsu H, Kataoka H, et al (1997) Administration of corticosterone alters intracellular localization of brain-derived neurotrophic factor-like immunoreactivity in the rat brain. Neurosci Lett 226:115–118
17. Nakajima M, Furukawa S, Hayashi K, et al (1993) Age-dependent survival-promoting activity of vitamin K on cultured CNS neurons. Dev Brain Res 73:17–23
18. Awatsuji H, Furukawa Y, Furukawa S, et al (1993) Interleukin-2 as a trophic factor for supporting the survival of neurons cultured from various regions of fetal rat brain. J Neurosci Res 35:305–311
19. Miwa T, Furukawa S, Nakajima K, et al (1997) Lipopolysaccharide enhances expression of brain-derived neurotrophic factor in cultured rat microglia. J Neurosci Res 50:1023–1029
20. Furukawa S, Kamo I, Akazawa S, et al (1983) A highly sensitive enzyme immunoassay for mouse beta nerve growth factor. J Neurochem 40:734–744
21. Kontula K, Anderson LC, Paavonen T, et al (1980) Glucocorticoid receptors and glucocorticoid sensitivity of human leukemic cells. Int J Cancer 26:177–183
22. Smith MA, Makino S, Kvetnansky R, et al (1995) Stress and glucocorticoids affect the expression of brain-derived neurotrophic factor and neurotrophin-3 mRNAs in the hippocampus. J Neurosci 15:1768–1777
23. Lindholm D, Carroll P, Tzimagiogis G, et al (1996) Autocrine-paracrine regulation of hippocampal neuron survival by IGF-1 and the neurotrophins BDNF, NT-3, and NT-4. Eur J Neurosci 8:1452–1460
24. Beck T, Lindholm D, Castren E, et al (1994) Brain-derived neurotrophic factor protects against ischemic cell damage in rat hippocampus. J Cereb Blood Flow Metab 14:689–692

Neurotrophins Regulate the Function of Microglia

KAZUYUKI NAKAJIMA, ETSUKO IKOMA, SHIZUYO HONDA, and SHINICHI KOHSAKA

Summary. Although the physiological role of neurotrophins in neuronal development and survival has been extensively investigated, their role in glial cell function remains to be elucidated. In the present study, we investigated the effects of neurotrophins on cultured microglia from newborn rat brain. All the neurotrophins tested, nerve growth factor (NGF), brain-derived neurotrophic factor (BDNF), neurotrophin-3 (NT-3), and neurotrophin-4 (NT-4), increased the secretion of plasminogen and specific activity of acid phosphatase, but suppressed the release of constitutively produced nitric oxide (NO) from microglia. Western blotting revealed that cultured microglia express Trk A, B, and C, and low-affinity NGF receptor, LNGFRp75. Neurotrophin was found to phosphorylate Trk A and B. Furthermore, neurotrophins caused an activation of transcription factor, NF-κB. These results indicate that the neurotrophin family regulate the function of microglia through Trks or LNGFRp75-mediated signal transduction.

Key words. Microglia—Neurotrophin—Plasminogen—Nitric oxide—Acid phosphatase

Introduction

The neurotrophins are a family of protein factors composed of at least five structurally related members: nerve growth factor (NGF), brain-derived neurotrophic factor (BDNF), neurotrophin-3 (NT-3), neurotrophin-4/5 (NT-4/5), and NT-6. These proteins not only regulate neuronal function but also support the survival and enhance the growth of various types of neurons in both the peripheral nervous system (PNS) and the central nervous system (CNS) [1]. Neurotrophins such as NGF, BDNF, and NT-3 were originally shown to be produced in target neurons or tissues in both the PNS and the CNS [2], but later they were reported to be produced in glial cells as well. Astrocytes have been found to express mRNAs and proteins of the neurotrophin

---

Department of Neurochemistry, National Institute of Neuroscience, 4-1-1 Ogawa-higashi, Kodaira, Tokyo 187-8502, Japan family, including NGF [3]. These findings further suggest that the neurotrophins produced by glial cells act on glial cells in an autocrine or paracrine fashion in addition to neurons. However, information on the significance of neurotrophins in the regulation of glial function has been quite limited.

In the present study, as a means of testing the hypothesis that neurotrophins regulate microglial function in the CNS, we biochemically examined the ability of four different neurotrophins to regulate secretory products and characteristic enzyme activities of cultured microglia.

Materials and Methods

Preparation of Microglia and Stimulation with Neurotrophins

Microglia were obtained from primary cell cultures of the cerebral hemispheres of neonatal rats, as described previously [4]. In brief, after 10–20 days in primary cell culture, microglia were prepared as floating cells with an 8-character orbital shaker (80 rpm, 3 min). Aliquots of the microglial cell suspension (1.5–2 $\times$ 10^5) were transferred to each well of a 12-well plate (Coster, Cambridge, MA, USA) and allowed to attach at 37°C. After 30 min, unattached cells, if any, were removed by rinsing three times with serum-free DMEM. After the final rinse, 1 ml of serum-free DMEM was poured into each well, and the plate was allowed to stand in 10% CO_2 for 3–4 h before stimulation. Each neurotrophin was added to the well to a final concentration of 0–100 ng/ml, and the cultures were maintained in 10% CO_2. After the microglial culture had been maintained for 24 h, the conditioned medium was removed and centrifuged at 1500 rpm for 10 min to remove any floating materials. The supernatant was used as microglial-conditioned medium. Aliquots of 10 μl $\times$ 3 were taken for the determination of nitric oxide (NO), and the 750 μl of remaining supernatant was concentrated with Ultrafree C3-TTK (30-kDa cut; Millipore) and freeze-dried to determine the amount of plasminogen (PGn). The microglial cells on the dishes were rinsed twice with saline and sonicated with 80 μl of cold saline. The lysate was collected as a cell homogenate and maintained at −30°C until used.

Detection of Plasminogen by Western Blot

The freeze-dried conditioned medium was solubilized with 50 μl of nonreducing Laemmli's sample solution (62.5 mM Tris HCl, pH 6.8, 2% sodium dodecyl sulfate [SDS], 5% glycerol), and 20 μl was subjected to SDS-polyacrylamide gel electrophoresis (PAGE) with 10%–20% gradient gel under nonreducing conditions. The proteins were transferred to Immobilon-P (Millipore) with a semidry-type transblotting apparatus (ATTO, Tokyo, Japan). The Immobilon was incubated with a concentration of 2 μg/ml of polyclonal anti-rat PGn antibody (IgG) in TNB solution (10 mM Tris HCl, pH 7.5, 150 mM NaCl, 3% bovine serum albumin [BSA]) for 1 h at room temperature. After washing with TN solution (10 mM Tris HCl, pH 7.5, 150 mM NaCl), the membrane was incubated with horse radish peroxidase- (HRP-) conjugated anti-rabbit IgG (1:200) for 1 h at room temperature. The antigen–antibody complex was detected by staining with 0.02% diaminobenzidme (DAB) and 0.01% H_2O_2.

Measurement of Nitric Oxide in the Conditioned Medium

Nitric oxide (NO) is a short-lived gas and immediately changes to NO_2 in aqueous solution after production. The amounts of NO in microglial-conditioned medium were monitored by the amount of NO_2, a breakdown product of NO, using an exclusive NO analyzing system (ENO-10; EICOM, Kyoto, Japan). At various times during culture, 10 μl of culture medium was directly applied to the analyzer with a column that enables separation of NO_2 and NO_3 (NO-PAK; 9.6 mm × 50 mm). The NO_3 is further reduced to NO_2 in a reducing column (NO-RED, NORTIP). The original and reduced NO_2 could be detected individually, and the amount of NO_2 was calculated by using sodium nitrite as the standard.

Determination of Acid Phosphatase

Acid phosphatase (EC 3.1.3.2) activity was essentially measured according to the method previously described [5]. A reaction mixture containing 100 mM citrate buffer (pH 4.8), 10 mM *p*-nitrophenyl phosphate, 0.5% bovine serum albumin (BSA), and cell homogenate in a total volume of 0.25 ml was incubated at 37°C for 20 min. The reaction was stopped by addition of 0.75 ml of 0.03 N NaOH, and absorbance at 405 nm was determined. Activity was calculated using *p*-nitrophenol as the standard.

Western Blotting for Detection of Trks and LNGFp75

The microglia were allowed to adhere to the culture dishes (75 cm^2; Nunclon, Denmark) at high density and rinsed with saline twice. The cells were then harvested with a rubber policeman and recovered as a precipitate by low-speed centrifugation (1500 × *g*, 10 min). The cell precipitate was solubilized with RIPA buffer (10 mM Tris HCl, pH 7.5, 150 mM NaCl, 0.5% deoxychorate [DOC], 0.1% Triton ×100) and centrifuged at 100 000 × *g* for 30 min to remove insoluble material. The supernatant was removed and an aliquot was removed to determine the amount of proteins. An equal volume of reducing Laemmli's SDS sample solution (125 mM Tris HCl, pH 6.8, 5% 2-mercaptoethanol, 4% SDS, 10% glycerol) was then added to the cell extract, and it was heated at 90°C for 3 min. The protein (20–30 μg) was separated by electrophoresis on 7.5% (for Trks) or 10%–20% (for p75) polyacrylamide gel as described, and then transferred to an Immobilon membrane. The blot was blocked with STw/TNB (1% skim milk and 0.01%Tween 20 in TNB solution) for 1 h, and incubated with Trk antibody (1 μg/ml) in STw/TNB at 4°C overnight. After rinsing with Tw/TN three times and finally with TN solution once, the membrane was incubated with HRP-conjugated antirabbit IgG (1 : 1000) for 1 h at room temperature. The antigen–antibody complex was detected using an ECL chemiluminescence system.

Immunocytochemistry for Localization of p65 Subunit of NF-κB

The microglia were seeded on glass coverslips as described, washed with serum-free DMEM, and exposed to each neurotrophin (200 ng/ml) in the same medium for 2 h. The cells were stained essentially according to the method of Tozawa et al. (1995) and Sakurada et al. (1996). In brief, the cells were fixed in 4% paraformaldehyde/

PBS for 10 min, permeabilized by 0.5% Triton X-100/PBS for 10 min, and then incubated with anti-p65 antibody (3 μg protein/ml) at 4°C overnight. The bound antibody was detected by incubating with FITC-conjugated goat anti-rabbit antibody for 1 h.

Results

Morphology and Proliferation of Cultured Microglia

Cultured microglia were exposed to various concentrations (20, 50, and 100 ng/ml) of each neurotrophin (NGF, BDNF, NT-3, NT-4) for 24 h in serum-free medium. The morphology of the cells appeared to change slightly, but the cell number did not seem to change in response to the addition of any of the neurotrophins within 24 h, so long as the microglia were maintained in serum-free medium. Actually, we could not detect a significant difference between cell number in the control and neurotrophin-stimulated wells. As a result, we concluded that none of the neurotrophins used here expressed any proliferative activity on the cultured microglia under serum-free conditions.

Effect on the Acid Phosphatase Activity of Microglia

The effects of the neurotrophins on the enzyme activity of cultured microglia were investigated under the conditions described earlier. Acid phosphatase and 5′-nucleotidase were chosen as characteristic enzymes of cultured microglia. All the neurotrophins significantly increased the specific activity of acid phosphatase in

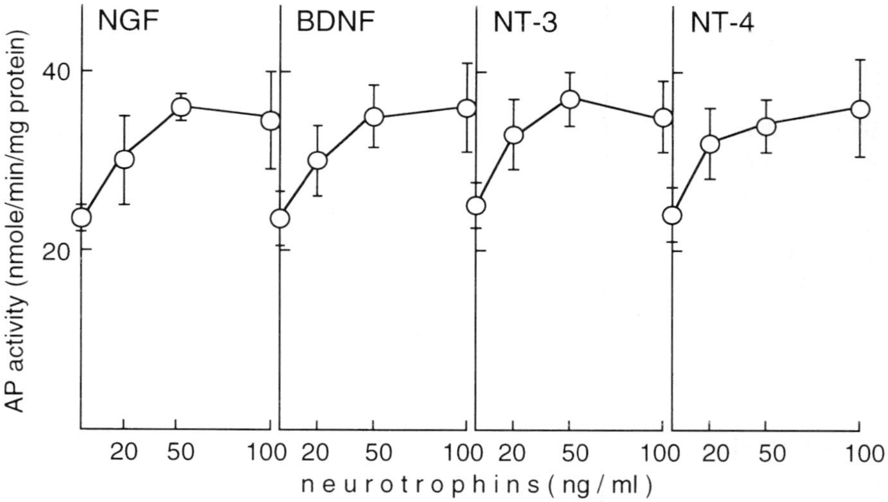

FIG. 1. Effect of neurotrophins on acid phosphatase of cultured microglia. Microglial cells exposed to each neurotrophin were collected and homogenized. The specific activity of acid phosphatase was determined in triplicate assays using a 10-μl sample of each homogenate. The values shown are means ± SD of three or four experiments. *AP*, acid phosphatase

a dose-dependent manner (Fig. 1). The effect of all four neurotrophins was quite comparable in potency, and appeared to be saturated at about 50 ng/ml.

Effect on PGn Secretion

As an effect of neurotrophins on the secretory products of microglia, secretion of PGn was examined by Western blotting (Fig. 2). In the control culture medium, about 3–4 ng of PGn was detected per well. Addition of neurotrophins to the culture increased the amounts of PGn secreted into the medium roughly dose dependently (Fig. 2A).

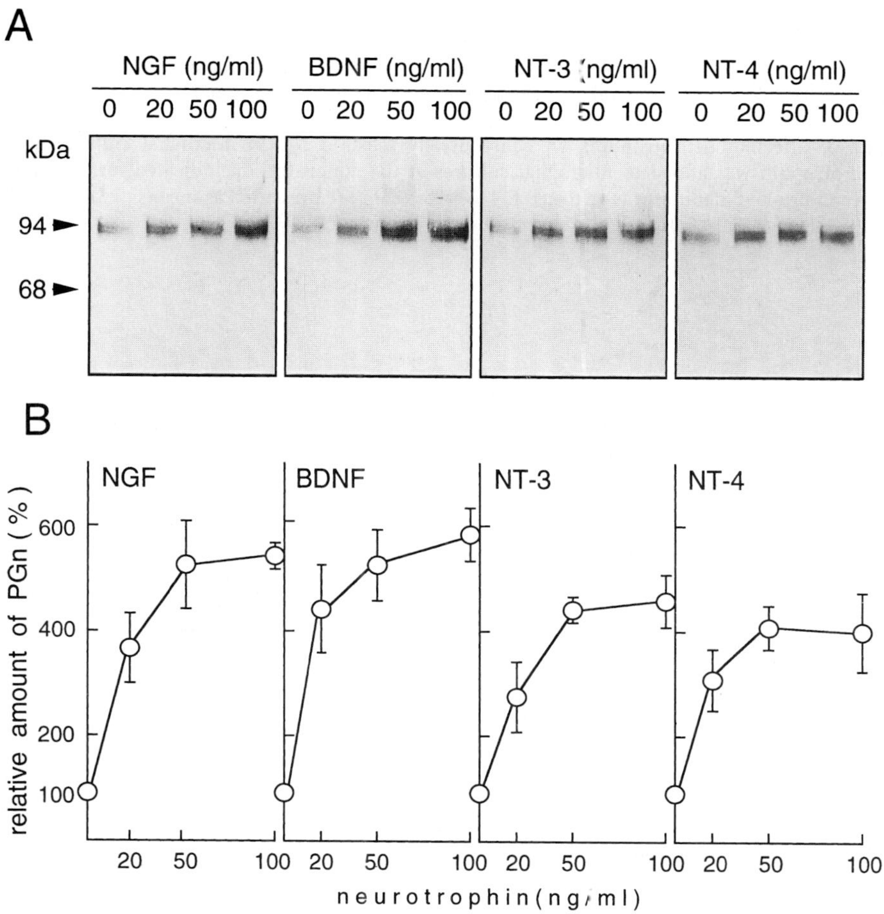

FIG. 2A,B. Effect of neurotrophins on plasminogen (PGn) secretion by cultured microglia. **A** Analysis of secretion of PGn by Western blotting. Microglial-conditioned medium prepared 24 h after exposure to the indicated concentration of each neurotrophin was analyzed for PGn by Western blotting as described in Materials and Methods. Representative effects are shown. **B** Quantification of the effects of neurotrophins on PGn secretion. The results of Western blotting as described were analyzed with densitometer, normalized, and quantified from five or six independent experiments. Values are means ± SD

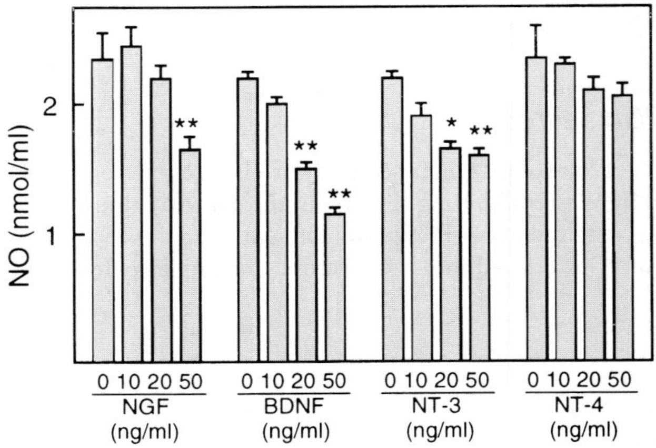

FIG. 3. Effect of neurotrophins on constitutively released NO in microglial culture. Each neurotrophin was added to microglial cultures at the concentration indicated for 24h, and the conditioned medium was analyzed for NO. The data are presented as means ± SE ($n = 4$). *, $P < 0.05$; **, $P < 0.01$, calculated according to one-way ANOVA

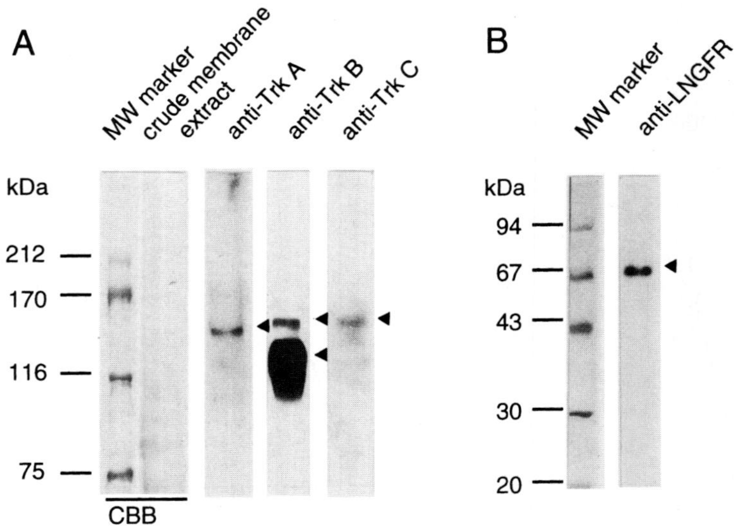

FIG. 4A,B. Detection of Trks and LNGFRp75 on cultured microglia by Western blotting. Extract (30 μg protein for Trks and 20 μg protein for LNGFRp75) of the crude membrane fraction of cultured microglia was applied to SDS-PAGE using 7.5% (for Trks) or 10%–20% gradient gel (for LNGFRp75), and Western blotted using specific anti-Trk antibodies (**A**) or anti-p75 antibody (**B**) as described in Materials and Methods. The protein profiles of the crude membrane extract and the molecular weight marker are shown at the *left*. *CBB*, Coomassie brilliant blue staining

The increasing effect of neurotrophins was quantified with a densitometer by determining the intensity of the immunoreactive bands (Fig. 2B). Neurotrophins in the 50–100 ng/ml dose range increased the release of PGn four- to sixfold compared to the control, showing a saturated state at about 50 ng/ml.

Effect on the Release of NO from Microglial Cultures

The effect of neurotrophins on NO production in cultured microglia was also examined. Basically, cultured microglia constitutively produce and release small amounts of NO, and the amount was increased by lipopolysaccharide (LPS) in a dose-dependent and a time-dependent manner (data not shown). The effect of neurotrophins was examined in regard to the constitutive production of NO in microglia. The neurotrophins decreased the amount of NO in the conditioned medium dose dependently (Fig. 3). BDNF had a stronger effect than the other neurotrophins.

Biochemical Detection of the Trk Family and LNGFRp75 in Cultured Microglia

To determine whether these effects of neurotrophins are mediated by their receptors, the expression of Trks on cultured microglia was analyzed by Western blotting. As shown in Fig. 4A, Trk A and Trk C were present as a single band at around 140 kDa and 150 kDa, respectively. Trk B was detected at about 150 kDa, although other smaller molecular forms were also recognized by anti-Trk B antibody at about 120–130 kDa. The staining of Trk A at 140 kDa, and of Trk B and Trk C at 150 kDa, suggests the presence of the full length of each Trk in cultured microglia. The heavily stained band by anti-Trk B antibody around 120–130 kDa may be attributed to the difference in glycosylation. LNGFRp75 was also detected in the same sample used for Trks detection by monoclonal antibody specific for LNGFRp75 (Fig. 4B). These results indicate the coexistence of Trk receptor and LNGFRp75 in the same microglial preparation.

To determine the functional involvement of Trks, we analyzed the tyrosine phosphorylation of Trks by treatment with neurotrophin (50 ng/ml). NGF stimulation induced a phosphorylation of 140-kDa protein and BDNF phosphorylated both 120–130 kDa protein and 150 kDa protein (data not shown).

Activation of NF-κB by Neurotrophins

To investigate the possibility that the biological responces of cultured microglia to a relatively high dose of neurotrophins are also mediated by LNGFRp75-associated signaling, we determined the activation of NF-κB, a transcription factor, because neurotrophin binding to LNGFRp75 can activate NF-κB in Schwann cells [6]. We have examined immunocytochemically the localization of NF-κB by using an antibody to p65 subunit of NF-κB (Fig. 5). In nonstimulated microglia, NF-κB was found mainly in the cytoplasm. However, when microglia were treated with NGF (200 ng/ml) for 2 h, p65 was found to be translocated to the nucleus in many cells. BDNF and NT-3 were also effective to a similar degree on the translocation of p65 to nucleus so far as microglia are stimulated with a concentration of 200 ng/ml (data not shown).

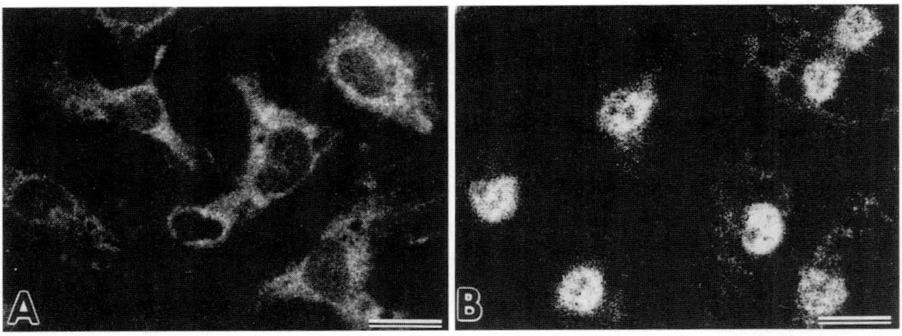

Fig. 5A,B. Nuclear translocation of NF-κB (p65). Cultured microglia on 24-well plates were exposed to each neurotrophin (200 ng/ml) for 2 h and immunocytochemically stained by anti-p65 (subunit of NF-κB) antibody as described in Materials and Methods. Typical examples of nonstimulated (**A**) and NGF-stimulated microglia (**B**) are shown in photomicrographs by laser scanning confocal microscope (Molecular Dynamics, Tokyo, Japan). BDNF or NT-3 also provided similar results. *Bar* 10 μm

Discussion

To clarify the role of neurotrophins in microglial function, in this study we biochemically examined the effect of NGF, BDNF, NT-3, and NT-4/5 on the secretory products and cellular enzymes of cultured microglia. We have been convinced that it is important to identify the biochemical changes that occur in microglia when they are exposed to certain factors, because such changes are believed to be deeply involved in cellular function. The application of neurotrophins to cultured microglia resulted in slight morphological changes but essentially had no effect on the proliferation of cultured microglia under serum-free conditions. Elkabes et al. [7] recently reported that NT-3 induced microglial proliferation and enhanced survival when cultured in 10% FCS-containing medium. However, they also observed that the actual number of microglia cells did not change when cultured in serum-free medium, consistent with our findings in this study.

Interestingly, neurotrophins caused biochemical changes in cultured microglia, as shown in our study. PGn is a secretory product that has been identified in cultured microglia [8], and we have demonstrated that PGn has a neurotrophic effect on neocortical explants [4] and mesencephalic dopaminergic neurons [9]; PGn therefore has been considered to be a functionally important molecule for intercellular interaction between microglia and neurons. By contraries the neurotrophins suppressed NO levels in microglial-conditioned medium. NO, synthesized from L-arginine and NO synthase (NOS), has been concluded to be an intercellular messenger with many important functions in the brain. On the other hand, NO has also been shown to possess neurotoxic activity, impairing mitochondrial function and the activity of other iron/sulfur-containing enzymes, and causing DNA damage. In this study, we showed that neurotrophins can downregulate the constitutively producing NO levels in microglial cultures.

Apart from their effect on secretory products, neurotrophins were found to increase the cellular enzyme. Acid phosphatase activity in the microglia was enhanced by all the neurotrophins. Acid phosphatase is a lysozomal enzyme, and is a characteristic enzyme of cultured microglia, with much higher specific activity in them than in other glial cells such as astrocytes or oligodendrocytes [10]. This enzyme appears to indicate the strength of the cell's phagocytic activity or survivability.

These responses of cultured microglia to neurotrophins strongly suggested the presence of specific receptors. As might be expected from their biological activities, cultured microglia have been demonstrated to express the functional Trk receptor family (Trk A, Trk B, and Trk C). However, regarding Trk B, smaller molecular forms with a molecular mass of 120–130kDa were recognized by anti-Trk B antibody in addition to a reasonable band for Trk B. In addition to these high-affinity receptors, the low-affinity NGF receptor LNGFRp75, which binds to all neurotrophins with similar affinity, has also been identified in microglia by Western blotting. A potentiality of LNGFRp75 to induce sphingomyelin turnover [11] and to cause activation of transcription factor NF-κB [6] has suggested that LNGFRp75 is functional receptor to produce intracellular signals despite the lack of kinase domain. We showed here that neurotrophins (NGF, BDNF, NT-3) have the ability to activate NF-κB in microglia. If activation of NF-κB is mediated by LNGFRp75, our results seem to be reasonable because all neurotrophins show similar effects on microglia.

In any event, these results clearly indicate that cultured microglia essentially have the ability to express neurotrophin receptors, including the Trks family and LNGFRp75, and that they may be able to respond to neurotrophins added exogenously. Thus, microglia expressing Trks and LNGFRp75 may be functionally regulated by neurotrophins produced by glial cells or other cells at lesion sites or in the developmental stage of the brain. It will be of particular interest in future studies to examine whether microglia alter cell function through neurotrophin-Trk/LNGFRp75 signal transduction in vivo.

References

1. Lindsay R, Wiegard S, Alter C, Distefano PS (1994) Neurotrophic factors: from molecule to man. Trends Neurosci 17:182–189
2. Maisonpierre PC, Belluscio L, Friedman B, Alderson RF, Wiegard ST, Furth ME, Lindsay RM, Yancopoluos GD (1990) NT-3, BDNF, and NGF in the developing rat nervous system: parallel as well as reciprocal patterns of expression. Neuron 5:501–509
3. Furukawa S, Furukawa Y, Satoyoshi E, Hayashi K (1986) Synthesis and secretion of nerve growth factor by mouse astroglial cells in culture. Biochem Biophys Res Commun 136:57–63
4. Nakajima K, Nagata K, Hamanoue M, Takemoto N, Kohsaka S (1993) Microglia-derived elastase produces a low-molecular-weight plasminogen that enhances neurite outgrowth in rat neocortical explant cultures. J Neurochem 61:2155–2163
5. Nakajima K, Nagata K, Hamanoue M, Takemoto N, Shimada A, Kohsaka S (1993) Plasminogen-binding protein associated with the plasma membrane of cultured embryonic rat neocortical neurons. FEBS Lett 333:223–228
6. Carter BD, Kaltschmidt C, Kaltschmidt B, Offenhauser N, Bohm-Matthaei R, Baeuerle PA, Barde Y-A (1996) Selective activation of NF-κB by nerve growth factor through the neurotrophin receptor p75. Science 272:542–545

7. Elkabes S, DiCicco-Bloom EM, Black IB (1996) Brain microglia/macrophages express neurotrophins that selectively regulate microglial proliferation and function. J Neurosci 16:2508–2521
8. Nakajima K, Tsuzaki N, Nagata K, Takemoto N, Kohsaka S (1992) Production and secretion of plasminogen in cultured rat brain microglia. FEBS Lett 308:179–182
9. Nagata K, Nakajima K, Kohsaka S (1993) Plasminogen promotes the development of rat mesencephalic dopaminergic neurons in vitro. Dev Brain Res 75:31–37
10. Nakajima K, Kohsaka S (1993) Functional roles of microglia in the brain. Neurosci Res 17:187–203
11. Dobrowsky RT, Werner MH, Castellino AM, Chao MV, Hannun YA (1994) Activation of sphingomyelin cycle through the low affinity neurotrophin receptor. Science 265:1596–1599

Neurotrophins and CNTF: Endogenous Mechanisms of Trafficking and Therapeutic Potential

Ronald M. Lindsay

Polypeptide neurotrophic factors such as the neurotrophin family (NGE, BDNF, NT-3 and NT-4) and the neuroactive cytokine CNTF were first identified as potent survival factors for developing nerve cells in cell culture assays, using primary cultures of various classes of peripheral and central nervous system neurons. It is now clear in some cases that such pharmacological effects of these factors identified in cell culture, accurately predict the precise physiological role of these molecules in vivo. This has been best confirmed through generation of null mutant mice lacking one or other of the neurotrophins; such mice display selective loss of FNS and CNS neurons that are consistent with the specificity of NGF, BDNF and NT-3 determined in vitro. Furthermore, it is now well established that neurotrophic factors are not only required during development, but also play critical roles in the differentiation and maintenance of the phenotype of mature neurons. We have been interested both in basic aspects of the biology of neurotrophic factors and the therapeutic potential of such factors in neurodegenerative diseases.

As first postulated, the neurotrophic hypothesis embodies the idea that neurotrophic factors are made in target tissues where specific classes of innervating neurons compete for precisely limited supplies of the particular factor that will promote their survival during development and their maintenance of function in the adult. Receptor mediated retrograde axonal transport has been established as the most plausible mechanism whereby a neurophic factor in a distant peripheral tissue or central neuronal target can influence the survival and maintenance of afferent neurons. While there is abundant evidence that NGF and NT-3 may act endogenously by retrograde transport in the PNS, I will present evidence that suggests that the predominant role of BDNF is not as a retrogradely signalling effector moledule, but is rather as an afferent supplied effector molecular delivered by anterograde transport.

The potent action of neurotrophins and CNTF as survival factors for developing neurons in vitro, has suggested that pharmacological doses of such factors may have neuroprotective and neuroregenerative effects in vivo. I will summarise examples of in vivo studies in animal models that demonstrate the therapeutic potential of the neurotrophins in peripheral neuropathies (sensory neurons and motor neurons) and of CNTF in Huntington's disease, a degenerative disease that involves loss of GABAergic neurons of the striatum.

Regeneron Pharmaceuticals, Tarrytown, NY10591 USA

Genes Involved in Programmed Cell Death of Cerebellar Granule Neurons

K. Suzuki, K. Sakai, and T. Koike

Summary. Cerebellar granule cells isolated from 7-day-old rats and cultured in normal medium underwent apoptotic cell death, but remained healthy under depolarizing conditions with high K^+ (25 mM) or in the presence of brain-derived neurotrophic factor (BDNF). We have searched for genes that are upregulated during cell death by differential hybridization. One of the five genes, *gcd-10*, which consists of two transcripts, 2.3 kbp and 1.6 kbp was induced during cell death. Cyclin D1 and c-*jun*, both essential mediators of apoptosis, were found to be upregulated in our culture system. The cyclin D1 gene remained un-stimulated during the cell death of granule neurons following combined serum deprivation and low K^+ shift, suggesting that our culture system is suitable for further analysis of molecular events of cell death in cerebellar granule neurons.

Key words. Cerebellum—Granule cells—Apoptosis—Gene regulation—Death cascade

Introduction

During development of the cerebellum, granule cell precursors continue to divide and form a layer of the external granular layer (EGL). The young granule cells with their biochemical maturation then leave the EGL, extend their parallel fiber axon, and migrate centripetally down into the internal granular layer (IGL) where they make synaptic contacts to the developing dendrites of the Purkinje cells [1]. Target-derived neurotrophic factors play a decisive role in regulating the survival of these innervating neurons [2]. There is circumstantial evidence that brain-derived neurotrophic factor (BDNF) and insulin-like growth factor-I (IGF-I) are involved in this function. Cerebellar granule cells in the EGL undergo apoptosis at 5–9 days postnatal [3], although much less is known about cell death at the IGL, where the astrocyte is the major cell type undergoing cell death during this period [4].

Graduate Program in Biological Sciences, Faculty of Science, Hokkaido University, N10 W8, Kita-ku, Sapporo 060-0810, Japan

Programmed cell death (PCD) in the nervous system is subject to elaborated gene regulation to execute the active death program, which consists of the four phases of activation, propagation, commitment, and execution of cell death [2]. To elucidate the molecular mechanism underlying PCD of cerebellar granule cells, we have examined the pattern of gene expression in PCD of granule cells in vitro [5–7] and compared it to the pattern of gene expression in other systems.

Materials and Methods

Equal amounts of total RNA (20 μg) were loaded per lane onto formaldehyde-denaturing gel and separated by electrophoresis. The RNAs were transferred to a Hybond-N+ membrane (Amersham, Tokyo, Japan). Hybridization was performed as previously described [6]. DNA probes were labeled with [^{32}P]dCTP to a specific activity of 1–1.5 × 106 dpm/ng by random priming (Amersham Pharmacia Biotech, Tokyo, Japan). The rat cyclin D1 (446 bp) and G3PDH (983 bp) probes were prepared from PCR products subcloned into each plasmid vector as described [6].

Primers used for PCR reaction of cyclin D1 were as follows: 5′ (5′-CTGCATGTTCGTGGCCTCTAAG-3′) and 3′ (5′-GGGCTTCAATCTGTTCCTGGCA-3′), corresponding to nucleotides 440–461 and 864–885, respectively, of the rat cyclin D1 cDNA clone. For G3PDH, the primers used for PCR reaction were as described previously [6]. The cDNA fragment of rat c-*jun* probe was kindly gifted by Dr. Nakabeppu (Kyushu University). The Bio-imaging Analyzer (BAS2000; Fuji, Tokyo, Japan) was used for northern blot analysis.

Results and Discussion

Cerebellar granule cells, isolated from 7-day-old rats and cultured in normal medium, began to degenerate and decrease their numbers progressively within 4–5 days in vitro (DIV); their cell bodies became irregular and shrunken in shape, and the neurite network was disintegrated. Finally, more than 90% of the neurons died about 7 DIV. In contrast, they remained healthy under depolarizing conditions with high K^+ (25 mM) or in the presence of BDNF [5,6]. These neurons die actively, requiring protein and RNA synthesis, with morphological features of apoptosis.

We have searched for genes that are specifically induced during cell death using differential hybridization, with a subtracted cDNA probe used to screen a cDNA library from 5 DIV mRNA where granule cells began to die. One of the five genes, *gcd-10*, consists of two transcripts, 2.3 and 1.6 kbp, and was induced during cell death (Fig. 1). Database searches revealed that *gcd-10* is homologous to *Mus musculus* retinoic acid-inducible E3 protein mRNA. However, the full-length cDNA of *gcd-10* remains to be isolated and characterized. This gene was also induced during apoptosis of the granule neurons, which were maintained in 30 mM K^+ following their replacement to a 5 mM K^+ medium (a low K^+ shift) [8].

Northern blot analysis was also made on several genes known to be upregulated during programmed cell death. We found that the cyclin D1 gene was strikingly upregulated, while only a modest induction of c-*jun* occurred, which was persistent during cell death (Fig. 1). Under depolarizing conditions, the expression of cyclin D1

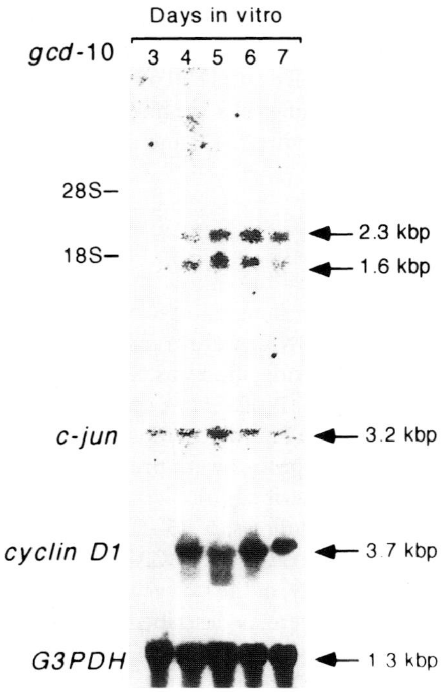

FIG. 1. Alterations of the mRNA level of *gcd-10*, c-*jun*, *cyclin D1*, and *G3PDH* in programmed cell death (PCD) of cerebellar granule cells in vitro. Hybridization procedures and probes used here are described in Materials and Methods. Equal amounts of total RNA per lane were loaded, which was confirmed by detecting 18S and 28S rRNA bands in methylene blue-staining membranes. The putative *gcd-10* transcripts were detected as two bands, 2.3 and 1.6 kbp. The experiment was repeated twice with similar results

was inhibited significantly (data not shown). In contrast, we have failed to detect the expression of c-*fos*. The expression of p75, the low-affinity neurotrophin receptor, was evaluated by RT-PCR, and was found to be downregulated at 3–5 DIV (data not shown). Furthermore, trkB was also upregulated during PCD of granule cells, as described in our previous report [6]. Immunohistochemical analysis revealed that enhanced expression of the cyclin D1 gene was observed in the cerebellar granule neurons themselves as well as in the microglia (data not shown). This finding is consistent with the fact that the microglial genes are also upregulated in response to neuronal cell death under our culture conditions [7].

The activation and propagation phases in PCD consist of convergent pathways to the commitment point. Genes whose expression was induced before the commitment point of cycloheximide were identified in PCD of cultured sympathetic neurons following nerve growth factor (NGF) deprivation [9]. One of these, c-*jun*, was identified as a key player in the cell death cascade; microinjection of neutralizing antibodies to c-Jun or a dominant negative c-jun construct suppressed PCD of sympathetic neurons [9,10]. The cyclin D1 gene has been shown to be an essential mediator of apoptosis in sympathetic neurons following NGF deprivation [11], and overexpression of cyclin D1 induces the apoptosis of neuroblastoma cell lines [12]. These findings suggest that there are common mediators such as c-*jun* and cyclin D1 in PCD in the central nervous system as well. Interestingly, *gcd-10* was undetectable during PCD of cultured sympathetic neurons deprived of NGF (data not shown).

In conclusion, *gcd-10*, cyclin D1, and c-*jun* were upregulated during PCD of cerebellar granule cells in vitro. The cyclin D1 gene remained un-stimulated during the cell death of granule neurons following combined serum deprivation and low K^+ shift [13], suggesting that our culture system is suitable for further analysis of molecular events of cell death in cerebellar granule neurons.

References

1. Burgoyne RD, Cambray-Deakin MA (1988) The cellular neurobiology of neuronal development: the cerebellar granule cell. Brain Res 472:77–101
2. Johnson EM Jr, Greenlund LJ, Akins PT, Hsu CY (1995) Neuronal apoptosis: current understanding of molecular mechanisms and potential role in ischemic brain injury. J Neurotrauma 12:843–852
3. Wood KA, Dipasquale B, Youle RJ (1993) In situ labeling of granule cells for apoptosis-associated DNA fragmentation reveals different mechanisms of cell loss in developing cerebellum. Neuron 11:621–632
4. Krueger BK, Burne JF, Raff MC (1995) Evidence for large-scale astrocyte death in the developing cerebellum. J Neurosci 15:3366–3374
5. Koike T (1991) Neuronal survival of cerebellar granule cells in vitro is regulated by levels of intracellular calcium. Soc Neurosci Abstr 17:1499
6. Suzuki K, Koike T (1997) Brain-derived neurotrophic factor suppresses programmed death of cerebellar granule cells through a post-translational mechanism. Mol Chem Neuropathol 29:101–124
7. Tanaka S, Suzuki K, Watanabe M, et al (1997) Up-regulation of a new microglial gene (mrf-1) in response to programmed neuronal cell death. Soc Neurosci Abstr 23:1988
8. D'Mello SR, Galli C, Ciotti T, et al (1993) Induction of apoptosis in cerebellar granule neurons by low potassium: inhibition of death by insulin-like growth factor I and cAMP. Proc Natl Acad Sci USA 90:10989–10993
9. Estus S, Zaks WJ, Freeman RS, et al (1994) Altered gene expression in neurons during programmed cell death: identification of c-jun as necessary for neuronal apoptosis. J Cell Biol 127:1717–1727
10. Ham J, Babij C, Whitfield J, et al (1995) A c-Jun dominant negative mutant protects sympathetic neurons against programmed cell death. Neuron 14:927–939
11. Freeman RS, Estus S, Johnson EM Jr (1994) Analysis of cell cycle-related gene expression in postmitotic neurons: selective induction of cyclin D1 during programmed cell death. Neuron 12:343–355
12. Kranenburg O, van der Eb AJ, Zantema A (1996) Cyclin D1 is an essential mediator of apoptotic neuronal death. EMBO J 15:46–54
13. Miller TM, Johnson EM Jr (1996) Metabolic and genetic analyses of apoptosis in potassium/serum-deprived rat cerebellar granule cells. J Neurosci 16:7487–7495

Retrograde Transport of Endogenous NT-3 in Rat Sciatic Nerve

Atsumi Nitta, Takayoshi Jin-nouchi, Toshio Asami, Kaori Toyoda, Mayuko Hino, and Shoei Furukawa

Summary. To address active transport of neurotrophin, nerve growth factor (NGF), brain-derived neurotrophic factor (BDNF), and neurotrophin-3 (NT-3) in the peripheral nerves, we examined their levels of protein and mRNA in the sciatic nerve of adult rats following transection, using enzyme immunoassays and the reverse transcription polymerase chain reaction method, respectively. NT-3 protein increased 1 day after transection only in the distal segment next to the transection site and returned to the original level 2 days later. This was considered to reflect accumulation of NT-3 transported from the periphery toward the neuronal cell bodies, because the NT-3 mRNA level was not changed in any sciatic segments during the experimental period. In contrast, an increase in BDNF protein was observed simultaneously in both distal and proximal stumps 3 days after transection. BDNF mRNA was elevated in the same stumps 1 day before, suggesting that BDNF was produced within the transected stumps. These observations demonstrate that NT-3, like NGF, is retrogradely transported in the sciatic nerve but that BDNF is not, which suggests that NT-3 plays particular roles in the conveyance of trophic signals from target organs to neurons.

Key words. Neurotrophin—Brain-derived neurotrophic factor (BDNF)—Neurotrophin-3 (NT-3)—Enzyme immunoassay (EIA)—Reverse transcription polymerase chain reaction (RT-PCR)

Nerve growth factor (NGF), brain-derived neurotrophic factor (BDNF), and neurotrophin-3 (NT-3) are members of the neurotrophin family proteins that regulate the development and maintenance of some specific functions in different populations of peripheral and central neurons [1–4]. The survival of some neurons depends on the retrograde transport of trophic molecules, because such neurons die when this transport is interrupted [5]. NGF is a well-known neurotrophic factor that is trans-

Laboratory of Molecular Biology, Gifu Pharmaceutical University, 5-6-1 Mitahora-higashi, Gifu 502-8585, Japan ported retrogradely from the peripheral end organs to the neuronal cell bodies where its characteristic biochemical effects are produced [6,7]. Biologically active exogenous BDNF and NT-3 can be also retrogradely transported by a much broader category of neuronal populations in the peripheral and central nervous systems [8]. However, there are few reports showing axonal transport of endogenous BDNF or NT-3, probably because of a lack of reliable methods to quantify small amounts of these factors. Enzyme immunoassay (EIA) methods sensitive enough to detect endogenous BDNF or NT-3 [9,10] would be a potent tool to address this issue. In the present work, we developed specific EIA systems with a sensitivity as low as 5 pg/ml for BDNF and NT-3 and examined changes in these proteins in lesioned sciatic nerves.

All the experiments with animals were carried out according to the guidelines for animal experimentation of Gifu Pharmaceutical University. Male Wistar rats (Nippon SLC, Shizuoka, Japan) were used. The animals were transected at the indicated site (Fig. 1A) [6]. On days 1, 2, and 3 after surgery, the sciatic nerve was removed from each of six animals and cut into six segments (numbered from 1 to 6; Fig. 1A), each 3 mm in length. Each segment was added to 100 μl of homogenizing buffer [6], pulse-sonicated for 30s, and centrifuged at 10 000 × *g* for 30 min. The supernatant was then mixed vigorously with 100 μl of chloroform and centrifuged at 20 000 × *g* for 15 min, after which the aqueous phase was collected and used for the EIA measurement.

The BDNF or NT-3 mRNA level was estimated by use of a reverse transcription polymerase chain reaction (RT-PCR) method as described [11–13]. The density of the amplified products was analyzed by image analysis software (Mac Bass 3000; Fuji, Tokyo, Japan) run in a Macintosh system and was expressed as a ratio to the density of the β-actin cDNA product amplified from an identical RNA sample.

The detection limit of our EIA system for both BDNF and NT-3 was as low as 5 pg/ml. No cross-reactivities were observed among other neurotrophins tested. Levels of NGF, BDNF, and NT-3 in each segment of the normal rat sciatic nerve are shown in Fig. 1B, Fig. 1C, and Fig. 1D, respectively. The level of NGF was comparable to that found previously [6,7]. The recoveries of exogenously added BDNF (61.8 pg/ml) and NT-3 (104.9 pg/ml) following disruption of the sciatic segment were 71.6% and 75.7%, respectively. The values of NGF (recovery, 89.5%), BDNF, and NT-3 content thus obtained were expressed without correction, in view of the practical recovery of the exogenous neurotrophins, whose level was comparable with the endogenous one in the normal sciatic nerve (data not shown).

The time course of changes in NGF, BDNF, and NT-3 levels after the transection are shown in Fig. 1B–D. The NGF level in the distal part nearest to the transection site (segment no. 4) was dramatically increased at 1–2 days, and decreased at 3 days, after the transection. This observation agreed well with previous data [6,7]. No significant change in BDNF content could be found within 2 days after the transection (Fig. 1C); however, a significant increase was observed simultaneously both in distal and proximal sites 3 days after the transection (Fig. 1C). As shown in Fig. 1C, the NT-3 level was markedly increased 1 day after transection, but only in segment no. 4, and returned to the original level 1 or 2 days later. No significant changes were found in the other segments during the experimental period of 3 days.

The time course of the change in NGF mRNA expression is presented in Fig. 2A. In all distal segments (segment nos. 4–6), the NGF mRNA level was increased 2 and 3

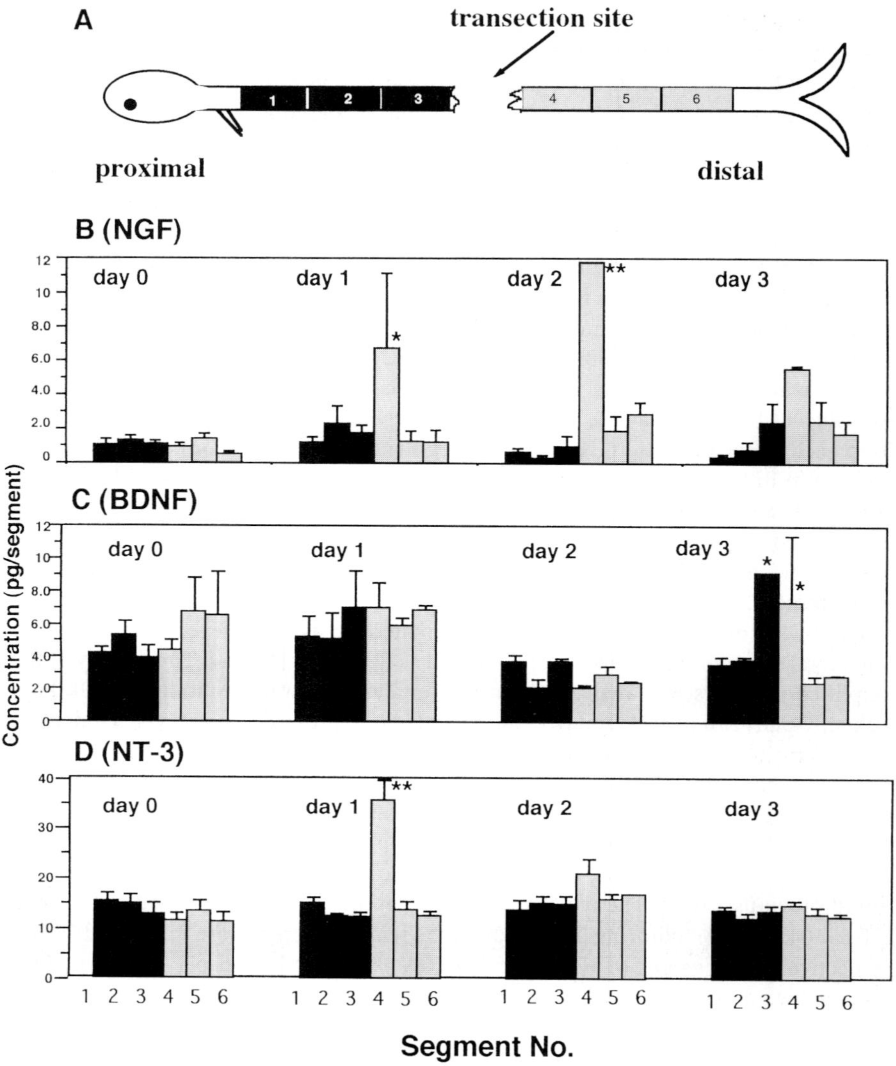

FIG. 1. **A** Schematic representation of transection site and sites used for determination of neurotrophin (NT) protein and mRNA in the sciatic nerve. The site of transection was between segments 3 and 4. Segments were numbered 1 to 6 from the central to the peripheral portion. **B–D** Time course of level of nerve growth factor (NGF), brain-derived neurotrophic factor (BDNF), or NT-3 protein in the sciatic nerve after its transection between segments 3 and 4. The sciatic nerves were transected at day 0, collected 1, 2, and 3 days later, and cut into 3-mm-long segments. Contents of NGF (**B**), BDNF (**C**), and NT-3 (**D**) were measured by the respective EIA systems. The values given represent the mean ± SE of four rats. *, $P < 0.05$; **, $P < 0.01$ vs. control level in each corresponding segment (Tukey's test)

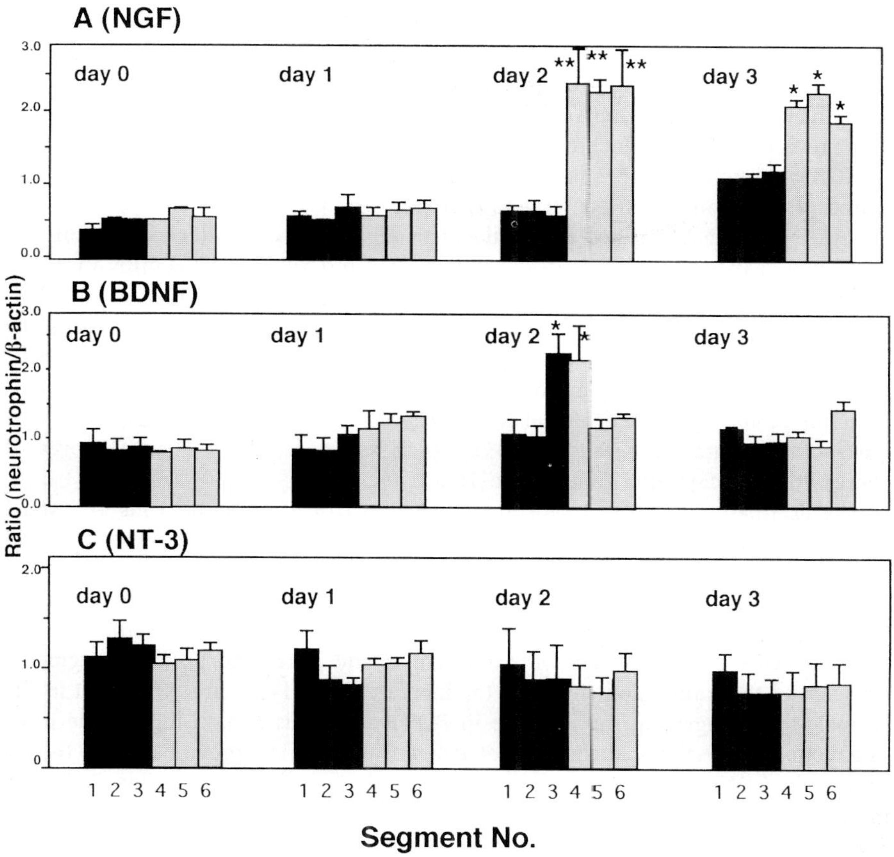

FIG. 2A–C. Time course of level of NGF (**A**), BDNF (**B**), and NT-3 (**C**) mRNA in the sciatic nerve after its transection. Experimental procedures were the same as described in the legend of Fig. 1. Level of mRNA in each segment was estimated by the RT-PCR method using corresponding primers. The values given represent the mean ± SE of four rats. *, $P < 0.05$ vs control level in each corresponding segment (Tukey's test)

days after transection, without any increment at 1 day. The BDNF mRNA level was increased in both proximal and distal stumps next to the transection site (segment no. 3 and 4) only 2 days after transection (Fig. 2B). In contrast, the NT-3 mRNA level remained completely unchanged in all segments during the experimental period (Fig. 2C).

In this study we obtained direct evidence that endogenous NT-3 is retrogradely transported in the rat sciatic nerve after transection, because the NT-3 protein level was dramatically increased in the distal stump adjacent to the transection site at the time immediately after transection of the sciatic nerve without an alteration of the mRNA level (Figs. 1D and 2C). The pattern of changes in the NT-3 protein content is similar to that of NGF shown in the current and previous studies [6,7]. The rapid increment of NGF protein in the distal segment is considered to reflect the accumulation of NGF transported by a still-functioning intact axonal transport mechanism from the periphery, in light of on unchanged mRNA level at the time and site of protein elevation. The pattern of change in NGF mRNA is also consistent with results of a previous northern blot study [15].

NGF is thought to be produced by nonneuronal cells, such as Schwann cells and fibroblast cells, on day 2 after transection [14]. The present results supported this, because NGFmRNA was increased in the distal segments 2 days after transection. The pattern of change in mRNA differed between NGF and NT-3, as NT-3 mRNA was not changed even within 3 days after transection. Our observations demonstrate that NT-3 production is kept constant following transection in nonneuronal cells within the sciatic nerves. Therefore, we can conclude that the transient elevation of NT-3 in the distal stump was caused by an interruption and accumulation of NT-3 transported from the periphery to the neuronal cell bodies. NT-3 is thought to be produced in the gastrocnemius muscle, because NT-3 mRNA is reported to be constantly expressed there within 14 days after transection [16].

The destinations of NT-3 transport probably involve motor neurons and some neurons in the dorsal root ganglion (DRG), because these cells express trkC and p75 mRNA [16,17] and their axons are included in the sciatic nerves. In contrast to NT-3, it is not clear, from our results, whether BDNF is transported, irrespective of direction, in the rat sciatic nerve. Increases in BDNF protein and its mRNA could be observed in the same segments adjacent to the distal and proximal segments adjacent to the transection site, with the increase in the latter preceding that in the former, thus suggesting the increase in BDNF mRNA is responsible for de novo production within the sciatic nerve, not axonal transport. In a previous study [16], no change in BDNF mRNA was observed within 3 days after sciatic transection. A marked elevation of mRNA was found only after 7 days. The transient increase in BDNF mRNA detected in the present study may be attributed to the use of the sensitive RT-PCR method for measurement. Physiologically active ^{125}I-labeled BDNF was transported retrogradely in the rat sciatic nerve following injection into the sciatic nerve [8].

The transport system of neurotrophins was recently studied in the developing visual system [18]. This apparent discrepancy may be explained as follows. (1) Endogenous and exogenous BDNF may have different types of transport systems; (2) they are transported by an identical mechanism, but BDNF transport in the sciatic nerve is at too low a level to detect; and (3) different transport systems may function depending on neuronal types and/or degree of neuronal differentiation.

In this study, NT-3 was shown to be retrogradely transported in the sciatic nerve, like NGF, suggesting that NT-3 plays some role in the conveyance of trophic signals from organs to neurons such as motoneurons and proprioceptive sensory neurons in adult rats. BDNF, however, does not appear to have such a role.

Acknowledgments. The authors are grateful to Sumitomo Pharmaceutical Ltd. and Amgen for the supply of recombinant human BDNF and NT-3, respectively. This work was supported, in part, by grants in aid (09280231 and 09780727) for science research from the Ministry of Education, Science, Sports and Culture of Japan.

References

1. Barde YE (1989) Trophic factors and neuronal survival. Neuron 2:1525–1534
2. Kang H, Schuman EM (1995) Long-lasting neurotrophin-induced enhancement of synaptic transmission in the adult hippocampus. Science 267:1658–1662
3. Lohof AM, I NY, Poo MM (1993) Potentiation of developing neuromuscle synapses by the neurotrophins NT-3 and BDNF. Nature (Lond) 363:350–353
4. Thoenen H (1991) The changing scene of neurotrophic factors. Trends Neurosci 14:165–170
5. Linden R (1994) The survival of developing neurons: a review of affect control. Neuroscience 58:671–682
6. Ishikawa R, Nishokori K, Furukawaet S, et al (1992) Injury-induced reduction of acidic fibroblast growth factor levels in the distal parts of rat sciatic nerve. Neurosci Lett 135:113–116
7. Korsching S, Thoenen H (1983) Quantitive demonstration of the retrograde axonal transport of endogenous nerve growth factor. Neurosci Lett 39:1–4
8. DiStefano PS, Friedman B, Radziejewski C, et al (1992) The neurotrophins BDNF, NT-3, and NGF display distinct patterns of retrograde axonal transport in peripheral and central neurons. Neuron 8:983–993
9. Katoh-Semba R, Kaisyo Y, Shintani A, et al (1996) Tissue distribution and immunocytochemical localization of neurotrophin-3 in brain and peripheral tissue of rats. J Neurochem 66:330–337
10. Nawa H, Carnahan J, Gall C (1995) BDNF protein measured by a novel enzyme immunoassay in normal brain and after seizure: partial disagreement with mRNA levels. Eur J Neurosci 7:1527–1535
11. Miwa T, Furukawa S, Nakajima K, et al (1997) Lipopolysaccharide enhances synthesis of brain-derived neurotrophic factor in cultured rat microglia. J Neurosci Res 50:1023–1029
12. Nitta A, Murakami Y, Furukawa Y, et al (1994) Oral administration of idebenone induces nerve growth factor in the brain and improves learning and memory in basal forebrain-lesioned rats. Naunyn-Schmiedeberg's Arch Pharmacol 349:401–407
13. Awatsuji H, Furukawa Y, Hirota M, et al (1992) Interleukin-4 and -5 as modulators of nerve growth factor synthesis/secretion in astrocytes. J Neurosci Res 34:539–545
14. Heumann R, Korching S, Bandtlow C, et al (1987) Changes of nerve growth factor synthesis in nonneuronal cells in response to sciatic nerve transection. J Cell Biol 104:1623–1631
15. Meyer M, Matsuoka I, Wetmotr C, et al (1992) Enhanced synthesis of brain-derived neurotrophic factor in the lesioned peripheral nerve; different mechanisms are responsible for the regulation of BDNF and NGF mRNA. J Cell Biol 119:45–54
16. Funakoshi H, Frisen J, Barbany G, et al (1993) Differential expression of mRNA for neurotrophins and their receptors after the sciatic nerve transection. J Cell Biol 123:455–465
17. McMahon SB, Armanini MP, Ling LH, et al (1994) Expression and coexpression of Trk receptors in subpopulations of adult primary neurons projecting to identified peripheral targets. Neuron 12:1161–1171
18. Bartheld CS, Byers MR, Williams R, et al (1996) Anterograde transport of neurotrophins and axodendritic transfer in the developing visual system. Nature (Lond) 379:830–833
19. Furukawa S, Kamo I, Furukawa Y, et al (1983) A highly sensitive enzyme immunoassay for mouse b nerve growth factor. J Neurochem 40:734–744
20. Kaechi K, Ikegami R, Nakamura N, et al (1995) 4-Methylcathechol, an inducer of nerve growth factor synthesis, enhances perioheral nervous generation across nerve gaps. J Pharmacol Exp Ther 272:1300–1304

Expression of *c-fms* Proto-oncogene/ M-CSF Receptor in Postnatal Purkinje Cells: Possible Function of M-CSF in the Central Nervous System

Shin-ichi Murase[1], Kazuo Motoyoshi[2], and Yokichi Hayashi[3]

Summary. The *c-fms* proto-oncogene is known to encode the receptor for macrophage colony-stimulating factor (M-CSF, also called CSF-1), and is expressed mainly in the mononuclear phagocyte lineage. In the central nervous system, nonneuronal cells such as microglia, astrocytes, and oligodendrocytes express *c-fms* especially when cultured, while the localization of *c-fms*-positive cells in vivo has not been extensively examined. Recently, M-CSF production has been confirmed in cultured cerebellar neurons, and *c-fms* expression has been confirmed also in Purkinje cells of both mouse and rat cerebella. M-CSF not only directly promotes neuronal survival after ischemic damage but also supports in vitro survival of hippocampal neurons in an indirect manner. These findings, taken together, imply that *c-fms* expression can be regarded as a developmental marker for Purkinje cells, and that M-CSF possibly acts as a neurotrophic factor for neurons directly and indirectly.

Key words. Integrin—M-CSF—Purkinje cell—Neurotrophin—*c-fms*

Introduction

Neurons require neurotrophic factors (neurotrophins) for their survival, maturation, and differentiation during development. Several lymphokines are known as neurotrophic factors that mediate intercellular signaling between neuronal cells. Interleukins 1, 3, and 6 act on some type of neurons. Nerve growth factor (NGF) is also known as a lymphokine that mediates intercellular signaling between immunocompetent cells, and its cellular receptor trk has have been studied in terms of the immunological and neuronal functions [1–5]. Therefore, the nervous system appears to share some humoral trophic factors with the immune system.

---

[1] Department of Anatomy, School of Medicine, Keio University, 35 Shinanomachi, Shinjuku-ku, Tokyo 160-8582, Japan
[2] Third Department of Internal Medicine, National Defense Medical College, 3-2 Namiki, Tokorozawa, Saitama 359-0042, Japan
[3] Department of Cell and Molecular Biology, Nagano College of Nursing, Akoho 1694, Komagane, Nagano 399-4117, Japan

Several colony-stimulating factors (CSF) promote proliferation of either macrophages or granulocytes. Macrophage colony-stimulating factor (M-CSF) has been defined on the basis of its biological activities in stimulating specific subsets of leukocytes, because it is essential for the development and the proliferation of macrophages/monocytes and microglia.

In osteopetrotic mutant (op/op) mice that are devoid of active M-CSF because of a frameshift mutation [6,7], decreased numbers of macrophage lineage cells are observed. The op/op mice are characterized by lack of teeth, decreased body weight, and skeletal deformities caused by the lack of osteoclasts, whereas no reports have been published on defects in formation of other tissues including the central nervous system.

Interest in investigating the role of neuronal functions of M-CSF and *c-fms* using op/op mice arose later, however, and appreciation of the importance of this system has emerged only recently.

Molecular Properties of M-CSF and *c-fms*

M-CSF was found as an urinary glycoprotein capable of stimulating macrophage colony formation [8], and its gene structure was revealed in 1987 [9]. Two types of M-CSF, a soluble form and a proteoglycan form (PG-M-CSF), have been found [10–12]. Soluble M-CSF is a homodimer composed of 43-kDa subunits. PG-M-CSF is further classified into homodimer and heterodimer types [10,13]. The subunit of the homodimer type is a polypeptide consisting of the 43-kDa subunit with 180 residues added to the C terminus, and this long polypeptide is further linked to a glycosaminoglycan. Thus, the final molecular weight of the subunit is 150–200 kDa. The heterodimer type of PG-M-CSF is composed of one 43-kDa subunit and one 150 to 200-kDa subunit (Fig. 1). The PG-M-CSF binds to collagen type V, and is not detected in the blood [11,14].

In contrast, *c-fms*, the receptor for M-CSF encoded by the *c-fms* proto-oncogene, is a membranous glycoprotein [15], and is mainly expressed in the mononuclear phagocyte lineage, trophoblasts and decidual cells, and B lymphocytes. The extracellular domain of the *c-fms* protein has five immunoglobulin domains and thus it belongs to the immunoglobulin superfamily. The intracellular domain of the *c-fms* protein has four tyrosine phosphorylation sites (Tyr697, Tyr706, Tyr721, Tyr807) and a kinase insert. Phosphorylated *c-fms* binds phospatidylinositol 3-kinase when Tyr721 is phosphorylated, or binds Grb-2 (growth factor receptor-bound protein 2) when Tyr697 is phosphorylated [16,17]. Thus, the signals mediated through *c-fms* might be linked to Ras or MAP kinase cascades.

Expression of M-CSF and *c-fms* in the Central Nervous System

In the central nervous system, M-CSF is mainly produced in astrocytes and microglia, and is regarded as the most effective growth factor for microglial proliferation [18–22]. Astrocytes and oligodendroglia express *c-fms* [23]; however, the functional roles

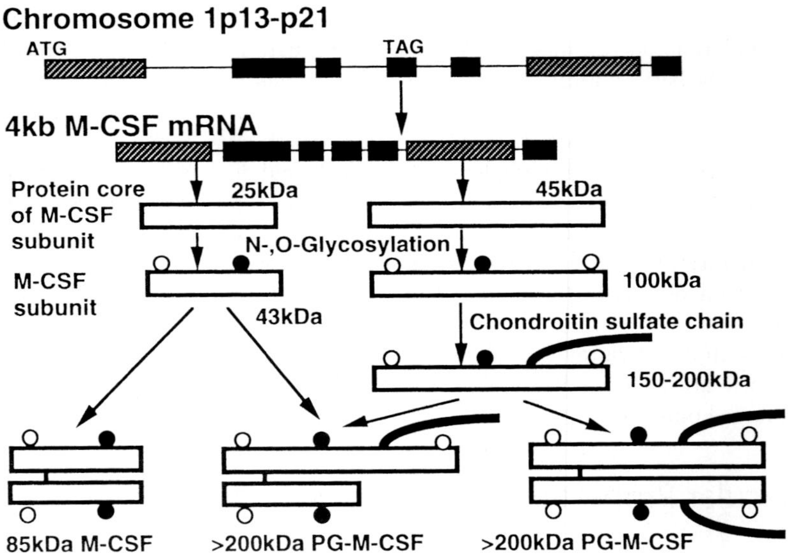

FIG. 1. The biosynthetic process of macrophage colony-stimulating factor (M-CSF) in a given cell. The human M-CSF gene is located on chromosome 1q13-p21, which allows a 4-kb mRNA encoding M-CSF to be transcribed. Two distinct subunits of M-CSF are resultantly synthesized to form three types of M-CSF dimeric molecules

of M-CSF in relation to these glia remain to be elucidated. One simple explanation is that M-CSF may play roles in the cytokine network among these cells scattered throughout the nervous system.

Very recently, cerebellar neurons also have been shown to produce M-CSF in vitro [24]. This finding strongly indicates that cellular communication through the M-CSF take place not only between glial cells, but also between glial and neuronal cells.

Rescue of Neurons by M-CSF After Ischemic Damage

Berezovskaya and co-workers focused the point to examine neuronal dependence on M-CSF in the process of recovery from ischemic damage in op/op mice [25–27]. They found that exogenous M-CSF enhanced neuronal survival as well as enhancing microglial and astroglial activities in vivo. Therefore, M-CSF functions as a neurotrophic factor for damaged neurons. Although M-CSF may function in a similar manner in cases of neuronal damage resulting from events such as inflammation, infectious disease, aging, and neuronal degenerative diseases, there is no direct information available to support this idea.

Postnatal Expression of *c-fms* in Purkinje Cells

As cerebellar neurons produce M-CSF in vitro [24], we suspect that there is a type of cells expressing M-CSF receptor/*c-fms* in the cerebellum. As expected, Purkinje cells of mouse and rat were found to be immunopositive for *c-fms* (Murase and Hayashi, in

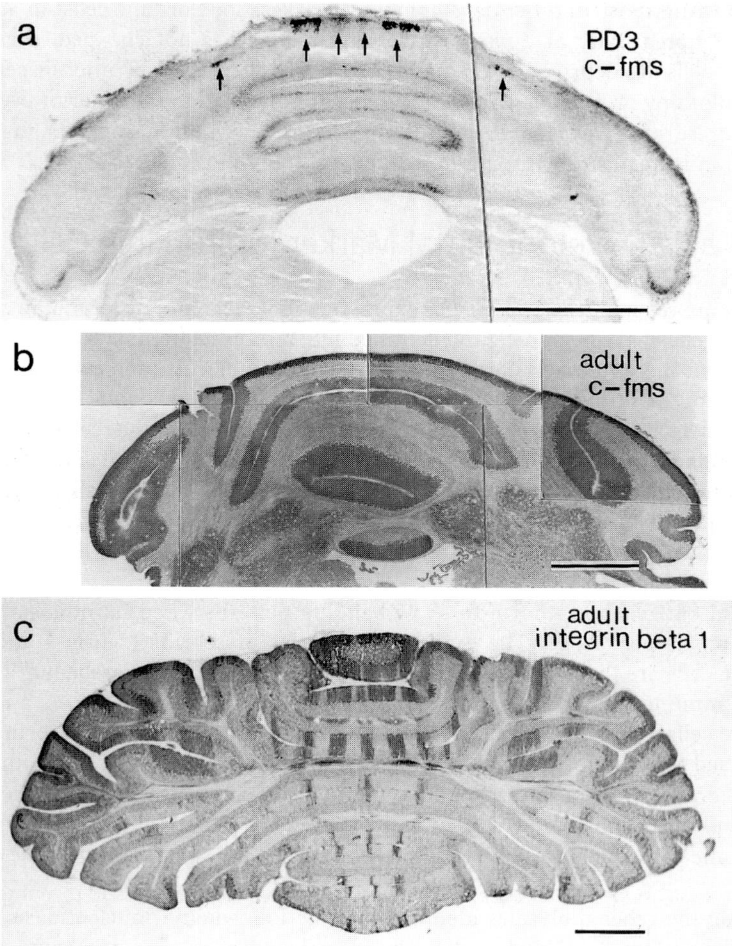

FIG. 2a–c. Distribution of *c-fms* immunoreactivity in murine cerebella. Horizontal sections from mouse cerebellum of postnatal day 3 (**a**), an adult mouse (**b**), and an adult rat (**c**) were prepared and processed for immunostaining using anti-*c-fms* (**a** and **b**) and anti-integrin β1 antibody (**c**). The *c-fms*-positive cells were arranged symmetrically (*arrows* in **a**). All Purkinje cells showed *c-fms* immunoreactivities in the adult cerebellum (**b**). The expression of integrin β1 was suppressed in some Purkinje cells, so that the positive Purkinje cells in the adult rat were organized into parasagittal bands interposed by negative cells throughout the vermis and hemispheres (**c**). Note the different parasagittal pattern of immunopositive Purkinje cells comparing the two antigens. *Bars*, 1 mm

manuscript). Interestingly, the distribution shows unique patterns in postnatal development. The *c-fms*-positive Purkinje cells first appear in the caudal cerebellum at postnatal day (PD) 0 of the mouse in a parasagittal manner, and most Purkinje cells gradually become positive by PD 6 (Fig. 2a). Subsequent to PD 7 and in adult mice, *c-fms* is expressed in most Purkinje cells (Fig. 2b). The parasagittal pattern observed until PD 5 is different from those of zebrins and integrin β1 subunit (Fig. 2c), and different from those of other developmental markers for Purkinje cells. In addition, the immunoreactivity of c-*fms*-positive Purkinje cells is not changed, even when Purkinje cells are deprived of their climbing fiber innervation by inferior cerebellar pedunculotomy or transplantation of cerebellar anlagen into the anterior eye chamber (Murase and Hayashi, in manuscript). This observation suggests that c-*fms* expression is intrinsically programmed in developing Purkinje cells.

c-*fms* as a Developmental Marker for Purkinje Cells

It is also interesting to examine c-*fms* expression in weaver and reeler mutants, where mutated genes have been recently identified [28,29]. In the homozygous weaver cerebellum, most granule cells die at the interface of the molecular and external granular layer without emitting an axon during the first and second postnatal weeks. In mice with this defect, only a few parallel fibers are evident [30,31]. In the reeler cerebellum, malpositioned Purkinje cells are present in the granular layer or white matter [32–34]. Despite the absence of the parallel fiber inputs to the Purkinje cells, the mutant cerebellum possesses c-*fms*-positive Purkinje cells (Murase and Hayashi, in manuscript). Thus, c-*fms* expression in Purkinje cells is independent of the climbing or parallel fiber inputs.

Several antigens are known to show banding patterns when immunostaining of cerebellum is performed. The expression patterns of zebrins (zebrin I and II) in Purkinje cells are thoroughly studied parasagittal stripes in the cerebellum [35–41]. Zebrin-immunoreactive Purkinje cells first appear caudally at PD 6, and most Purkinje cells gradually become positive by PD 12. The expression of zebrin is then suppressed in some of these cells, so that the positive Purkinje cells in the adult become organized into parasagittal bands interposed by negative cells throughout the vermis and hemispheres. Recently, the expression pattern of integrin β1 subunit in Purkinje cells has been observed and found to be the same as that of the zebrins [42].

Among the other molecules identified are the following: 5′-nucleotidase [43,44], acetylcholinesterase [45], glutamic acid decarboxylase [46], cytochrome oxidase [47,48], motilin [46,49], and synaptophysin [35,50]. However, none of these show a pattern of expression comparable to that of c-*fms*, even though these molecules were stained under the same conditions to compare their parasagittal pattern in the cerebellum. Therefore, the banding pattern of c-*fms* could be new and unique as a developmental marker for Purkinje cells.

Possible Function of c-*fms* Expression in the Purkinje Cells

Very recently, the spatial pattern of neuronal activity evoked by peripheral stimulation showed the parasagittal bands corresponding with the parasagittal compartments revealed by immunostaining with anti-zebrin II [51]. It is possible that the c-*fms* bands also reflect the functional parasagittal compartments; however, no experimental evidence has been obtained up to this point. There is another possi-

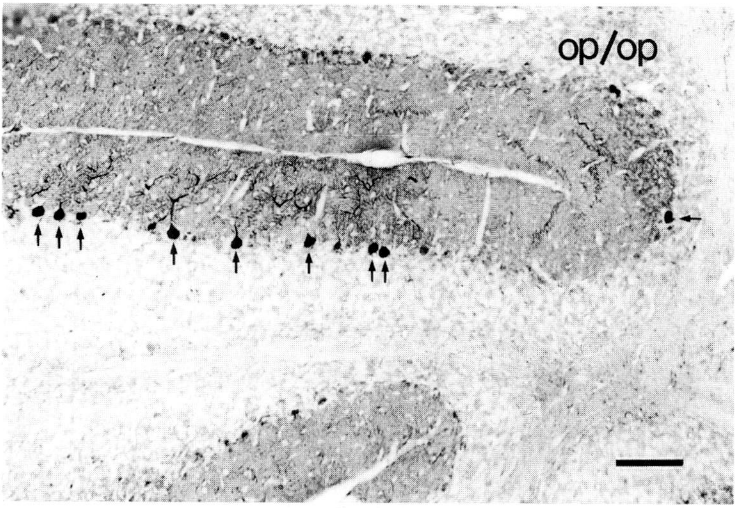

FIG. 3. Partial absence of Purkinje cells in osteopetrotic mutant (op/op) cerebellum. Horizontal sections from mouse cerebellum were prepared and processed for immunostaining as described in Fig. 1 using anti-calbindin antibody. Note that surviving Purkinje cells are visualized as immunopositive cells (*arrows*). *Bar*, 100 μm

bility that M-CSF signaling through c-*fms* would function to promote neuronal survival.

To check this question, we surveyed the cerebella of op/op mutant mice that are devoid of active M-CSF. We found that the number of calbindin-positive Purkinje cells in a given cerebellum was reduced substantially during the initial 4–5 weeks of the postnatal period, and the remaining Purkinje cells were both c-*fms*- and calbindin positive (Fig. 3). These data strongly suggest that the survival of Purkinje cells is dependent on M-CSF after weaning, but the underlying molecular mechanism remains unclear.

Effect of the M-CSF-c-*fms* System on Neuronal Survival

After ischemic damage has occurred, M-CSF functions as a neurotrophic factor, as damaged neurons and their neighbors become immunopositive for c-*fms* [25,26]. However, the majority of other neurons are immunonegative for c-*fms*. It is therefore interesting to test whether the M-CSF-c-*fms* system may exert some biological activity toward the neurons immunonegative for c-*fms*. To examine this, we utilized a rat hippocampal neuronal culture and incubated these neurons in the presence or absence of conditioned medium (CM) prepared by Purkinje cell culture (Hayashi et al., unpublished observation). When CM was prepared by incubating Purkinje cells with M-CSF, the hippocampal neurons survived considerably well compared with those incubated in CM without M-CSF (Fig. 4). However, treatment of these cells with M-CSF or its antibody did not reproduce a similar effect (data not shown). These data suggest that M-CSF signaling through c-*fms*, while it occurs in either Purkinje cells or

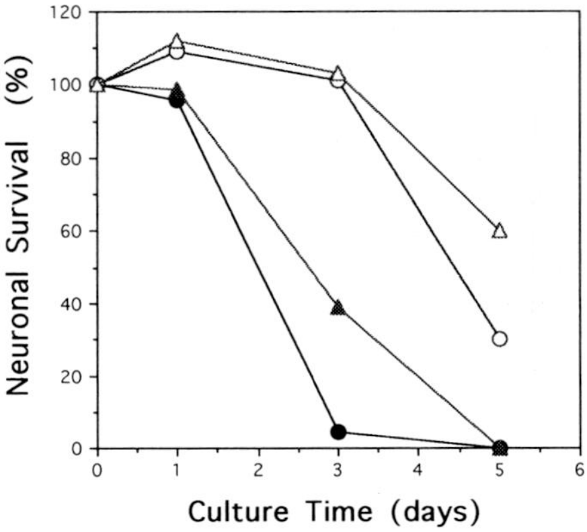

FIG. 4. Stimulatory effect of M-CSF-treated Purkinje cell culture on neuronal survival. Purkinje cell cultures were prepared and maintained in the presence or absence of M-CSF (1000 U/ml) for 5 days in DMEM/Ham F12 medium with N2 supplements as described previously [52]. The resultant culture media were recovered as conditioned media (CM). Hippocampal neurons were prepared from 18-day-old rat embryos and plated onto 24-multiwell plates at a concentration of 10^4 cells/ml for 3 days [53]. Various amounts of each CM were dispensed into the neuronal culture, and the survival rate of the neurons was determined up to 5 days as the percentage relative to the number of neurons initially plated. *Open* and *solid symbols* denote CM prepared by Purkinje cell culture with and without M-CSF, respectively. *Circles* denote cell cultures without addition of M-CSF; *triangles* denote further addition of 1000 U/ml of M-CSF after the CM was dispensed

other type of cells such as microglia or astroglia, leads to enhanced synthesis of other neurotrophins, which eventually results in elevating the survival of hippocampal neurons. Although this M-CSF-inducible neurotrophic activity has not been identified, it is possibly involved not only in neuronal development but also in the process of neuronal repair after damage induced by events such as inflammation, ischemia, aging, and neuronal degenerative disease.

References

1. Ehrhard PB, Erb P, Graumann U, et al (1993) Expression of nerve growth factor and nerve growth factor receptor tyrosine kinase Trk in activated CD4-positive T-cell clones. Proc Natl Acad Sci USA 90:10984–10988
2. Otten U, Scully JL, Ehrhard PB, et al (1994) Neurotrophins: signals between the nervous and immune systems. Prog Brain Res 103:293–305
3. Ehrhard PB, Erb P, Graumann U, et al (1994) Expression of functional trk tyrosine kinase receptors after T cell activation. J Immunol 152:2705–2709
4. Melamed I, Kelleher CA, Franklin RA, et al (1996) Nerve growth factor signal transduction in human B lymphocytes is mediated by gp140trk. Eur J Immunol 26:1985–1992
5. Torcia M, Bracci-Laudiero L, Lucibello M, et al (1996) Nerve growth factor is an autocrine survival factor for memory B lymphocytes. Cell 85:345–356
6. Wiktor-Jedrzejczak W, Bartocci A, Ferrante A Jr, et al (1990) Total absence of colony-stimulating factor 1 in the macrophage-deficient osteopetrotic (*op/op*) mouse. Proc Natl Acad Sci USA 87:4828–4832

7. Yoshida H, Hayashi S, Kunisada T, et al (1990) The murine mutation osteopetrosis is in the coding region of the macrophage colony stimulating factor gene. Nature (Lond) 345:442–444
8. Motoyoshi K, Takaku F, Mizoguchi H, et al (1978) Purification and some properties of colony-stimulating factor from normal human urine. Blood 52:1012–1020
9. Wong GG, Temple PA, Leary AC, et al (1987) Human CSF-1: Molecular cloning and expression of 4-kb cDNA encoding the human urinary protein. Science 235:1504–1508
10. Suzu S, Ohtsuki T, Yanai N, et al (1992) Identification of a high molecular weight macrophage colony-stimulating factor as a glycosaminoglycan-containing species. J Biol Chem 267:4345–4348
11. Suzu S, Ohtsuki T, Makishima M, et al (1992) Biological activity of a proteoglycan form of macrophage colony-stimulating factor and its binding to type V collagen. J Biol Chem 267:16812–16815
12. Price LK, Choi HU, Rosenberg L, et al (1992) The predominant form of secreted colony stimulating factor-1 is a proteoglycan. J Biol Chem 267:2190–2199
13. Ohtsuki T, Suzu S, Nagata N, et al (1992) A human osteoblastic cell line, MG-63, produces two molecular types of macrophage-colony-stimulating factor. Biochim Biophys Acta 1136:297–301
14. Suzu S, Yanai N, Sato-Somoto Y, et al (1991) Characterization of macrophage colony-stimulating factor in body fluids by immunoblot analysis. Blood 77:2160–2165
15. Hampe A, Shamoon BM, Gobet M, et al (1989) Nucleotide sequence and structural organization of the human FMS proto-oncogene. Oncogene Res 4:9–17
16. Reedijk M, Liu X, van der Geer P, et al (1992) Tyr721 regulates specific binding of the CSF-1 receptor kinase insert to PI 3′-kinase SH2 domains: a model for SH2-mediated receptor-target interactions. EMBO J 11:1365–1372
17. van der Geer P, Hunter T (1993) Mutation of Tyr697, a GRB2-binding site, and Tyr721, a PI 3-kinase binding site, abrogates signal transduction by the murine CSF-1 receptor expressed in Rat-2 fibroblasts. EMBO J 12:5161–5172
18. Hao C, Guilbert LJ, Fedoroff S, et al (1990) Production of colony-stimulating factor-1 (CSF-1) by mouse astroglia in vitro. J Neurosci Res 27:314–323
19. Théry C, Hetier E, Evrard C, et al (1990) Expression of macrophage colony-stimulating factor gene in the mouse brain during development. J Neurosci Res 26:129–133
20. Frei K, Nohava K, Malipiero UV, et al (1992) Production of macrophage colony-stimulating factor by astrocytes and brain macrophages. J Neuroimmunol 40:189–195
21. Lee SC, Liu W, Roth P, et al (1993) Macrophage colony-stimulating factor in human fetal astrocytes and microglia. Differential regulation by cytokines and lipopolysaccharide, and modulation of class II MHC on microglia. J Immunol 150:594–604
22. Liu W, Brosnan CF, Dickson DW, et al (1994) Macrophage colony-stimulating factor mediates astrocyte-induced microglial ramification in human fetal central nervous system culture. Am J Pathol 145:48–53
23. Sawada M, Itoh Y, Suzumura A, et al (1993) Expression of cytokine receptors in cultured neuronal and glial cells. Neurosci Lett 160:131–134
24. Nohava K, Malipiero U, Frei K, et al (1992) Neurons and neuroblastoma as a source of macrophage colony-stimulating factor. Eur J Immunol 22:2539–2545
25. Berezovskaya O, Maysinger D, Fedoroff S (1995) The hematopoietic cytokine, colony-stimulating factor 1, is also a growth factor in the CNS: congenital absence of CSF-1 in mice results in abnormal microglial response and increased neuron vulnerability to injury. Int J Dev Neurosci 13:285–299
26. Berezovskaya O, Maysinger D, Fedoroff S (1996) Colony stimulating factor-1 potentiates neuronal survival in cerebral cortex ischemic lesion. Acta Neuropathol (Berl) 92:479–486
27. Fedoroff S, Berezovskaya O, Maysinger D (1997) Role of colony stimulating factor-1 in brain damage caused by ischemia. Neurosci Biobehav Rev 21:187–191

28. D'Arcangelo G, Miao GG, Chen SC, et al (1995) A protein related to extracellular matrix proteins deleted in the mouse mutant reeler. Nature (Lond) 374:719–723
29. Patil N, Cox DR, Bhat D, et al (1995) A potassium channel mutation in weaver mice implicates membrane excitability in granule cell differentiation. Nat Genet 11:126–129
30. Rakic P, Sidman RL (1973) Weaver mutant mouse cerebellum: defective neuronal migration secondary to abnormality of Bergmann glia. Proc Natl Acad Sci USA 70:240–244
31. Rakic P, Sidman RL (1973) Organization of cerebellar cortex secondary to deficit of granule cells in weaver mutant mice. J Comp Neurol 152:133–161
32. Rakic P, Sidman RL (1972) Synaptic organization of displaced and disoriented cerebellar cortical neurons in reeler mice. J Neuropathol Exp Neurol 31:192
33. Mariani J, Crepel F, Mikoshiba K, et al (1977) Anatomical, physiological and biochemical studies of the cerebellum from Reeler mutant mouse. Philos Trans R Soc Lond Biol 281:1–28
34. Heckroth JA, Goldowitz D, Eisenman LM (1989) Purkinje cell reduction in the reeler mutant mouse: a quantitative immunohistochemical study. J Comp Neurol 279:546–555
35. Hawkes R, Colonnier M, Leclerc N (1985) Monoclonal antibodies reveal sagittal banding in the rodent cerebellar cortex. Brain Res 333:359–365
36. Hawkes R, Leclerc N (1987) Antigenic map of the rat cerebellar cortex: the distribution of parasagittal bands as revealed by monoclonal anti-Purkinje cell antibody mabQ113. J Comp Neurol 256:29–41
37. Leclerc N, Gravel C, Hawkes R (1988) Development of parasagittal zonation in the rat cerebellar cortex: MabQ113 antigenic bands are created postnatally by the suppression of antigen expression in a subset of Purkinje cells. J Comp Neurol 273:399–420
38. Brochu G, Maler L, Hawkes R (1990) Zebrin II: a polypeptide antigen expressed selectively by Purkinje cells reveals compartments in rat and fish cerebellum. J Comp Neurol 291:538–552
39. Hawkes R, Brochu G, Gravel C, et al (1992) Zebrins: molecular markers of compartmentation in the cerebellum. In: Llinás R, Sotelo C (eds) The cerebellum revisited. Springer, Berlin, pp 22–55
40. Tano D, Napieralski JA, Eisenman LM, et al (1992) Novel developmental boundary in the cerebellum revealed by zebrin expression in the lurcher (Lc/+) mutant mouse. J Comp Neurol 323:128–136
41. Ahn AH, Dziennis S, Hawkes R, et al (1994) The cloning of zebrin II reveals its identity with aldolase C. Development (Camb) 120:2081–2090
42. Murase S, Hayashi Y (1996) Expression pattern of integrin β1 subunit in Purkinje cells of rat and cerebellar mutant mice. J Comp Neurol 375:225–237
43. Scott TG (1963) A unique pattern of localization within the cerebellum. Nature (Lond) 200:793
44. Scott TG (1964) A unique pattern of localization within the cerebellum of the mouse. J Comp Neurol 122:1–8
45. Marani E, Voogd J (1977) An acetylcholinesterase band-pattern in the molecular layer of the cat cerebellum. J Anat 124:335–345
46. Chan-Palay V, Nilaver G, Palay SL, et al (1981) Chemical heterogeneity in cerebellar Purkinje cells: existence and coexistence of glutamic acid decarboxylase-like and motilin-like immunoreactivities. Proc Natl Acad Sci USA 78:7787–7791
47. Hess DT, Voogd J (1986) Chemoarchitectonic zonation of the monkey cerebellum. Brain Res 369:383–387
48. Leclerc N, Doré L, Parent A, et al (1990) The compartmentalization of the monkey and rat cerebellar cortex: zebrin I and cytochrome oxidase. Brain Res 506:70–78
49. Nilaver G, Defendini R, Zimmerman EA, et al (1982) Motilin in the Purkinje cell of the cerebellum. Nature (Lond) 295:597–598

50. Leclerc N, Beesley PW, Brown I, et al (1989) Synaptophysin expression during synaptogenesis in the rat cerebellar cortex. J Comp Neurol 280:197–212
51. Chen G, Hanson CL, Ebner TJ (1996) Functional parasagittal compartments in rat cerebellar cortex: an in vivo optical imaging study using neutral red. J Neurophysiol (Bethesda) 76:4169–4174
52. Messer A (1989) Cell-specific enrichment of primary monolayer cultures of postnatal rodent cerebellum. In: Shahar A, de Vellis J, Vernadakis A, et al (eds) A dissection and tissue culture manual of the nervous system. Liss, New York, pp 195–197
53. Nakajima M, Furukawa S, Hayashi K, et al (1993) Age-dependent survival-promoting activity of vitamin K on cultured CNS neurons. Dev Brain Res 73:17–23

Part 4
Adhesion Molecules

Cadherin-Mediated Adhesion in Central Nervous System Architecture

Wei-Song Shan[1], Kazuyoshi Tamura[2], Hidekazu Tanaka[1], Larry Shapiro[3], and David R. Colman[1]

Summary. Cadherins are calcium-dependent adhesion molecules that function in all tissues of epithelial derivation, including the brain. About two dozen cadherins have been identified in brain tissue, and of these, N-cadherin has been definitively localized to central nervous system synaptic junctional complexes, where it is believed to link up and lock in pre- and postsynaptic membranes. In addition, N-cadherin has been shown to mediate neurite outgrowth, a phenomenon that requires relatively weak adhesive bonds between interacting membrane surfaces. X-ray crystallographic analyses of cadherin structure have yielded testable models that may explain how cadherins can function in neurite outgrowth and synapse formation, two phenomena which require very different adhesive strengths.

Key words. Cadherins—Synapse formation—X-ray crystallography—Adhesion—Neurite outgrowth

We now recognize sets of surface molecules that, in various combinations, act together to generate membrane adhesion in nervous tissue. Modulatable adhesion is an extremely important functional feature that underlies all formative operations which occur in the brain; the best studied examples are axonal outgrowth, myelin sheath formation, and cell migration. It is really quite extraordinary that the same sets of molecules which function to organize simpler tissues also orchestrate the immensely elaborate assembly of central nervous tissue. Recent studies have revealed that one set of related molecules are important in all these organizational phenomena, but only recently has it been recognized that they may be important in synapse formation in the CNS as well. These are the *cadherins*, which are calcium-dependent homophilic adhesion molecules that in epithelial cells generate adherens junctions and desmo-

[1] Brookdale Center for Molecular and Developmental Biology, Mount Sinai School of Medicine, One Gustave L. Levy Place, New York, NY 10029, USA
[2] Department of Neurosurgery, Osaka University School of Medicine, 2-2 Yamadaoka Suita, Osaka 565-0871, Japan
[3] Program in Structural Biology, Mount Sinai School of Medicine, One Gustave L. Levy Place, New York, NY 10029, USA

somes. X-ray crystallographic analysis coupled with site-directed mutagenesis and expression experiments are beginning to yield an atomic-level picture of how cadherins function as adhesive moieties to join two cells together, or in certain special cases, to join one cell to itself.

Cadherins are now recognized as a large superfamily of glycoproteins. The classic cadherins, which include N-, E-, R, and P-cadherin, have in common a primary structure that consists of an amino-terminal ectodomain, a short transmembrane domain, and a conserved carboxy terminal cytoplasmic domain by which cadherins interact with cytoplasmic proteins. The ectodomain comprises five regions termed extracellular regions (EC) 1 to 5 of about 108 amino acids each. The ectodomain is made rigid by the presence of calcium ions articulating between the EC domains.

The structural similarities between adherens junctions and the central nervous system synaptic junctional complex have been noted previously [1], and so it should come as no surprise that the same cadherins that mediate adherens junctional integrity in epithelia might also operate to link up and lock in, adhesively, pre- and postsynaptic membranes in the CNS [2–4]. For example, N-cadherin immunoreactivity can be readily demonstrated in some but not all synaptic junctional complexes throughout the brain, while E-cadherin can be detected at other synaptic endings [3]. Because the brain is known to express at least two dozen cadherins [5,6], each of which presumably operates homophilically, it is logical to suggest that the different adhesive specificities of individual cadherins may provide a mechanism by which synaptic specificity is ultimately laid in [3,7].

The molecular basis for interactions within the classic cadherin subfamily are now being studied in our laboratory. X-ray crystallographic analysis of the EC1 domain of N-cadherin led to a model in which classic cadherins protrude from the cell surface in lateral pairs (Fig. 1; see [8]). They thus create a dimer interface, separate from the putative cell adhesion interface, which consists of part of a beta strand of one mol-

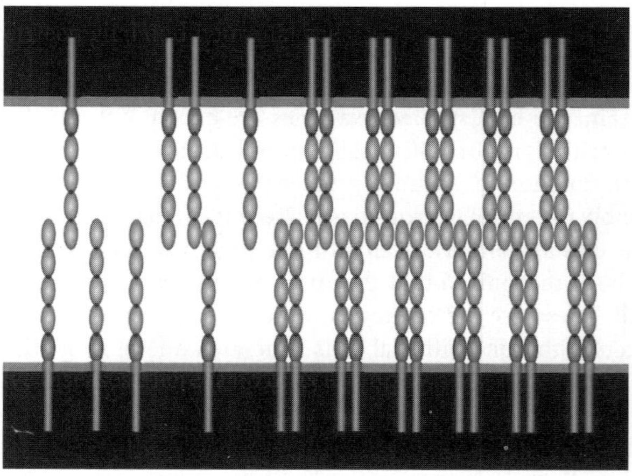

FIG. 1. Cadherins may exist on cell surfaces as monomers and dimers. Cadherins are likely to exist in dynamic equilibrium between monomeric (yielding weak or no adhesion) and strand dimeric (strongly adhesive) forms. Strand dimers intercalate from opposite cell surfaces, which is the structural basis for the adhesion zipper

ecule fitting into a hydrophobic receptor pocket in the partner molecule, yielding a "strand dimer." According to this model, lateral dimerization is dependent on the presence of a tryptophan in position 2 of EC1, whose hydrophobic side chain intercalates into the receptor pocket. Interestingly, this tryptophan is conserved in the classic cadherins, and the receptor pocket comprises the HAV sequence that some consider to be important for cell adhesion [9]. It should be stressed that the best evidence to date indicates that cadherins exist on cell surfaces in monomeric and strand dimeric forms. It is likely that monomeric cadherins are either extremely weak in terms of adhesive capacity or cannot mediate adhesion at all; in fact, classic cadherins might require the ability to form strand dimers to function in cell adhesion, and it is furthermore possible that regulation of cadherin adhesion occurs through a mechanism involving dynamic changes between the monomeric and dimeric forms.

Site-directed mutagenesis experiments to be described elsewhere confirm that the formation of strand dimers is essential to cadherin adhesion [12], and in fact studies from other groups have already provided strong evidence that the strand dimer is the active form that mediates cadherin adhesion [10,11]. The strand dimer model allows each cadherin pair to present adhesive interfaces in ideal position to engage an identically coupled pair of cadherins emanating from an opposing cell surface. Repetition of this cadherin functional unit generates what we have termed the "adhesion zipper," which is a linear, clustered, and oriented set of cadherins that together yield tightly adhering cadherin dimeric units [8]. This zipper depends on the integrity of the strand dimer, and this idea fits well with the concept of a regular arrangement of cadherins on cell surfaces; the zipper structure is also suggested in electron microscopic images of cadherin-mediated junctions.

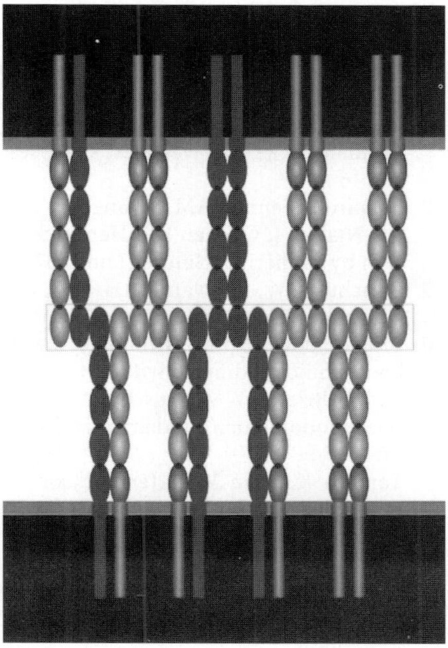

Fig. 2. Strand dimers theoretically can be heterodimers that nevertheless mediate homophilic adhesion. In this figure, two cadherins with different adhesive specificities are depicted. Because of the conservation of the tryptophan in position 2 in EC1 and the hydrophobic receptor pocket on the partner strand, it is conceivable that two classic cadherins with different specificities may nevertheless form a strand dimer. In such a scenario, the adhesive specificities remain uniquely homophilic

One intriguing implication of the fact that the two components of the strand dimer (the tryptophan in position 2 and the hydrophobic receptor pocket) are conserved elements among the classic cadherins is that the strand dimer need not always be composed of *homodimeric* cadherin molecules. It is possible that the conserved tryptophan in certain of the classic cadherins might be able to be inserted into hydrophobic pockets in adjacent strands, which however display different adhesive faces with markedly different specificities (Fig. 2). This would be particularly important in organizing a tissue such as the brain where it is likely that several cadherins are coexpressed by the same cell. Strand dimers composed of heterodimers would not be expected a priori to interfere with the adhesion zipper assembly and would yield a cadherin superfamily with vast combinatorial properties, allowing for exquisite modulation of adhesive interactions between neural cells. Clearly, the strand dimer model is an attractive one that provides a framework in which we can understand stable intermembrane associations such as those that are found in epithelial junction in other tissues and synaptic junctions in the CNS.

References

1. Peters A, Palay SL, Webster H de F (1991) The fine structure of the nervous system: neurons and their supporting cells. Oxford University Press, New York
2. Yamagata M, Herman JP, Sanes JR (1995) Lamina-specific expression of adhesion molecules in developing chick optic tectum. J Neurosci 15:4556–4571
3. Fannon AM, Colman DR (1996) A model for central synaptic junctional complex formation based on the differential adhesive specificities of the cadherins. Neuron 17:423–434
4. Uchida N, Honjo Y, Johnson KR, Wheelock MJ, Takeichi M (1996) The catenin/cadherin adhesion system is localized in synaptic junctions bordering transmitter release zones. J Cell Biol 135:767–779
5. Suzuki S, Sano K, Tanihara H (1991) Diversity of the cadherin family: evidence for eight new cadherins in nervous tissue. Cell Regul 2:261–270
6. Sano K, Tanihara H, Heimark RL, Obata S, Davidson M, St. John T, Taketani S, Suzuki S (1993) Protocadherins: a large family of cadherin-related molecules in central nervous system. EMBO J 12:2249–2256
7. Colman DR (1997) Neurites, synapses and cadherins reconciled. Mol Cell Neurosci 10:1–6
8. Shapiro L, Fannon AM, Kwong PD, Thompson A, Lehmann MS, Grübel G, Legrand JF, Als-Nielsen J, Colman DR, Hendrickson WA (1995) Structural basis of cell-cell adhesion by cadherins. Nature (Lond) 374:327–337
9. Blaschuk OW, Sullivan R, David S, Pouliot Y (1990) Identification of a cadherin cell adhesion recognition sequence. Dev Biol 139:227–229
10. Brieher WM, Yap AS, Gumbiner BM (1996) Lateral dimerization is required for the homophilic binding activity of C-cadherin. J Cell Biol 135:487–496
11. Yap A, Brieher WM, Pruschy M, Gumbiner BM (1997) Lateral clustering of the adhesive ectodomain: a fundamental determinant of cadherin function. Curr Biol 7:308–315
12. Tamura K, Shan WS, Hendrickson WA, Colman DR, Shapiro L (1998) Structure-function analysis of cell adhesion by neural (N-) cadherin. Neuron 20:1153–1163

Auditory, Somatosensory, and Other Restricted Circuits Are Delineated by Cadherin-6 Expression in Perinatal Mouse Brains

Takayoshi Inoue[1,2], Taro Tanaka[1], Sachihiro C. Suzuki[1], and Masatoshi Takeichi[1]

Summary. Multiple subtypes of the cadherin cell–cell adhesion molecules are differentially expressed in developing and maturing brain. Cadherin-6 (cad6) is a member of the type II classic cadherin subfamily, and recent studies of cad6 mRNA expression in certain areas of postnatal mouse brains have shown its regional distribution. We extended the analysis of cad6 mRNA expression to the entire brain, and also to parts of the peripheral nervous system (PNS), by use of both embryonic and postnatal brains. The results demonstrated that cad6 expression delineates restricted cortical areas and nuclei, and also in cranial ganglia, constituting the auditory, somatosensory, and certain other neuronal circuits in postnatal brains. Not all neurons belonging to these systems were cad6 positive, indicating that its expression delineates only subdivisions of them. In embryonic brain, most of these cad6-positive areas were detected even in E14.0 samples, suggesting that this expression pattern of cad6 is established during embryonic stages. We also confirmed that cad6 expression occurred in neuronal cell populations, although it was also expressed by nonneuronal cells. These findings support the idea that cadherin-mediated specific cell–cell adhesion is involved in interneuronal connections to form neural circuits.

Key words. Cadherin—Synapse—Neural network—Auditory system—Somatosensory system

Introduction

Neural development includes several aspects in which the regulation of cell adhesiveness may play crucial roles. A simple example is the compartmentation of early neural tube into neuromeres: in this aspect, the chemical differences among neuromeres should produce cell lineage restrictions [1,2], and such differences could be partly explained by the selective cell adhesiveness [3,4]. Another sophisticated example is

[1] Department of Biophysics, Faculty of Science, Kyoto University, Kitashirakawa, Sakyo-ku, Kyoto 606-01, Japan
[2] Division of Biochemistry and Cellular Biology, National Institute of Neuroscience, National Center of Neurology and Psychiatry, 4-1-1 Ogawa-higashi, Kodaira, Tokyo 187-8502, Japan

found in later stages of neural development: cells sharing similar functions selectively aggregate to form the nucleus, or are sorted into specific layers, and subsequently neuronal cells in these nuclei or specific layers are connected to each other. These aspects, in which a correctly guided growth cone should finally select a proper target from thousands of cells, also involve selective cell adhesiveness. Thus, cell-surface molecules, especially the cell adhesion molecules, are most likely to hold a key to understand still-veiled mechanisms of neural development.

Cadherins are cell adhesion receptors that give cells the subclass specific adhesiveness: cells that express a particular cadherin subclass selectively adhere to those which express the same cadherin subclass [5]. During early embryogenesis, the expression pattern of cadherins changes dynamically corresponding to morphogenetic events, and the fact that the ectopic expression [6,7] or the dysfunction of cadherins [8–15] perturbed histogenesis suggests that cadherins play a central role by their characteristic selective adhesiveness.

Thus far, more than 20 members of classic cadherins have been cloned, and they are now tentatively subdivided into type I and type II groups by their amino acid sequence similarities [16–18]. As for neural development, it has been revealed that the expression pattern of type I classic cadherins is well correlated to the aspects described here. For instance, R-cadherin expression delineated some compartments of brain and restricted nucleus in mice [19] and chickens [20–22]. We recently demonstrated that type II classic cadherins were also expressed in specific rhombomeres [23] as well as a restricted set of brain nuclei [24] of the prenatal brain. Moreover, within the postnatal forebrain, a unique combination of three type II cadherins delineated some cortical areas and thalamic subdivisions [25]. In light of the fact that the classic cadherins were localized at synaptic junctions [26–28], we further examined a mouse type II classic cadherin, cadherin-6 (m-cad6) in the present study to reveal how cadherins are involved in neural circuit formation.

Results and Discussion

As previously reported [23], m-cad6 cDNA was originally cloned by means of RT-PCR from mouse newborn brains, and from a mouse neuroblastoma × rat glioma hybrid cell line NG108–15 that can be induced to differentiate into neuron like cells in vitro [29–31]. In the process of cloning, we noticed that the number of m-cad6 PCR clones amplified from neuronally differentiated NG108–15 (Fig. 1B) were greater than those from pre-differentiated NG108–15 (Fig. 1A). Consistently, we confirmed that the expression level of this cadherin was upregulated through the neuronal induction of NG108–15 cells by northern blot analysis (Fig. 1C). In vivo, the expression of m-cad6 mRNA was restricted in the brain and kidneys of postnatal day 1 (P1) mice (Fig. 1D). Throughout brain development, m-cad6 mRNA was detected from embryonic day 14 (E14) to P28 (Fig. 1E). As for the expression level, it increased from E14 to P1 and was maintained until P28, compared with beta-actin expression (Fig. 1E). On the contrary, the expression level of mouse N-cadherin assessed in the same blot decreased during development (Fig. 1E). These results from in vitro and in vivo suggested that m-cad6, rather than N-cadherin, could be expressed in neuronal cells at differentiated stages.

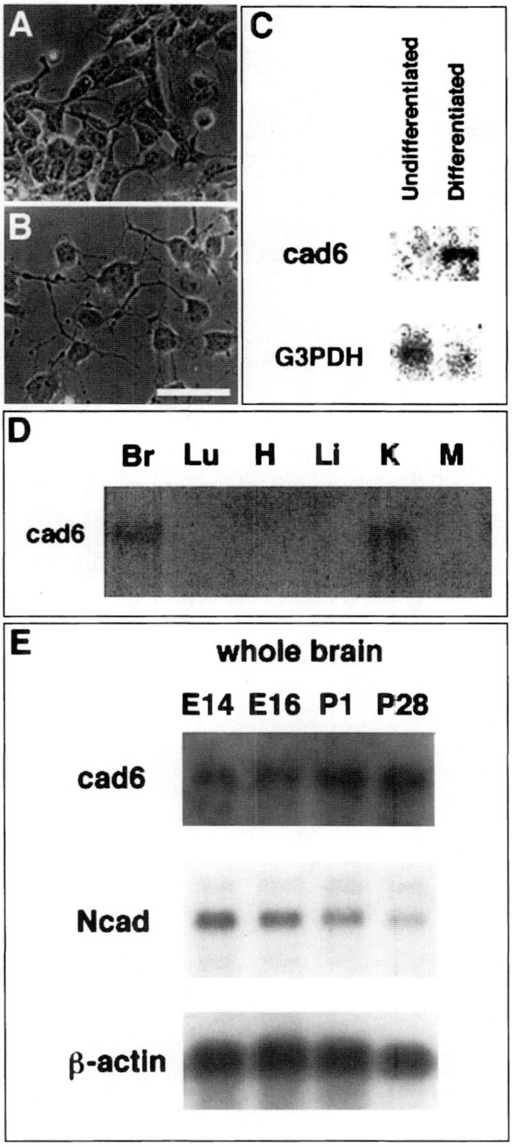

FIG. 1A–E. Cad6 expression in NG108–15 cells and northern blot analyses of cad6 and N-cadherin mRNA expression. **A** Undifferentiated NG108–15 cells. **B** Neuronal cells that appeared 5 days after the addition of the reagents for neuronal induction. **C** Northern blotting for *cad6*, and *G3PDH*, a control housekeeping gene marker, in cells as prepared in **A** and **B**. A reduced amount of mRNA was purposely loaded in the *right lane*, as compared to the *left lane*. *Bar* 20 μm. **D** Cad6 expression in various tissues of P1 mouse. Only the brain and kidney express *cad6* messages. *Br*, brain; *Lu*, lung; *H*, heart; *Li*, liver; *K*, kidney; *M*, muscle. **E** Cad6 and N-cadherin (*Ncad*) expression during development. Note that a similar level of cad6 expression is maintained during the E16–P28 development or is even increased during this period, whereas N-cadherin expression tends to decrease during the postnatal stages

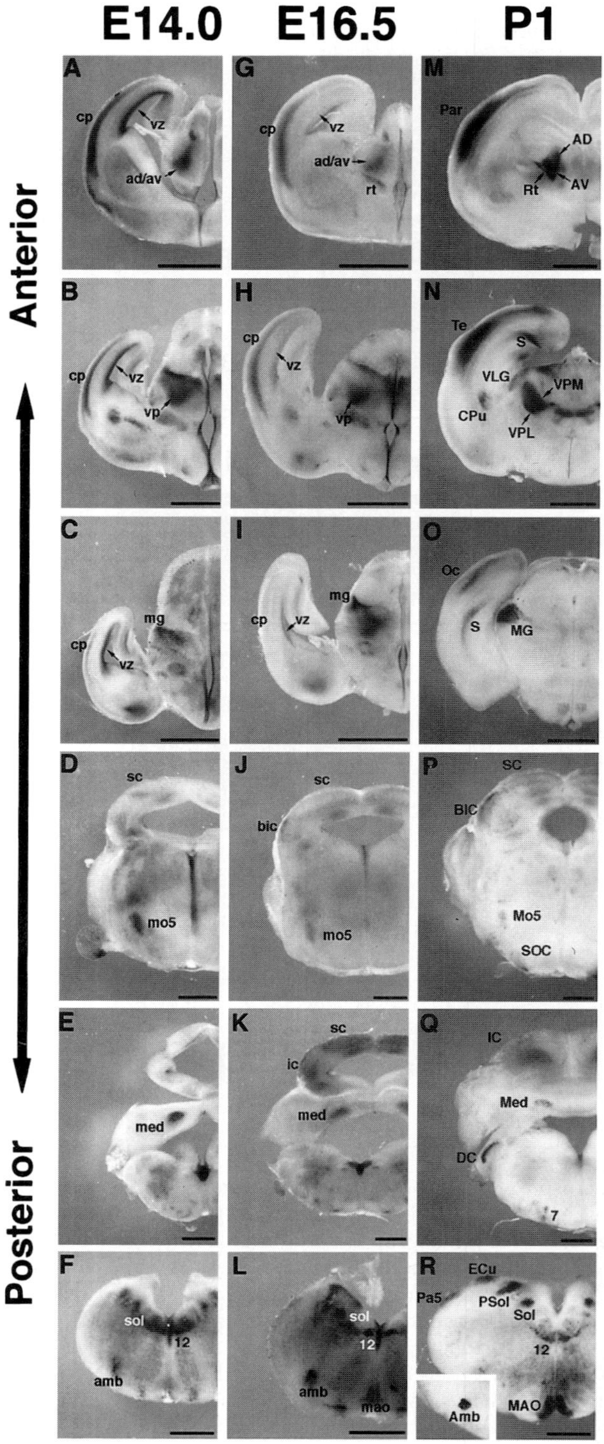
E14.0
E16.5
P1
Anterior
Posterior
A
cp
vz
ad/av
G
cp
vz
ad/av
rt
M
Par
AD
Rt
AV
B
cp
vz
vp
H
cp
vz
vp
N
Te
S
VLG
VPM
CPu
VPL
C
mg
cp
vz
I
mg
cp
vz
O
Oc
S
MG
D
sc
mo5
J
sc
bic
mo5
P
SC
BIC
Mo5
SOC
E
med
K
sc
ic
med
Q
IC
Med
DC
7
F
sol
12
amb
L
sol
12
amb
mao
R
ECu
Pa5
PSol
Sol
12
Amb
MAO

Then, we made frontal brain slices of various developmental stages and mapped the detailed expression pattern of m-cad6 using in situ hybridization technique. As summarized in Fig. 2, cad6 was revealed to be expressed in restricted cortical areas and nuclei, constituting the auditory, somatosensory, and certain other neuronal circuits in postnatal brains: these include the auditory cortex (Te), medial geniculate body (MG), inferior colliculus (IC), and dorsal cochlear nucleus (DC) for the auditory system and the somatosensory cortex (Par), ventroposterior thalamic nucleus (VPL/M), spinal trigeminal nucleus (Sp5), and external cuneate nucleus (ECu) for the somatosensory system. At higher magnifications of nuclei or cortices, we found that cad6 signals were always localized in the cytoplasm of large, round cells, the morphology of which was characteristic of neurons (data not shown; see also [32]). To further confirm the neuronal expression of cad-6, we prepared monolayer cultures of parietal cortical cells derived from E18.0 cerebrums, and doubly stained them for cad6 mRNA and MAP-2, a dendrite marker, after 2 days. The results showed that most cells expressing cad6 were also positive for MAP-2 immunoreactivity (data not shown; see also [32]). This provided an evidence that a large population of cad6-positive cells in the parietal cortex are neurons. We performed similar analysis for other regions of prenatal brains, and obtained essentially the same results. As for the peripheral nervous system (PNS), cranial ganglion belonging to auditory or somatosensory system also contained cad6-positive neuronal cells at the postnatal stage [32].

Thus, cad6 expression delineated specific neuronal circuits from PNS to central nervous system (CNS) at P1 stage (Fig. 3). Not all neurons belonging to these systems were cad6 positive, suggesting that its expression delineates only subdivisions of them. Furthermore, the ventral cochlear nucleus (VC), for example, a major relay nucleus in auditory system, had few cad6-positive cells, indicating only a subpathway of the system was cad6 positive. Consistently, the superior olivary complex (SOC), which receives axons from VC and send axons to IC, had only weak staining for cad6 (Fig. 2P, Fig. 3). In embryonic brains, most of these cad6-positive areas were detected

---

FIG. 2A–R. Cad6 mRNA distribution during brain development. *E14.0*, *E16.5*, and *P1* brains were frontally sliced, and their cad6 expression was examined by whole-mount in situ hybridization. As indicated, the *left*, *middle*, and *right* columns are for E14.0, E16.5, and P1 brains, respectively. In each column, different levels of the brain are shown, with more anterior portions in *upper panels* and posterior portions in *lower panels* as indicated. Slices of similar level are arranged in each horizontal set of the panel. *Insert* in **R** shows part of a more anterior slice. Developing nuclei in E14.0 and 16.5 samples are indicated by abbreviated names depicted with small characters: *AD*, anterodorsal thalamic nucleus; *Amb*, ambiguus nucleus; *AV*, anteroventral thalamic nucleus; *BIC*, nucleus of the brachium of the inferior colliculus; *cp*, cortical plate; *CPu*, caudate-putamen; *DC*, dorsal cochlear nucleus; *ECu*, external cuneate nucleus; *IC*, inferior colliculus; *MAO*, medial accessory olive; *Med*, medial cerebellar nucleus; *MG*, medial geniculate nucleus; *Mo5*, motor trigeminal nucleus; *Oc*, occipital cortex; *Pa5*, paratrigeminal nucleus; *Par*, parietal cortex; *PSol*, parasolitary nucleus; *Rt*, reticular thalamic nucleus; *S*, subiculum; *SC*, superior colliculus; *SOC*, superior olivary complex; *Sol*, nucleus of the solitary tract; *Te*, temporal cortex; *VLG*, ventral lateral geniculate nucleus; *vp*, developing ventral posterior thalamic nucleus; *VPL*, ventral posterolateral thalamic nucleus; *VPM*, ventral posteromedial thalamic nucleus; *vz*, ventricular zone; *7*, facial nucleus; *12*, hypoglossal nucleus. *Bars* 1 mm (**A–C**, **G–I**, **M–O**); 0.5 mm (**D–F**, **J–L**, **P–R**)

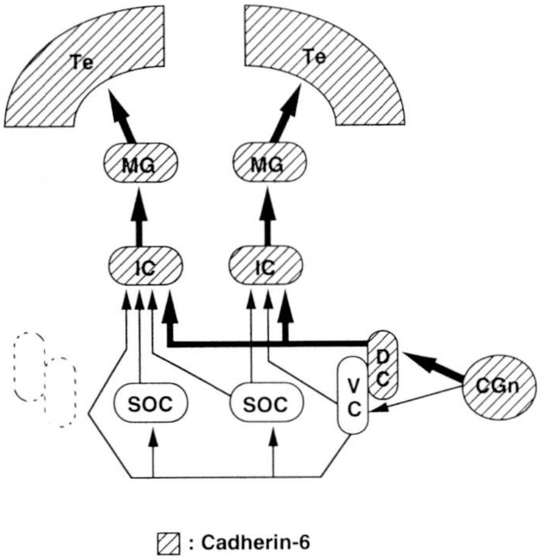

FIG. 3. An example of the connection systems delineated by m-cad6 expression. Cad6-positive elements in the auditory system are marked by *oblique shading*; the *bold arrows* indicate the presumptive pathways with this cadherin. *CGn*, spiral ganglion; *VC*, ventral cochlear nucleus. For other abbreviations, see legend for Fig. 2

even in E14.0 samples (Fig. 2A–F), and the emergence of these cad6 signals seemed to coincide with neuronal differentiation processes of these areas [33,34].

All these findings support the hypothesis that cadherins play important roles in some aspects of the later neural development by their characteristic selective adhesiveness. First, m-cad6 expression delineated restricted nuclei and cortices from the E14.0 to P1 brain. Considering that neighboring cells expressed other sets of type II cadherin subclasses in P2 brain [25], the differential expression pattern of cadherins could be also expected at embryonic stages, and it would be possible to segregate cells within a specific nucleus, or to sort cells into specific cortical layers and areas at postnatal stages. It might otherwise be required to maintain such brain structures. Second, m-cad6 mRNA expression delineated specific neuronal circuits. In other words, m-cad6 mRNA was expressed at both pre- and postsynaptic groups of neuronal cells within specific circuits. As the numbers of synaptic junctions containing chicken N-cadherin were smaller than those delineated by alpha N- and beta-catenin that usually associated with various types of cadherins [27], it is natural to assume many types of cadherins including m-cad6 are localized at synaptic junctions to generate or maintain the correct neuron–neuron connection. Consistently, our recent results using the confocal microscope revealed that m-cad6 proteins were sometimes colocalized with synaptotagmin, a synapse marker, in P5 brains (data not shown; see also [32]). Third, m-cad6 expression in the present study was found only in subdivisions or subpathways of neuronal circuits (Fig. 3). This arouses a possibility that other types of cadherins also delineate other specific pathways as m-cad6 did, and such combinatorial expression pattern of cadherins enables a growth cone to select a correct target from the thousands of cells in the nervous system. Supporting this idea, it was shown that the expression of mouse cadherin-8 and -11 was found in nuclei among other restricted systems [25] and that chicken N-cadherin actually plays a role in laminar-specific connectivity [35].

Of course, other classes of cell adhesion molecules and various receptors for chemoattractants or chemorepellents at the cell surface [36,37] are also synergestically important to construct precise neuronal networks. Above all such known molecules, however, the expression pattern of a single molecular family, cadherins, and their characteristic adhesive specificities well explain the complex mechanisms of the neuronal connectivity in the entire nervous system beyond species. This strongly suggests the unique and indispensable roles of cadherins during neural development. To test this hypothesis, detailed mapping of other multiple types of cadherin in the brain as well as their functional analysis using gene knockout techniques would be required in future study.

References

1. Wilkinson DG (1993) Molecular mechanisms of segmental pattering in the vertebrate hindbrain and neural crest. BioEssays 15:499–505
2. Puelles L, Rubenstein JLR (1993) Expression patterns of homeobox and other putative regulatory genes in the embryonic mouse forebrain suggests a neuromeric organization. Trends Neurosci 16:472–479
3. Lumsden A (1990) The cellular basis of segmentation in the developing hindbrain. Trends Neurosci 13:329–335
4. Wingate RJT, Lumsden A (1996) Persistence of rhombomeric organization in the postsegmental hindbrain. Development (Camb) 122:2143–2152
5. Nose A, Nagafuchi A, Takeichi M (1988) Expressed recombinant cadherins mediate cell sorting in model system. Cell 54:993–1001
6. Detrick RJ, Dickey D, Kintner CR (1990) The effect of N-cadherin misexpression on morphogenesis in *Xenopus* embryos. Neuron 4:493–506
7. Fujimori T, Miyatani S, Takeichi M (1990) Ectopic expression of N-cadherin perturbs histogenesis in *Xenopus* embryos. Development (Camb) 110:97–104
8. Matsunaga M, Hatta K, Takeichi M (1988) Role of N-cadherin cell adhesion molecules in the histogenesis of neural retina. Neuron 1:289–295
9. Hirai Y, Nose A, Kobayashi S, et al (1989) Expression and role of E- and P-cadherin adhesion molecules in embryonic histogenesis. Development (Camb) 105:263–277
10. Barami K, Kirschenbaum B, Lemmon V, et al (1994) N-cadherin and Ng-CAM/8D9 are involved serially in the migration of newly generated neurons into the adult songbird brain. Neuron 13:567–582
11. Larue L, Ohsugi M, Hirchenhain J, et al (1994) E-cadherin null mutant embryos fail to form a trophectoderm epithelium. Proc Natl Acad Sci USA 91:8263–8267
12. Riethmacher D, Brinkmann V, Birchmeier C (1995) A targeted mutation in the mouse E-cadherin gene results in defective preimplantation development. Proc Natl Acad Sci USA 92:855–859
13. Hermiston ML, Gordon JI (1995) *In vivo* analysis of cadherin function in the mouse intestinal epithelium: essential roles in adhesion, maintenance of differentiation, and regulation of programmed cell death. J Cell Biol 129:489–506
14. Uemura T, Oda H, Kraut R, et al (1996) Zygotic *Drosophila* E-cadherin expression is required for the processes of dynamic epithelial cell rearrangement in the *Drosophila* embryo. Genes Dev 10:659–671
15. Radice GL, Rayburn H, Matsunami H, et al (1997) Developmental defects in mouse embryos lacking N-cadherin. Dev Biol 181:64–78
16. Suzuki S, Sano K, Tanihara H (1991) Diversity of the cadherin family: evidence for eight new cadherins in nervous tissue. Cell Regul 2:261–270
17. Ranscht B (1994) Cadherins and catenins: interaction and function in embryonic development. Curr Opin Cell Biol 6:740–746
18. Takeichi M (1995) Morphogenetic roles of classic cadherins. Curr Opin Cell Biol 7:619–627

19. Matsunami H, Takeichi M (1995) Fetal brain subdivisions defined by R- and E-cadherin expressions: evidence for the role of cadherin activity in region-specific, cell-cell adhesion. Dev Biol 172:466–478
20. Redies C, Engelhart K, Takeichi M (1993) Differential expression of N- and R-cadherin in functional neuronal systems and other structures of the developing chicken brain. J Comp Neurol 333:398–416
21. Ganzler SI, Redies C (1995) R-cadherin expression during nucleus formation in chick forebrain neuromeres. J Neurosci 15:4157–4172
22. Arndt K, Redies C (1996) Restricted expression of R-cadherin by brain nuclei and neural circuits of the developing chicken brain. J Comp Neurol 373:373–399
23. Inoue T, Chisaka O, Matsunami H, et al (1997) Cadherin-6 expression transiently delineates specific rhombomeres, other neural tube subdivisions, and neural crest subpopulations in mouse embryo. Dev Biol 183:183–194
24. Kimura Y, Matsunami H, Takeichi M (1996) Expression of cadherin-11 delineates boundaries, neuromeres, and nuclei in the developing mouse brain. Dev Dyn 206:455–462
25. Suzuki SC, Inoue T, Kimura Y, et al (1997) Neuronal circuits are subdivided by differential expression of type-II classic cadherins in postnatal mouse brains. Mol Cell Neurosci 9:433–447
26. Yamagata M, Herman JP, Sanes JR (1995) Lamina-specific expression of the adhesion molecules in developing optic tectum. J Neurosci 15:4556–4571
27. Uchida N, Honjo Y, Johnson KR, et al (1996) The catenin/cadherin adhesion system is localized in synaptic junctions bordering transmitter release zones. J Cell Biol 135:767–779
28. Fannon AM, Colman DR (1996) A model for central synaptic junctional complex formation based on the differential adhesive specificities of the cadherins. Neuron 17:423–434
29. Nelson P, Christian C, Nirenberg M (1976) Synapse formation between clonal neuroblastoma × glioma hybrid cells and straited muscle cells. Proc Natl Acad Sci USA 73:123–127
30. Nirenberg M, Wilson S, Higashida H, et al (1983) Modulation of synapse formation by cyclic adenosine monophosphate. Science 222:794–799
31. Han H-Q, Nichols RA, Rubin MR, et al (1991) Induction of formation of presynaptic terminals in neuroblastoma cells by synapsin IIb. Nature (Lond) 349:697–700
32. Inoue T, Tanaka T, Suzuki SC, et al (1998) Cadherin-6 in the developing mouse brain: expression along restricted connection systems and synaptic localization suggest a potential role in neuronal circuitry. Dev Dyn 211:338–351
33. Bayer SA, Altman J (1995) Neurogenesis and neuronal migration. In: The rat nervous system, 2nd edn. Academic Press, San Diego, pp 1041–1078
34. Bayer SA, Altman J (1995) Principles of neurogenesis, neuronal migration and neural circuit formation. In: The rat nervous system, 2nd edn. Academic Press, San Diego, pp 1079–1098
35. Inoue A, Sanes JR (1997) Laminar-specific connectivity in the brain: regulation by N-cadherin, neurotrophins, and glycoconjugates. Science 276:1428–1431
36. Tessier-Lavigne M, Goodman CS (1996) The molecular biology of axon guidance. Science 274:1123–1133
37. Chiba A, Keshishian H (1996) Neural pathfinding and recognition: roles of cell adhesion molecules. Dev Biol 180:424–432

Cadherins in CNS Wiring

Masatoshi Takeichi, Tadashi Uemura, Yoichi Iwai, Takayoshi Inoue, Sachihiro C. Suzuki, Taro Tanaka, and Naoshige Uchida

Cadherins are a family of homophilic cell–cell adhesion molecules. More than 20 subtypes of this molecular family have been identified. Specific, homophilic interaction of each subtype leads cells to adhere to a restricted group of other cells, i.e., those expressing the same cadherin subtypes. This activity of cadherins has been implicated for selective cell association events. We are interested in how the cadherin-mediated specific adhesion is involved in CNS morphogenesis. Our recent findings indicate that this group of adhesion molecules are essential for axon patterning, and also suggest that they may be involved in synapse and neural network formation.

We identified a novel *Drosophila* cadherin, *D*N-cadherin, which is expressed by axons of CNS. *D*N-cadherin null mutants are embryonic lethal, and our analysis of their phenotypes demonstrated that axonal patterning is severely disorganized; e.g., axons were defasciculated, and migration of some growth cones was misoriented. These findings indicate that *D*N-cadherin is crucial for axon–axon interactions. Some alleles of the *D*N-cadherin mutants can grow to the adult stage, but interestingly they exhibit potentially neuronal disorders, such as convulsion of legs, suggesting that this cadherin is also involved in wiring for the adult nervous system.

In maturing vertebrate brains, cadherins and associated proteins, the catenins, become to be localized at synaptic junctions, bordering the "active zones." Many cadherin subtypes are detected in the brains, each being expressed by a specific subset of neurons that are synaptically connected with one another. For example, cadherin-6 is expressed by neurons constituting subpathways of the auditory, somatosensory, and other restricted circuits; and other cadherins delineate different circuits. These findings suggest that cadherin-mediated adhesion is involved in linking neurons so as to generate their networks. In conclusion, cadherins likely play multiple roles in brain morphogenesis, from axon–axon to axon–target interactions.

Department of Biophysics, Faculty of Science, Kyoto University, Kitashirakawa, Sakyo-ku, Kyoto 606-8502, Japan

References

Uchida N, Honjo Y, Johnson KR, Wheelock MJ, Takeichi M (1996) The catenin/cadherin adhesion system is localized in synaptic junctions bordering the transmitter release zones. J Cell Biol 135:767–779

Iwai Y, Usui T, Hirano S, Steward R, Takeichi M, Uemura T (1997) Axon patterning requires *D*N-cadherin, a novel neuronal adhesion receptor, in the *Drosophila* embryonic CNS. Neuron 19:77–89

Suzuki SC, Inoue T, Kimura Y, Tanaka T, Takeichi M (1997) Neuronal circuits are subdivided by differential expression of type-II classic cadherins in postnatal mouse brains. Mol Cell Neurosci 9:433–447

Inoue T, Tanaka T, Suzuki SC, Takeichi M (1998) Cadherin-6 in the developing mouse brain: expression along restricted connection systems and synaptic localization suggest a potential role in neuronal circuitry. Dev Dyn 211:338–351

OCAM, a Novel Member of NCAM Family, and Zone-to-Zone Projection of Olfactory Axons

KENSAKU MORI[1,2] and YOSHIHIRO YOSHIHARA[1,2]

Zone-to-zone projection of olfactory and vomeronasal sensory axons is a basic principle for organization of the mammalian olfactory system. Chemoreceptor expression zones of the olfactory and vomeronasal sensory epithelia are topographically and functionally mapped onto zonally arranged glomeruli in the main and accessory olfactory bulbs, respectively. We report here the structure and possible function of OCAM (R4B12 antigen), an axonal surface glycoprotein expressed by subsets of both olfactory and vemeronasal axons in a zone-specific manner. OCAM is a novel cell adhesion molecule belonging to the NCAM family within the immunoglobulin superfamily. As revealed by binding studies, OCAM has a homophilic binding activity. In the main olfactory system, OCAM mRNA was expressed selectively by sensory neurons in the odorant-receptor expression zones II, III, and IV. OCAM expressing olfactory axons projected into the glomeruli in zones II, III, and IV of the main olfactory bulb. In the accessory olfactory system, OCAM mRNA was expressed by vomeronasal sensory neurons in the apical pheromone receptor expression zone. OCAM-positive vomeronasal axons projected selectively into glomeruli in the rostral zone of the accessory olfactory bulb. In contrast, NCAM was expressed by all the olfactory and vomeronasal axons.

These results demonstrate a precisely coordinated zonal expression of chemoreceptors and OCAM in both main and accessory olfactory systems, and suggest that OCAM may play important roles in selective axon fasciculation and in the formation and maintenance of zone-to-zone projection of the primary olfactory axons. We discuss also possible functional roles of NCAM family cell adhesion molecules, which include NCAM, OCAM, apCAM, and fasciclin II.

References

1. Yoshihara Y, Kawasaki M, Tamada A, et al. (1997) OCAM: a new member of the neural cell adhesion molecule family related to zone-to-zone projection of olfactory and vomeronasal axons. J Neurosci 17:5830–5842

---

[1]Laboratory for Neuronal Recognition Molecules, Frontier Research Program, RIKEN, 2-7 Hirosawa, Wako, Saitama 351-0198, Japan

[2]Department of Neuroserences, Osaka Bioscience Institute, Furuedai, Suita, Osaka 565, Japan

2. von Campenhausen H, Yoshihara Y, Mori K (1997) OCAM reveals segregated mitral/tufted cell pathways in developing accessory olfactory bulb. Neuro Report 8:2607–2612
3. Yoshihara Y, Mori K (1997) Basic principles and molecular mechanisms of olfactory axon pathfinding. Cell Tissue Res 290:457–463
4. Mori K, Yoshihara Y (1995) Molecular recognition and olfactory processing in the mammalian olfactory system. Prog Neurobiol 45:585–619
5. Yoshihara Y, Katoh K, Mori K (1993) Odor stimulation causes disappearance of R4B12 epitope on axonal surface molecule of olfactory sensory neurons. Neuroscience 53:101–110
6. Mori K, Fujita SC, Imamura K, Obata K (1985) Immunohistochemical study of subclasses of olfactory nerve fibers and their projections to the olfactory bulb in the rabbit. J Comp Neurol 242:214–229
7. Mori K, Imamura K, Fujita SC, Obata K (1985) Projections of two subclasses of vomeronasal nerve fibers to the accessory olfactory bulb in the rabbit. Neuroscience 20:259–278

Cell Signaling and Cell Adhesion Molecule-Mediated Neurite Outgrowth

Paula J. Green[1], Patrick Doherty[2], and Frank S. Walsh[1]

Summary. During development the ability of neurons to extend axons to and innervate appropriate targets is fundamental for the correct establishment of a highly complex pattern of neural pathways. Thus, the molecules involved in controlling nerve growth and guidance play key roles in these processes, and the understanding of these molecular mechanisms is one of the major challenges facing developmental neurobiology. The cell adhesion molecules (CAMs) NCAM, L1, and N-cadherin have key roles in axonal growth, and the molecular basis of this response has been extensively studied. It is clear from experiments using a variety of different approaches that there is a critical involvement of the fibroblast growth factor (FGF) receptor in mediating the downstream responses of these CAMs. Activation of the FGF receptor is both necessary and sufficient to account for the ablility of the CAMs to stimulate axonal growth.

Key words. Cell adhesion molecules—Axon growth—Signal transduction—Fibroblast growth factor receptor—Tyrosine kinase

Introduction

A highly motile structure present at the tip of a developing axon, known as the growth cone, is responsible for guiding the axon to its correct target to allow for appropriate synapse formation. Thus, the ultimate direction and rate of growth of a developing axon involves the integration of all signals detected by the finger-like filopodia on the growth cone as it senses cues present in its environment [1].

Considerable advances in understanding have been made in identifying the molecules involved in these processes. These may be diffusible long-range cues, for example, netrins and semaphorins [2], or contact-dependent short-range cues, e.g.,

[1] Department of Neuroscience, SmithKline Beecham Pharmaceuticals, New Frontiers Science Park, Harlow, Essex, CM19 5AW, UK

[2] Department of Experimental Pathology, UMDS, Guy's Hospital, London Bridge, London, SE1 9RT, UK

cell adhesion molecules (CAMs) [3], Eph receptors, and their ligands, the ephrins [2,4]. Molecules in both categories can be inhibitory or growth promoting for neurons, with some molecules acting as both depending on the type of neuron, e.g., myelin-associated glycoprotein (MAG) and the netrins [3]. Some families of molecules may also have both short- and long-range-acting members; e.g., the semaphorins have both cell-surface and secreted forms [2]. Recent advances in this area include the demonstration that some CAMs, such as L1, NCAM, and N-cadherin, can activate the fibroblast growth factor (FGF) receptor [5] (see following), the identification of deleted in colorectal cancer (DCC) as a receptor for netrins [6,7], the elucidation of the roles of Roundabout and Commissureless in controlling neurons crossing the midline [8], and the identification of the neuropilins as receptors for the semaphorins [9,10]. However, the events occuring downstream from these ligand–receptor interactions are as yet largely unknown apart from the CAM–FGF receptor interaction, which is the focus of this review.

Cell Adhesion Molecules

The three cell adhesion molecules L1, NCAM, and N-cadherin are expressed both on the surface of the developing axon and on the non-neuronal cells that the growth cone contacts, in a manner consistent with a role in axonal growth and regeneration. The key role these molecules have in neuronal growth has been highlighted recently with the first example of a genetic disease associated with a neural CAM; this was the identification of severe neurological deficits in humans being associated with mutations in the L1 gene [11–13].

L1 and NCAM are members of the immunoglobulin superfamily while N-cadherin is a member of the cadherin superfamily. They bind primarily in a homophilic manner, N-cadherin requiring the presence of calcium for binding, and mediate cell–cell interactions. A wide range of heterophilic interactions have also been described, particularly for L1 and NCAM. All three CAMs have a role in neurite outgrowth, L1 also functions in axonal fasiculation and guidance [3], and N-cadherin in neural cell migration [14], while NCAM is thought to have a role in synaptic plasticity associated with learning and memory [15].

Role of CAMs in Axonal Growth

Evidence for the role of these CAMs in axonal growth has come from a range of studies. Antibody perturbation experiments initially gave an indication that these CAMs could promote axonal growth [16–18], which was then confirmed by other approaches. Cells lacking endogenous CAMs were transfected with the CAM of interest, which resulted in enhanced neurite outgrowth, demonstrating that physiological levels of these CAMs were able to stimulate neurite outgrowth [19–26]. Further work showing that soluble forms of L1 and NCAM can also promote neurite outgrowth helped to establish that the growth response to CAMs can be separated from their adhesive role and that activation of second-messenger cascades is vital to their function [27].

Second-Messenger Cascades Involved in CAM-Mediated Neurite Outgrowth

Involvement of nonreceptor tyrosine kinases has been suggested for the CAMs, L1 and NCAM. The nonreceptor tyrosine kinase $pp60^{c\text{-}src}$ and $pp59^{fyn}$ are thought to modulate the function of NCAM and L1 [28,29], but it is unclear if they interact directly with the CAMs and whether their activation is sufficient to mimic CAM responses. More recently, coprecipitation experiments have shown that L1 coprecipitates with another nonreceptor tyrosine kinase, $p90^{rsk}$ [30], and Shc, an adaptor protein and substrate for tyrosine kinases, coprecipitates with N-cadherin [31]. Although $p90^{rsk}$ was shown to be able to selectively phosphorylate L1 on serine 1152, again it is unclear if this is a downstream event involved in the neurite outgrowth signaling cascade or if it is regulating the ligand-binding properties of L1. However, there is good experimental evidence for the involvement of receptor tyrosine kinases in the downsteam events of CAM-mediated neurite outgrowth.

A series of pharmacological studies on CAM-stimulated neurite outgrowth revealed the possible involvement of a tyrosine kinase in the axonal growth response. A tyrosine kinase inhibitor and a phospholipase Cγ inhibitor and other downstream inhibitors such as diacyl glycerol lipase inhibitors and calcium channel antagonists were able to block the CAM neurite outgrowth response [32]. Also, agents that can activate this cascade, for example, arachidonic acid, were able to mimic the CAM response in a nonadditive manner. These experiments led to the suggestion that the receptor tyrosine kinase involved in the CAM neurite outgrowth response may be the FGF receptor. The FGF receptors are a family of tyrosine kinase receptors comprised of four distinct but related members [33]. This suggestion was also based on the identification of a CAM homology domain within the FGF receptor, a region that shows homology to each of the CAMs, L1, NCAM, and N-cadherin [26]. Subsequent experiments have confirmed the critical role of this receptor in the CAM-mediated neurite outgrowth response.

Evidence That the FGF Receptor Is Required for CAM Neurite Outgrowth Responses

In both cerebellar granule cells and PC12 cells, antibodies that inhibit FGF receptor function inhibited axonal growth responses stimulated by NCAM, L1, and N-cadherin [23] (J. L. Saffell, unpublished, 1997), and in retinal ganglion cells these antibodies inhibited L1 and NCAM but not N-cadherin responses, as there was no outgrowth response to N-cadherin in these cells [34]. High levels of FGF, which desensitize the FGF receptor signal transduction cascade, also cross-desensitized the growth responses stimulated by the CAMs [24]. A soluble form of L1, an L1-Fc chimera, and FGF stimulated tyrosine phosphorylation on a common set of proteins [23], and both L1-Fc and NCAM-Fc chimeras can stimulate tyrosine phosphorylation of the FGF receptor [5], a hallmark of receptor activation. This latter activation of the FGF receptor by the CAMs can be inhibited if a dominant-negative form of the FGF receptor, which lacks a functional tyrosine kinase domain, is expressed. Finally, using the dominant-negative form of the FGF receptor, either expressed in primary neurons from transgenic mice or stably transfected into PC12 cells, is sufficient to inhibit neurite outgrowth stimulated by L1, NCAM, and N-cadherin with no loss in the basal integrin-dependent neurite outgrowth or nerve growth factor responsiveness of these neurons [5]. These experiments show that by mimicking FGF receptor tyrosine kinase activation it is possible to mimic CAM-mediated neurite outgrowth responses.

Conclusion

These results show that CAMs have key roles in mediating axonal growth and that a functional FGF receptor is required to mediate CAM-stimulated neurite outgrowth. It remains unclear as to whether the interaction between the FGF receptor and the CAMs is direct or indirect, but what is clear is that activation of the FGF receptor is both necessary and sufficient to account for CAM responses and that CAMs can activate the FGF receptor second-messenger cascade. It will be important to determine the precise mechanisms involved and the role of the putative second-messenger cascades that the CAMs may activate and to determine how activation of this cascade by CAMs is translated into neurite outgrowth. One possible link between the cell surface and the cytoskeleton is the observation that addition of soluble CAMs to isolated growth cones leads to an enhancement of phosphorylation of GAP-43 [35].

Finally, these experiments show the importance of regulated tyrosine phosphorylation as a fundamental signaling process controlling growth cone behavior, but dynamic regulation of this process requires phosphatase activity as well. The role of phosphatases in the regulation of neuronal growth and guidance is only just recently being recognized, and any future studies must address not only the roles of tyrosine kinases but of phosphatases in these responses.

References

1. Rehder V, Kater SB (1996) Filopodia on neuronal growth cones: multifunctional structures with sensors and motor capabilities. Semin Neurosci 8:81–88
2. Tessier-Lavigne M, Goodman CS (1996) The molecular biology of axon guidance. Science 274:1123–1133
3. Walsh FS, Doherty P (1997) Neural cell adhesion molecules of the immunoglobulin superfamily: role in axon growth and guidance. Annu Rev Cell Dev Biol 13:425–456
4. Orioli D, Klein R (1997) The Eph receptor family: axonal guidance by contact repulsion. Trends Genet 13:354–359
5. Saffell JL, Williams EJ, Mason IJ, et al (1997) Expression of a dominant negative FGF receptor inhibits axonal growth and FGF receptor phosphorylation stimulated by CAMs. Neuron 18:231–242
6. Keino-Masu K, Masu M, Hinck L, et al (1996) Deleted in colorectal cancer (DCC) encodes a netrin receptor. Cell 87:175–185
7. Kolodziej PA, Timpe LC, Mitchell KJ, et al (1996) Frazzled encodes a *Drosophila* member of the DCC immunoglobulin superfamily and is required for CNS and motor axon guidance. Cell 87:197–204
8. Chien C-B (1998) Why does the growth cone cross the road? Neuron 20:3–6
9. Kolodkin AL, Levengood DV, Rowe EG, et al (1997) Neuropilin is a Semaphorin III receptor. Cell 90:753–762
10. He Z, Tessier-Lavigne M (1997) Neuropilin is a receptor for the axonal chemorepellent Semaphorin III. Cell 90:739–751

11. Jouet M, Rosenthal A, Armstrong G, et al (1994) X-linked spastic paraplegia (SPG 1), MASA syndrome and X-linked hydrocephalus result from mutations in the L1 gene. Nat Genet 7:402–407
12. Wong EV, Kenwrick S, Willems P, et al (1995) Mutations in the cell adhesion molecule L1 cause mental retardation. Trends Neurosci 18:168–172
13. Bateman A, Jouet M, MacFarlane J, et al (1996) Outline structure of the human L1 cell adhesion molecule and the sites where mutations cause neurological disorders. EMBO J 15:6050–6059
14. Takeichi M (1995) Morphogenetic roles of classical cadherins. Curr Opin Cell Biol 7:619–627
15. Doherty P, Fazeli MS, Walsh FS (1995) The neural cell adhesion molecule and synaptic plasticity. J Neurobiol 26:437–446
16. Tomaselli KJ, Reichardt RF, Bixby JL (1986) Distinct molecular interactions mediate neuronal process outgrowth on non-neuronal cell surfaces and extracellular matrix. J Cell Biol 103:2659–2672
17. Bixby JL, Pratt RL, Lilien J, et al (1987) Neurite outgrowth on muscle cell surfaces involves extracellular matrix receptors as well as Ca^{2+}-dependent and independent cell adhesion molecules. Proc Natl Acad Sci USA 84:2555–2569
18. Chang S, Rathjen FG, Raper JA (1987) Extension of neurites on axons is impaired by antibodies against specific neural cell surface glycoproteins. J Cell Biol 104:355–362
19. Matsunaga M, Hatta K, Nagafuchi A, et al (1988) Guidance of optic nerve fibers by N-cadherin adhesion molecules. Nature (Lond) 334:62–64
20. Doherty P, Cohen J, Walsh FS (1990) Neurite outgrowth in response to transfected NCAM changes during development and is modulated by polysialic acid. Neuron 5:209–219
21. Doherty P, Fruns M, Seaton P, et al (1990) A threshold effect of the major isoforms of NCAM on neurite outgrowth. Nature (Lond) 343:464–466
22. Williams EJ, Doherty P, Turner G, et al (1992) Calcium influx into neurons can solely account for cell-contact dependent neurite outgrowth stimulated by transfected L1. J Cell Biol 119:885–892
23. Williams EJ, Furness J, Walsh FS, et al (1994) The production of arachidonic acid can account for calcium channel activation in the second messenger pathway underlying neurite outgrowth stimulated by NCAM, N-cadherin and L1. J Neurochem 62:2124–2131
24. Williams EJ, Walsh FS, Doherty P (1994) Tyrosine kinase inhibitors can differentially inhibit integrin-dependent and CAM-stimulated neurite outgrowth. J Cell Biol 124: 1029–1037
25. Williams EJ, Furness J, Walsh FS, et al (1994) Characterisation of the second messenger pathway underlying neurite outgrowth stimulated by FGF. Development (Camb) 120:1685–1693
26. Williams EJ, Furness J, Walsh FS, et al (1994) Activation of the FGF receptor underlies neurite outgrowth stimulated by L1, N-CAM, and N-cadherin. Neuron 13:583–594
27. Doherty P, Williams EJ, Walsh FS (1995) A soluble chimeric form of the L1 glycoprotein stimulates neurite outgrowth. Neuron 14:1–20
28. Ignelzi MA, Miller DR, Soriano P, et al (1994) Impaired neurite outgrowth of src-minus cerebellar neurons on the cell adhesion molecule L1. Neuron 12:873–884
29. Beggs HE, Soriano P, Maness PF (1994) NCAM dependent neurite outgrowth is inhibited in neurons from fyn-minus mice. J Cell Biol 127:825–833
30. Wong EV, Schaefer AW, Landreth G, et al (1996) Involvement of p90rsk in neurite outgrowth mediated by the cell adhesion molecule L1. J Biol Chem 271:18217–18223
31. Xu Y, Deng-Fu G, Davidson M, et al (1997) Interaction of the adaptor protein Shc and the adhesion molecule cadherin. J Biol Chem 272:13463–13466
32. Doherty P, Walsh FS (1996) CAM-FGF receptor interactions: a model for axonal growth. Mol Cell Neurosci 8:99–111

33. Green PJ, Walsh FS, Doherty P (1996) Promiscuity of fibroblast growth factor receptors. BioEssays 18:639–646
34. Brittis PA, Silver J, Walsh FS, et al (1996) FGF receptor function is required for the orderly projection of ganglion cell axons in the developing mammalian retina. Mol Cell Neurosci 8:120–128
35. Walsh FS, Meiri K, Doherty P (1997) Cell signalling and CAM-mediated neurite outgrowth. Soc Gen Physiol Ser 52:221–226

Development of the Neural Circuit in the Rat Cerebral Cortex

Koki Kawamura, Hitoshi Kawano, Tetsuya Fukuda, and Kyoji Ohyama

Summary. Expression of several molecules is involved in the formation of neural circuits. The neural cell adhesion molecules (NCAMs) L1 and TAG-1 are expressed in specific neurons of the embryonic brain and promote axonal growth, bundle formation, and axonal guidance. In the developing rat cerebral cortex, L1 and TAG-1 immunoreactions were specifically localized on thalamocortical axons and cortical efferent axons, respectively. L1-bearing thalamocortical axons preferentially traveled within the cortical subplate where neurocan, a chondroitin sulfate proteoglycan (CSPG), was specifically expressed. In contrast, axons immunoreactive for TAG-1 crossed the subplate and entered the intermediate zone, where another type of CSPG, phosphacan, was expressed. Although both TAG-1 and L1 have been reported to bind both neurocan and phosphacan in vitro, various patterns of interaction between NCAMs and CSPGs may play crucial roles in the formation of pathways in the rat forebrain. Molecular interactions among these elements at particular times and places are determined by genetic programming. Pax-6 is a transcription factor that is expressed in the developing eyes, nose, and brain. In the normal rat forebrain, Pax-6 was expressed in the primordial cerebral cortex and ventral thalamus. In the Pax-6 deficient mutant, thalamocortical axons took abnormal courses: they did not form the internal capsule and extended further ventrally. Our data indicate that Pax-6 regulates essential steps in forebrain development, especially in the formation of neural pathways.

Key words. Cortical development—Rat—Pax-6—L1—TAG-1

Introduction

During development of the mammalian cerebral cortex, postmitotic cells in the neuroepithelium of the lateral ventricle migrate toward the pial surface and form the cortical plate. Neurons in the superficial layer, designated as the preplate, are formed first, before cortical plate neurons. When cortical plate neurons migrate toward the

Department of Anatomy, School of Medicine, Keio University, 35 Shinanomachi, Shinjuku-ku, Tokyo 160-8582, Japan pial surface, the preplate is divided into two laminae: the marginal zone above and the subplate below the cortical plate [1,2]. Recent studies have shown that the preplate, marginal zone, and subplate play important roles in the formation of the cortical laminar structure and of afferent and efferent fiber connections in the cerebral cortex in cats [3–5], in rats [6,7], and in mice [8]. Furthermore, neural cell adhesion molecules (NCAMs) and extracellular matrix molecules that are expressed at specific times and places may be involved in cortical development. In this study, we examined immunohistochemically the localization of two groups of active intercellular molecules in the cerebral cortex of normal rat fetuses. One group is chondroitin sulfate proteoglycans (CSPGs), which are abundantly expressed in the fetal brain [9,10], and the other group is NCAMs, which are localized on elongating axons, including thalamocortical fibers [11,12]. The present data suggest that the heterophilic interaction between CSPGs and NCAMs is involved in the pathfinding of thalamocortical axons.

We also examined the development of axonal pathways in the cerebral cortex of Pax-6-deficient rats. The Pax-6 gene, which encodes a transcription factor, is normally found in the eyes, olfactory epithelium, pituitary gland, forebrain, hindbrain, and spinal cord during development in mice [13,14]. A spontaneous mutation of the Pax-6 gene results in severe defects in regions where Pax-6 is specifically expressed in mice (Small eye, Sey) [15] and in rats (rat Small eye, rSey) [16]. Aberrant axonal trajectories in Pax-6-deficient animals have also been shown recently in the pretectum [17] and hindbrain [18]. At an early stage of development, the Pax-6 message is intensely expressed in the anlagen of the cerebral cortex and thalamus [13,14,19]. It is, therefore, likely that normal expression of Pax-6 regulates the formation of cortical pathways. The present results indicate that Pax-6 regulates essential steps in forebrain development, especially in the determination of specific neural pathways.

Results

Immunohistochemical Localization of NCAMs in the Fetal Rat Cerebral Cortex

Immunohistochemical localization of the NCAMs L1 and TAG-1 was examined in the fetal rat cerebral cortex. L1 immunoreactivity was first found in the cortical anlage at E16 and was localized on a few fibers coming from the internal capsule. These fibers were believed to be thalamocortical axons, because fibers with a similar trajectory were labeled when 1,1′-dioctadecyl-3,3,3′,3′-tetramethyl-indocarbocyanine perchlorate (DiI) was placed on the E16 dorsal thalamus. In the cortical anlage, L1-immunoreactive fibers were observed to run in the subplate. With advancing age, L1-immunoreactive fibers increased in number and gradually extended to the lower part of the cortical plate.

In contrast, TAG-1 immunoreactivity was first seen at E14 in the preplate, the outermost layer of the cortical anlage where differentiated neurons were first found. At E15, TAG-1-immunoreactive fibers were observed in the cortical anlage extending toward the ganglionic eminence. When DiI was placed onto the surface of the E15 cortex, the trajectory of anterogradely labeled axons was similar to that of TAG-1-

immunoreactive axons, indicating that TAG-1-immunoreactive fibers are corticofugal. After E16, TAG-1 immunoreactivity was localized in newly formed cortical plate neurons and in thick fiber bundles in the intermediate zone. TAG-1 immunoreactivity appeared to peak between E16 and E20, then markedly decreased.

Immunohistochemical Localization of CSPGs in the Fetal Rat Cerebral Cortex

In the present study, two kinds of brain-specific CSPGs, neurocan and phosphacan, were examined immunohistochemically in the fetal rat cerebral cortex. At E14, neurocan immunoreactivity was first detected in the pial surface of the cortical anlage. At E16, intense neurocan immunoreactivity was localized in the marginal zone (the outermost layer) and the subplate. Thereafter, neurocan immunoreactivity gradually extended to the lower part of the cortical plate and was localized throughout the cortical wall at birth. In contrast, phosphacan immunoreactivity was uniformly distributed in all layers of the cortical anlage from the earliest stage of cortical development (E14). The pattern of distribution of phosphacan immunoreactivity persisted until birth.

Comparative Localization of NCAMs and CSPGs in the Fetal Rat Cortex

During cortical development, L1-immunoreactive fibers and areas of neurocan immunoreactivity overlapped extensively. At E16, intense neurocan immunoreactivity was localized in the subplate, and at E17 L1-immunoreactive axons invaded the cortical anlage along the subplate (Fig. 1A). With advancing age, neurocan immunoreactivity extended to the lower part of the cortical plate, and L1-immunoreactive fibers proceeded into the lower cortical plate.

In contrast, immunoreactivities of TAG-1 and neurocan did not overlap in the developing cortex. At E16–E17, corticofugal axons immunoreactive for TAG-1 ran in the intermediate zone but not in the subplate, whereas neurocan immunoreactivity was intensely localized in the subplate (Fig. 1B). At later stages of development, neurocan immunoreactivity extended throughout the cortical wall, while TAG-1 immunoreactivity had sharply decreased.

TAG-1-immunoreactive axons entered phosphacan-rich regions in the cortical anlage. TAG-1-immunoreactive cortical efferent axons ran in the intermediate zone, and phosphacan immunoreactivity was seen throughout the cortical anlage, including the intermediate zone.

Development of Cerebral Cortex in the Pax-6-Deficient Rat Fetus

In the present study, $rSey^2$, which is a novel allele of rSey, showed a phenotype similar to that of rSey. In $rSey^2$ homozygotes, both the telencephalic and diencephalic anlagen were clearly underdeveloped. Although the cortical wall of $rSey^2/rSey^2$ was much

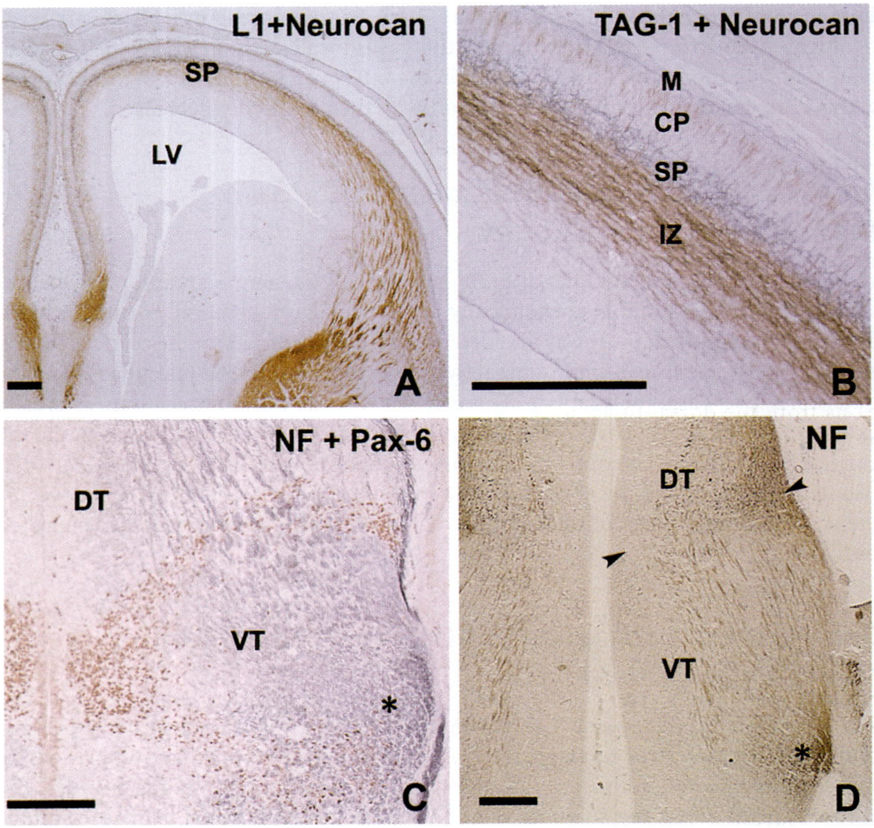

FIG. 1A–D. Photomicrographs showing the development of thalamocortical (**A,C**) and corticothalamic (**B**) fibers in normal E17 (**A,B**) and E16 (**C**) rats by double immunostaining in coronal sections. **A** L1-positive thalamic fibers (visualized with diaminobenzidine tetrahydrochloride [DAB] as *brown*) enter the subplate (*SP*) where neurocan (visualized with 4-chloro-1-naphthol as *blue/purple*) is localized. **B** TAG-1-positive corticothalamic fibers run in the intermediate zone (*IZ*) just below the neurocan-positive SP. **C** Neurofilament (NF) -positive fibers (*blue/purple*) from the dorsal thalamus (*DT*) traverse the dorsal band of the Pax-6 area (*brown*) but not the ventral Pax-6 band and course laterally (*asterisk*) before entering the internal capsule. **D** NF-immunoreactive axons from the DT of E16 $rSey^2/rSey^2$ traverse the zona limitans intrathalamica (*ZLI*, *arrowheads*) and proceed to the bottom (*asterisk*) of the ventral thalamus (*VT*). *LV*, lateral ventricle; *M*, marginal zone; *CP*, cortical plate. *Bars* 200 μm

thinner than that of normal SD rats, the lamination of the $rSey^2/rSey^2$ cortical anlage was observed at E16 by immunostaining of neurocan. In the cortical anlage of both normal and $rSey^2/rSey^2$ rats, neurocan was intensely expressed in the marginal zone and in the subplate.

In the cortices of both normal and mutant E17 rats, neurons in the cortical plate and fibers in the intermediate zone exhibited immunoreactivity for TAG-1. Thus, the course of corticofugal axons is normal in the cortical anlage of mutant rats.

Abnormal Formation of Thalamocortical Fiber Pathway in $rSey^2/rSey^2$

In the dorsolateral part of the cortical anlage, L1-immunoreactive thalamocortical axons were first detected at E16 running along the subplate in normal rats but not in mutant rats. When neurofilament (NF) was immunostained in forebrain sections from E16 normal rats, the trajectories of the thalamocortical pathway were clearly visible: NF-immunoreactive axons from the dorsal thalamus crossed the zona limitans intrathalamica (ZLI), which is the border between the dorsal and ventral thalami, and formed a compact bundle at the lateral part of the ventral thalamus. At more anterior levels, NF-immunoreactive axons from the dorsal thalamus changed their course to run within the anlage of the corpus striatum and then coursed up into the ventrolateral part of the cortical anlage. In age-matched $rSey^2/rSey^2$ homozygotes, axons from the dorsal thalamus followed an abnormal course in the diencephalon. NF-immunoreactive axons from the dorsal thalamus crossed the ZLI, then continued to the bottom of the ventral thalamus (Fig. 1D). At more anterior levels, they converged on the ventrolateral corner of the ventral thalamus and coursed up toward the ganglionic eminence to fan out in the basal region of the telencephalon.

In E16 normal rats, L1-immunoreactive axons were localized in the internal capsule and the cortical anlage. In the basal telencephalon of the mutant, axons that converged abnormally on the ventrolateral corner of the ventral thalamus were immunoreactive for L1, indicating that these abnormal fibers were of dorsal thalamic origin.

Immunohistochemical Localization of Pax-6 Protein in the Cerebral Cortex of Normal Rats

In the neocortex of normal rats, Pax-6-immunoreactive cells were localized in the ventricular zone throughout the embryonic period examined, although the number of immunoreactive cells gradually decreased with advancing age. Scattered immunoreactive cells were seen in the presumptive amygdala. In the middle level of the diencephalon of normal rats, Pax-6-immunoreactive cells in the mantle layer of the ventral thalamus were divided into two domains: one domain consisted of the cell cluster arranged along the ZLI and the other domain was in the ventral part. At E16, axons from the dorsal thalamus formed a thick bundle in the lateral part of the ventral thalamus between two Pax-6 domains (Fig. 1C). Thalamocortical axons in the internal capsule ran immediately dorsal to the amygdaloid Pax-6-immunoreactive cells.

Discussion

Neurocan and L1 are Potentially Important for Thalamocortical Pathway Formation

In the present study, neurocan immunoreactivity was shown to be specifically localized in the subplate of the fetal rat cortex. Bicknese et al. [7] showed that thalamocortical axons preferentially travel within the subplate where CSPGs are specifically expressed. Furthermore, Emerling and Lander [20] reported that in slices

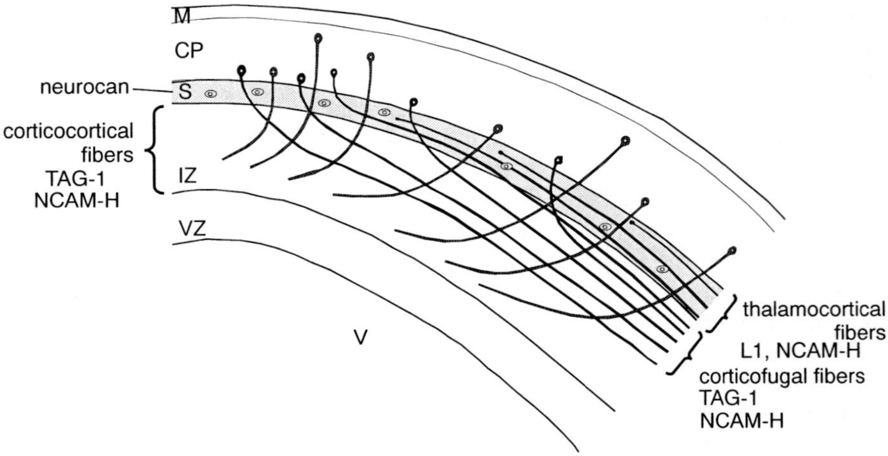

FIG. 2. Schematic diagram of the pathways of cortical axons in the E16 normal rat. L1-immunoreactive thalamocortical axons run in the subplate (*S*) where neurocan is specifically localized. In contrast, TAG-1-bearing corticofugal axons run in the intermediate zone where neurocan is scanty

of fetal mouse cerebral cortex, dissociated fetal thalamic neurons attached well to the intermediate zone and subplate but poorly to the cortical plate. When slices were treated with chondroitinase ABC, which degrades chondroitin sulfate side chains, the thalamic neurons attached less well to the subplate, suggesting that the specific affinity of the subplate for thalamic neurons may be attributed to chondroitin sulfate proteoglycans. Results of these studies strongly suggest that neurocan is a permissive substrate for both cell attachment and axonal growth of thalamic neurons.

In the present study, L1 was specifically localized on growing axons of the thalamocortical pathway from their initial stage of cortical innervation, as previously demonstrated in the mouse cortex [11,12]. That the growth-cone trajectories of L1-bearing thalamocortical axons in the present study were in close relation to areas of neurocan immunoreactivity suggests that L1-positive axons prefer neurocan-rich regions of the developing rat cortex and that the interaction between these molecules is involved in determining the thalamocortical pathway (Fig. 2).

L1 exerts its effects by homophilic interactions between L1 molecules and by heterophilic interactions between L1 and other molecules of the immunoglobulin superfamily [21–23]. Moreover, neurocan binds in vitro to L1 [24] and to its presumed chick homolog Ng-CAM [25]. Therefore, this heterophilic interaction supports the hypothesis that in the rat fetus neurocan and L1 mediate the determination of the pathways and growth of thalamic axons that invade the cortex.

TAG-1-Immunoreactive Fiber Pathways and CSPGs

As demonstrated in this study, immunoreactivity of TAG-1, a cell adhesion molecule, was localized on corticofugal and commissural axons, but not on thalamocortical afferent axons, from an early stage of development. As with L1, two groups of molecules that can bind to TAG-1 have been reported: adhesion molecules of the Ig superfamily, including TAG-1 itself [23,26–28], and extracellular matrix molecules, such as neurocan, phosphacan, tenascin-C, laminin, and collagen types I, II, IV, and VI [28]. The present histological results support the involvement of neurocan and phosphacan in the pathfinding of TAG-1-bearing cortical axons (see Fig. 2).

TAG-1-immunoreactive cortical axons do not invade the neurocan-rich subplate. As shown here, TAG-1-immunoreactive axons never overlapped with areas of neurocan immunoreactivity at any stage of cortical development. The spatiotemporal relationship between TAG-1 and neurocan agrees well with the present in vitro results that the process outgrowth of TAG-1-expressing PC12D cells was markedly inhibited on 1G2/neurocan substrate. Although TAG-1 binds to neurocan in vitro [28], neurocan may act as a repulsive and inhibitory substrate for axons of TAG-1-expressing cortical neurons and cause them to avoid neurocan-rich regions. Unlike that of neurocan, phosphacan immunoreactivity is distributed widely in the cortical anlage. During most stages of cortical development, phosphacan immunoreactivity detected with monoclonal antibodies is homogeneously distributed in all cortical layers [10,29,30]. Our histological results suggest that phosphacan acts as a permissive substrate for growth of TAG-1-bearing cortical axons, a conclusion consistent with findings of an in vitro study that phosphacan substrate promotes process outgrowth of TAG-1-expressing PC12D cells (manuscript in preparation).

Defects of Thalamocortical Pathway Formation in rSey^2/rSey^2 Rats

To our knowledge, this study is the first to demonstrate an aberrant trajectory of the thalamocortical tract in Pax-6-deficient animals. In the normal forebrain, both Pax-6 protein and mRNA are abundant in the olfactory bulb, neocortex, amygdala, hippocampus, and ventral thalamus [14]. Because the dorsal thalamic axons are misrouted in the ventral thalamus, the lack of Pax-6 expression in the ventral thalamus of mutants may be responsible (Fig. 3).

In the mantle layer of the ventral thalamus, Pax-6-immunoreactive cells were divided into dorsal and ventral clusters. The dorsal cluster was immediately below the ZLI, and the ventral cluster was distributed along the ventral border with the hypothalamus. In Pax-6-deficient mice, the expression patterns of such genes as Sonic hedgehog, Patched, Gli3, mSIM, and Dbx suggest that the ZLI is enlarged [31]. However, the present finding that thalamic axons of rSey^2/rSey^2 traversed the ZLI, as do those of the wild type, suggests that the Pax-6 localized in a band just below the ZLI (dorsal Pax-6 domain) does not play an essential role in the pathfinding of dorsal thalamic axons. In contrast, when dorsal thalamic axons encountered the ventral cluster of Pax-6-immunoreactive cells (ventral Pax-6 domain) they changed their direction to run laterally, suggesting the "negative" involvement of the ventral Pax-6 domain in the pathfinding of thalamocortical axons. In the ventral thalamus of rSey^2/rSey^2, dorsal thalamic axons traversed the region where the ventral Pax-6 domain would normally be in the wild type and reached the ventrolateral corner of the ventral thalamus.

A similar phenomenon was observed with a cluster of Pax-6-immunoreactive cells in the amygdala. In normal SD rats, when dorsal thalamic axons extended to the

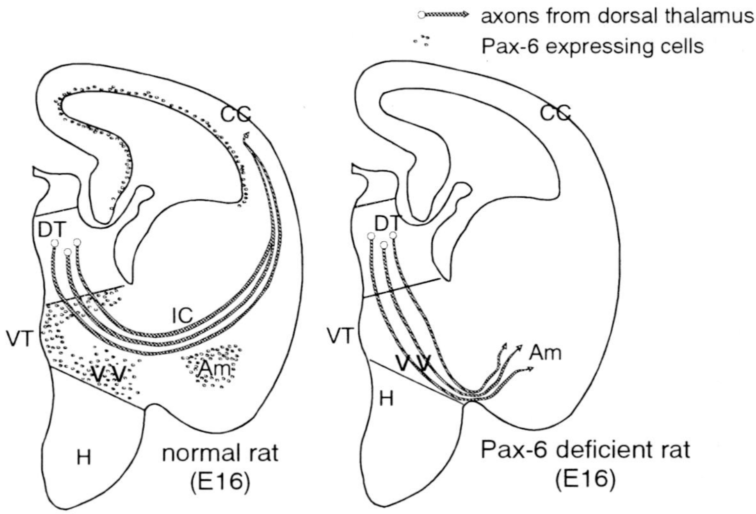

FIG. 3. Schematic diagram of the trajectory of developing thalamic axons. In the normal rat (*left side*), axons from the dorsal thalamus (*DT*) traverse the ZLI and form a bundle in the lateral side of the ventral thalamus (*VT*) between the two Pax-6 domains. These then continue in the internal capsule immediately dorsal to the amygdoloid Pax-6-immunoreactive domain and course up into the cortical anlage. In the Pax-6 mutant (*right side*), axons from the DT cross the ZLI, proceed to the ventrolateral corner of the VT, and fan out into the basal region of the telencephalon

ganglionic eminence they coursed up to the cortical anlage without entering the Pax-6-immunoreactive domain of the amygdala. In $rSey^2/rSey^2$, dorsal thalamic axons, which abnormally ran along the base of the brain, fanned out into the basal region of the telencephalon where the presumptive amygdala was located.

The present results raise the possibility that Pax-6-immunoreactive cells in the ventral thalamus and the amygdala act as "barriers" to dorsal thalamic axons. Because Pax-6 is a transcriptional regulator, its downstream target genes likely repulse the growth cones of thalamic axons. Although the genes of L1 and NCAM have been suggested to be targets of Pax-6 [32,33], distribution patterns of these NCAMs did not seem to be markedly affected in the forebrain of $rSey^2/rSey^2$. As yet unidentified molecules of a downstream target of Pax-6 may be involved in mediating the effects of Pax-6 in determining the trajectory of this projection.

Acknowledgments. This work was supported in part by grant-in-aid #07279101 (to K.K.) for scientific research on priority areas on "Functional Development of Neural Circuits," the Ministry of Education, Science, Sports and Culture of Japan.

References

1. Shatz CJ, Chun JJM, Luskin MB (1988) The role of the subplate in the development of the mammalian telencephalon. In: Peters A, Jones EG, (eds) Cerebral cortex, vol 7. Development and maturation of cerebral cortex. Plenum, New York, pp 35–58

2. Bayer SA, Altman J (1991) Neocortical development. Raven, New York
3. McConnell SK, Ghosh A, Shatz CJ (1989) Subplate neurons pioneer the first axon pathway from the cerebral cortex. Science 245:978–982
4. Shatz CJ, Ghosh A, McConnell SK, et al (1990) Pioneer neurons and target selection in cerebral cortical development. Cold Spring Harbor Symp Quant Biol 55: 469–480
5. Ghosh A, Antonini A, McConnell SK, et al (1990) Requirement for subplate neurons in the formation of thalamocortical connections. Nature (Lond) 347:179–181
6. De Carlos JA, O'Leary DDM (1992) Growth and targeting of subplate axons and establishment of major cortical pathways. J Neurosci 12:1194–1211
7. Bicknese AR, Sheppard AM, O'Leary DDM, et al (1994) Thalamocortical axons extend along a chondroitin sulfate proteoglycan-enriched pathway coincident with the neocortical subplate and distinct from the efferent path. J Neurosci 14:3500–3510
8. Rakic P, Caviness VS (1995) Cortical development: view from neurological mutants two decades later. Neuron 14:1101–1104
9. Oohira A, Matsui F, Watanabe E, et al (1994) Developmentally regulated expression of a brain specific species of chondroitin sulfate proteoglycan, neurocan, identified with a monoclonal antibody 1G2 in the rat cerebrum. Neuroscience 60:145–157
10. Miller B, Sheppard AM, Bicknese AR, et al (1995) Chondroitin sulfate proteoglycans in the developing cerebral cortex: the distribution of neurocan distinguishes forming afferent and efferent axonal pathways. J Comp Neurol 355:615–628
11. Fushiki S, Schachner M (1986) Immunocytological localization of cell adhesion molecules L1 and N-CAM and the shared carbohydrate epitope L2 during development of the mouse neocortex. Dev Brain Res 24:153–167
12. Chung W-W, Lagenaur CF, Yan Y, et al (1991) Developmental expression of neural cell adhesion molecules in the mouse neocortex and olfactory bulb. J Comp Neurol 314:290–305
13. Walther C, Gruss P (1991) *Pax-6*, a murine paired box gene, is expressed in the developing CNS. Development 113:1435–1449
14. Stoykova A, Gruss P (1994) Roles of *Pax*-genes in developing and adult brain as suggested by expression patterns. J Neurosci 14:1395–1412
15. Roberts RC (1967) *Small eyes*—a new dominant mutation in the mouse. Genet Res Camb 9:121–122
16. Matsuo T, Osumi-Yamashita N, Noji S, et al (1993) A mutation in the Pax-6 gene in rat *small eye* is associated with impaired migration of midbrain crest cells. Nat Genet 3:299–304
17. Mastick GS, Davis NM, Andrews GL, et al (1997) *Pax-6* functions in boundary formation and axon guidance in the embryonic mouse forebrain. Development (Camb) 124:1985–1997
18. Osumi N, Hirota A, Ohuchi H, et al (1997) *Pax-6* is involved in the specification of hindbrain motor neuron subtype. Development (Camb) 124:2961–2972
19. Stoykova A, Fritsch R, Walther C, et al (1996) Forebrain patterning defects in *Small eye* mutant mice. Development (Camb) 122:3453–3465
20. Emerling DE, Lander AD (1994) Laminar specific attachment and neurite outgrowth of thalamic neurons on cultured slices of developing cerebral neocortex. Development (Camb) 120:2811–2822
21. Lemmon V, Farr KL, Lagenaur C (1989) L1-mediated axon outgrowth occurs via a homophilic binding mechanism. Neuron 2:1597–1603
22. Kadmon G, Kowitz A, Altevogt P, et al (1990) The neural cell adhesion molecule N-CAM enhances L1-dependent cell-cell interactions. J Cell Biol 110:193–208
23. Felsenfeld DP, Hynes MA, Skoler KM, et al (1994) TAG-1 can mediate homophilic binding, but neurite outgrowth on TAG-1 requires an L1-like molecule and β1 integrins. Neuron 12:675–690
24. Friedlander DR, Milev P, Karthikeyan L, et al (1994) The neuronal chondroitin sulfate proteoglycan neurocan binds to the neural cell adhesion molecules Ng-CAM/L1/NILE

and N-CAM, and inhibits neuronal adhesion and neurite outgrowth. J Cell Biol 125:669–680
25. Grumet M, Flaccus A, Margolis RU (1993) Functional characterization of chondroitin sulfate proteoglycans of brain: interactions with neurons and neural cell adhesion molecules. J Cell Biol 120:815–824
26. Kuhn TB, Stoeckli ET, Condrau MA, et al (1991) Neurite outgrowth on immobilized axonin-1 is mediated by a heterophilic interaction with L1(G4). J Cell Biol 115:1113–1126
27. Stoeckli ET, Landmesser LT (1995) Axonin-1, Nr-CAM, and Ng-CAM, play different roles in the in vivo guidance of chick commissural neurons. Neuron 14: 1165–1179
28. Milev P, Maurel P, Häring M, et al (1996) TAG-1/Axonin-1 is a high-affinity ligand of neurocan, phosphacan/protein-tyrosine phosphatase-ζ/β, and N-CAM. J Biol Chem 271:15716–15723
29. Maeda N, Hamanaka H, Oohira A, et al (1995) Purification, characterization and developmental expression of a brain-specific chondroitin sulfate proteoglycan, 6B4 proteoglycan/phosphacan. Neuroscience 67:23–35
30. Fukuda T, Kawano H, Ohyama K, et al (1997) Immunohistochemical localization of neurocan and L1 in the formation of thalamocortical pathway of developing rats. J Comp Neurol 382:141–152
31. Grindley JC, Hargett LK, Hill RE, et al (1997) Disruption of PAX6 function in mice homozygous for the $Pax6^{Sey\text{-}1Neu}$ mutation produces abnormalities in the early development and regionalization of the diencephalon. Mech Dev 64:111–126
32. Chalepakis G, Wijnholds J, Giese P, et al (1994) Characterization of Pax-6 and Hoxa-1 binding to the promoter region of the neural cell adhesion molecule L1. DNA Cell Biol 13:891–900
33. Holst BD, Wang Y, Jones FS, et al (1997) A binding site for Pax proteins regulates expression of the gene for the neural cell adhesion molecule in the embryonic spinal cord. Proc Natl Acad Sci USA 94:1465–1470

The Role of Sialic Acid Binding in the Inhibition of Axonal Regeneration by Myelin-Associated Glycoprotein, MAG

Marie T. Filbin, Song Tang, Maria Elena deBellard, and Ying Jing Shen

Summary. The mammalian CNS does not regenerate after injury in part because of the presence of myelin-specific inhibitory molecules. One of these molecules, myelin-associated glycoprotein (MAG), is a potent inhibitor of axonal regeneration when presented to neurons expressed by cells as a substrate, or when added to growing neurons in solution. MAG is also a sialic acid-binding protein and binds neurons in a sialic acid-dependent manner. Moreover, desialylation of neurons not only partially reserves inhibition by MAG but also that by CNS myelin as a substrate. We have mapped the sialic acid-binding site on MAG to arginine 118 (R118). When mutated at R118, soluble MAG no longer binds to neurons and does not inhibit axonal regeneration. In contrast, when R118-mutated MAG expressed by cells is used as a substrate, inhibition is as potent as that of unmutated MAG. Therefore, we suggest that there are two recognition sites on MAG, a sialic acid-binding site and a distinct, neurite outgrowth inhibition site. Sialic acid binding of MAG may potentiate inhibition but alone is insufficient to effect inhibition of axonal regeneration.

Key words. Regeneration—Inhibition—MAG—Myelin—Sialic acid

Introduction

Myelin-associated glycoprotein (MAG) represents 0.1% and 1% of the total myelin protein of the peripheral nervous system (PNS) and the central nervous system (CNS), respectively [1,2]. Because of the early expression of MAG during development and because of its location in the first myelin wrap next to the axon, MAG has long been suggested to play a role both in the initiation of myelination and in the maintenance of mature myelin [1–4]. However, studies with the MAG knockout (MAG−/−) mouse would support the former suggestion but not the latter, as in these mice essentially normal myelin is formed that is subsequently lost with age [5,6]. In addition to its putative role in the formation and maintenance of myelin, MAG has been shown, like

Department of Biological Sciences, Hunter College of the City University of New York, 695 Park Avenue, New York, NY 10024, USA many other members of the immunoglobulin (Ig) gene superfamily (MAG contains five Ig-like domains in its extracellular sequences), to promote axonal growth, in this case from neonatal dorsal root ganglion (DRG) neurons [7]. However, more recently, we [8,9], and others [10,11] have shown that, depending on the age and type of neuron, MAG can either promote or inhibit axonal regrowth. That is, MAG is a bifunctional molecule because it promotes axonal growth from neonatal DRG neurons but inhibits outgrowth from adult DRG neurons and all other postnatal neurons tested to date [8,9]. Furthermore, it is well established that CNS myelin in vivo and in vitro is inhibitory for regeneration of axons of the adult mammalian CNS [12,13]. MAG in myelin has been suggested to contribute significantly to the inhibitory properties of this membrane when used as a substrate [10,23]. On the other hand, studies on regeneration in the MAG−/− mouse are controversial; one group reported not only an improvement in the length of regenerating axons compared to wild-type mice but also an increase in the number of axons regenerating [11], while another group reported no difference in regeneration [14]. Finally, a clue to the identity of the putative MAG neuronal receptor came when MAG was identified as a sialic acid-binding protein and included in a subfamily of the Ig superfamily termed the Siglecs (previously called Sialoadhesins) [15]. The next step was to assess whether neuronal sialic acid-dependent binding played a role in inhibition by CNS myelin and how MAG effected sialic acid binding to neurons.

Desialylation of Neurons Improves Neurite Outgrowth on Myelin

We have already established that MAG binds to neurons in a sialic acid-dependent manner [9,15] and that desialylation of neurons partially reverses the inhibitory effects of MAG-expressing Chinese ovary (CHO) cells [15]. In addition, studies using MAG−/− and MAG+/+ myelin as a substrate suggest that MAG contributes to the overall inhibitory properties of myelin [23]. To determine if a sialic acid-bearing component plays a role in the inhibition by myelin, neurons were desialylated before overnight culture on dishes coated with purified myelin. It was found that, when grown on CNS myelin as a substrate, neurites from desialylated neurons were about 100% longer than neurites from neonatal cerebellar neurons that had intact surface sialic acid residues (Fig. 1). Desialylation had no effect on neurite length from neurons grown on liver or gray matter as a substrate. This indicates that neuron–myelin sialic acid-dependent binding plays a role in the inhibitory properties of CNS myelin [23].

The Sialic Acid-Binding Site on MAG is Arg118

Two other members of the Ig subfamily of sialic acid-binding proteins, the Siglecs, are CD22 and sialoadhesin [15]. Recently, by site-directed mutagenesis, an arginine residue conserved in the first Ig domains of both CD22 and sialoadhesin was shown to be crucial for the sialic acid-dependent binding of each of these molecules [16,17]. This arginine is also conserved in MAG at Arg118. To determine if this conserved arginine

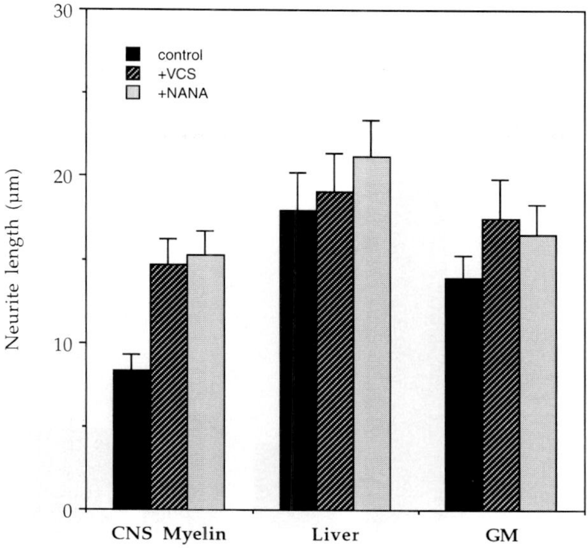

FIG. 1. The effect of desialylation on inhibition of neurite outgrowth by CNS myelin. Dissociated cerebellar neurons from PND3 rats were cultured overnight on wells coated with CNS myelin, liver, or gray matter (*GM*) at a concentration of 3 μg total protein/well before being fixed, immunostained for GAP43, and the neurite length measured. Results show the mean length of the longest neurite per cell (± SEM) for at least 200 individual neurons. *Black bars*, control neurons; *striped bars*, neurons were desialylated before the neurite outgrowth assay and sialidase was included in the cultures; *gray bars*, the sialic acid-bearing sugar, DDNANA, was included in the cultures at 1 mM

played a role in the sialic acid-dependent binding of MAG, the nucleotides coding for Arg118 were mutated such that they coded for either aspartate, a disruptive substitution (R118D), or for alanine, a nondisruptive substitution (R118A). The mutated MAG proteins were expressed as soluble Fc-chimeras (the extracellular domain of MAG fused to the Fc region of human IgG). Because a conformation-dependent MAG monoclonal antibody recognized both R118D- and R118A-MAG-Fc with the same affinity as it recognized unmutated MAG, it was concluded that these mutations in MAG did not affect its conformation [18].

The ability of R118D- and R118A-MAG-Fc, when coated onto a 96-well plate, to bind to neurons was tested in a manner similar to that which we had previously used to demonstrate the sialic acid-dependent binding of unmutated, soluble MAG-Fc to neurons [15]. Binding of neither R118D-MAG-Fc nor R118A-MAG-Fc could be detected in this binding assay (Fig. 2). Only background binding equivalent to that of a control Fc-chimera, MUC18-Fc, was observed. These results, together with the observation that a conformation-dependent MAG monoclonal antibody still recognizes the mutated proteins, suggests that the amino acid Arg118 is directly involved in the sialic acid-dependent binding of MAG [18].

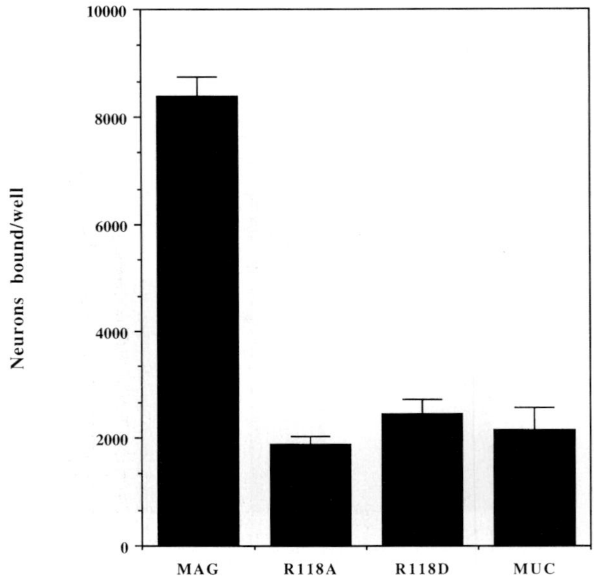

Fig. 2. The effect of mutation of R118 in MAG-Fc on binding to neurons. Dissociated PND4 cerebellar neurons, vitally labeled with calcein AM, were allowed to bind to MAG-Fc (*MAG*), R118A-MAG-Fc (*R118A*), R118D-MAG-Fc (*R118D*), or MUC18-Fc (*MUC*) immobilized at 10 μg/ml on a 96-well dish; 200 000 neurons were added to each well, and after incubation and washing, the number of cells bound was quantitated with a FluorImager. Results are from three experiments, each with 10 samples, and represent mean ± SE. (From [18], with permission)

Soluble, Mutated MAG-Fc Does Not Inhibit Axonal Regeneration

We had already shown that soluble MAG-Fc could inhibit axonal growth from cerebellar neurons on a variety of substrates in a specific, dose-dependent manner [19]. It is possible that R118D- and R118A-MAG-Fc could still bind to neurons but with an affinity too low to be measured in the binding assay but which would still effect inhibition of axonal growth. To test this possibility, R118D- or R118A-MAG-Fc were each added to cultures of cerebellar neurons growing on a substrate of the growth-promoting molecule L1 [19]. When added to cultures of cerebellar neurons at 25 μg/ml, neither R118D- nor R118A-MAG-Fc had any effect on neurite outgrowth; neurite length was the same as in the presence of the control chimera, MUC18-Fc (Fig. 3). In contrast, as observed previously [19], unmutated MAG-Fc at a similar concentration inhibited outgrowth by 50% (Fig. 3). These results indicate that the inhibition of axonal regeneration by soluble MAG-Fc is completely dependent on its ability to bind to neurons in a sialic acid-dependent manner [18].

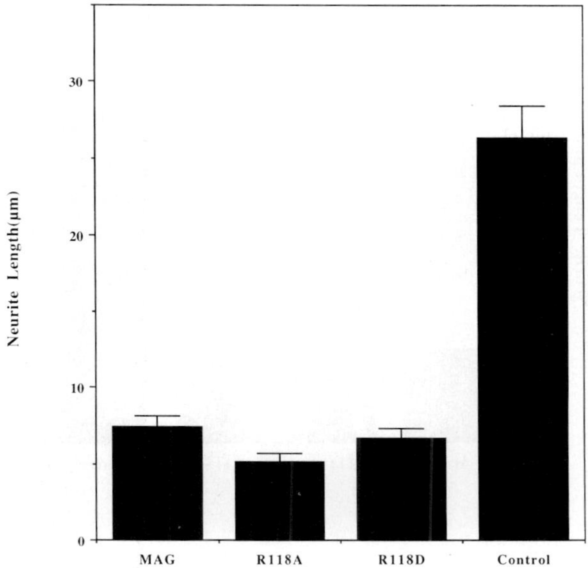

FIG. 3. The effect of mutation of R118 in MAG-Fc on inhibition of axonal growth. Dissociated PND 4 cerebellar neurons were incubated with MAG-Fc, R118A-MAG-Fc (R118A-Fc), R118D-MAG-Fc (R118D-Fc), or MUC18-Fc (MUC-Fc) at 25 µg/ml before being plated onto a substrate of L1-Fc immobilized on wells coated with anti-Fc antibody. Neurons were cultured for 18 h, fixed, and immunostained for GAP 43, and neurite length was measured. Results show the mean length of the longest neurite per cell (± SEM) for 100–200 individual neurons. (From [18], with permission)

Mutated MAG Expressed by Cells Does Inhibit Axonal Regeneration

The first demonstration that MAG could inhibit axonal regeneration was with MAG expressed by CHO cells [8]. Neurite length on MAG-expressing CHO cells was compared to neurite length on control CHO cells and inhibition on the MAG-expressing CHO cells was 70%–90%, depending on the type of neuron [8,9]. Subsequent to these studies we showed that MAG expressed by Schwann cells was also a potent inhibitor of axonal growth [23]. To test if mutated MAG when expressed by cells had any effect on axonal growth, the full-length R118D- and R118A-MAG cDNAs were individually transfected into CHO cells and a Schwann cell line that did not express MAG and which was permissive for axonal outgrowth [20,23]. A number of clones for each construct and for each cell type were chosen and expression of mutated MAG was characterized. For each clonal cell line, with the exception of Schwann cells transfected with R118D-MAG, mutated MAG reached the cell surface in quantities equivalent to cells expressing wild-type MAG, and mutated MAG was glycosylated to the same extent as wild-type MAG. For Schwann cells transfected with R118D-MAG, no

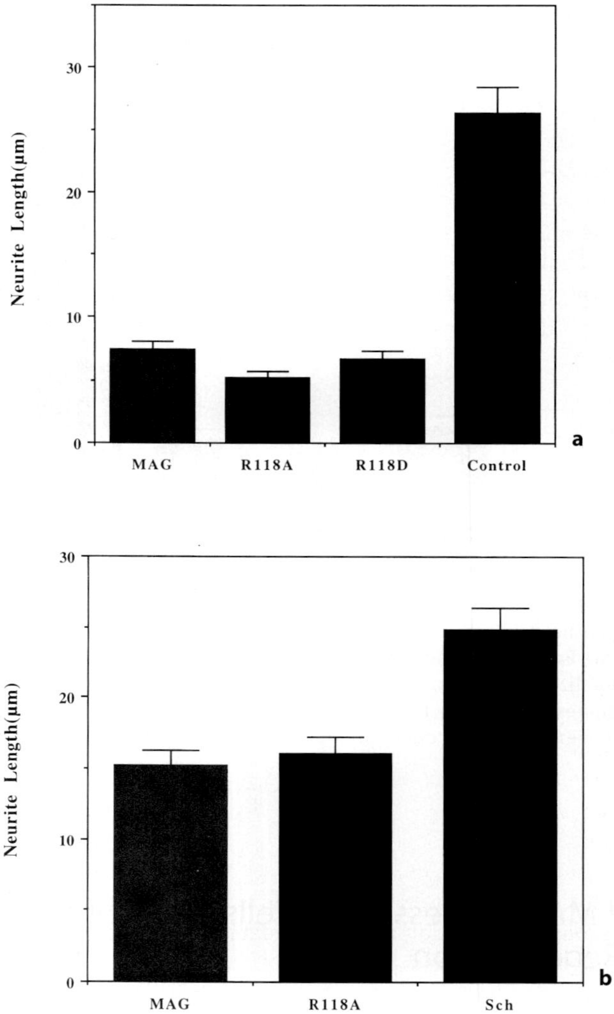

FIG. 4a,b. The effect of R118-mutated-MAG expressed by CHO or Schwann cells on neurite outgrowth from cerebellar neurons. Dissociated PND 4 cerebellar neurons were cultured for 18 h on confluent monolayers of (**a**) unmutated MAG-expressing (*MAG*), R118A-MAG-expressing (*R118A*), R118D-MAG-expressing (*R118D*), or control CHO cells or (**b**) MAG-expressing (*MAG*), R118A-MAG-expressing (*R118A*), or control Schwann cells not expressing MAG (*Sch*). Cocultures were fixed and immunostained for GAP 43, and neurite length was measured. Results show the mean length of the longest neurite per cell (± SEM) for 100–200 individual neurons. (From [18], with permission)

expression could be detected. It is possible that expression of R118D-MAG preferentially induces cell death in Schwann cells but not in the more robust CHO cells. However, neurite outgrowth assays were conducted with CHO cells expressing R118D- or R118A-MAG and with Schwann cells expressing R118A-MAG. Surprising, we found that the mutated forms of MAG, expressed in either CHO or Schwann cells, inhibited axonal regeneration to the same extent as wild-type MAG expressed by the same cells (Fig. 4). The observation that R118A-MAG expressed by Schwann cells is an effective inhibitor of regrowth indicates that the results we obtained are not an artifact of CHO cells which, under physiological conditions, never express MAG. These results, unlike the findings with soluble MAG-Fc, indicate that the ability of MAG expressed by cells to inhibit axonal regeneration is independent of its ability to bind neurons in a sialic acid-dependent manner [18,19].

Conclusions

The requirements for inhibition of axonal regeneration by MAG appear to differ depending on whether MAG is presented to the neuron in a soluble form compared to when it is expressed by cells. That is to say, soluble MAG-Fc appears to be entirely dependent on sialic acid-binding to effect inhibition while MAG expressed by cells does not require interaction via a neuronal sialic acid to inhibit neurite outgrowth. To reconcile these apparent contradictory observations, we propose a model of MAG interaction with neurons and its subsequent effect on axonal growth [18]. In this model we propose that MAG has two neuronal recognition sites, a sialic acid-binding site and a neurite outgrowth inhibition site (Fig. 5). In a soluble form, MAG is completely dependent on sialic acid binding to allow the inhibitory site to interact. In contrast, when mutated MAG is expressed by cells other cell adhesion molecules, expressed by either CHO or Schwann cells, interact with the neuron in a sialic acid-independent manner and allow the inhibition site to engage; therefore, when expressed by cells, MAG mutated at the sialic acid inhibition site is as effective an inhibitor of axonal regeneration as the unmutated MAG (Fig. 5). It is interesting that a two-site model of interaction has recently been proposed for another family of inhibitory molecules, the collapsin/semaphorin family [20,21]. In that study the inhibition site was mapped to a sequence of 70 amino acids within a region conserved in all family members, termed the semaphorin domain. A second site, which is involved in potentiating the inhibitory effects of collapsins/semaphorins, has been identified in the positively charged carboxy terminus of the molecule [22].

However, because we have shown previously that inhibition of neurite outgrowth by MAG expressed by CHO cells can be partially reversed by desialylation of the neurons before the assay [9], sialic acid binding, when allowed to occur, may potentiate the inhibitory effects of MAG. This notion is supported by our observation that desialylation of neurons before culture on purified myelin as a substrate results in improved axonal growth, suggesting that sialic acid binding influences the overall inhibitory properties of myelin.

In conclusion, we suggest that the amino acid Arg118, in the first Ig domain of MAG, is crucial for its sialic acid-dependent binding to neurons. The ability of MAG to bind neurons in a sialic acid-dependent manner appears to be imperative for

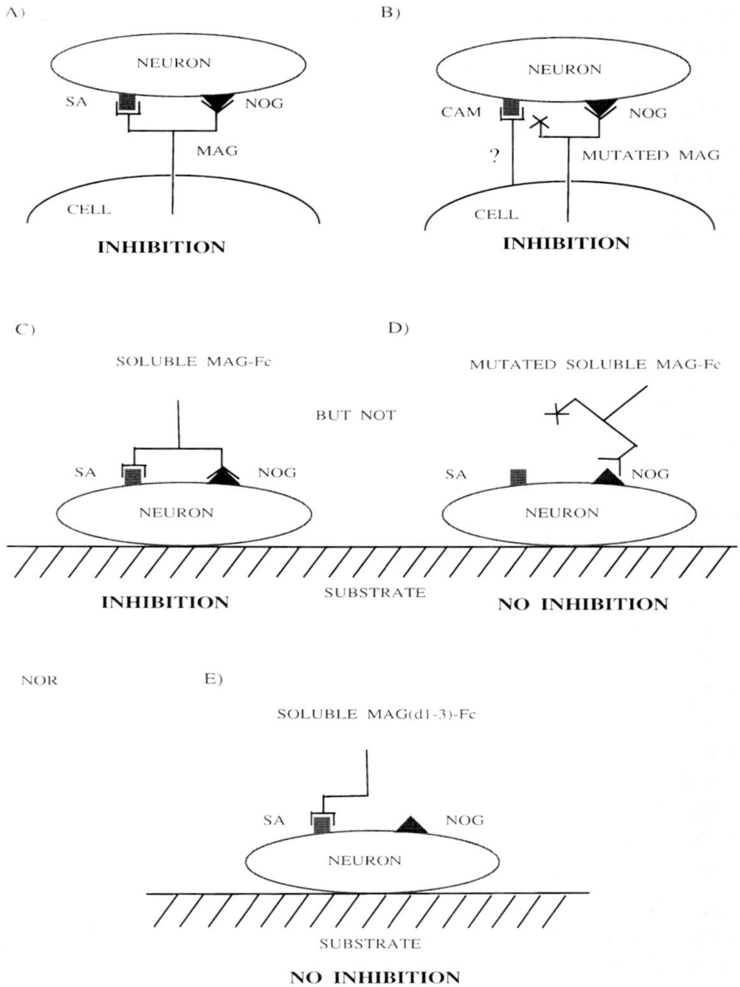

FIG. 5A–E. The sialic acid-binding site on MAG (R118) is critical for inhibition of axonal regeneration by soluble MAG-Fc but not MAG expressed by CHO cells. There are two recognition sites on MAG, a sialic acid-binding site (*SA, square symbol*) and a neurite outgrowth inhibition site (*NOG, triangular symbol*). When MAG is expressed by CHO cells, both sites engage the neuron and neurite outgrowth is inhibited (**A**). When MAG mutated at its sialic acid-binding site is expressed in CHO cells, another cell adhesion molecule (CAM) on the CHO cell surface engages the neuron along with the neurite outgrowth inhibition site and neurite outgrowth is still inhibited (**B**). When soluble MAG-Fc is added to neurons, both the sialic acid and the inhibition sites engage the neuron and neurite outgrowth is inhibited (**C**). However, when MAG-Fc mutated at the sialic acid-binding site is added to neurons, it cannot bind to neurons, and consequently the inhibition site cannot engage and there is no inhibition of neurite outgrowth (**D**). In contrast, when MAG(d1-3)-Fc is added to neurons, it binds via its sialic acid-binding site but does not inhibit axonal growth because the inhibition site is absent. (From [18], with permission)

soluble MAG to effect inhibition of axonal regeneration, while MAG expressed by cells seems somewhat independent of sialic acid binding in its inhibition.

Acknowledgments. We thank Dr. Lloyd Williams for his assistance with the image analysis. This work was supported by a grant from the National Multiple Sclerosis Society, the National Institutes of Health (NS 37060), and a core facility grant from the Research Centers for Minorities Institute-NIH. M.T.F. is a recipient of an Established Investigator Award from the American Heart Association, NY chapter, and holds the Marie L. Hesselbach Chair in Biology.

References

1. Quarles RH (1984) Myelin-associated glycoprotein in development and disease. Dev Neurosci 6:285–303
2. Trapp BD (1990) Myelin-associated glycoprotein, location and potential function. Ann NY Acad Sci 605:29–43.
3. Filbin MT (1995) Myelin-associated glycoprotein: a role in myelination and in the inhibition of axonal regeneration? Curr Opin Neurobiol 5:588–595
4. Filbin MT (1996) The muddle with MAG. Mol Cell Neurosci 8:84–92
5. Li C, Tropak MB, Gerlai R, Clapoff S, Abramow-Newerly W, Trapp B, Peterson A, Roder J (1994) Myelination in the absence of myelin-associated glycoprotein. Nature (Lond) 369:747–750
6. Montag D, Giese KP, Bartsch U, Martini R, Lang Y, Bluthmann H, Karthigasan J, Kirschner DA, Wintergerst ES, Nave KA, Schachner M (1994) Mice deficient for the myelin-associated glycoprotein show subtle abnormalities in myelin. Neuron 13:229–246
7. Johnson PW, Abramow-Newerly W, Seilheimer B, Sadoul R, Tropak MB, Arquint M, Dunn RJ, Schachner M, Roder JC (1989) Recombinant myelin associated glycoprotein confers neural adhesion and neurite outgrowth function. Neuron 3:377–385
8. Mukhopadhyay G, Doherty P, Walsh FS, Crocker PR, Filbin MT (1994) A novel role for myelin-associated glycoprotein as an inhibitor of axonal regeneration. Neuron 13:757–767
9. DeBellard ME, Tang S, Mukhopadhyay G, Shen Y, Filbin MT (1996) Myelin-associated glycoprotein inhibits axonal regeneration from a variety of neurons via interaction with a sialoglycoprotein. Mol Cell Neurosci 7:89–101
10. McKerracher L, David S, Jackson DL, Kottis V, Dunn RJ, Braun PE (1994) Identification of myelin-associated glycoprotein as a major myelin-derived inhibitor of neurite outgrowth. Neuron 13:805–811
11. Li M, Shibata A, Li C, Braun PE, McKerracher L, Roder J, Kater SB, David S (1996) Myelin-associated glycoprotein inhibits neurite/axon growth and causes growth cone collapse. J Neurosci Res 46:404–414
12. Johnson AR (1993) Contact inhibition in the failure of mammalian CNS axonal regeneration. BioEssays 15:807–813
13. Schwab ME, Kapfhammer JP, Bandtlow CE (1993) Inhibitors of neurite growth. Annu Rev Neurosci 16:565–596
14. Bartsch L, Bandtlow CE, Schnell L, Bartsch S, Spillmann AA, Rubin BP, Hillenbrand R, Montag D, Schwab ME, Schachner M (1995) Lack of evidence that the myelin-associated glycoprotein is a major inhibitor of axonal regeneration in the CNS. Neuron 15:1375–1381
15. Kelm S, Pelz A, Schauer R, Filbin MT, Tang S, de Bellard M-E, Schnaar RL, Mahoney JA, Hartnell A, Bradfield P, Crocker PR (1994) Sialoadhesin, myelin-associated glyco-

protein and CD22 define a new family of sialic acid-dependent adhesion molecules of the immunoglobulin superfamily. Curr Biol 4:965–972
16. van der Merwe A, Crocker PR, Vinson M, Barclay AN, Schauer R, Kelm S (1996) Localization of the putative sialic acid-binding site on the immunoglobulin superfamily cell-surface molecule CD22. J Biol Chem 271:9273–9280
17. Vinson M, van der Merwe A, Kelm S, May A, Jones EY, Crocker PR (1996) Characterization of the sialic acid-binding site in sialoadhesin by site-directed mutagenesis. J Biol Chem 271:9267–9272
18. Tang S, Shen Y, Mukhopadhyay G, DeBellard ME, Crocker P, Filbin MT (1997) Myelin-associated glycoprotein, MAG, recognizes neurons via two epitopes, a sialic acid binding site at Arg118 and a distinct neurite inhibition epitope. J Cell Biol 138:1355–1366
19. Tang S, Woodhall RW, Shen YJ, DeBellard ME, Saffell JL, Doherty P, Walsh FS, Filbin MT (1997) Soluble myelin-associated glycoprotein (MAG) found in vivo inhibits axonal regeneration. Mol Cell Neurosci 9:333–346
20. Porter SL, Glaser L, Bunge RP (1987) Release of autocrine growth factor by primary and immortalized Schwann cells. Proc Natl Acad Sci USA 83:7768–7771
21. Luo Y, Raible D, Raper JA (1993) Collapsin: a protein in brain that induces the collapse and paralysis of neuronal growth cones. Cell 75:217–227
22. Koppel AM, Feiner L, Kobayashi H, Raper JA (1997) A 70 amino acid region within the semaphorin domain activates specific cellular response of the semaphorin family members. Neuron 19:531–537
23. Shen Y, Debellard ME, Salzer JL, Roder J, Filbin MT (1998) Myelin-associated glycoprotein (MAG) in myelin and expressed by schwann cells inhibits axonal regeneration and branching. Mol Cell Neurosci (in press)

Function of the Cytoplasmic Domain of the Cell Adhesion Molecule L1 in Normal and Abnormal Brain Development

Hiroyuki Kamiguchi[1], Andrew W. Schaefer[1], Kristin E. Long[1], Mami Yamasaki[2], and Vance Lemmon[1]

Summary. L1 is a molecule that participates in axon growth in the nervous system and in neural cell migration in some brain regions. Several lines of evidence show that L1 is a multifunctional protein, capable of binding to different intracellular and extracellular partners. Because mutations of the L1 cytoplasmic domain in humans are sufficient to cause disruptions of major axon pathways, it is a reasonable conclusion that signals generated via the cytoplasmic domain or interactions with the cytoskeleton or endocytic machinery are required for normal axon growth. The demonstration that disruption of the L1 gene in mice produces a similar phenotype to human mutations indicates that targeted mutations of the L1 cytoplasmic domain in mice will allow careful functional analysis of this critical region of the protein.

Key words. Cell adhesion—Axon guidance—X-linked hydrocephalus—L1

Introduction

During development of the embryonic nervous system, migrating neurons and growing axons follow specific pathways to establish highly ordered structures and patterns of innervation. The interaction of migrating cells or neuronal growth cones with their microenvironment is an important component of proper brain development. There has been evidence for many years from many different types of experimental systems that cell adhesion molecules (CAMs) are important in development of the nervous system. Thus far, three major classes of cell-surface molecules have been identified that mediate binding of neural cells and axons to each other and to the extracellular matrix (ECM). These are cadherins and integrins and members of the immunoglobulin superfamily. Recently, it has become evident that mutations in one neural cell adhesion molecule, L1, produce devastating effects on development of the human nervous system.

[1] Department of Neurosciences, Case Western Reserve University, Cleveland, OH 44106-4975, USA

[2] Department of Neurosurgery, Osaka National Hospital, 2-1 Hoenzaka-cho, Chuo-ku, Osaka 540, Japan

Introduction to L1

L1 is a cell adhesion molecule found in very high concentrations on axons of projection neurons in the central nervous system (CNS) and the peripheral nervous system (PNS). This molecule has been described independently by several investigators. L1 was first found on the surface of rat PC12 cells, where its expression is regulated by NGF [1], hence the name NGF-inducible large external glycoprotein, or "NILE." Later, L1 was found on granule cells in the mouse cerebellum, being concentrated on their axons in the molecular layer [2], and on neurons in the chick embryo brain (neuron glial cell adhesion molecule, or Ng-CAM) [3–5]. While there has been confusion over the years as to whether L1 and Ng-CAM are true homologs, the recent work of Sonderegger and associates has shown that the original report of NgCAM sequence significantly underestimated the homology with L1 [6]. For this and other reasons it is clear that L1 and NgCAM are structural and functional homologs (see [7]).

Functional Studies of L1

L1 has been implicated in neural cell migration and in axon growth. Early studies found that antibodies to L1 disrupt migration of cerebellar granule cells from the external granule cell layer to the internal granule cell layer [8,9]. Subsequent studies expressing L1 in cell lines confirmed its ability to support migration of cerebellar granule cells [10,11]. Cell biological studies have also demonstrated that L1 is a particularly potent CAM in the regulation of axon behavior. Antibodies against L1 cause defasciculation of axons [5,12,13], and inhibit neurite growth on other neurites [14]. Purified L1 attached to a tissue culture dish is an excellent substrate for CNS and PNS axon growth in virto [15]. Chick and rat neurons are able to grow neurites on chick, mouse, rat, and human L1, indicating that the binding site of L1 has been conserved over a substantial range of evolution [7,15,16] despite significant changes in the amino acid sequence of the overall protein from one species to another.

While the most overt function of L1 appears to involve homophilic interactions [17] this molecule is also capable of interactions with other receptors on another cell, that is, by heterophilic *trans* binding. This additional role has been supported by studies on neuron-glial adhesion [18] and studies on neurite outgrowth on CNS glia [13]. Recently, direct evidence has shown that L1 can bind via a heterophilic interaction to axonin-1, a molecule present on axons [19], and that interaction with axonin-1 is crucial in L1-mediated neurite growth [6,20]. Axonin-1 is the chick homolog of TAG-1 in the mouse. There are also reports that L1 interacts with F3/F11/contactin and DM1/GRASP [21,22] in a heterophilic manner, although the data on these molecules are not as fully developed as that for axonin-1. Perhaps the most intriguing conclusion at this time is that axonin-1 binds to NgCAM in *cis* rather than *trans* and that this plays a crucial role in regulating L1-mediated neurite growth [6,23]. L1 has been reported to bind with high affinity to phosphacan/neurocan [24] and also bind integrins [25]. These data indicate that L1 has a complex set of interactions and raises the possibility that different L1-binding partners could produce different sequelae that change cell migration or growth cone behavior.

Structure of L1

Mammalian L1 is an integral membrane protein with about 1250 amino acids, a single transmembrane domain, and a cytoplasmic tail of 114 amino acids. The entire sequences for mouse L1 [26], rat L1 [10,27], and human L1 [7] are now available. L1 homologs have also been identified in the chick [16,28], goldfish and zebrafish [29,30], and *Drosophila* [31].

L1, like several other neuronal adhesion molecules such as NCAM and F11, is a member of the immunoglobulin superfamily (Ig). L1 has six Ig C2-like domains at the amino-termius and five fibronectin (Fn) type III-like domains adjacent to the cell membrane. The cytoplasmic domains of human L1 and mouse L1 are identical, suggesting an important function for this region because of its conservation. An alternatively spliced form of L1 is expressed in the PNS by non-myelinating Schwann cells and in neuroblastomas [32]. In the PNS variant, there are four amino acids (RSLE) deleted from the L1 cytoplasmic domain (L1CD) as well as a short spliced-out sequence in the amino-terminal region.

Biochemical studies have revealed that L1 has a molecular weight on SDS-PAGE of about 200 kDa, depending on the species [5,18,33,34]. Of the total, about 138 kDa is from polypeptide with the remaining resulting from carbohydrate side chains. Following immunopurification of L1 from detergent-solubilized brain membranes, a major portion of the material is found as two peptides with SDS-PAGE molecular weights of 135 and 85 kDa, which represent the amino-terminal and the carboxy-terminal portions, respectively, of the intact molecule. These two fragments are probably present in the intact tissue because of endogenous protease activity and not proteolysis during purification. The 135- and 85-kDa peptides interact with each other via a noncovalent interaction that persists even after endogenous proteolytic cleavage [34].

Bacterially expressed fusion proteins have been used to identify L1 domains important in promoting neurite growth. The second Ig domain is necessary and sufficient to mediate L1 homophilic binding and neurite growth [35]. However, the sixth Ig domain, which contains an RGD sequence, is also capable of promoting neurite growth for some but not all neurons, perhaps by binding to an integrin [36].

L1, Second-Messenger Systems, and the Cytoskeleton

Like several other CAMs and substrate adhesion molecules (SAMs), L1 has been implicated in intracellular signaling. Treatment of PC12 cells with anti-L1 antibodies, but not antiliver membrane antibodies, leads to an increase in intracellular pH and a decrease in intracellular Ca^{2+} [37]. Antibodies to L1, but not N-cadherin, inhibit endogenous phosphorylation of tubulin in growth cones [38]. Also, Ca^{2+} channel blockers inhibit L1-mediated neurite growth, suggesting Ca^{2+} is an essential component of the cascade initiated by L1–L1 binding [39]. Doherty and his associates have performed an extensive set of experiments suggesting that L1 signals are mediated via the fibroblast growth factor (FGF) receptor [40]. An alternative pathway involving *src* has been identified by Maness and associates [41]. Finally, we have found that components of a MAPK signaling pathway, raf, ERK2, and p90rsk, associate with L1, and ERK2 and p90rsk phosphorylate the L1CD and influence L1-mediated neurite growth [42,43]. L1 is also phosphorylated by casein kinase II at a site likely to regulate L1 trafficking [44].

L1 has been found to interact with the actin cytoskeleton. First, Bennet and associates found that ankyrin binds to a highly conserved segment of 10 amino acids near the C-terminal end of L1 and two related molecules, neurofascin and NrCAM [45]. Recently Bennett's group has found that the tyrosine in the ankyrin-binding region of neurofascin is phosphorylated and that this negatively regulates ankyrin binding [46]. Because this tyrosine is in a highly conserved region in the L1 family of CAMs, it probably regulates L1–ankyrin interactions as well. Subsequently, Stallcup and associates found that a second highly conserved region of the L1CD, immediately adjacent to the transmembrane region, is necessary and sufficient for colocalization with actin stress fibers in a cell line [47]. These experiments show that adhesion is not the only function of L1 and demonstrate that L1 may be able to influence growth cone and cell migration behavior via different mechanisms, such as by the initiation of different intracellular second-messenger cascades or binding to the actin cytoskeleton.

Trafficking of L1

The neuronal form of L1 contains an alternatively spliced sequence of RSLE that follows a tyrosine. The resulting sequence, *Y*RS*LE*, conforms to a tyrosine-based sorting signal, YxxL, that mediates interactions with the machinery that sorts membrane proteins into different membrane compartments. If the YRSLE sequence is altered by mutating the tyrosine or by alternative splicing to the nonneuronal form of L1, L1 is not targeted to neural axons and growth cones but remains in the neuronal cell body [48]. The YRSL mediates clathrin-dependent endocytosis of L1 by binding AP-2, an adaptor that links membrane proteins to the clathrin-based endocytic machinery. Mutations of the YRSLE prevent L1:AP-2 interactions and alter the endocytic pathways taken by L1. Immunohistochemical studies have revealed that L1 colocalizes with Eps15, another component of clathrin endocytic machinery, at the rear of growth cones [49]. This provides a mechanism whereby the L1 can be removed from the growth cone, permitting advancement of the extending axon. Internalization of L1 via a clathrin-based mechanism is also likely to be important in neural plasticity because a similar mechanism is involved in synaptic sprouting in *Aplysia* [50–52].

Genetics of Human L1

In 1990, L1 was mapped to the X-chromosome band Xq28 [53]. This region is known to contain a number of different forms of X-linked mental retardation with overlapping clinical signs, including X-linked hydrocephalus (XLH), MASA syndrome (mental retardation, aphasia, spastic paraplegia, adducted thumbs), agenesis/dysgenesis of the corpus callosum (ACC), and X-linked spastic paraplegia (XSP). Subsequently it was shown that a family with human X-linked hydrocephalus had a mutation in the L1 gene [54]. This inherited disease is either lethal or results in severe retardation, corpus callosum and corticospinal tract hypoplasia, and adducted thumbs [55,56]. Subsequently, mutations in L1 have been found in more than 80 additional XLH, MASA, XSP, and ACC families (see following for details). Based on the common clinical findings associated with L1 mutations, the syndrome has been reclassified as CRASH syndrome for corpus callosum hypoplasia, retardation, adducted thumbs, spastic paraparesis, and hydrocephalus [57]. By examining the clinical literature we have found that mutations that affect only the L1CD are much less likely to cause severe malformations of the developing brain that lead to early death or severe hydrocephalus than mutations of the L1 extracellular domain (L1ED) [58]. However, mutations of the L1CD are equally likely to cause defects in axon guidance in the corticospinal tracts. This indicates that the L1CD is crucial for axon guidance or outgrowth in some projection systems but is not required for important developmental processes requiring L1-mediated adhesion, probably involving cell migration near the brain midline.

The L1 Knockout Mice

Two different groups have published results from independent mice lines in which the L1 gene has been disrupted [57,58]. There are a number of similarities with the human mutations. Depending on the mouse strain, significant hydrocephalus is found. The carrier females are poor breeders, and L1 knockout offspring appear at a frequency approximately one-half that expected by Mendalian genetics. Nonmyelinating Schwann cells do not establish normal interactions with axons; they exhibit an immature phenotype surrounding multiple axons within a single lacuna. It is not yet known if this results from a defect in the axons or the Schwann cells. The corticospinal tract, which is located in a different part of the spinal cord in mice than in humans, is developmentally retarded and diminished in size. The establishment of a reliable animal model for CRASH syndrome indicates that it will be possible to analyze the function of the L1CD using in vivo systems, opening the possibility for direct testing of the hypothesis that the L1CD is critical for axon guidance, while L1-mediated adhesion is involved in morphogenetic movements that form important midline structures such as the thalamus and anterior vermis.

Acknowledgment. The work was supported by NIH grants EY-5285 and NS-34252 to V.L.

References

1. McGuire JC, Greene LA, Furano AV (1978) Nerve growth factor stimulates incorporation of fucose or glucosamine into an external glycoprotein in cultured rat PC12 pheochromocytoma cells. Cell 15:357–365
2. Rathjen FG, Schachner M (1984) Immunocytological and biochemical characterization of a new neuronal cell surface component (L1 antigen) which is involved in cell adhesion. EMBO J 3:1–10
3. Grumet M, Hoffman S, Edelman GM (1984) Two antigenically related neuronal cell adhesion molecules of different specificities mediate neuron-neuron and neuron-glia adhesion. Proc Natl Acad Sci USA 81:267–271
4. Lemmon V, McLoon S (1986) The appearance of an L1-like molecule in the chick primary visual pathway. J Neurosci 6:2987–2994

5. Rathjen FG, Wolff JM, Frank R, et al (1987) Membrane glycoproteins involved in neurite fasciculation. J Cell Biol 104:343–353
6. Buchstaller A, Kunz S, Berger P, et al (1996) Cell adhesion molecules NgCAM and axonin-1 form heterodimers in the neuronal membrane and cooperate in neurite outgrowth promotion. J Cell Biol 135:1593–1607
7. Hlavin ML, Lemmon V (1991) Molecular structure and functional testing of human L1CAM: an interspecies comparison. Genomics 11:416–423
8. Lindner J, Rathjen FG, Schachner M (1983) L1 mono- and polyclonal antibodies modify cell migration in early postnatal mouse cerebellum. Nature (Lond) 305:427–430
9. Chuong C-M, Crossin KL, Edelman GM (1987) Sequential expression and differential function of multiple adhesion molecules during the formation of cerebellar cortical layers. J Cell Biol 104:331–342
10. Miura M, Kobayashi M, Asou H, et al (1991) Molecular cloning of cDNA encoding the rat neural cell adhesion molecule L1—two L1 isoforms in the cytoplasmic region are produced by differential splicing. FEBS Lett 289:91–95
11. Miura M, Asou H, Kobayashi M, et al (1992) Functional expression of a full-length cDNA coding for rat neural cell adhesion molecule-L1 mediates homophilic intercellular adhesion and migration of cerebellar neurons. J Biol Chem 267:10752–10758
12. Stallcup WB, Beasley L (1985) Involvement of the nerve growth factor-inducible large external glycoprotein (NILE) in neurite fasciculation in primary cultures of rat brain. Proc Natl Acad Sci USA 82:1276–1280
13. Drazba J, Lemmon V (1990) The role of cell adhesion molecules in neurite outgrowth on Muller cells. Dev Biol 138:82–93
14. Chang S, Rathjen FG, Raper J (1987) Extension of neurites on axons is impaired by antibodies against specific neural cell surface glycoproteins. J Cell Biol 104:355–362
15. Lagenaur C, Lemmon V (1987) An L1-like molecule, the 8D9 antigen, is a potent substrate for neurite extension. Proc Natl Acad Sci USA 84:7753–7757
16. Burgoon MP, Grumet M, Mauro V, et al (1991) Structure of the chicken neuron-glial cell adhesion molecule, Ng-CAM: origin of the polypeptides and relation to the Ig superfamily. J Cell Biol 112:1017–1029
17. Lemmon V, Farr K, Lagenaur C (1989) L1-mediated axon outgrowth occurs via a homophilic binding mechanism. Neuron 2:1597–1603
18. Grumet M, Hoffman S, Chuong C-M, et al (1984) Polypeptide components and binding functions of neuron-glia cell adhesion molecules. Proc Natl Acad Sci USA 81:7989–7993
19. Kuhn TB, Stoeckli ET, Condrau MA, et al (1991) Neurite outgrowth on immobilized axonin-1 is mediated by a heterophilic interaction with L1(G4). J Cell Biol 115:1113–1126
20. Stoeckli ET, Ziegler U, Bleiker AJ, et al (1996) Clustering and functional cooperation of Ng-CAM and axonin-1 in the substrate contact area of growth cones. Dev Biol 177:15–29
21. DeBernardo AP, Chang S (1996) Heterophilic interactions of DM-GRASP: GRASP-NgCAM interactions involved in neurite extension. J Cell Biol 133:657–666
22. Brümmendorf T, Hubert M, Treubert U, et al (1993) The axonal recognition molecule F11 is a multifunctional protein: specific domains mediate interactions with Ng-CAM and restrictin. Neuron 10:711–727
23. Rader C, Kunz B, Lierheimer R, et al (1996) Implications for the domain arrangement of axonin-1 derived from the mapping of its NgCAM binding site. Embo J 15:2056–2068
24. Friedlander DR, Milev P, Karthikeyan L, et al (1994) Neuronal chondroitin sulfate proteoglycan neurocan binds to the neural cell adhesion molecules ng-CAM/L1/Nile and N- CAM, and inhibits neuronal adhesion and neurite outgrowth. J Cell Biol 125:669–680

25. Montgomery AMP, Becker JC, Siu CH, et al (1996) Human neural cell adhesion molecule L1 and rat homologue NILE are ligands for integrin αVβ3. J Cell Biol 132:475–485
26. Moos M, Tacke R, Scherer H, et al (1988) Neural adhesion molecule L1 as a member of the immunoglobulin superfamily with binding domains similar to fibronectin. Nature (Lond) 334:701–703
27. Prince JT, Alberti L, Healy PA, et al (1991) Molecular cloning of NILE glycoprotein and evidence for its continued expression in mature rat CNS. J Neurosci Res 30:567–581
28. Grumet M, Mauro V, Burgoon MP, et al (1991) Structure of a new nervous system glycoprotein, Nr-CAM, and its relationship to subgroups of neural cell adhesion molecules. J Cell Biol 113:1399–1412
29. Tongiorgi E, Bernhardt RR, Schachner M (1995) Zebrafish neurons express two L1-related molecules during early axonogenesis. J Neurosci Res 42:547–561
30. Bastmeyer M, Ott H, Leppert CA, et al (1995) Fish E587 glycoprotein, a member of the L1 family of cell adhesion molecules, participates in axonal fasciculation and the age-related order of ganglion cell axons in the goldfish retina. J Cell Biol 130:969–976
31. Bieber AJ, Snow PM, Hortsch M, et al (1989) *Drosophila* neuroglian: a member of the immunoglobulin superfamily with extensive homology to the vertebrate neural adhesion molecule L1. Cell 59:447–460
32. Takeda Y, Asou H, Murakami Y, et al (1996) A noneuronal isoform of cell adhesion molecule L1: tissue-specific expression and functional analysis. J Neurochem 66:2338–2349
33. Sadoul R, Kirchhoff F, Schachner M (1989) A protein kinase activity is associated with and specifically phosphorylates the neural cell adhesion molecule L1. J Neurochem 53:1471–1478
34. Prince JT, Milona N, Stallcup WB (1989) Characterization of a partial cDNA clone for the NILE glycoprotein and identification of the encoded polypeptide domain J Neurosci 9:1825–1834
35. Zhao X, Siu CH (1995) Colocalization of the homophilic binding site and the neuritogenic activity of the cell adhesion molecule L1 to its second Ig-like domain. J Biol Chem 270:29413–29421
36. Yip PM, Zhao X, Montgomery AM, et al (1998) The Arg-Gly-Asp motif in the cell adhesion molecule L1 promotes neurite outgrowth via interaction with the α5β3 integrin. Mol Biol Cell 9:277–290
37. Schuch U, Lohse MJ, Schachner M (1989) Neural cell adhesion molecules influence second messenger systems. Neuron 3:13–20
38. Atashi JR, Klinz SG, Ingraham CA, et al (1992) Neural cell adhesion molecules modulate tyrosine phosphorylation of tubulin in nerve growth cone membranes. Neuron 8:831–842
39. Williams EJ, Doherty P, Turner G, et al (1992) Calcium influx into neurons can solely account for cell contact-dependent neurite outgrowth stimulated by transfected L1. J Cell Biol 119:883–892
40. Saffell JL, Williams EJ, Mason IJ, et al (1997) Expression of a dominant negative FGF receptor inhibits axonal growth and FGF receptor phosphorylation stimulated by CAMs. Neuron 18:231–242
41. Ignelzi MA, Miller DR, Soriano P, et al (1994) Impaired neurite outgrowth of src-minus cerebellar neurons on the cell adhesion molecule L1. Neuron 12:873–884
42. Wong EV, Schaefer A, Landreth G, et al (1996) Involvement of p90rsk in neurite outgrowth mediated by the cell adhesion molecule L1. J Biol Chem 271:18217–18223
43. Schaefer A, Wong E, Landreth G, et al (1997) MAPK signal cascade components c-raf and erk2 associate with the neural cell adhesion molecule L1. Society for Neuroscience, New Orleans
44. Wong EV, Schaefer AW, Landreth G, et al (1996) Casein kinase II phosphorylates the neural cell adhesion molecule L1. J Neurochem 66:779–786

45. Davis JQ, Bennett V (1994) Ankyrin binding activity shared by the neurofascin/L1/NrCAM family of nervous system cell adhesion molecules. J Biol Chem 269:27163–27166
46. Garver TD, Ren Q, Tuvin S, et al (1997) Tyrosine phsophorylation at a site highly conserved in the L1 family of cell adhesion molecules abolishes ankyrin binding and increases lateral mobility of neurofascin. J Cell Biol 137:703–714
47. Dahlin-Huppe K, Berglund EO, Ranscht B, et al (1997) Mutational analysis of the L1 neuronal cell adhesion molecule identifies membrane-proximal amino acids of the cytoplasmic domain that are required for cytoskeletal anchorage. Mol Cell Neurosci 9:144–156
48. Kamiguchi H, Lemmon V (1998) A neuronal form of the cell adhesion molecule L1 contains a tyrosine based signal required for sorting to the axonal growth cone. J Neurosci 18:3749–3756
49. Kamiguchi K, Long K, Pendergast M, et al (1998) The neural cell adhesion molecule L1 interacts with the AP-2 adaptor and is endocytosed via the clathrin-mediated pathway. J Neurosci 18:5311–5321
50. Mayford M, Barzilai A, Keller F, et al (1992) Modulation of an NCAM-related adhesion molecule with long-term synaptic plasticity in *Aplysia*. Science 256:638–644
51. Peter N, Aronoff B, Wu F, et al (1994) Decrease in growth cone neurite fasciculation by sensory or motor cells in vitro accompanies downregulation of aplysia cell adhesion molecules by neurotransmitters. J Neurosci 14:1413–1421
52. Bailey CH, Chen M, Keller F, et al (1992) Serotonin-mediated endocytosis of apCAM: an early step of learning-related synaptic growth in *Aplysia*. Science 256:645–649
53. Djabali M, Mattei MG, Nguyen C, et al (1990) The gene encoding L1, a neural adhesion molecule of the immunoglobulin family, is located on the X-chromosome in mouse and man. Genomics 7:587–593
54. Rosenthal A, Jouet M, Kenwrick S (1992) Aberrant splicing of neural cell adhesion molecule L1 messenger RNA in a family with X-linked hydrocephalus. Nat Genet 2:107–112
55. Willems PJ, Brouwer OF, Dijkstra I, et al (1987) X-linked hydrocephalus. Am J Med Genet 27:921–928
56. Winter RM, Davies KE, Bell MV, et al (1989) MASA syndrome: further clinical delineation and chromosomal localisation. Hum Genet 82:367–370
57. Fransen E, Lemmon V, Van Camp G, et al (1995) CRASH syndrome: a clinical spectrum of corpus callosum hypoplasia, retardation, adducted thumbs, spastic paraparesis and hydrocephalus due to mutations in one single gene, L1. Eur J Hum Genet 3:273–284
58. Yamasaki M, Thompson P, Lemmon V (1997) CRASH syndrome: mutations in the L1 gene correlate with severity of the disease. Neuropediatrics 28:175–178
59. Dahme M, Bartsch U, Martini R, et al (1997) Disruption of the mouse L1 gene leads to malformations of the nervous system. Nat Genet 17:346–349
60. Cohen N, Taylor J, Scott L, et al (1997) Errors in corticospinal axon guidance in mice lacking the neural cell adhesion molecule L1. Curr Biol 8:26–33

Functional Analysis of Cell Adhesion Proteins, P0 and L1

KEIICHI UYEMURA[1], KUNIO Kitamura[2], JAN SEDZIK[1], TAKAHITO YAZAKI[1], HIROAKI ASOU[1], and YASUO TAKEDA[1]

Summary. Cell adhesion proteins of the immunoglobulin superfamily play important roles during development of the nervous system. The cells expressing a peripheral myelin P0 protein showed strong homophilic adhesion. This adhesion is inhibited by addition of the glycopeptide of P0 protein, and therefore the oligosaccharide moiety is likely involved in this process. In coculture of P0-expressing cells, neurons showed neurite outgrowth, which was specifically inhibited by monoclonal antibody against the sequence 68–79 of P0. Several point mutations of human P0 were reported in Charcot-Marie-Tooth disease type 1B, but they were not directly related to the possible sites of cell adhesion and neurite outgrowth. We developed a useful method to purify P0 protein by immobilized metal affinity chromatography for crystallization. The whole L1 protein mediates homophilic cell adhesion, neurite extension, and neuronal migration. Complete L1 is expressed exclusively in neurons, but its isoform, lacking several amino acids, is found in nonneuronal cells. Recently, a new frameshift mutation of the L1 gene was found in a Japanese family with hereditary hydrocephalus, and abnormal expression of L1 was observed in human disorders of neuronal migration.

Key words. Cell adhesion molecule—Immunoglobulin superfamily—P0—L1

Introduction

Certain groups of proteins play essential roles for development, maintenance, and regeneration in the nervous system of vertebrates. Among them, the superfamily of immunoglobulin is a representative group of cell adhesion proteins [1]. In this chapter, we review recent studies on P0 and L1 proteins and focus attention on their physiological roles and involvement in neurological diseases.

[1] Department of Physiology, School of Medicine, Keio University, 35 Shinanomachi, Shinjuku-ku, Tokyo 160-8582, Japan
[2] Saitama Medical School, Irumagun, Saitama 350-0495, Japan

Structure and Function of P0

In mammals, P0 protein is expressed exclusively in peripheral nerve myelin [2]. Amino acid sequences of P0 in rat and human were deduced from cDNA, but bovine sequence was determined from the protein. P0 contained an extracellular domain, a membrane-spanning segment, and an intracellular domain. The extra-cellular domain of P0 shows homology to the variable domain of immunoglobulin, and it is localized in the intraperiod line of peripheral myelin where the two opposed external surfaces of the Schwann cell membranes adhere tightly and form the myelin sheath. Several posttranslational modifications of P0 were reported, such as glycosylation, phosphorylation, sulfation, and acylation.

Crystallization of the recombinant extracellular domain of P0 has been reported, but there is no report on crystallization of the whole P0 protein. We developed an effective purification method of bovine P0 protein suitable for crystallization [3]. Previously, we established a standard procedure of purification as follows: (1) myelin preparation, (2) delipidation, and (3) gel filtration on Sephadex G200 with 2% SDS. In the new method, the myelin membrane is solubilized by 0.5% SDS and applied in succession to three columns: Cu-immobilized metal ions affinity, Con-A agarose, and hydroxyapatite. The preliminary trial of crystallization of purified P0 was successful.

To analyze the functions of P0, we established a P0-expressing C6 cell line by transfecting them with P0 cDNA. When rat dorsal root ganglion cells or cortical neurons were incubated on P0-expressing C6 cells, neurite extensions were remarkably promoted as compared to control experiments [4]. The neurite extensions were inhibited by the addition of P0 antibodies recognizing the protein sequence 68–79 but not by the P0 glycopeptide. The P0-expressing cells formed large aggregates. Addition of the P0 glycopeptide showed 85% inhibition of the cell aggregation, suggesting that the peptide-containing oligosaccharide is the most important for the adhesion. The peptide of P0 responsible for adhesion is clearly different from the peptide 68–79 for neurite outgrowth (Fig. 1). Based on biochemical analysis of exoglycosidase digestion, Smith degradation, and gas chromatography-mass spectrometry analysis, we postulated the unique structure of the P0 oligosaccharide as a hybrid type with possible several microheterogeneities [1]. It is interesting that one of the oligosaccharide subtypes contains a HNK-1 epitope of glucuronic acid and sulfate. The possibility remains that the HNK-1 epitope is involved in the adhesive activity.

P0 and Hereditary Neuropathies

The chromosomal locus of the P0 gene was determined by the fluorescent in situ hybridization method as q22–23 of chromosome 1. As determined by linkage studies, this corresponds well to the locus of Charcot-Marie-Tooth disease type 1B. Seven point mutations (I30M, Y82C, D90E, K96E, R98H, R98C, and K131R) and one deletion (S63-) of P0 in Charcot-Marie-Tooth disease 1B and two point mutations (S63C and G167R) in Dejerine-Sottas disease were reported [5,6]. Two mutations, S63C in the cytoplasmic domain and G167R in the transmembrane segment, induces more severe disease, suggesting a stronger functional perturbation by these mutations. In the case of P0 knockout mouse, only the null P0 homozygote mouse showed hypomyelination

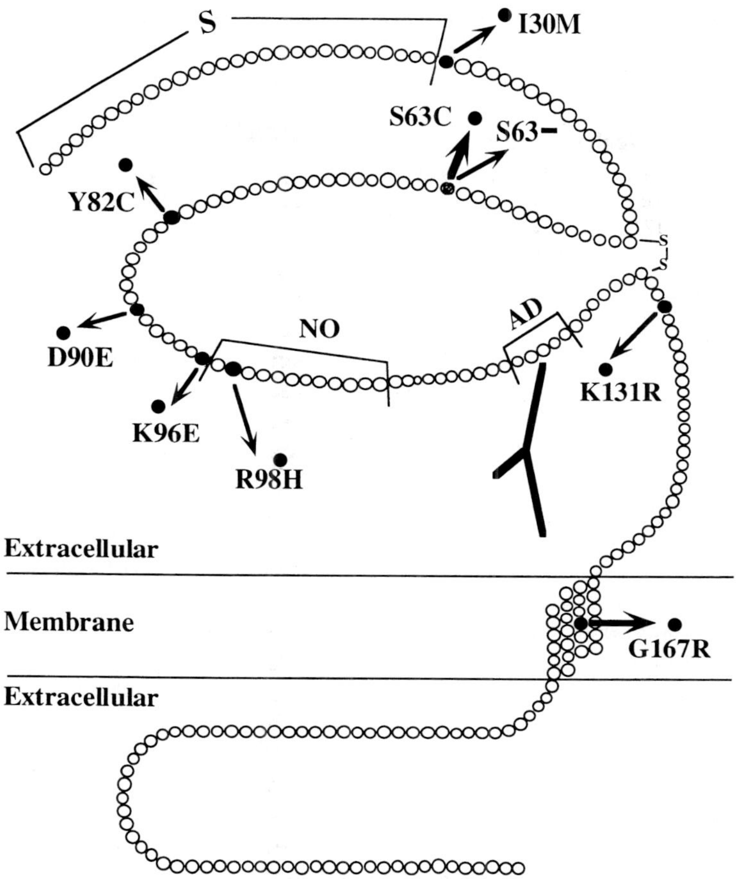

FIG. 1. Structure of P0 protein: signal sequence (*S*); active sites for neurite outgrowth (*NO*); cell adhesion (*AD*). The locations of point mutations that induce Charcot-Marie-Tooth disease type IB (*thin arrows*) and Dejerine-Sottas disease (*thick arrows*) are indicated

and severe clinical symptoms. The heterozygote expressing 50% of normal P0 have developed normally at least in the younger stages. In human, Charcot-Marie-Tooth disease 1B is an dominant neuropathy containing 50% of normal P0 with equivalent mutated P0. Mutated P0 may severely disturbs formation or maintenance of the myelin.

PASII/PMP22 Protein

PASII were purified from peripheral myelin by a procedure of gel filtration with high concentration of SDS. The sequence analysis of amino-terminal and carbohydrate-attaching peptide revealed that PASII is not a degradation product of P0 and differs from P0 [7]. Recently PMP22 protein, which was down-regulated after nerve damage, was characterized. Homology search revealed that bovine PASII is a homolog of rat PMP22 and mouse growth arrest-specific protein 3. These highly hydrophobic proteins contain four transmembrane segments including a signal-like domain. In the case of PASII/PMP22, major mutation of gene duplication (50% increase) induces Charcot-Marie-Tooth disease type 1A, and gene deletion (50% decrease) causes hereditary neuropathy of pressure palsy [6]. This evidence indicate that the constant level of PASII/PMP22 mRNA/protein expression is necessary for myelination.

Functional Analysis of L1 Protein

The complete L1 amino acid sequences of mouse, rat, and human were deduced. The L1 is a large 200-kDa molecule and belongs to the immunoglobulin superfamily. It contained six immunoglobulin domains and five fibronectin type III domains.The dissociated neurons of a young rat cerebellar cortex showed strong adhesion to the L1-expressing cells transfected with L1 cDNA [8]. In the case of coculture with the neuronal aggregates from a young cerebellum, the L1-expressing cells noticeably promoted both neurite outgrowth and migration of neurons from the aggregate. These results confirmed the roles of L1 for cell adhesion, neurite outgrowth, and neuronal migration in the nervous system.

In the course of sequence analysis of L1, we found an isoform of L1 lacking several amino acids in the cytoplasmic domain. We examined distribution in the nervous system of the L1 isoform by the PCR method. While rat and mouse brains contain the complete L1, the short isoform of L1 is found in only their sciatic nerves. Neurons have complete L1, but Schwann cells express exclusively the short L1 isoform [9]. Interestingly, both oligodendrocytes and astrocytes express a small amount of the short L1 isoforms. These results indicated that L1 in nonneuronal cells is a short form and differs from L1 in neurons. The deletion segment of four amino acids in the intracellular domain contains two possible sites of phosphorylation by casein kinase I and II. The deletion of RSLE in L1 cytoplasm may influence the migration ability of L1-expressing cells [10], but not a cell aggregation activity. Further studies revealed that the nonneuronal isoform has an additional deletion of five amino acids in the amino-terminal regions [9]. The deletions correspond to exon 2 and 27, which indicates existence of different alternative splicing systems between neurons and nonneuronal cells. Another interesting point is a cleavage site of L1 in the middle of the third fibronectin type III domain. Probably 140- and 80-kDa L1 are produced from the 200-kDa L1 by kexin-like proprotein convertases [11].

X-linked hydrocephalus, MASA syndrome (mental retardation, aphasia, shuffling gait, adducted thumbs), and spastic paraplegia type 1 are caused by mutations of the L1 gene [7]. A new frame shift mutation of the L1 gene was recently identified in a Japanese family of hereditary hydrocephalus [12]. In this case, the deletion of five nucleotides in the third immunoglobulin domain causes a shift of the reading frame and introduces a premature stop codon, which results in expression of a truncated protein. The relationship between the L1 mutation and various phenotypes is an interesting area for future study.

In human neuronal migration disorders and malformations such as Zellweger syndrome, thanatophoric dysplasia, and hemimegalencephaly, three types of abnormal expression of L1 were observed: (1) high expression associated with neuroaxonal overgrowth, (2) delayed expression with neuronal dysmaturation and dysmyelinogenesis, and (3) no expression in brain injury [13,14]. Impaired expression of L1 in the extrinsic nerve fibers in Hirschsprung's disease was also reported [15]. Concerning neuronal migration, partial sequence homology was found between the KALIG-1 protein and the L1 peptides in the fibronectin domains. It is suggested that the KALIG-1 is a protein responsible for Kallmann syndrome with hypogonadism and anosmia caused by defects in migration of olfactory neurons and LH-RH neurons. It is likely that consensus sequence common to these two proteins are related to cell migration.

Conclusions

Studies on structure and function of neural cell adhesion proteins have progressed greatly. Involvement of these proteins in various neurological diseases has been reported. Although gene mutations of the proteins are clearly demonstrated, their relations to each phenotype are not yet well established. Further studies are desirable to elucidate how these mutations induce functional changes. This information will be useful to understand the functional aspects of these proteins.

References

1. Uyemura K, Asou H, Yazaki T, Takeda Y (1996) Cell adhesion proteins of immunoglobulin superfamily in nervous system. Essays Biochem 31:37–48
2. Uyemura K, Kitamura K, Miura M (1992) Structure and molecular biology of P0 protein. In: Martenson RE (ed) Myelin: biology and chemistry. CRC Press, Boca Raton, FL, pp 481–508
3. Sedzik J, Kotake Y, Uyemura K (1998) Purification of PASII/PMP22—an extremely hydrophobic glycoprotein of PNS myelin membrane. Neuro Report 9:1595–1600
4. Yazaki T, Miura M, Asou H, et al (1994) Peripheral P0 protein mediates neurite outgrowth of cortical neurons in vitro and axonal regeneration in vivo. Neurosci Lett 176:13–16
5. Hayasaka K, Himoro M, Sato W, et al (1993) Charcot-Marie-Tooth neuropathy type 1B is associated with mutation of the myelin P0 gene. Nat Genet 5:31–34
6. Uyemura K, Takeda Y, Asou H, et al (1994) Neural cell adhesion proteins and neurological disease. J Biochem 116:1187–1192
7. Kitamura K, Sakamoto Y, Suzuki M, et al (1981) Microheterogeneity of carbohydrate in P0 protein from bovine peripheral nerve myelin membrane. In: Yamakawa T, Osawa T, Hnada S (eds) Glycoconjugates. Japan Scientific Society Press, Tokyo, pp 273–274
8. Miura M, Asou H, Kobayashi M, et al (1992) Functional expression of a full length cDNA encoding for rat neural cell adhesion molecule L1 mediates homophilic intercellular adhesion and migration of cerebellar neurons. J Biol Chem 267:10752–10758
9. Takeda Y, Asou H, Murakami Y, et al (1996) A nonneuronal isoform of cell adhesion molecule L1: tissue-specific expression and functional analysis. J Neurochem 66:2338–2349
10. Asou H, Takeda Y, Itoh K, et al (1998) The RSLE sequence in the cytoplasmic domain of cell adhesion molecule L1 is involved in cell migration on L1 substrates. In: this volume, pp 267–272

11. Nagahama M, Yoshida I, Uyemura K, et al (1998) Proteolytic cleavage of the neural cell adhesion molecule L1 by subtilsin-like proprotein convertases. In: this volume, pp 261–266
12. Takechi T, Tohyama J, Kurashige T, et al (1996) A deletion of five nucleotides in the L1CAM gene in a Japanese family with X-linked hydrocephalus. Hum Genet 97:353–356
13. Tsuru A, Mizuguchi M, Uyemura K, et al (1997) Immunohistochemical expression of cell adhesion molecule L1 in hemimegalencephaly. Pediatr Neurol 16:45–49
14. Tsuru A, Mizuguchi M, Uyemura K, et al (1997) Abnormal expression of cell adhesion molecule L1 in migration disorders: a developmental immunohistochemical study. Clin Neuropathol 16:122–126
15. Ikawa H, Kawano H, Takeda Y, et al (1997) Impaired expression of neural cell adhesion molecule L1 in the extrinsic nerve fibers in Hirschsprung's disease. J Pediatr Surg 32:542–545

Signal-Mediated Sorting of the Cell Adhesion Molecule L1: Role of Its Cytoplasmic Domain

Hiroyuki Kamiguchi and Vance Lemmon

Summary. The neural cell adhesion molecule L1 plays a crucial role in the formation of major axonal tracts such as the corpus callosum and corticospinal tract. L1 is preferentially expressed in the axons and growth cones, and its expression on the cell surface is likely to be dynamically altered to regulate L1-dependent adhesivity. Tyr^{1176} in the L1 cytoplasmic domain is adjacent to a neuron-specific alternatively spliced sequence, RSLE (Arg-Ser-Leu-Glu). The resulting sequence of YRSL conforms to a tyrosine-based consensus motif (YxxL) for sorting of integral membrane proteins into specific cellular compartments. It has been demonstrated that the YRSL sequence functions as a sorting signal that mediates intracellular L1 trafficking, such as sorting into axons and internalization from the plasma membrane. We review recent advances in understanding the functions of this sorting signal and its interaction with clathrin-associated adaptor proteins.

Key words. Neural cell adhesion molecule—L1—Sorting—Endocytosis—Tyrosine-based signal

Introduction

The neural cell adhesion molecule L1 plays a variety of important processes during nervous system development, including neuronal migration [1,2], neurite growth [3], and neurite fasciculation [4]. The fact that humans and mice with L1 mutations have hydrocephalus and major axonal tract defects confirms the importance of this molecule [5–7]. Interestingly, mutations that affect only the L1 cytoplasmic domain (L1CD) also produce abnormal axon tracts, but not severe forms of hydrocephalus [8], suggesting that intracellular signals or protein interactions involving the L1CD are key factors for proper axon growth and guidance.

The amino acid sequence of the L1CD is identical among mammalian species [9] and highly conserved among L1 subfamily members such as NrCAM and neurofascin [10]. Near the middle of the L1CD is an alternatively spliced RSLE (Arg-Ser-Leu-Glu)

Department of Neurosciences, Case Western Reserve University, Cleveland, OH 44106-4975, USA sequence [11,12], which is expressed in neurons but not in the other L1-expressing cells such as Schwann cells and lymphocytes [13]. The RSLE sequence is preceded by a tyrosine residue in the L1 subfamily members. The resulting sequence of YRSL conforms to the tyrosine-based sorting signal YxxØ, where x is any amino acid and Ø is an amino acid with a bulky hydrophobic side chain [14]. The tyrosine-based signal has been shown to mediate a variety of sorting events that target signal-bearing proteins into specific cellular compartments [15]. The tyrosine and Ø residues are critical for the signals to be active in protein sorting, whereas the importance of the x residues is less well defined [16,17].

In this chapter, we review recent advances in understanding intracellular L1 trafficking mediated by the tyrosine-based sorting signal in its cytoplasmic domain.

Clathrin-Mediated Endocytosis of L1

Cell–cell and cell–extracellular matrix interactions mediated via cell adhesion molecules (CAMs) are dynamically regulated during initial nervous system development as well as subsequent synaptic remodeling. One important strategy is to control the amount of CAMs on the cell surface by regulating their recycling from the plasma membrane. Support for this idea comes from studies in *Aplysia*, where synaptic growth requires downregulation of apCAM expression on the cell surface by endocytosis [18,19]. Furthermore, integrin expression on migrating cells is dynamically regulated by internalization at the rear followed by forward transport and recycling to the leading edge [20,21].

Clathrin-coated vesicles are the best-characterized vehicles for rapid endocytosis of plasma membrane proteins [22]. The formation of clathrin coats at the plasma membrane depends on the adaptor complex AP-2 [23], which is composed of two large chains (α-adaptin and either β1- or β2-adaptin), one medium chain (μ2), and one small chain (σ2). A specific interaction between tyrosine-based signals and the clathrin-associated AP-2 is believed to concentrate the signal-bearing molecules in clathrin-coated areas [15,24]. The medium chain, μ2, of AP-2 is responsible for interacting with the tyrosine-based signal in plasma membrane proteins to be endocytosed [25,26].

It has recently been shown that L1 associates with AP-2 in rat brain and that this interaction occurs via the YRSL sequence of L1 and the μ2 chain of AP-2 [27]. In L1-transfected 3T3 cells, L1 can be endocytosed from the plasma membrane. Endocytosed L1 colocalizes with markers for clathrin-mediated internalization, such as the transferrin receptor (TfR). If the YRSL sequence is altered by mutating the tyrosine or by alternative splicing to the nonneuronal form of L1, these forms of L1 do not colocalize with TfR. These observations indicate that the YRSL mediates endocytosis of L1, most likely by interacting with AP-2. In neuronal growth cones migrating on an L1 substrate, colocalization of L1 with Eps15, another marker for the clathrin endocytic pathway, is restricted to the rear of growth cones [27]. This observation suggests a mechanism in which a front-versus-rear asymmetry in L1 expression can be generated by internalizing L1 preferentially at the rear of growth cones. This strategy is regarded as important to create a gradient of adhesivity from the front to the rear, which is required for anterograde cell migration [28].

In adult brain, L1 is expressed in the hippocampus and cerebellum, regions where synaptic remodeling is a critical and ongoing process [29]. Indeed, L1 is implicated in hippocampal long-term potentiation [30]. Considering that clathrin-mediated endocytosis of apCAM is associated with learning-related synaptic growth, L1 endocytosis might also play a role in remodeling of synaptic connections.

Further studies are needed to clarify what mechanism is involved in regulating the L1 and AP-2 interaction to regulate clathrin-dependent endocytosis of L1. It has been speculated that the critical tyrosine (Tyr^{1176}) in the sorting signal YRSL in the L1CD may be subject to phosphorylation [31]. Given that the critical tyrosine has to be in a nonphosphorylated state for the signals to be active in sorting [16,17], the phosphorylation of Tyr^{1176} may be involved in regulating L1 endocytosis.

Axonal Sorting of L1

During the period when axons elongate toward their targets, their tips form the growth cone that interacts with environmental cues to produce directed axon growth. The growth cone is considered as the major site of action of L1, where, in response to extracellular ligand binding, L1 generates intracellular signals and interacts with the cytoskeleton to regulate axon growth [32]. L1 is transported directly to the axonal growth cone and inserted into the plasma membrane [33]. Subsequently, L1 spreads from the growth cone into the axonal shaft, as the axon elongates by preferentially adding new membrane components into the growth cone [34]. As a result, L1 is localized in the axonal growth cone and axonal shaft. This raises the question as to what molecular signal L1 uses to be sorted into transport vesicles destined for the axonal growth cone.

The mechanism of protein sorting to distinct plasma membrane compartments is well characterized in polarized epithelial cells [22,35]. The cytoplasmic domain of some plasma membrane proteins contains tyrosine-based sorting signals that mediate basolateral sorting. At the trans-Golgi network, the clathrin-associated AP-1 adaptor recognizes the tyrosine-based signals and sort the signal-bearing proteins into a specific population of transport vesicles destined for the basolateral plasma membrane. Accumulating evidence has suggested that epithelial apical and neuronal axonal membranes, as well as epithelial basolateral and neuronal somatodendritic membranes, may be equivalent domains in terms of protein sorting [36,37]. However, several proteins that contradict this hypothesis have also been reported [36,38–40]. Given the existence of nervous tissue-specific isoforms of adaptor complex subunits [41,42], it is most likely that neurons and epithelial cells utilize similar but distinct molecules to segregate proteins into different transport vesicles.

Recently, the tyrosine-based signal, YRSL, in the L1CD has been shown to mediate axonal sorting of L1 in dorsal root ganglion (DRG) neurons [43]. When DRG neurons are transfected with cDNA coding for wild-type full-length L1 or L1 truncated immediately after the RSLE sequence, both forms of L1 are expressed in the axonal growth cone and to a lesser degree in the cell body. However, if the YRSL sequence is altered by mutating the tyrosine or by alternative splicing to the nonneuronal form of L1, expression of these L1 forms is restricted to the cell body and the proximal axonal shaft. These observations demonstrate that the YRSL sequence is required for sorting

L1 to the axonal growth cone. However, an adaptor-like molecule that interacts with tyrosine-based signals and mediates axonal sorting has not yet been identified.

Conclusion

Significant advances have been made in understanding the molecular mechanisms of L1 sorting. A particularly interesting aspect is the demonstration that the tyrosine-based sorting signal, YRSL, depends on the alternative splicing only found in neuronal form of L1. Because this sorting signal is found in other L1 subfamily members, this mechanism is likely to have general functions in sorting CAMs.

Acknowledgment. This work was supported by N.I.H. grants Ey-5285 and NS-34252 to V.L.

References

1. Lindner J, Rathjen FG, Schachner M (1983) L1 mono- and polyclonal antibodies modify cell migration in early postnatal mouse cerebellum. Nature (Lond) 305:427–430
2. Barami K, Kirschenbaum B, Lemmon V, et al (1994) N-cadherin and Ng-CAM/8D9 are involved serially in the migration of newly generated neurons into the adult songbird brain. Neuron 13:567–582
3. Lagenaur C, Lemmon V (1987) An L1-like molecule, the 8D9 antigen, is a potent substrate for neurite extension. Proc Natl Acad Sci USA 84:7753–7757
4. Stallcup WB, Beasley L (1985) Involvement of the nerve growth factor-inducible large external glycoprotein (NILE) in neurite fasciculation in primary cultures of rat brain. Proc Natl Acad Sci USA 82:1276–1280
5. Wong EV, Kenwrick S, Willems P, et al (1995) Mutations in the cell adhesion molecule L1 cause mental retardation. Trends Neurosci 18:168–172
6. Dahme M, Bartsch U, Martini R, et al (1997) Disruption of the mouse L1 gene leads to malformations of the nervous system. Nat Genet 17:346–349
7. Cohen NR, Taylor JSH, Scott LB, et al (1997) Errors in corticospinal axon guidance in mice lacking the neural cell adhesion molecule L1. Curr Biol 8:26–33
8. Yamasaki M, Thompson P, Lemmon V (1997) CRASH syndrome: Mutations in L1CAM correlate with severity of the disease. Neuropediatrics 28:175–178
9. Hlavin ML, Lemmon V (1991) Molecular structure and functional testing of human L1CAM: an interspecies comparison. Genomics 11:416–423
10. Hortsch M (1996) The L1 family of neural cell adhesion molecules: old proteins performing new tricks. Neuron 17:587–593
11. Reid RA, Hemperly JJ (1992) Variants of human L1 cell adhesion molecule arise through alternate splicing of RNA. J Mol Neurosci 3:127–135
12. Miura M, Kobayashi M, Asou H, et al (1991) Molecular cloning of cDNA encoding the rat neural cell adhesion molecule L1—two L1 isoforms in the cytoplasmic region are produced by differential splicing. FEBS Lett 289:91–95
13. Takeda Y, Asou H, Murakami Y, et al (1996) A nonneuronal isoform of cell adhesion molecule L1: tissue-specific expression and functional analysis. J Neurochem 66:2338–2349
14. Trowbridge IS, Collawn JF, Hopkins CR (1993) Signal-dependent membrane protein trafficking in the endocytic pathway. Annu Rev Cell Biol 9:129–161

15. Marks MS, Ohno H, Kirchhausen T, et al (1997) Protein sorting by tyrosine-based signals: adapting to the Ys and wherefores. Trends Cell Biol 7:124–128
16. Boll W, Ohno H, Songyang Z, et al (1996) Sequence requirements for the recognition of tyrosine-based endocytic signals by clathrin AP-2 complexes. EMBO J 15:5789–5795
17. Ohno H, Fournier MC, Poy G, et al (1996) Structural determinants of interaction of tyrosine-based sorting signals with the adaptor medium chains. J Biol Chem 271:29009–29015
18. Mayford M, Barzilai A, Keller F, et al (1992) Modulation of an NCAM-related adhesion molecule with long-term synaptic plasticity in *Aplysia*. Science 256:638–644
19. Bailey CH, Chen M, Keller F, et al (1992) Serotonin-mediated endocytosis of apCAM: an early step of learning-related synaptic growth in *Aplysia*. Science 256:645–649
20. Lawson MA, Maxfield FR (1995) Ca^{2+}- and calcineurin-dependent recycling of an integrin to the front of migrating neutrophils. Nature (Lond) 377:75–79
21. Palecek SP, Schmidt CE, Lauffenburger DA, et al (1996) Integrin dynamics on the tail region of migrating fibroblasts. J Cell Sci 109:941–952
22. Mellman I (1996) Endocytosis and molecular sorting. Annu Rev Cell Dev Biol 12:575–625
23. Gallusser A, Kirchhausen T (1993) The β1 and β2 subunits of the AP complexes are the clathrin coat assembly components. EMBO J 12:5237–5244
24. Kirchhausen T, Bonifacino JS, Riezman H (1997) Linking cargo to vesicle formation: receptor tail interactions with coat proteins. Curr Opin Cell Biol 9:488–495
25. Ohno H, Stewart J, Fournier MC, et al (1995) Interaction of tyrosine-based sorting signals with clathrin-associated proteins. Science 269:1872–1875
26. Rapoport I, Miyazaki M, Boll W, et al (1997) Regulatory interactions in the recognition of endocytic sorting signals by AP-2 complexes. EMBO J 16:2240–2250
27. Kamiguchi H, Long KE, Pendergast M, et al (1998) The neural cell adhesion molecule L1 interacts with the AP-2 adaptor and is endocytosed via the clathrin-mediated pathway. J Neurosci 18:5311–5321
28. Lauffenburger DA, Horwitz AF (1996) Cell migration: a physically integrated molecular process. Cell 84:359–369
29. Brümmendorf T, Rathjen FG (1994) Cell adhesion molecules 1: immunoglobulin superfamily. Protein Profile 1:963–1108
30. Lüthl A, Laurent JP, Figurov A, et al (1994) Hippocampal long-term potentiation and neural cell adhesion molecule L1 and NCAM. Nature (Lond) 372:777–779
31. Heiland PC, Hertlein B, Traub O, et al (1996) The neural adhesion molecule L1 is phosphorylated on tyrosine and serine residues. Neuroreport 7:2675–2678
32. Kamiguchi H, Lemmon V (1997) Neural cell adhesion molecule L1: signaling pathways and growth cone motility. J Neurosci Res 49:1–8
33. Vogt L, Giger RJ, Ziegler U, et al (1996) Continuous renewal of the axonal pathway sensor apparatus by insertion of new sensor molecules into the growth cone membrane. Curr Biol 6:1153–1158
34. Craig AM, Wyborski RJ, Banker G (1995) Preferential addition of newly synthesized membrane protein at axonal growth cones. Nature (Lond) 375:592–594
35. Matter K, Mellman I (1994) Mechanisms of cell polarity: sorting and transport in epithelial cells. Curr Opin Cell Biol 6:545–554
36. Craig AM, Banker G (1994) Neuronal polarity. Annu Rev Neurosci 17:267–310
37. Scannevin RH, Murakoshi H, Rhodes KJ, et al (1996) Identification of a cytoplasmic domain important in the polarized expression and clustering of the Kv2.1 K^{+} channel. J Cell Biol 135:1619–1632
38. Ferreira A, Caceres A, Kosik KS (1993) Intraneuronal compartments of the amyloid precursor protein. J Neurosci 13:3112–3123

39. Haass C, Koo EH, Teplow DB, et al (1994) Polarized secretion of β-amyloid precursor protein and amyloid β-peptide in MDCK cells. Proc Natl Acad Sci USA 91:1564–1568
40. Higgins D, Burack M, Lein P, et al (1997) Mechanisms of neuronal polarity. Curr Opin Neurobiol 7:599–604
41. Pevsner J, Volknandt W, Wong BR, et al (1994) Two rat homologs of clathrin-associated adaptor proteins. Gene 146:279–283
42. Ball CL, Hunt SP, Robinson MS (1995) Expression and localization of α-adaptin isoforms. J Cell Sci 108:2865–2875
43. Kamiguchi H, Lemmon V (1998) A neuronal form of the cell adhesion molecule L1 contains a tyrosine-based signal required for sorting to the axonal growth cone. J Neurosci 18:3749–3756

Proteolytic Cleavage of the Neural Cell Adhesion Molecule L1 by Subtilisin-Like Proprotein Convertases

Masami Nagahama[1], Ichiro Yoshida[1], Keiichi Uyemura[2], Akihiko Tsuji[1], and Yoshiko Matsuda[1]

Summary. Subtilisin-like proprotein convertases (SPCs) are a family of proteases that are responsible for the processing of many precursors at sites containing multiple basic residues. The region-specific expression of SPCs during the early development of the brain implicates this class of proteases in the functional plasticity of the nervous system. L1 is a Ca^{2+}-independent neural cell adhesion molecule that may play a critical role in some aspects of axonal guidance. L1 has several molecular forms including 200, 140, and 80 kDa (L1-200, L1-140, and L1-80). L1-140/80 are thought to be generated by proteolysis of L1-200 at its extracellular region, which contains the multiple basic recognition motif for SPCs. To demonstrate the involvement of SPCs in this process, we have coexpressed L1 with the SPCs in HEK293 cells. Results demonstrated that SPC1 (furin) is most active at cleaving L1-200 with additional cleavage of the generated products. SPC6 (PC6) and SPC7 (PC8/LPC) are less active than SPC1; however, L1-200 was correctly processed into L1-140/80. These results suggest that SPCs are responsible for the proteolytic processing of L1 observed in the developing brain and thus play a regulatory role by modulating biological activity of L1 during brain development and neural regeneration.

Key words. Cell adhesion molecule—Convertase—L1—Processing—SPCs

Introduction

Most biologically active peptides, such as hormones, growth factors, and neuropeptides, are generated from inactive precursors by endoproteolytic processing at sites of paired basic amino acid residues. Several different convertases have been identified that catalyze this maturation of proproteins. Among them, the best-characterized class of convertases are subtilisin-like proprotein convertases (SPCs), which are a family of proteins with homology to the yeast Kex2 and the bacterial

[1] Department of Biological Science and Technology, Faculty of Engineering, The University of Tokushima, 2-1 Minamijosanjima, Tokushima 770-8506, Japan

[2] Department of Physiology, School of Medicine, Keio University, 35 Shinanomachi, Shinjuku-ku, Tokyo 160-8582, Japan

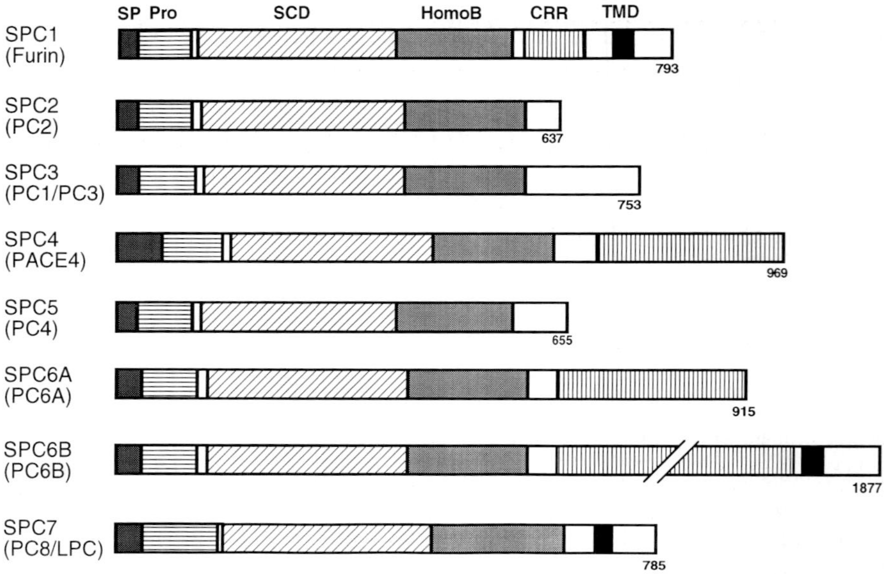

FIG. 1. Structure of mammalian subtilisin-like proprotein convertases (SPCs). Seven distinct members of the SPC family have been identified in mammalian species. SPC4, SPC5, and SPC6 have isoforms that are generated via alternative splicing. The various subregions are designated as follows: signal peptide (*SP*), proregion (*Pro*), subtilisin-like catalytic domain (*SCD*), Homo B domain (*HomoB*), cysteine-rich region (*CRR*), transmembrane domain (*TM*)

subtilisin serine proteases that cleave C-terminally to a multiple basic motif [1–3]. To date, seven mammalian SPCs have been identified, including SPC1/furin, SPC2/PC2, SPC3/PC1/PC3, SPC4/PACE4, SPC5/PC4, SPC6/PC5/PC6, and SPC7/PC8/LPC, some having isoforms generated via alternative splicing (Fig. 1). SPC2 and SPC3 are restricted in expression to endocrine and neuroendocrine cells and are active in the regulated secretory pathway. Expression of SPC5 is highly restricted to the testis. In contrast, SPC1, SPC4, SPC6, and SPC7 are expressed in many tissues and are active in the constitutive secretory pathway.

L1 is a Ca^{2+}-independent cell adhesion molecule (CAM) that is expressed in the nervous system and is involved in cell–cell adhesion, neurite outgrowth, neurite fasciculation, neuronal cell migration, and myelination [4,5]. L1 is a member of a closely related subfamily of the immunoglobulin superfamily that includes chick Ng-CAM and Nr-CAM. These cell-surface glycoproteins have extracellular regions containing six immunoglobulin-like domains and five fibronectin type III repeats, and are similar to each other in amino acid sequence, particularly in their cytoplasmic regions. These molecules are also posttranslationally cleaved in vivo at comparable sites in the middle of the third fibronectin type III repeats that contain unique multiple basic processing motifs (Fig. 2). In rat brain, the major form of L1 is 200 kDa (L1-200), but lower molecular mass forms are found including 140 and 80 kDa [6,7]. These smaller components are generated by proteolysis of L1-200 yielding the amino-

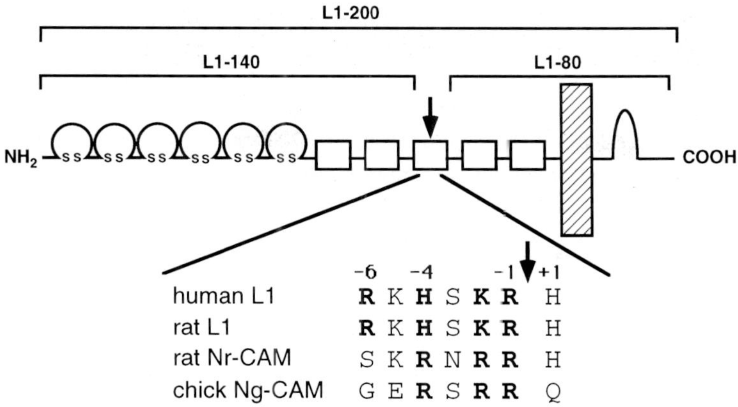

FIG. 2. Domain structure of L1 and its conserved processing site. The six immunoglobulin-like domains (*loops*), five fibronectin type III repeats (*open rectangles*), and transmembrane region (*vertical striped rectangles*) are shown. The 140- and 80-kDa components (L1-140 and L1-80) are generated from the ~200-kDa component (L1-200) by cleavage in the third fibronectin type III repeat, indicated by the *arrow(s)*. The amino acid sequence around the putative cleavage site is shown in comparison with other related molecules. The basic residues conforming to the consensus cleavage signal for SPCs [11–13] are shown by *bold letters*

terminal 140-kDa extracellular fragment (L1-140) and the 80-kDa transmembrane fragment (L1-80). The ratio of these different molecular forms of L1 changes during development among different regions of the brain [8]; therefore, regulation of the conversion of L1-200 into L1-140/80 is a potential mechanism by which the biological activity of this adhesion molecule might be modulated.

The fact that putative cleavage sites of the L1 subfamily are conserved and conform to the consensus sequences for SPC recognition, and also that the expression of SPCs is spatiotemporally regulated in the brain [9,10], suggests that SPCs may be good candidates for processing enzymes of L1. In an attempt to identify the enzyme(s) involved in conversion of L1-200 into L1-140/80, we have coexpressed human or rat L1 with known mammalian SPCs in HEK293 cells.

Results and Discussion

Recent studies indicate that protein precursors that have a triad of basic residues arranged in an Arg-X-Lys/Arg-Arg motif can be cleaved on the C-terminal side of the consensus sequence by a family of processing proteases, subtilisin-like proprotein convertases (SPCs) [1–3,11]. Sequence rules established for precursor cleavage within the constitutive secretory pathway by SPCs indicate important roles for the basic residues at positions-1, -2, -4, and -6 relative to the scissile bond (that is, P_1, P_2, P_4, and P_6, respectively) [12,13]. Because the putative processing site of L1 contains the sequence Arg-Lys-His-Ser-Lys-Arg, which conforms to the SPC cleavage consensus site, and similar sequences are present in all L1 and its related molecules characterized so far, it was hypothesized that one or more members of the SPC family are responsible for such a cleavage. To test the possibility that SPCs convert L1-200 into L1-140/80, HEK293 cells were cotransfected with L1 cDNA and each SPC cDNA (Fig. 3a). When human L1 cDNA was transfected in HEK293 cells, a major band corresponding to L1-200 was detected in the membrane fraction. The minor band that migrates slightly faster than L1-200 corresponded to an N-glycosylation intermediate that is still in the endoplasmic reticulum, as demonstrated by its endoglycosidase H sensitivity (data not shown). When L1 was cotransfected with SPC6A, SPC6B, and SPC7, L1-200 was processed to L1-140/80. SPC1 also processed L1-200; however, an additional cleavage was observed. The same results were also obtained when rat L1 was examined (data not shown).

To assess the role of amino acid residues at the putative cleavage site, amino acid substitutions were generated and analyzed as described. For cleavage by SPCs, Arg in the P_1 position has been shown to be necessary, and a basic residue in the P_4 position is also critical, preferably Arg [11,12]. Thus Arg^{863} located in P_1 of the putative cleavage site of L1 was replaced by Gln (RQ^{-1} mutant), and His^{860} in P_4 was replaced by Arg

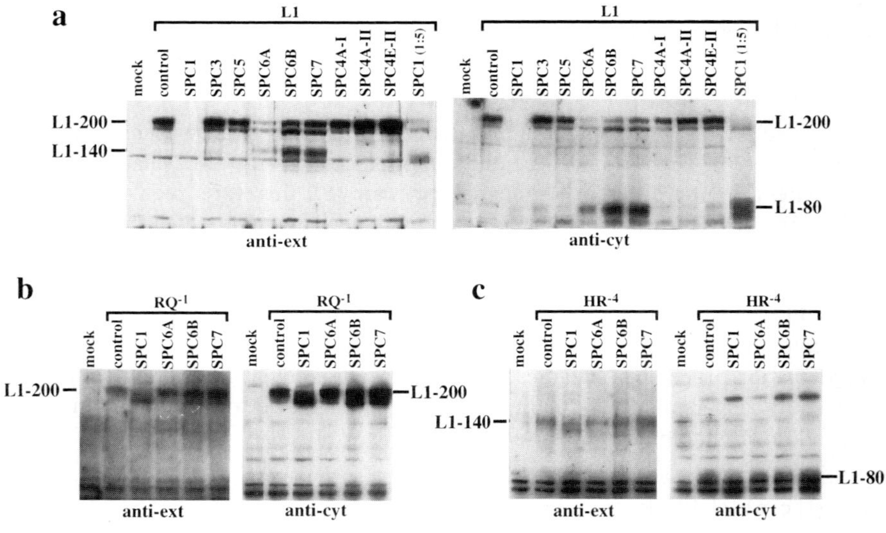

FIG. 3. **a** Conversion of L1-200 into L1-140/80 by subtilisin-like proprotein convertases (SPCs) in HEK293 cells. HEK293 cells were either transfected with control plasmid (mock) or cotransfected with equal amount of human L1 expression vector and one SPC expression vector. SPC1 expression vector was also cotransfected with a ratio of 1 : 5 to L1 expression vector. Total membrane fractions were prepared, and L1 products were analyzed by western blotting using polyclonal antibodies against extracellular (anti-ext) or cytoplasmic region (anti-cyt) of L1. **b** Processing of L1 cleavage mutant, RQ^{-1}, by SPCs in HEK293 cells. The mutation was introduced into the coding sequence of rat L1 by site-directed mutagenesis to create the substitution of -1Arg → Gln. The expression construct harboring the mutation was transiently cotransfected with an equal amount of control vector or expression vectors for SPCs. **c** Processing of L1 cleavage mutant, HR^{-4}, by SPCs in HEK293 cells. The -4His → Arg mutation was introduced into the coding sequence of rat L1 and the effects were analyzed as described

(HR^{-4} mutant). Consistent with the important role of Arg in P_1 for cleavage by SPCs, the Arg → Gln mutation at P_1 (RQ^{-1}) completely blocked the conversion of L1-200 into L1-140/80 (Fig. 3b). In contrast, the His → Arg mutant at P_4 (HR^{-4}) underwent more efficient processing, and even the endogenous SPC-like activity in HEK293 cells was sufficient for complete cleavage of L1-200 (Fig. 3c). When these L1 mutants and SPC1 were cotransfected, cleavage products other than L1-140/80 were detected, indicating the presence of additional cleavage site(s) for SPC1 in the L1 structure under these experimental conditions. Thus, within the putative cleavage sequence, two positions (His^{860}, Arg^{863}) play an important role, suggesting that these residues are part of the protease recognition site.

The abilities of L1 to support multiple functions during the neural development might be under the control of mechanisms such as differential posttranslational processing. The regulated cleavage of L1 may enhance the heterogeneity and hence the specificity of signals involved in axonal guidance during development. Regional and development-specific expression of specific processing proteases may provide a way to modulate the functions of L1. The results described here suggest that L1-140 and L1-80 are the results of regulated processing of L1-200, and that SPC1, SPC6, and SPC7 may be responsible for this process.

Although these convertases are expressed ubiquitously, the time and extent of their expression differs among tissues or cells. In addition, their specific sequence requirements for substrate recognition and their intracellular distributions are also different. Which convertase(s) is responsible for the L1 modulation in physiological circumstances remains to be elucidated. Finally, the functional significance of the L1 polypeptide components also remains to be determined. Proteolytic cleavage of L1 may reduce L1-mediated cell binding. Alternatively, diffusion of L1 fragments may have local cellular consequences. Our results show that the HR^{-4} mutant of L1 is expressed as a fully proteolytically processed form (L1-140/80) in cells, suggesting the potential of this mutant in studies on the functional role of the proteolytic modification.

References

1. Barr PJ (1991) Mammalian subtilisins: the long-sought dibasic processing endoproteases. Cell 66:1–3
2. Steiner DF, Smeekens SP, Ohagi S, et al (1992) The new enzymology of precursor processing endoproteases. J Biol Chem 267:23435–23438
3. Nakayama K (1997) Furin: a mammalian subtilisin/Kex2p-like endoprotease involved in processing of a wide variety of precursor proteins. Biochem J 327:625–635
4. Hortsch M (1996) The L1 family of neural cell adhesion molecules: old proteins performing new tricks. Neuron 17:587–593
5. Burden-Gulley SM, Pendergast M, Lemmon V (1997) The role of cell adhesion molecule L1 in axonal extension, growth cone motility, and signal transduction. Cell Tissue Res 290:415–422
6. Faissner A, Teplow DB, Kubler D, et al (1985) Biosynthesis and membrane topography of the neural cell adhesion molecule L1. EMBO J 4:3105–3113
7. Sadoul K, Sadoul R, Faissner A, et al (1988) Biochemical characterization of different molecular forms of the neural cell adhesion molecule L1. J Neurochem 50:510–521
8. Liljelund P, Ghosh P, van den Pol AN (1994) Expression of the neural axon adhesion molecule L1 in the developing and adult rat brain. J Biol Chem 269:32886–32895

9. Akamatsu T, Daikoku S, Nagamune H, et al (1997) Developmental expression of a novel Kexin family protease, PACE4E, in rat olfactory system. Histochem Cell Biol 108:95–103
10. Zheng M, Seidah NG, Pinter JE (1997) The developmental expression in the rat CNS and peripheral tissues of protease PC5 and PACE4 mRNAs: comparison with other proprotein processing enzymes. Dev Biol 181:268–283
11. Hosaka M, Nagahama M, Kim WS, et al (1991) Arg-X-Lys/Arg-Arg motif as a signal for precursor cleavage catalyzed by furin within the constitutive secretory pathway. J Biol Chem 266:12127–12130
12. Watanabe T, Nakagawa T, Ikemizu J, et al (1992) Sequence requirements for precursor cleavage within the constitutive secretory pathway. J Biol Chem 267:8270–8274
13. Watanabe T, Murakami K, Nakayama K (1993) Positional and additive effects of basic amino acids on processing of precursor proteins within the constitutive secretory pathway. FEBS Lett 320:215–218

The RSLE Sequence in the Cytoplasmic Domain of Cell Adhesion Molecule L1 Is Involved in Cell Migration on L1 Substrates

Hiroaki Asou[1,2], Yasuo Takeda[1,2], Kouichi Itoh[3], and Keiichi Uyemura[1]

Summary. L1 has two alternatively spliced exons, one that codes for part of the extreme N-terminus region and the other which codes for four amino acids (Arg-Ser-Leu-Glu, RSLE) in the cytoplasmic region. To investigate the functional difference of L1 alternatively spliced forms, we have established cell lines transfected with either a cDNA lacking the 12-bp coding for the RSLE amino acids (L4-2 cells) or complete/full-length cDNA (L3-1 cells). A western blot analysis revealed that the expression levels of L1 in L3-1 and L4-2 cells were similar to those obtained with adhesion assays. The ability of L3-1 cells to migrate from aggregates on to a monolayer of L1-expressing cells, however, was found to be much greater than that of L4-2 cells. Similarly, when L1 fragments were used as a substrate, L3-1 cells migrated over larger areas of the dish compared to L4-2 cells, even though both cell types were capable of migration over other substrates (i.e., poly L-lysine and fibronetin.). These results indicate that the RSLE sequence in the L1 cytoplasmic domain is important in the regulation of L1-dependent cell migration. This process is likely to be important in neural cell migration during brain development and in axon growth and guidance.

Key words. L1cs—cell migration activity—L4-2—L3-1

Introduction

Cell adhesion molecules (CAMs) have been widely implicated to play important roles in a diverse set of process that include the migration of neuroblasts, neurite outgrowth, fasciculation, and axon guidance to the target cells [1]. However, little is known about either the underlying associated molecular mechanism or the nature of interactions that may exist between these CAMs. L1, a transmembrane protein that

[1] Department of Physiology, School of Medicine, Keio University, 35 Shinanomachi, Shinjuku-ku, Tokyo 160-8582, Japan
[2] Tokyo Metropolitan Institute of Gerontology, 35-2 Sakae-cho, Itabashi-ku, Tokyo 173-0015, Japan
[3] Tokyo Metropolitan Institute of Medical Science, Honkomagome, Bunkyo-ku, Tokyo 113-8613, Japan

belongs to the immunoglobulin (Ig) superfamily of cell adhesion molecules found mainly in the nervous system, is a developmentally regulated molecule which may play important roles in some aspects of axonal guidance and cell migration [2].

We have previously reported that both L cell transfectant, which expresses rat L1, and substrate-purified brain L1 could promote the migration of cerebellar neuron in vitro, suggesting that the L1 molecule has a critical role in cerebellar neuronal migration in vivo [3,4]. An L1 isoform (L1cs) with a small deletion, i.e., an isoform that lacks four sequential amino acid residues (RSLE) in the cytoplasmic domain and which is generated by alternative splicing form of a single gene, is the predominant isoform in nonneuronal tissues. To elucidate the molecular mechanism associated with neuronal migration mediated by L1, we have transfected complete rat L1 cDNA (L1) and its cytoplasmic short isoform (L1cs) into mouse fibroblast L cells. In this chapter, we describe the functions of the alternatively spliced form of L1.

Materials and Methods

Transformants expressing the full length of rat L1 (pBactL3) and the cytoplasmic short form of L1 (pBactL4) were established according to the method of Miura et al. [4]. To visualize tracks of L1-transfected L cells moving on the L1 substrate-coated coverglass, and to assess the motility of L cells without the anomaly of cell division, a short-term cell migration assay was used according to the method of Woodley et al. [5]. Briefly, glass coverslips (thickness, 13 nm) were coated with either poly L-lysine plus the fusion protein of the L1-IgC2 (I–VI) domain or L1-FN III (I–V) domain or fibronectin (Boehringer, Mannheim, Germany). Transfectants (5×10^4/ml) were plated on each coverglass in Dulbecco's modified Eagle's medium (DMEM) +2% calf serum (CS). Unattached cells were then gently removed and incubated in an incubator at 37°C for 20 h with 95% CO_2/5% O_2, then washed with PBS and fixed in 2% paraformaldehyde in PBS. To quantify cell migration on L1-coated coverglasses, anti-L1-IgC2 (I–V), anti-L1-FN III (I–V) antibodies, and antifibronectin monoclonal antibody (Boehringer) were added after fixation and examined with an inverted fluorescence microscope (Nikon, Tokyo, Japan). The percentage of the area of each field occupied by phagokinetic tracks was calculated using computer-assisted image analysis (Hamamatsu Photonics, Hamamatsu, Japan). The percentage of total field area taken up by tracks was termed the migration index (% of area).

Results

The rate of L3-1 cell migration, estimated at regular intervals after cell seeding, was three- to fourfold higher than that of L4-2 cells on the same L1 substrate. The mean area of spreaing by L3-1 cells after 20 h of cultivation was $16250 \pm 2630\,\mu mm^2$, as summarized in Fig. 1A. In contrast, both L3-1 cells and L4-2 cells showed the same migratory behavior on other substrates (e.g., fibronectin and bovme serum albumm, BSA) (Fig. 1B). These results suggested that the differences in migratory behavior are specific for L1-mediated migration.

Additional studies of the migration activity of L3-1 and L4-2 cells were conducted with each coculture assay system. When DiO-labeled reaggregates of L3-1 were cocultured on a feeder layer of either L3-1 or L4-2 cells, the DiO-labeled L3-1 cells

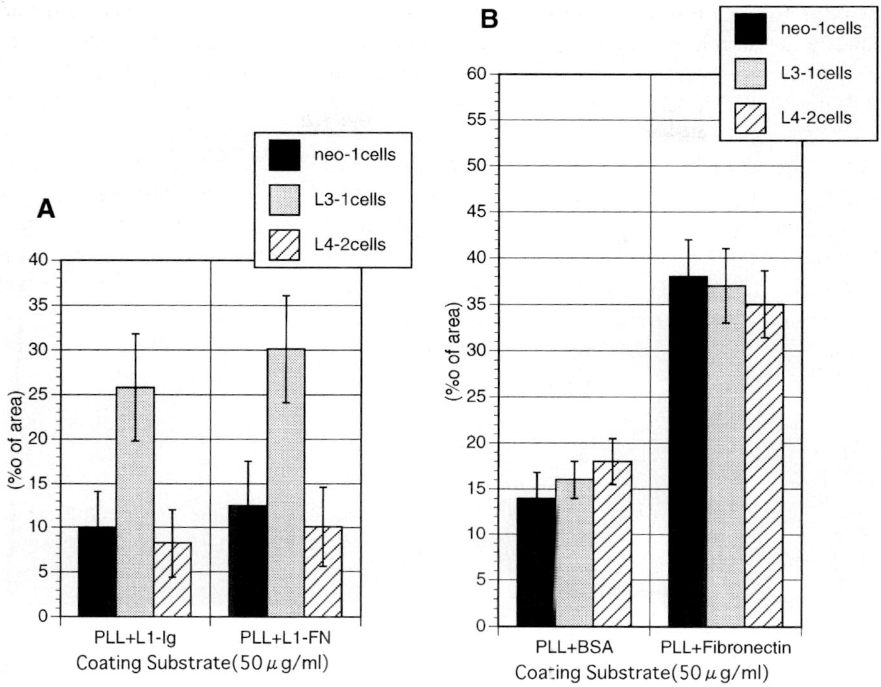

FIG. 1A,B. Migratory activity of L3-1 and L4-2 cells. L3-1 and L4-2 cells dispersed on coverslips coated with recombinant polypeptide fragments of domain from IgC2 (I–VI) (50 μg/ml) and FN III (I–V) (50 μg/ml) of L1 (**A**), and BSA (50 μg/ml) and fibronectin (50 μg/ml) (**B**). The migration of L3-1 and L4-2 cells was analyzed as described in Methods. Values are expressed as the mean ± SEM for three independent experiments and are represented in the Y-axis as percentage of total field area

migrated rapidly from the reaggregates to the outer perimeter (Fig. 2). Similar effects were observed with cerebellar neurons cultured on L3-1 cells [3,4]. Feeder layers of L3-1 and L4-2 cells were similarly potent in stimulating the migration of L3-1 cells. The average numbers of migrants (moving less than 50 μm) in coculture with L3-1 cells and L4-2 cells were 95 ± 34 and 88 ± 26, respectively (Table 1; Fig. 2). Surprisingly, when DiO-labeled reaggregates of L4-2 cells were cocultured with either of these cell types, no differences were observed. In this case, the average numbers of migrants (moving less than 50 μm) were only 12 ± 10 in coculture with L3-1 and only 8 ± 6 in coculture with L4-2 (Table 1; Fig. 3).

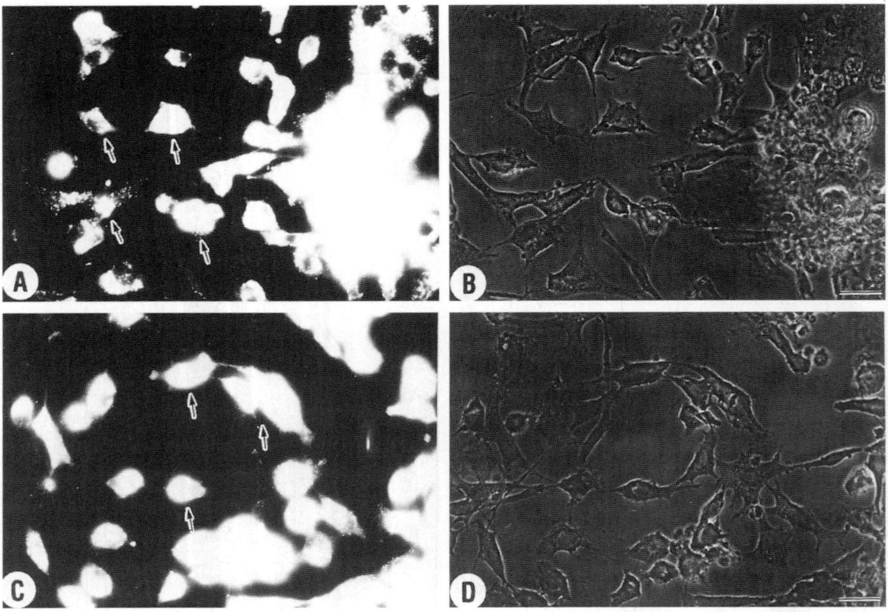

FIG. 2A–D. Migration of L3-1 cells cocultured with L1-transfected cells. Aggregation of L3-1 cells that were labeled with DiO (molecular probes) were overlaid on unlabeled L3-1 cells (**A,B**) and L4-2 cells (**C,D**) as feeder cells. After incubation for 20h, fluorescence (**A,C**) and phase-contrast (**B,D**) photomicrographs were taken with a Nikon-FX microscope. *Arrows* show the migrating DiI-labeled L3-1 cells from the reaggregation cultures. *Bars* 24.2 μm

TABLE 1. Migration of L1-transfected cells (L3-1/L4-2) from reaggregation cultures in coculture with L3-1 cells or L4-2 cells

Culture combination	Numbers of migrating L1-transfected cells			
	<50 μm	50–100 μm	100–150 μm	>150 μm
L3-1/L3-1 cells	95 ± 34	56 ± 18	4 ± 3	0
L3-1/L4-2 cells	88 ± 26	46 ± 26	3 ± 2	0
L4-2/L4-2 cells	12 ± 10	0	0	0
L4-2/L3-1 cells	8 ± 6	0	0	0

Migration of L3-1 cells is promoted by coculture with L1-tansfected L3-1 cells or L4-2 cells. The number of migrating transformants and their distance from the margin of the reaggregates for three independent aggregates in the same cultures were measured as described previously.
Values are means ± SEM for three independent experiments.

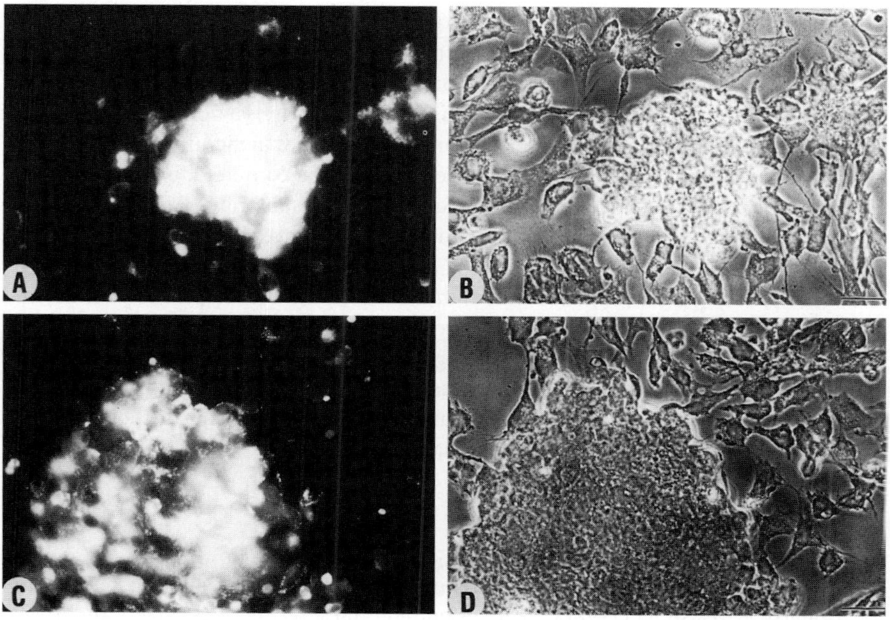

FIG. 3A–D. Migration of L4-2 cells cocultured with L1-transfected cells. Dissociated L4-2 cells were incubated with 15 μg/ml of DiO (molecular probes) for 2 h under a constant rotation of 80 rpm. Reaggregated L4-2 cells were then seeded on coverslips that were coated with unlabeled L3-1 cells (**A**,**B**) or unlabeled L4-2 cells (**C**,**D**). The cells were grown under subconfluent conditionns for 20 h. **A** and **C**, fluorescent photographs taken with a BG3 filter; **B** and **D**, phase-contrast photographs. *Bars* 31.2 μm

Discussion

The molecular mechanisms of cell migration triggered by L1 have not been elucidated. Cells transfected with L1 containing the RSLE sequence in the cytoplasmic domain (L3-1 cells) migrate out from cell aggregates, whereas cells transfected with L1 lacking the RSLE sequence (L4-2 cells) do not migrate. The L1 expressed by L4-2 cells lacks one of the putative phosphorylation sites acted on by casein kinase II [6]. The deletion of RSLE may cause a functional change in L1 because of the lack of phosphorylation sites. This is an interesting finding, which could help to explain the presence of alternatively splice forms of L1.

Recently, using the reverse transcription polymerase chain reaction (RT-PCR) and the total RNA purified from various tissues an cells, we have found another alternative splicing of the extracellular region of L1. This form lacked a 15-bp sequence corresponding to a KGHHV amino acid sequence starting at position 27 from the translation initiation site. Comparison of the two L1 PCR products, one lacking a 15-bp sequence in the extracellular region and one lacking a 12-bp sequence in the cytoplasmic region, showed that alternative splicing in both regions occurred in the same way [7]. Both truncated isoforms of L1 are major components of L1 in the peripheral nervous system. Further analysis revealed that Schwann cells contain only the truncated forms, while neural cells contain exclusively the full length of L1. Astrocytes and oligodendrocytes derived from rat brain primary cultures also express a low level of the truncated L1 isoform, indicating that the truncated form is a nonneuronal isoform [8]. The phosphorylation of RSLE and other serine residues in vicinity of RSLE may be related to the cytoskelton cascade in cells expressing the complete form of L1, which mediates the triggering of cell migration: kinase and its substrates trigger a signaling cascade similar to that which occurs with the growth factor receptor and tyrosine phosphorylation [9].

Recently, Wong et al. [10] have shown that CK II immunoprecipitates with L1, and that the phosphorylation of L1 by CK II requires the presence of the RSLE exon. This finding as well as our present findings suggest that the phosphorylation-dependent function of the L1 molecule in neuronal cells might play an important role in their migration. Further experiments on the function of the RSLE site in the cytoplasm domain of L1 are now in progress in our laboratory.

References

1. Edelman GM, Crossin KL (1991) Cell adhesion molecules: implications for a molecular histology. Annu Rev Biochem 60:155–190
2. Doherty P, Walsh FS (1992) Cell adhesion molecules, second messengers and axonal growth. Curr Opin Neurobiol 2:595–601
3. Asou H, Miura M, Kobayashi M, Uyemura K (1992) The cell adhesion molecule L1 has a specific role in neural cell migration. NeuroReport 3:481–484
4. Miura M, Asou H, Kobayashi M, Uyemura K (1992) Functional expression of a full-length cDNA coding for rat neural cell adhesion molecule L1 mediates homophilic intercellular adhesion and migration of cerebellar neurons. J Biol Chem 267:10752–10758
5. Woodley DT, Bachmann PM, O'Keefe EJ (1988) Laminin inhibits human keratinocyte migration. J Cell Physiol 136:140–146
6. Miura M, Kobayashi M, Asou H, Uyemura K (1991) Molecular cloning of cDNA encoding the rat neural cell adhesion molecule L1. FEBS Lett 289:91–95
7. Takeda Y, Asou H, Miura M, Kobayashi M, Uyemura K (1994) An isoform of cell adhesion molecule L1 expressed in Schwann cell. J Neurochem 62(suppl):S80
8. Takeda Y, Asou H, Murakami Y, Uyemura K (1995) Cell specific expression and functional analysis of a truncated isoform of L1 cell adhesion molecule. J Neurochem 65(suppl):S144
9. Williams EJ, Furrness J, Walsh FS, Doherty P (1994) Activation of the FGF receptor underlies neurite outgrowth stimulated by L1, NCAM, and N-cadherin. Neuron, 13:583–594
10. Wong EV, Kenwrick S, Willems P, Lemmon V (1995) Mutations in the cell adhesion molecule L1 cause mental retardation. Trends Neurosci 18:168–172

Mutations in L1CAM Correlate with the Severity of Developmental Defects in Brain Development

Mami Yamasaki[1] and Vance Lemmon[2]

Summary. X-linked hydrocephalus, MASA syndrome—an acronym for mental retardation, aphasia, shuffling gait, and adducted thumbs, and certain forms of X-linked spastic paraplegia and agenesis of corpus callosum are caused by mutations in the gene for the neural cell adhesion molecule L1. A comparison of existing case reports with molecular genetic analysis reveals a striking correlation between the type of mutation in the L1CAM gene and the severity of the disease. Mutations that produce truncations in the extracellular domain of the L1 protein are more likely to produce severe hydrocephalus, grave mental retardation, or early death than point mutations in the extracellular domain or mutations affecting only the cytoplasmic domain of the protein. While less severe than extracellular truncations, point mutations in the extracellular domain do produce more severe neurological problems than mutations in only the cytoplasmic domain.

Key words. X-linked hydrocephalus—MASA syndrome—L1CAM

Introduction

The first report of a mutation in the L1CAM gene associated with a neurological disorder (X-linked hydrocephalus, XLH) [1] appeared in 1992. Since then, 77 human L1CAM gene mutations have been discovered in patients with either X-linked hydrocephalus (McKusick 307000), MASA syndrome (McKusick 303350), X-linked spastic paraplegia syndrome (SPG1) (McKusick 312900), or X-linked agenesis of the corpus callosum (ACC) [2]. X-linked hydrocephalus was originally described in 1949 [3], and later characterized by hydrocephalus, mental deficiency, spastic paraplegia, and flexion deformity of the thumbs. In 1974, Bianchine and Lewis reported a novel X-linked disorder characterized by mental retardation, aphasia, shuffling gate, and adducted thumbs and proposed the acronym "MASA" to name the syndrome [4].

[1] Department of Neurosurgery, Osaka National Hospital, 2-1-14 Hoenzaka, Chuo-ku, Osaka 540-0006, Japan

[2] Department of Neurosciences, Case Western Reserve University, 2109 Adelbert Rd., Cleveland, OH 44106-4975, USA

Based on the extensive genetic studies of Kenwrick and associates and Willems and colleagues (see [5] for review), it is clear that these syndromes should be considered as a single clinical entity with varying severity all resulting from mutations in the L1CAM gene.

While it is clear that mutations in the L1CAM gene cause a recurring set of developmental anomalies, it has been unclear whether different types of mutations influence the severity of the disease. Recently we reviewed the clinical descriptions of 129 individuals from 34 families in which mutations in the L1CAM gene are known from molecular studies and have correlated the severity of the disease with the type of L1CAM mutation [5]. Here we report on an additional 10 families with a total of 24 new patients [6–10]. The conclusions reported in our original paper are supported with the addition of the new data.

Presently 77 different mutations have been detected in 83 families. These are listed on the L1 World Wide Web site maintained by Willems and associates (http://hgins.uia.ac.be/dnalab/l1.html) [2]. Forty-four families having known mutations in L1CAM have been described in the clinical literature, detailing a total of 153 individuals. A complete listing of the individual cases examined in this chapter can be found at http://neurowww.cwru.edu/Lemmon/. Sufficient information is available to analyze severity of hydrocephalus, survival after the first year, degree of mental retardation, and presence or absence of adducted thumbs.

Classification Schemes for Clinical Signs

Hydrocephalus was categorized into three basic groups: severe, moderate, or absent. A total of 147 patients were analyzed. Mental retardation was analyzed for 83 patients that survived for more than 1 year and for whom there was sufficient description to allow categorization. Mental retardation was classified into two groups, grave (IQ below 50) and modest (IQ above 50). To define survivorship, patients were divided into two groups based on whether they died within the first year. A total of 148 patients were analyzed. The presence or absence of adducted thumbs was analyzed in 133 patients.

Classification of Mutations

L1CAM mutations can be arranged into three groups based on how they would affect the expression or structure of the protein (Fig. 1a). The first class of mutation includes missense mutations, nonsense mutations, frameshifts, duplications, and deletions that affect only the cytoplasmic domain of L1 at the C-terminal end of the protein. We analyzed 34 patients from 5 families with different mutations in this class. The second class consists of missense point mutations in the extracellular domain; 79 patients from 22 families in this category have published descriptions. The third class of mutation includes nonsense or frameshift mutations that produce a premature stop codon in the extracellular domain of L1, which would result in failure of the L1 to remain associated with the cell surface because of loss of its transmembrane domain. We analyzed 40 patients from 18 families with different mutations in this group.

The basic findings are summarized in Fig. 1b. Hydrocephalus showed a significant variation between the three classes. Among 34 patients in class 1, 6% showed

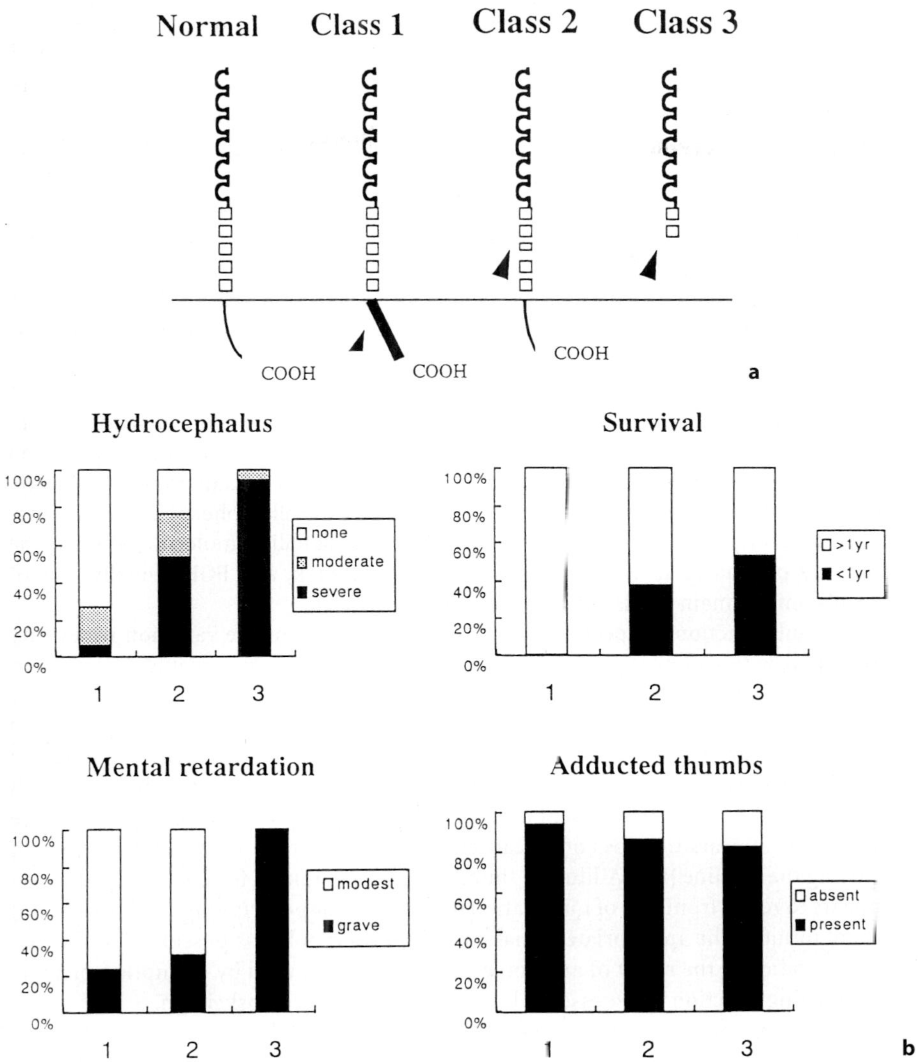

FIG. 1. **a** Structure of L1. Normal L1 has 6 Ig domains, 5 fibronectin (FN) type III domains, a short transmembrane region, and a 114-amino-acid cytoplasmic domain. *Class 1* mutations disrupt only the cytoplasmic domain. *Class 2* consists of missense point mutations in the extracellular domain. *Class 3* mutations result in a stop codon that causes a truncation in the extracellular domain. **b** Correlation of class of L1CAM mutation (*1*, *2*, *3*; see **a**) with clinical signs showing the percentage of patients with varying degrees of hydrocephalus, survival, mental retardation, and presence of adducted thumb

severe hydrocephalus, 21% had moderate hydrocephalus, and 73% did not have hydrocephalus. Of 79 patients in class 2, 53% showed severe hydrocephalus, 22% moderate hydrocephalus, and 24% had no hydrocephalus; 32 patients (94%) with class 3 mutations showed severe hydrocephalus and 2 patients had moderate hydrocephalus. The likelihood of survival showed a similar pattern. In class 1, all patients survived more than a year. For patients in class 2, 37% died within the first year of life while 52% of class 3 patients died within the first year. Mental retardation showed significant variation between the three classes. Of 34 patients in class 1, 24% showed grave mental retardation while 76% had modest retardation. In class 2, there were 39 patients and 30% showed grave mental retardation, while all 10 patients in class 3 showed grave mental retardation. Adducted thumbs showed much less difference between the three groups; in classes 1, 2, and 3, 94%, 86%, and 81%, respectively, of the patients presented with adducted thumbs.

L1 is a cell adhesion molecule (CAM) that promotes cell migration and axon growth (see [11] for review). Part of this activity is accounted for by L1 binding to ligands on adjacent cells. Known ligands include L1 itself, two other Ig superfamily adhesion molecules, TAG-1/axonin-1 and F3/F11, a proteoglycan called phosphacan, and integrins. L1 also promotes axon growth by acting as a signaling molecule, altering the activities of kinases such as the casein kinase II, p90rsk, and FGFr. Finally, the L1 cytoplasmic domain binds the cytoskeleton via ankyrin.

The multifunctional aspect of L1 most likely accounts for the variation in severity associated with different L1CAM mutations. Class 1 mutations would be expected to disrupt L1-mediated signaling and cytoskeleton interactions but not necessarily L1-mediated adhesion [11]. In contrast to the high incidence of spastic paraplegia and adducted thumbs, found equally in all classes, very few instances of severe hydrocephalus occurred in class 1. Spastic paraplegia is found in all individuals with L1 mutations related to the absent or diminished corticospinal tract easily seen in the medulla. It appears that loss of L1 causes a specific failure of corticospinal axons in crossing the midline [12]. Adducted thumbs is also a common finding. The adduction appears to result from loss of innervation of the extensor pollicis longus (EPL) and not loss of input to the appropriate spinal motor neuron pool. It is possible that the loss of innervation is the result of an axon guidance failure caused by disruption of some L1 signaling function or an essential interaction with the cytoskeleton.

The degree and progression of ventricle dilatation is highly variable. Our analysis showed that loss of cell-surface expression of L1 and disruption of L1-mediated adhesion correlates with a very high incidence of severe hydrocephalus. In the normal brain, axon–axon adhesion generates tension in the white matter that is sufficient to balance with hydrostatic pressure of the cerebral spinal flow. However, loss of adhesion would allow axons to slip along each other, resulting in increased brain compliance. This mechanism might underlie the ventricular dilatation in the L1 mutations syndrome. Class 2 mutations could prevent either L1 homophilic or heterophilic binding and might also disrupt *cis* interactions that are important in L1-mediated neurite outgrowth. Finally these mutations could also alter the confirmation of L1. Class 3 mutations would lose all these L1 functions and be predicted to have the most severe phenotypes.

The inter- and intrafamilial variability in phenotype in families with L1CAM mutations has been thought to be wide, and our analysis of interfamilial variability is consistent with this. However, the intrafamilial variability has been overestimated. There are three families with substantial intrafamilial variability, but many other families show a fairly consistent phenotype. This raises the possibility that intrafamilial variation may depend on genetic background or some other non-L1-associated factor. It is interesting that a background influence on severity of hydrocephalus has been recently found in L1 knockout mice [13].

Conclusions

Mutations in L1 produce a complex clinical entity that shows variation in severity between different affected families and even within given families. A careful review of the literature shows, however, that a strong correlation exists between the type of mutation in the L1CAM gene and the severity of the disease.

Acknowledgments. This work was supported by grants from NINDS and NEI to V. Lemmon.

References

1. Rosenthal A, Jouet M, Kenwrick S (1992) Aberrant splicing of neural cell adhesion molecule L1 messenger RNA in a family with X-linked hydrocephalus. Nat Genet 2:107–112
2. Fransen E, Vits L, Van Camp G, et al (1996) The clinical spectrum of mutations in L1, a neuronal cell adhesion molecule. Am J Med Genet 64:73–77
3. Bickers DS, Adams RD (1949) Hereditary stenosis of the aqueduct of Sylvius as a cause of congenital hydrocephalus. Brain 72:246–262
4. Bianchine JW, Lewis RC (1974) The MASA syndrome: a new heritable mental retardation syndrome. Clin Genet 5:298–306
5. Yamasaki M, Thompson P, Lemmon V (1997) CRASH syndrome: mutations in the L1 gene correlate with severity of the disease. Neuropediatrics 28:175–178
6. Gu SM, Orth U, Zankl M, et al (1997) Molecular analysis of the L1CAM gene in patients with X-linked hydrocephalus demonstrates eight novel mutations and suggests non-allelic heterogeneity of the trait. Am J Med Genet 71:336–340
7. MacFarlane JR, Du JS, Pepys ME, et al (1997) Nine novel L1 CAM mutations in families with X-linked hydrocephalus. Hum Mutat 9:512–518
8. Okamoto N, Wada Y, Kawabata H, et al (1997) A novel mutation in L1CAM gene in a Japanese patient with X-linked hydrocephalus. Jpn J Hum Genet 41:431–437
9. Okamoto N, Wada Y, Goto M, et al (1997) Hydrocephalus and Hirschprung's disease in a patient with a mutation of L1CAM. J Med Genet 34:670–671
10. Mizugishi K, Yanagisawa M, Kondo I, et al (1997) Gene analysis of the neural cell adhesion molecule L1 (L1CAM) in a Japanese family with X-linked hydrocephalus. Acta Pediatr Jpn 101:1383–1387
11. Kamiguchi H, Lemmon V (1997) Neural cell adhesion molecule L1: signaling pathways and growth cone motility. J Neurosci Res 49:1–8
12. Cohen N, Taylor J, Scott L, et al (1997) Errors in corticospinal axon guidance in mice lacking the neural cell adhesion molecule L1. Curr Biol 8:26–33
13. Dahme M, Bartsch U, Martini R, et al (1997) Disruption of the mouse L1 gene leads to malformations of the nervous system. Nat Genet 17:346–349

Assessment of P_0-Mediated Adhesion in *Xenopus* Oocytes

Mika Yoshida

Summary. Protein zero (P_0), the major integral protein in the peripheral nerve myelin, belongs to the immunoglobulin superfamily and is known to function as a homophilic adhesion molecule. The adhesive event mediated by P_0 is essential for normal myelin formation and maintenance, as point mutations and the null allele of the P_0 gene show various degrees of hypomyelination. Many point mutations of P_0 have been shown to be associated with certain human peripheral neuropathies, such as Charcot-Marie-Tooth disease type 1B, Dejerine-Sottas syndrome, and congenital hypomyelination. To test the adhesive properties of mutated P_0 observed in these neuropathies, I adapted the paired *Xenopus laevis* oocyte expression system, which has been utilized to investigate intercellular gap junction formation between two oocytes expressing various connexins. In a single oocyte, a wild-type P_0GFP chimeric protein was targeted correctly to the oocyte cell surface. When two oocytes expressing wild-type P_0 were coupled, protein was distributed to and heavily concentrated at cell–cell contact sites. In initial studies, a point mutation (G138R) that is found in Dejerine-Sottas syndrome was engineered, and the resultant mutated P_0GFP mRNA was injected into two oocytes. The mutated P_0GFP was accumulated mostly intracellularly, and when two oocytes were coupled, this mutated P_0 did not show strong surface distribution and was instead retained intracellularly. The studies presented here show that the paired *Xenopus laevis* oocyte expression system will be extremely useful for evaluating the adhesive properties of various adhesion molecules.

Key words. Protein zero (P_0)—P_0 mutation—*Xenopus* oocyte—Paired oocyte—GFP

Introduction

Protein zero (P_0) is the major myelin integral protein that mediates adhesion of Schwann cell plasma membrane surfaces to form compact myelin in the peripheral nervous system (PNS) of vertebrates. P_0 is a member of the immunoglobulin gene superfamily that contains a single variable-like region in its extracellular segment

---

The Brookdale Center for Developmental and Molecular Biology, Mount Sinai School of Medicine, One Gustave L. Levy Place, Box 1126, New York, NY 10029, USA

[1,2], and functions as a homophilic adhesion molecule through interactions of its extracellular domains [3–5]. X-ray crystallographic studies suggest that P_0-mediated adhesion is via tryptophan side-chain interaction with cell bilayers and P_0 tetramer associations in the extracellular space [6]. A null mutation of the P_0 gene showed impaired membrane compaction in peripheral myelin and a severe hypomyelination phenotype, revealing the importance of P_0 in the formation and maintenance of PNS myelin [7]. Moreover, point mutations in the human P_0 gene have been shown to be associated with certain human peripheral neuropathies, such as Charcot-Marie-Tooth type 1B (CMT1B), Dejerine-Sottas syndrome (DSS), and congenital hypomyelination (CH).

To evaluate the adhesive properties of adhesion molecules such as cadherins, the aggregation assay for stable transfectants is an established and widely used method [8–10], but it is extremely laborious to make stable cell lines for a large number of different point mutations. Dahl et al. [11] utilized the paired *Xenopus laevis* oocyte expression system to investigate the ability of different connexins to form the junctional channels. When two oocytes injected with certain connexin combinations were coupled together, channel activity through gap junctions was observed [11,12]. Inspired by these studies, we tested if the paired *Xenopus* oocyte expression system could be applied to evaluate the adhesive properties of P_0 and engineered mutants representing the various human peripheral neuropathies.

In this study, wild-type P_0 and a representative mutated P_0, which is found in DSS patients, were used. First, to visualize protein localization, we made a P_0 fusion construct with green fluorescent protein (GFP). When P_0 carrying GFP at the C-terminus was transfected to HeLa cells, P_0GFP intensely accumulated at the cell–cell contact site, indicating that P_0GFP fusion protein is correctly localized and is likely to have retained adhesive properties. Next, mRNAs made by in vitro transcription were injected into *Xenopus* oocytes and the two oocytes were coupled together. In single oocytes, wild-type P_0GFP protein was uniformly distributed at the cell surface. When two oocytes injected with wild-type P_0 were coupled together, P_0 protein was redistributed and intensely concentrated at cell–cell contact sites. However, P_0GFP bearing a missense mutation in the transmembrane domain (G138R), which is found in DSS patients [13], was only partially targeted to the plasma membrane and a substantial amount was retained in the cytoplasmic compartment. Furthermore, when two oocytes injected with this mutated P_0 were coupled, the protein did not show the typical surface redistribution of the wild-type P_0 and was not concentrated at cell–cell borders. With further development, we are confident that the coupled *Xenopus laevis* oocyte system will be valuable in evaluating the properties of various adhesion molecules.

Materials and Methods

Wild-Type and Mutated P_0 Constructs

Rat P_0 was subcloned into pEGFP-N1 vector (Clontech, Palo Alto, CA) to be fused with GFP. The point mutation (G138R) corresponding to the human peripheral neuropathy, DSS, was introduced into an expressible P_0 cDNA by the method based on the polymerase chain reaction (PCR).

HeLa Cell Transfection

HeLa cells were transiently transfected with P_0GFP by lipofection using DOTAP (Boerhinger-Mannheim, Germany). At 48 h after transfection, cells were fixed with 4% paraformaldehyde and processed for microscopy.

Injection and Pairing of Xenopus Oocytes

To make P_0 mRNA suitable for oocyte expression, constructs were subcloned into the oocyte transcription vector pGEMHE [14] and transcribed in vitro. Oocytes were collected from *Xenopus laevis* females [15] and injected with 40 nl of in vitro transcribed mRNA at a concentration of 500 ng/μl. After injection, oocytes were incubated for 15–30 min in hypertonic solution to strip their vitelline membrane. Oocytes were manually paired and incubated for 24–48 h at 19°C. The sections of single and paired oocytes were made by cryosectioning.

Results and Discussion

P_0GFP Is Targeted to the Plasma Membrane of Transfected HeLa Cells and Accumulates at Cell–Cell Borders

The great advantage to using GFP is that immunostaining is not needed after sectioning. This is important especially for sections of oocyte pairs, which tend to distort during immunostaining procedures. On the other hand, it is possible that GFP, whose molecular size is 27 kDa, interrupts the normal function of the P_0 molecule. To examine P_0 adhesive properties after being fused with GFP, we performed the transfections in HeLa cells. The P_0 without GFP is expressed at cell–cell contact sites (Fig. 1A), as previously described in HeLa cells [3,16]. P_0 tagged with GFP is also expressed on the cell surface and intensely accumulates at cell–cell borders when two expressors contact each other (Fig. 1B). P_0 expression is further confirmed by P_0 antibody labeling

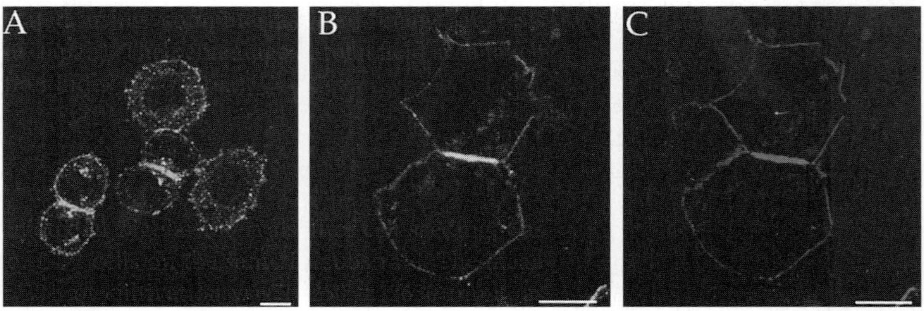

FIG. 1A–C. HeLa cells transiently transfected with P_0. **A** Confocal images of HeLa cells transfected with wild-type P_0. P_0 appears to be concentrated at the cell–cell interface. **B** and **C** HeLa cells transfected with P_0 tagged with green fluorescent proterm (GFP). The fusion protein observed by *green* fluorescence was distributed on the cell surface and intensely accumulated at the cell–cell border (**B**). P_0 expression was further comfirmed by P_0 antibody labeling (**C**). *Bars* 10 μm

(Fig. 1C). Although subtle changes in adhesivity may not be detectable, these transient transfection observations indicate that fusion with GFP does not change appreciably the adhesive properties of the P_0 molecule.

Wild-Type P_0 Is Uniformly Expressed at the Xenopus Oocyte Cell Surface, Whereas Mutated P_0 Is Retained Mainly in Cytoplasm

Oocytes injected with the control mRNA bearing only GFP coding sequence express GFP exclusively in nucleus (data not shown). When mRNA of the wild-type P_0GFP was injected into oocytes, protein distribution is strikingly changed: P_0GFP is uniformly inserted into the oocyte plasma membrane (Fig. 2A), which is also labeled by P_0

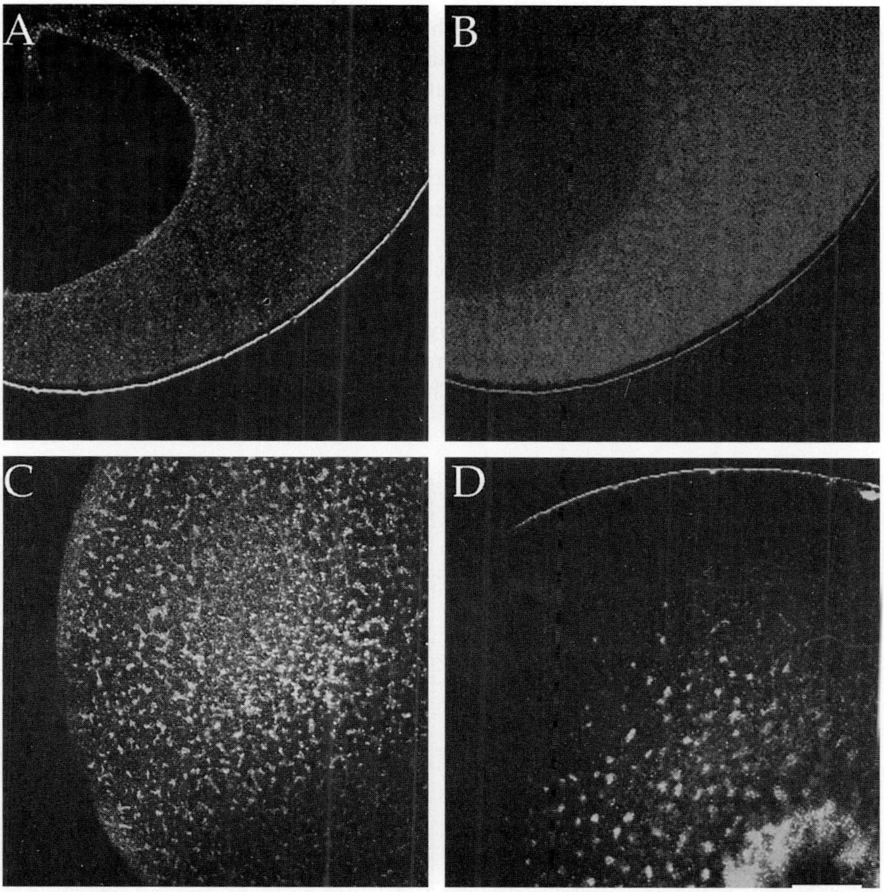

FIG. 2A–D. *Xenopus* oocytes injected with GFP and P_0GFP. *Xenopus* oocytes were injected with mRNA and allowed to incubate for 48 h. Wild-type P_0GFP was distributed on the cell surface, shown by *green* fluorescence (**A**) and P_0 antibody labeling (**B**). In contrast, P_0 bearing a point mutation in the transmembrane domain was mainly distributed in the oocyte cytoplasm (**C**) and only partially at the plasma membrane (**D**). *Bar* 100 μm antibody (Fig. 2B), showing that P_0GFP protein is correctly targeted to the oocyte cell surface. However, when a point mutation in the transmembrane domain of P_0, no. 138 glycine to arginine (which is found in DSS patients; [13]), is introduced, the distribution of P_0GFP is dramatically altered: substantial amounts of P_0GFP accumulate in the cytoplasm (Fig. 2C), although a proportion of the protein reaches the plasma membrane (Fig. 2D). It has been thought that this transmembrane missense mutation would make P_0 unstable within the plasma membrane [13]. Consistent with this idea, the mutated P_0 seems to change its targeting properties, is not correctly inserted in the plasma membrane, and is instead retained in the cytoplasm.

Wild-Type P_0, But Not Mutated P_0, Shows Close Membrane Apposition Between Two Oocytes

Homologous pairs of oocytes injected with either connexin32 or 43 mRNAs showed an accumulation of the specific antibody staining at the region of the cell–cell contact [12]. To investigate if a similar accumulation of protein, which is likely to reflect adhesive function, occurs for P_0, two oocytes were injected with wild-type P_0 mRNA and then coupled to each other. A striking redistribution of the P_0GFP to the cell–cell apposition region was revealed: P_0 is intensely accumulated at the border of two oocytes and reveals close membrane apposition (Fig. 3A), showing that P_0 behaves in a similar manner in *Xenopus* oocytes as in HeLa cells. However, the mutated P_0 does not show such a redistribution and no accumulation of mutated proteins was observed at cell–cell borders (Fig. 3B).

It is highly likely that wild-type P_0 is redistributed and accumulates at cell–cell contact sites because it is adhesive, and that mutated P_0 is not because the adhesive function is destroyed in this mutation. The redistribution of the connexins to the

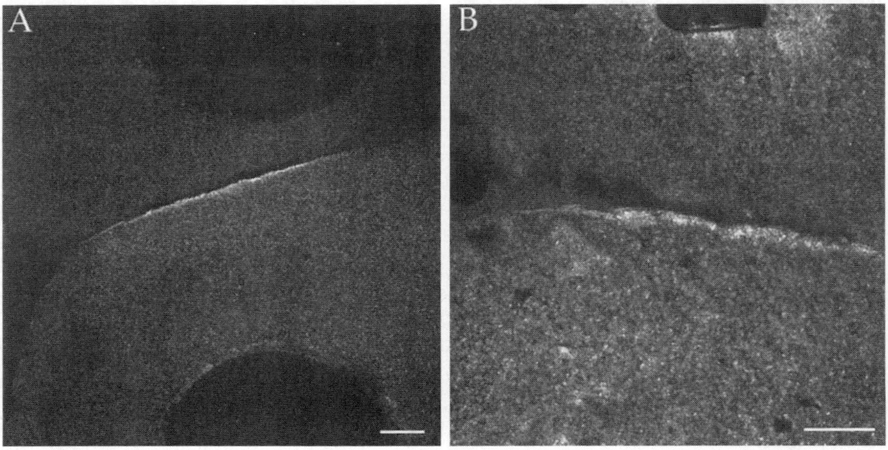

FIG. 3A,B. Coupling of two *Xenopus* oocytes injected with wild-type or mutated P_0GFP. Wild-type P_0GFP accumulates at cell–cell borders and shows the close membrane apposition (**A**). However, mutated P_0GFP did not show such apposition between two oocytes; instead, GFP fluorescence was mainly detected in the cytoplasm (**B**). *Bars* 100 μm appositional area is solely related to cell–cell contact and is not dependent on gap junction formation [12]. In this regard, it is possible that P_0 redistribution at cell–cell contact sites is not dependent on adhesive function but on cell–cell contact. Although further studies to confirm the results of the paired oocyte expression, such as adhesion assays for cell lines, are necessary, this system may be a useful new method for evaluating the altered adhesive properties of engineered P_0 molecules, and further, may be applicable to test this function in other adhesion molecules.

Acknowledgments. The author thanks Dr. David R. Colman for his suggestions and support. This work is supported by an advanced postdoctoral fellowship from the National Multiple Sclerosis Society (FA1230-A-1) to M.Y. and by a grant from the NIH (NS20147) to D.R.C. This is manuscript number 262 from the Brookdale Center for Developmental and Molecular Biology.

References

1. Lai C, Brow MA, Nave KA, Noronha AB, Quarles RH, Bloom FE, Milner RJ, Sutcliffe JG (1987) Two forms of 1B236/myelin-associated glycoprotein, a cell adhesion molecule for postnatal neural development, are produced by alternative splicing. Proc Natl Acad Sci USA 84:4337–4341
2. Lemke G, Lamar E, Patterson J (1988) Isolation and analysis of the gene encoding peripheral myelin protein zero. Neuron 1:73–83
3. D'Urso D, Brophy PJ, Staugaitis SM, Gillespie CS, Frey AB, Stempak JG, Colman DR (1990) Protein zero of peripheral nerve myelin: biosynthesis, membrane insertion, and evidence for homotypic interaction. Neuron 4:449–460
4. Filbin MT, Walsh FS, Trapp BD, Pizzey JA, Tennekoon GI (1990) Role of myelin P_0 protein as a homophilic adhesion molecule. Nature (Lond) 344:871–872
5. Schneider-Schaulies J, von Brunn A, Schachner M (1990) Recombinant peripheral myelin protein P_0 confers both adhesion and neurite outgrowth-promoting properties. J Neurosci Res 27:286–297
6. Shapiro L, Doyle JP, Hensley P, Colman DR, Hendrickson WA (1996) Crystal structure of the extracellular domain from P_0, the major structural protein of peripheral nerve myelin. Neuron 17:435–449
7. Giese KP, Martini R, Lemke G, Soriano P, Schachner M (1992) Mouse P_0 gene disruption leads to hypomyelination, abnormal expression of recognition molecules, and degeneration of myelin and axons. Cell 71:565–576
8. Takeichi M (1977) Functional correlation between cell adhesive properties and some cell surface proteins. J Cell Biol 75:464–474
9. Nose A, Nagafuchi A, Takeichi M (1988) Expressed recombinant cadherins mediate cell sorting in model systems. Cell 54:993–1001
10. Murphy-Erdosh C, Yoshida CK, Paradies N, Reichardt LF (1995) The cadherin-binding specificities of B-cadherin and LCAM. J Cell Biol 129:1379–1390
11. Dahl G, Miller T, Paul D, Voellmy R, Werner R (1987) Expression of functional cell-cell channels from cloned rat liver gap junction complementary DNA. Science 236:1290–1293
12. Swenson KI, Jordan JR, Beyer EC, Paul DL (1989) Formation of gap junctions by expression of connexins in *Xenopus* oocyte pairs. Cell 57:145–155
13. Hayasaka K, Himoro M, Sawaishi Y, Nanao K, Takahashi T, Takada G, Nicholson GA, Ouvrier RA, Tachi N (1993) De novo mutation of the myelin P_0 gene in Dejerine-Sottas disease (hereditary motor and sensory neuropathy type III). Nat Genet 5:266–268

14. Liman ER, Tytgat J, Hess P (1992) Subunit stoichiometry of a mammalian K^+ channel determined by construction of multimeric cDNAs. Neuron 9:861–871
15. Soreq H, Seidman S (1992) *Xenopus* oocyte microinjection: from gene to protein. Methods Enzymol 207:225–265
16. Doyle JP, Stempak JG, Cowin P, Colman DR, D'Urso D (1995) Protein zero, a nervous system adhesion molecule, triggers epithelial reversion in host carcinoma cells. J Cell Biol 131:465–482

Characterization of Binding of Myelin-Associated Glycoprotein with Neuronal Cells

Yoshihiro Matsuda, Hisami Koito, and Chigusa Shimizu

Summary. The structural requirements for myelin-associated glycoprotein (MAG) to bind with neuronal cells were examined. A soluble chimera protein, MAG-Fc, consisting of the extracellular region of mouse MAG (containing 5 immunoglobulin (Ig) domains) and the Fc region of human IgG, was prepared, and bound to neurons derived from mouse embryonal carcinoma P19 cells. As it appeared that the binding affinity of MAG-Fc by itself was not sufficient for detection, binding was sustained by cross-linking with dithio-bis(succimidylpropionate) (DSP). This binding was significantly diminished in the presence of anti-MAG antibodies. By using deletion chimera mutants, the area necessary for binding was shown to be the third Ig domain. Although MAG has the RGD sequence, a well-known cell adhesion motif of fibronectin, between the first and second Ig domains, this did not appear to play a critical role because an RGD-containing peptide failed to affect the MAG binding in the competition analyses. 3′-Sialyllactose inhibited the binding of MAG, but 6′-sialyllactose did not, suggesting involvement of α 2,3-linked sialic acid in the binding. Several species of gangliosides, such as GM1, GD1a, and GD1b, enhanced the MAG binding, which suggests their backing action during the binding of MAG with neurons.

Key words. Myelin-associated glycoprotein—Immunoglobulin superfamily—Embryonal carcinoma P19 cells—Cell adhesion—Sialic acid

Introduction

Myelin-associated glycoprotein (MAG) is a member of the immunoglobulin superfamily and considered to be a cell adhesion molecule linking myelin to axon membranes. MAG consists of two isoforms, L-MAG and S-MAG, the former being thought to play a critical role in the initial stage of myelination, and the latter being important for maintenance of myelin sheaths. However, little information has so far been available with respect to the binding mechanism of MAG and neurons. In the present

Department of Immunology, National Institute of Neuroscience, National Center of Neurology and Psychiatry, 4-1-1 Ogawa-higashi, Kodaira, Tokyo 187-8502, Japan study, using various MAG derivatives, we analyzed structural requirements of MAG for the binding.

Materials and Methods

Cell Culture

Mouse embryonal carcinoma P19 cells were maintained in α-minimum essential medium (α-MEM) containing 10% fetal calf serum (FCS), and subcultured every other day, as described by McCarrick and Andrews [1]. The P19 cells were directed to differentiate into the cells of neuronal phenotype by culture in aggregation in the presence of 0.7 μM retinoic acid, as previously described [2]. Then, the cells were dispersed with 0.025% trypsin and 1 mM EDTA, suspended in DMEM/Ham's F12 (1:1) containing 5% FCS, and plated in poly-L-lysine-coated 96-well multiplates at a density of 4×10^4 cells/well. After 2 days, the medium was exchanged for a serum-free defined medium consisting of DMEM/F12 supplemented with insulin (5 μg/ml), sodium selenite (30 nM), transferrin (50 μg/ml), hydrocortisone (10 nM), triiodothyronine (30 nM), and glucose (6 g/l), and the cells extending plenty of long neurites following 3 or 4 culture days were used for the MAG-binding assay.

Preparation of MAG-Fc Chimera Molecules

A DNA construct consisting of a cDNA fragment encoding the extracellular region of mouse MAG and a genomic DNA fragment of the Fc region of human immunoglobulin G1 was prepared and inserted into an expression vector, BCMGSneo, as previously reported [3]. Then, the plasmid was transfected into mouse myeloma P3U1 cells by electroporation, and only those transfectant cell clones that were found to be efficiently secreting the chimera protein into the culture medium were selected. The chimera protein containing the whole extracellular region of MAG (five Ig domains, 1–504 amino acids) was named MAG(D1-5)-Fc, and those containing the first three Ig domains (1–331 amino acids) and two Ig domains (1–249 amino acids) were named MAG(D1-3)-Fc and MAG(D1-2)-Fc, respectively.

Binding of MAG to Neuronal Cells

The culture medium of the P19 cell-derived neurons was removed from the 96-well multiplates and replaced by Hanks balanced salt solution (HBSS, 50 μl) containing MAG-related molecules. After incubation at room temperature for 40 min, an equal volume (50 μl) of 1 mM dithio-bis(succimidylproprionate) (DSP) dissolved in 10% dimethylsulfoxide (DMSO) in HBSS was added directly on the reaction medium and allowed to stand for 20 min. Then, the cells were washed with HBSS three times and fixed with 4% paraformaldehyde in PBS.

The binding of MAG was detected with conventional immunocytochemical procedures using biotinylated anti-MAG antibodies and peroxidase-conjugated streptoavidin. For cell staining, diaminobenzidine (DAB) was used as substrate. For quantitative determination, on the other hand, 2,2′-azino-di-(3-ethyl-benzthiazoline sulfonate) (ABTS) was used as a substrate for peroxidase, and absorbance at 415 nm was measured with a microplate reader.

Results and Discussion

It has been demonstrated that MAG binds to neuronal cells. Poltorak et al. [4] and Johnson et al. [5] showed that MAG-containing liposomes bound to neurites of dorsal root ganglion (DRG) and spinal cord neuronal cultures, and anti-MAG antibodies inhibited it. In this study, aiming to identify the MAG countermolecule on the neuronal surface, we tried to observe the binding of free MAG to neurons. As shown in Fig. 1, a soluble MAG-Fc chimera protein bound to P19-derived neurons with the aid of DSP. The neurite networks and the outlines of neuronal cells were clearly stained in Fig. 1a, but no significant binding was detected in the absence of DSP (Fig. 1c). When the MAG-Fc was pretreated with anti-MAG antibody, the binding to neurons was

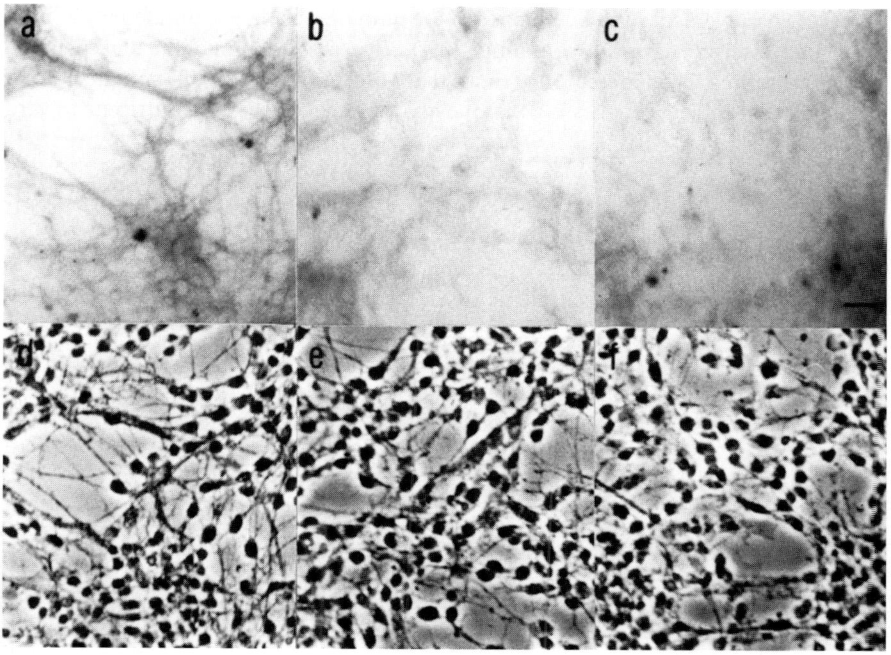

FIG. 1a–f. MAG (myelin-associated glycoprotein) binding to neurons with the aid of DSP: 10 μg/ml MAG(D1-5)-Fc was added on the neurons derived from P19 cells, which had been cultured for 4 days following the treatment for differentiation. The binding was detected by immunocytochemistry (**a–c**), and the corresponding phase-contrast fields are shown (**d–f**). The binding could be clearly seen when sustained with dithio-bis(succimidylpropionate) (DSP) (**a**), and is significantly inhibited when the MAG derivative is pretreated with 50 μg/ml anti-MAG monoclonal antibody (2B6) (**b**). No specific staining was found in the absence of DSP (**c**). *Bar* in **c**, 100 μm

diminished (Fig. 1b), whereas anti-human IgG(Fc) antibody or nonimmune immunoglobulins did not affect the binding (data not shown), indicating that this staining of the cells was dependent on the specific MAG binding.

Next, MAG's structural requirements for the binding were examined. MAG has five Ig domains in its extracellular region. Fahrig et al. [6] demonstrated by electron microscopic analysis that MAG was bent with the third Ig domain at one end, and Meyer-Franke et al. [7] showed with a panel of anti-MAG monoclonal antibodies and MAG peptide fragments that domain III may be important for binding to the neuronal cell surface. In this study, several Ig domain-deleted mutant molecules were prepared and compared for the binding to neurons. As shown in Fig. 2, MAG(D1-3)-Fc had a similar degree of binding to MAG(D1-5)-Fc. Anti-MAG antibodies inhibited the binding in both cases. On the other hand, MAG(D1-2)-Fc did not show any significant binding. These results also suggest that the third Ig domain is important for binding to neurons. In addition, MAG(D1-5), a MAG derivative that does not have the Fc portion, showed potent binding, indicating no involvement of the Fc portion in the binding.

MAG contains the RGD sequence, a well-known cell adhesion motif of fibronectin. Thus, any involvement of this sequence in the binding was then examined by competition analysis using a peptide containing RGD. An excess amount of the RGD peptide, of a molar ratio approximately 1000 times that of MAG(D1-5)-Fc, did not affect MAG binding at all (Table 1), suggesting that the RGD sequence does not contribute toward

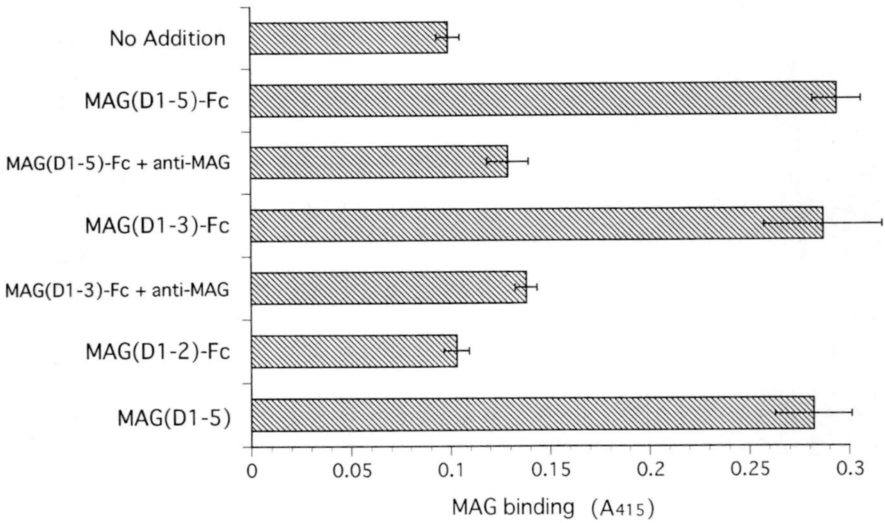

FIG. 2. Dependence of MAG binding to neurons on Ig domain numbers. MAG-Fc chimera proteins containing various numbers of Ig domains (10 μg/ml) were added to P19-derived neurons, followed by DSP. The MAG binding was quantified immunocytochemically using ABTS as a substrate; absorbances at 415 nm of 20-min reactions at room temperature are shown. To ascertain that increases in the absorbance are ascribed to the MAG binding, 50 μg/ml rabbit anti-MAG antibodies were added. Results represent mean ± SE from three experiments, each in triplicate. MAG(D1-5) is a derivative that contains only the extracellular region of MAG

TABLE 1. Effects of RGD peptides and sialyl oligosaccharides on myelin-associated glycoprotein (MAG) binding

G*RGD*SP peptide (100 μg/ml)	No effect
G*RGE*SP peptide (100 μg/ml)	No effect
3-Sialyllactose (500 μg/ml)	Inhibition (30%)
6-Sialyllactose (500 μg/ml)	No effect

The binding of 10 μg/ml MAG (D1-5) to P19-derived neurons was examined in the presence of the substances indicated.

the binding at all. On the other hand, DeBellard et al. [8] reported that a variety of neurons bound to immobilized MAG-Fc, and this binding was abolished when the neurons were treated with trypsin and sialidase. This suggests that the MAG-binding molecule on the neuronal surface may be a sialoglycoprotein. In our competition analyses, 2,3-sialyllactose inhibited MAG binding, whereas 2,6-sialyllactose did not, which is coincident with the data in this report [8].

Recently, it was reported that gangliosides might be involved in recognition and binding of the MAG-mediated cell–cell interactions [9,10]. Therefore, the effects of several gangliosides were investigated. Interestingly, gangliosides enhanced the binding of MAG in our experimental system. Twice as much MAG binding was observed in the presence of GM1, GD1a, and GD1b, but GT1b had no effect (data not shown). These results suggest that some species of gangliosides might support the binding, presumably because the association of gangliosides with MAG stabilizes the conformation necessary for binding.

References

1. McCarrick JW, Andrews PW (1992) Embryonal carcinoma cells and embryonic stem cells as models for neuronal development and function. In: Wood JN (ed) Neuronal cell lines: a practical approach. Oxford University Press, Oxford, pp 77–104
2. Matsuda Y, Koito H, Yamamoto H (1997) Induction of myelin-associated glycoprotein expression through neuron-oligodendrocyte contact. Dev Brain Res 100:110–116
3. Matsuda Y, Okitsu A, Sato S, et al (1996) Soluble myelin-associated glycoprotein-immunoglobulin G1 chimera protein promotes neurite outgrowth from mouse cerebellar neurons. Neurosci Lett 205:87–90
4. Poltorak M, Sadoul R, Keilhauer G, et al (1987) Myelin-associated glycoprotein, a member of the L2/HNK-1 family of neural cell adhesion molecules, is involved in neuron-oligodendrocyte and oligodendrocyte-oligodendrocyte interaction. J Cell Biol 105:1893–1899
5. Johnson PW, Abramow-Newerly W, Seilheimer B, et al (1989) Recombinant myelin-associated glycoprotein confers neural adhesion and neurite outgrowth function. Neuron 3:377–385
6. Fahrig T, Probstmeier R, Spiess E, et al (1993) Functional topography of the myelin-associated glycoprotein. I. Mapping of domains by electron microscopy. Eur J Neurosci 5:1118–1126
7. Meyer-Franke A, Tropak MB, Roder JC, et al (1995) Functional topography of the myelin-associated glycoprotein. II. Mapping of domains on molecular fragments. J Neurosci Res 41:311–323

8. DeBellard M-E, Tang S, Mukhopadhyay G, et al (1996) Myelin-associated glycoprotein inhibits axonal regeneration from a variety of neurons via interaction with a sialoglycoprotein. Mol Cell Neurosci 7:89–101
9. Yang LJ-S, Zeller CB, Shaper NL, et al (1996) Gangliosides are neuronal ligands for myelin-associated glycoprotein. Proc Natl Acad Sci USA 93:814–818
10. Collins BE, Yang LJ-S, Mukhopadhyay G, et al (1997) Sialic acid specificity of myelin-associated glycoprotein binding. J Biol Chem 272:1248–1255

Characterization and Developmental Regulation of Proteoglycan-Type Protein Tyrosine Phosphatase ζ/RPTPβ Isoforms

Taeko Nishiwaki, Nobuaki Maeda, and Masaharu Noda

Summary. Protein tyrosine phosphatase ζ (PTPζ/RPTPβ) is a receptor-like protein tyrosine phosphatase specifically expressed in the brain. Alternative splicing produces three isoforms of this molecule: PTPζ-A, the full-length form of PTPζ; PTPζ-B, the short form of PTPζ, in which a large part of the extracellular domain is deleted; and PTPζ-S, an extracellular variant. PTPζ-A and -S have been demonstrated to be chondroitin sulfate proteoglycans, but little is known about PTPζ-B. In this study, we identified PTPζ-B from rat brain as a chondroitin sulfate proteoglycan and compared its characteristics with the other isoforms. The level of PTPζ-A expression was maintained during prenatal periods and decreased markedly after birth. PTPζ-S was expressed in a similar manner, although the postnatal decrease was gradual. In contrast, relatively constant amounts of PTPζ-B were observed from embryonic day 13 (E13) through adulthood. In contrast to PTPζ-A and -S, which were constantly expressed as chondroitin sulfate proteoglycans, a substantial amount of non-proteoglycan form of PTPζ-B was detected at E13–E15. L cells transfected with PTPζ-A and -B cDNA indicated that they were colocalized with F-actin and cortactin in cell cortical structures. These results show that PTPζ-A and -B are differentially regulated during development and suggest that they may be involved in the reorganization of the cytoskeleton.

Key words. Brain development—Protein tyrosine phosphatase—Chondroitin sulfate proteoglycan—Alternative splicing—Subcellular localization

Introduction

Protein tyrosine phosphatase ζ (PTPζ/RPTPβ) is a receptor-like protein tyrosine phosphatase specifically expressed in the brain [1–5] and suggested to play roles in neuronal cell migration, neurite extension, and development of specific neural connections like the cerebellar mossy fiber system. PTPζ consists of an N-terminal car-

Division of Molecular Neurobiology, National Institute for Basic Biology, and Department of Molecular Biomechanics, The Graduate University for Advanced Studies, 38 Nishigonaka, Myodaiji-cho, Okazaki, Aichi 444-8585, Japan bonic anhydrase-like domain, a fibronectin type III domain, a large cystein-free, serine-glycine-rich region, a transmembrane segment, and two intracellular tyrosine phosphatase domains [1,2]. PTPζ-A and -S have been demonstrated to be chondroitin sulfate proteoglycans [3,6,7]. Here, we identified PTPζ-B also as a chondroitin sulfate proteoglycan and examined its developmental change in chondroitin sulfate modification, expression, and subcellular localization.

Results

We applied CsCl density gradient centrifugation, which separates proteoglycans from normal proteins, to postnatal day 0 (P0) rat brains [3]. PTP activity was recovered from proteoglycan fractions (fraction numbers 1–8) (Fig. 1). Aliquots of fraction numbers 4–6 were subjected to immunoprecipitation with the anti-6B4 proteoglycan, which recognizes all these isoforms (Fig. 2A). When the immunoprecipitate was treated with chondroitinase ABC, 220-, 300-, and 380-kDa core protein bands were detected by anti-6B4 proteoglycan (Fig. 2B). Previously, we demonstrated that 300- and 380-kDa core proteins correspond to those of PTPζ-S and PTPζ-A, respectively [3]. Chondroitin sulfate proteoglycan with 220-kDa core protein was recognized by anti-RPTPβ, which reacts with the D2 domain, but not by antiserum 31–5, the epitope of which is present only on PTPζ-A and -S (Fig. 2A), indicating that the proteoglycan with 220-kDa core protein is PTPζ-B.

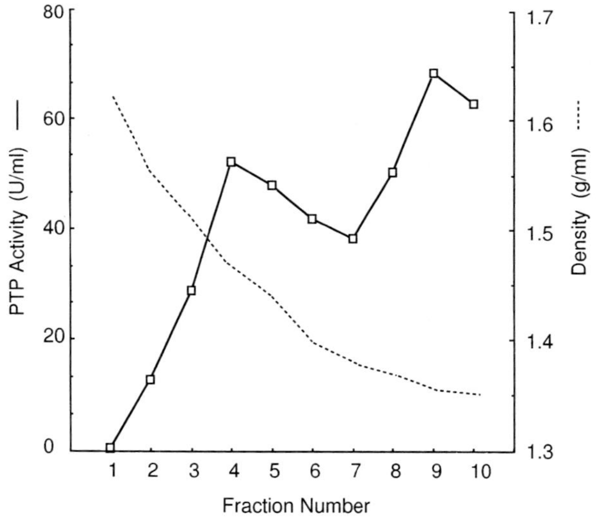

Fig. 1. Purification of proteoglycan-type protein tyrosine phosphatase (PTPs). CHAPS extract of the postnuclear fraction from P0 rat brain was applied to a DEAE-Toyopearl column, and the proteins were eluted with 0.6 M NaCl after washing with 0.25 M NaCl. The eluted proteins were fractionated by CsCl density gradient centrifugation. The sample was fractionated into 10 tubes, and the density (*dotted line*) and PTP activity (*open squares*) of each fraction were measured. [^{32}P]phosphotyrosine-labeled Raytide was used as a substrate for the assay of PTP activity

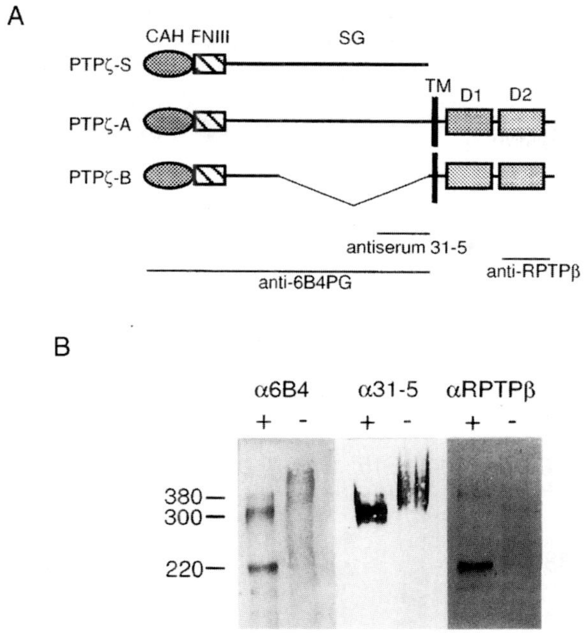

FIG. 2**A,B.** Identification of three PTPζ isoforms in the proteoglycan fraction. **A** Schematic representation of PTPζ isoforms and antigenic regions of the antibodies. *CAH*, carbonic anhydrase-like domain; *FNIII*, fibronectin type III domain; *SG*, serine-glycine-rich region (an 853 amino acid sequence is deleted in rat PTPζ-B); *TM*, transmembrane segment; *D1*, *D2*, tyrosine phosphatase domain. **B** The immunoprecipitate with anti-6B4 proteoglycan antibody contained three PTPζ isoforms. Aliquots of the CsCl gradient fractions (numbers 4–6) from P0 rat brain were applied to immunoprecipitation with anti-6B4 proteoglycan, and the immunoprecipitate was analyzed by immunoblotting using anti-6B4 proteoglycan (α6B4), antiserum 31-5 (α31-5), and anti-RPTPβ (αRPTPβ). The positions of the core protein bands of PTPζ-A (380), -S (300), and -B (220) are shown on the *left* (kDa). +, chondroitinase ABC-treated samples; −, nontreated samples

Developmental expression of PTPζ isoforms was examined in the brains from embryonic day 13 (E13) to P52. Expression of PTPζ-S increased from E13 to the perinatal period and gradually decreased thereafter (Fig. 3A). Expression of PTPζ-A was maintained from E13 to P0, and became almost undetectable after P20 (Fig. 3B). In contrast, little change was observed in the expression level of PTPζ-B during development (Fig. 3B). PTPζ-A and -S were constantly expressed as proteoglycans; however, non-proteoglycan-type PTPζ-B was detected from E13 to P0. After P0, almost all this isoform was expressed as chondroitin sulfate proteoglycan. Furthermore, it was revealed that PTPζ-B was not modified with Le^X, HNK-1 epitope, or keratan sulfate, although PTPζ-A and PTPζ-S were modified with these carbohydrates (data not shown). These results indicate that the expression and carbohydrate modification of PTPζ are developmentally regulated in an isoform-specific manner.

Next, we prepared L cell transfectants stably expressing cDNA-derived PTPζ-A and -B, and examined the subcellular localization of each PTPζ isoform. Indirect immun-

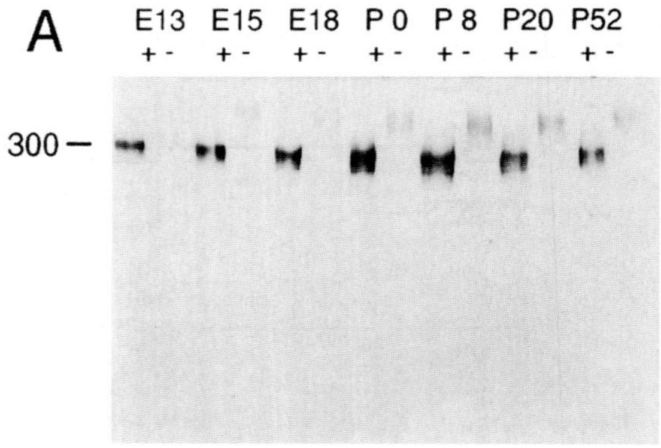

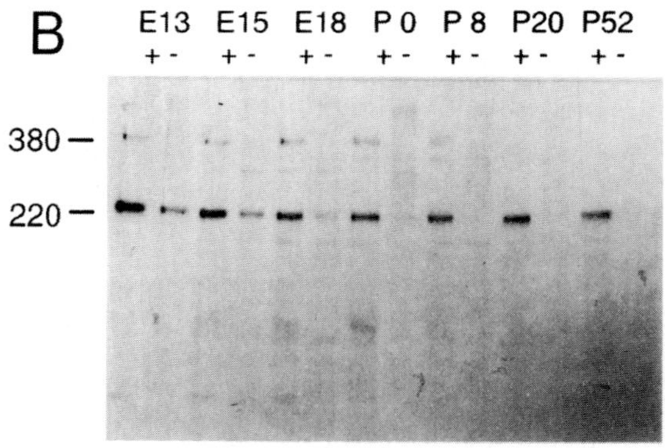

FIG. 3. **A** The brain homogenates prepared from E13 to P52 rats were solubilized and then analyzed by immunoblotting using antiserum 31-5 before (−) or after (+) chondroitinase ABC digestion. **B** Postnuclear fractions prepared from E13 to P52 rats were solubilized and subjected to immunoprecipitation with anti-6B4 proteoglycan. The immunoprecipitates were analyzed by immunoblotting using anti-RPTPβ before (−) or after (+) chondroitinase ABC digestion. The positions of the core protein bands of PTPζ-A (380), -S (300), and -B (220) are shown on the *left* (kDa).

FIG. 4**A–H.** Subcellular localization of PTPζ. L cell transfectants expressing PTPζ-B were double immunostained with anti-6B4 proteoglycan (**A**) and anti-cortactin (**B**), or anti-6B4 proteoglycan (**C**) and FITC-phalloidin (**D**). PTPζ-B and cortactin were colocalized at the processes (*arrows* in **A** and **B**) and ruffling membranes (*arrowheads* in **A** and **B**). PTPζ-B was distributed along F-actin at the cell periphery (*arrows* in **C** and **D**), but was not observed on stress fibers (*arrowhead* in **D**). When rabbit preimmune serum was used instead of the primary antibodies, only weak background stainings were observed (**E**). In the PTPζ-A-producing cells, processes (*arrows*) and ruffling membranes (*arrowhead*) were stained with anti-6B4 proteoglycan, although strong immunoreactivity was also observed inside the cell (**F**). Cortical neurons from E16 rats were cultured overnight on poly-L-lysine-coated coverslips, then fixed and double stained with anti-6B4 proteoglycan (**G**) and anti-cortactin (**H**). Anti-6B4 proteoglycan epitope and cortactin were colocalized at the growth cone (*arrowheads*), filopodial processes (*small arrows*), and the cell periphery (*large arrows*). Bars, 20 μm

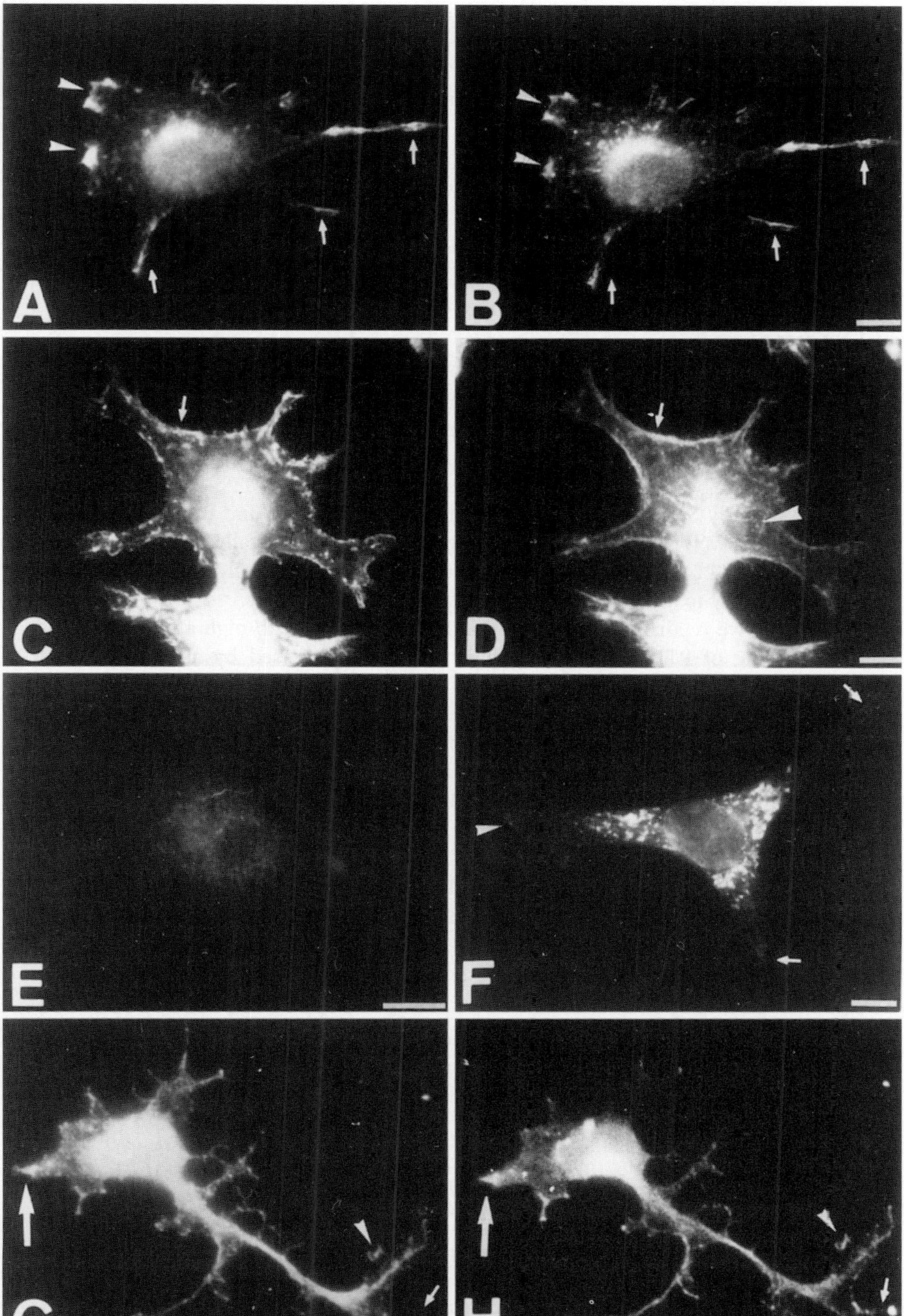
A
B
C
D
E
F
G
H

ofluorescence revealed that PTPζ-B and cortactin, an actin binding protein [8,9], were colocalized at the ruffling membranes and processes (Fig. 4A,B). PTPζ-B was distributed along the F-actin at the cell periphery but was not observed on stress fibers (Fig. 4C,D). PTPζ-A was also localized at the ruffling membranes and processes (Fig. 4F), although the signal was not so strong as the PTPζ-B-producing cells. Strong immunostaining was observed inside the cells, probably because of the accumulation of immature or processed products of PTPζ-A. Double immunofluorescence of cultured cortical neurons also revealed the colocalization of anti-6B4 proteoglycan epitope and cortactin at the growth cones, filopodial processes, and cell peripherals (Fig. 4G,H).

Discussion

This is the first demonstration that PTPζ-B is a chondroitin sulfate proteoglycan and that the expression and chondroitin sulfate modification of PTPζ isoforms are differentially regulated during brain development. The serine-glycine-rich region of PTPζ is considered to be the chondroitin sulfate attachment region. PTPζ-B still contains 10 Ser-Gly and Gly-Ser sequences in the corresponding region, and they may serve as attachment sites for chondroitin sulfate, which is developmentally regulated. On the other hand, PTPζ-A and -S contain additional 16 Ser-Gly and Gly-Ser sites, some of which are likely to attach chondroitin sulfate chains constantly.

Recently, we reported that PTPζ-S binds pleiotrophin with high affinity, and the total binding of PTPζ-S to pleiotrophin was also decreased by about 40% after chondroitinase digestion [10]. Binding of TAG-1 to PTPζ-S was also decreased by about 70% following chondroitinase ABC digestion [11]. These results indicate that chondroitin sulfate chains play significant roles for some ligand molecules in the regulation of binding to PTPζ. In a similar meaning it is noteworthy that PTPζ-B is not modified with Le^X, HNK-1 epitope, or keratan sulfate. Developmentally regulated modification of PTPζ-B with chondroitin sulfate might play a critical role in early prenatal development.

Using L cell transfectants producing PTPζ-A and -B, we demonstrated that both isoforms were localized at the cell cortical structures and colocalized with F-actin and cortactin. This raises the possibility that PTPζ acts as a receptor for extracellular molecules and regulates the organization of the cytoskeleton.

Acknowledgment. This work was supported by grants from the Ministry of Education, Science, Sports and Culture of Japan and from CREST of Japan Science and Technology Corporation.

References

1. Krueger NX, Saito H (1992) A human transmembrane protein-tyrosine phosphatase, PTPζ, is expressed in brain and has an N-terminal receptor domain homologous to carbonic anhydrases. Proc Natl Acad Sci USA 89:7417–7421
2. Levy JB, Canoll PD, Silvennoinen O, et al (1993) The cloning of a receptor-type protein tyrosine phosphatase expressed in the central nervous system. J Biol Chem 268:10573–10581

3. Maeda N, Hamanaka H, Shintani T, et al (1994) Multiple receptor-like protein tyrosine phosphatases in the form of chondroitin sulfate proteoglycan. FEBS Lett 354:67–70
4. Maurel P, Rauch U, Flad M, et al (1994) Phosphacan, a chondroitin sulfate proteoglycan of brain that interacts with neurons and neural cell-adhesion molecules, is an extracellular variant of a receptor-type protein tyrosine phosphatase. Proc Natl Acad Sci USA 91:2512–2516
5. Sakurai T, Friedlander DR, Grumet M (1996) Expression of polypeptide variants of receptor-type protein tyrosine phosphatase β: the secreted form, phosphacan, increases dramatically during embryonic development and modulates glial cell behavior in vitro. J Neurosci Res 43:694–706
6. Barnea G, Grumet M, Milev P, et al (1994) Receptor tyrosine phosphatase β is expressed in the form of proteoglycan and binds to the extracellular matrix protein tenascin. J Biol Chem 269:14349–14352
7. Shitara K, Yamada H, Watanabe K, et al (1994) Brain-specific receptor-type protein-tyrosine phosphatase RPTPβ is a chondroitin sulfate proteoglycan in vivo. J Biol Chem 269:20189–20193
8. Wu H, Parsons JT (1993) Cortactin, an 80/85-kilodalton $pp60^{src}$ substrate, is a filamentous actin-binding protein enriched in the cell cortex. J Cell Biol 120:1417–1426
9. Wu H, Reynolds AB, Kanner SB, et al (1991) Identification and characterization of a novel cytoskeleton-associated $pp60^{src}$ substrate. Mol Cell Biol 11:5113–5124
10. Maeda N, Nishiwaki T, Shintani T, et al (1996) 6B4 proteoglycan/phosphacan, an extracellular variant of receptor-like protein-tyrosine phosphatase ζ/RPTPβ, binds pleiotrophin/heparin-binding growth-associated molecule (HB-GAM). J Biol Chem 271:21446–21452
11. Milev P, Maurel P, Häring M, et al (1996) TAG-1/axonin-1 is a high-affinity ligand of neurocan, phosphacan/protein-tyrosine phosphatase-ζ/β, and N-CAM. J Biol Chem 271:15716–15723

Three-Dimensional Gels with Extracellular Matrix Containing Laminin Promote Migration of Cerebellar Granule Cells In Vitro

Junko Kimura-Kuroda[1], Hiroaki Asou[2], and Isao Nagata[1]

Summary. To establish suitable matrices for three-dimensional (3-D) neuronal migration in vitro, several kinds of gel matrices were examined in microexplant culture of mouse cerebellum. Massive granule cells migrated out from cerebellar explants in fibrin (FIB) and collagen type 1 (COL) gels that both contained laminin (LMN), but not from explants in FIB, COL, methylcellulose, and agarose gels alone. These results suggest that (1) both FIB and COL are suitable for constructing 3-D matrices for neurite extension and cell migration, and (2) LMN is essential for the promotion of granule cell migration in 3-D gels.

Key words. Cerebellar granule cell—Migration—Laminin—Three-dimensional culture—Extracellular matrix

Introduction

During central nervous system (CNS) development, neuroblasts migrate and extend their neuritic processes three-dimensionally (3-D) to appropriate sites in the brain. The establishment of a 3-D culture system is important for the analysis of neuronal migration and network formation. We have previously reported that in two-dimensional (2-D) culture laminin (LMN) promoted migration of cerebellar granule cells [1]. Recently it has been shown that olfactory neuroblasts migrate without the guidance of glia in a 3-D gel, Matrigel, which is composed of multiple extracellular matrix (ECM) components, including LMN, collagen type IV, and heparan sulfate proteoglycan (HSPG) [2]. In this study, to determine the ECM components utilized for 3-D migration of neuroblasts, several kinds of matrices were examined in the 3-D culture of cerebellar granule cells.

[1] Department of Cell Biology, Tokyo Metropolitan Institute for Neuroscience, 2-6, Musashidai, Fuchu, Tokyo 183, Japan

[2] Department of Neuro-cell Biology, Tokyo Metropolitan Institute of Gerontology, 35-2 Sakaecho, Itabashi-ku, Tokyo 173, Japan

Materials and Methods

Preparation of Gels

Fibrinogen from bovine plasma (FIB; Calbiochem, La Jolla, CA, USA), collagen type I from porcine tendon (COL; Nitta-Gelatin, Osaka, Japan), agarose type IV (AGA; Sigma, St. Louis, MO, USA), and methylcellulose (MET; Wako, Osaka, Japan) were used as a simple gel matrix. To make fibrin gel, fibrinogen was used after the removal of plasminogen using L-lysine [3]. Growth factor-reduced Matrigel (Collaborative Biomedical, Bedford, MA, USA) from Engelbreth-Holm-Swarm (EHS) mouse sarcoma was used as a gel matrix composed of multiple extracellular (basement membrane) matrix components. As shown in Table 1, each combination of gel matrices or additional components was mixed with a serum-free, hormone-supplemented medium as previously described [1]. Additional components were LMNCB (Collaborative Biomedical), LMNBT (Biomedical Technologies, Stoughton, MA, USA), HSPG (Sigma), and collagen type IV (Nitta-Gelatin), all of which were the main components of Matrigel. The former four substances were obtained from EHS mouse sarcoma, and collagen type IV was from bovine crystalline lens. Entactin/nidogen (Sigma) and ultrapure LMN (Collaborative Biomedical) were used for a comparative study.

Microexplant Culture

Small pieces of gray matter of the cerebellum (4- to 5-day-old BALB/C mice) were prepared as previously described [1] and immersed in each gel. After 2-day culture, microexplants were observed and photographed by phase-contrast microscopy. Migration distance of granule cells from the explant was measured in six randomly chosen fields (0.53 mm^2). Each value is the mean from two independent experiments.

Analysis of LMNs Used

Purity of LMNs used was examined by SDS-PAGE and Western blot by anti-LMN (Cappel, Aurora, OH, USA) and anti-entactin (Biogenesis, Sandown, NH, USA) antibodies. Biological activity of LMN was determined by cell attachment and neurite extension assays with cerebellar granule and PC12 cells [4].

Results

Characteristics of LMNs

Western blot analysis showed that entactin was detected in both LMNBT and LMNCB, while not in ultrapure LMN (data not shown). When the percent of attached granule cells to a poly-L-lysine-coated dish (0.1 mg/ml) was 100%, those cells attached to LMNBT and LMNCB were 105% and 95%, respectively, but those attached to ultrapure LMN were 15%. Neurite extension assays indicated that these LMNs with entactin are more effective than ultrapure LMN without entactin. We confirmed that entactin alone showed activity on neither neurite outgrowth nor the migration of granule cells in both 2-D and 3-D cultures. In PC12 cells, however, both cell attachment and neurite extension activities of ultrapure LMN were almost the same as the other LMNs. In the

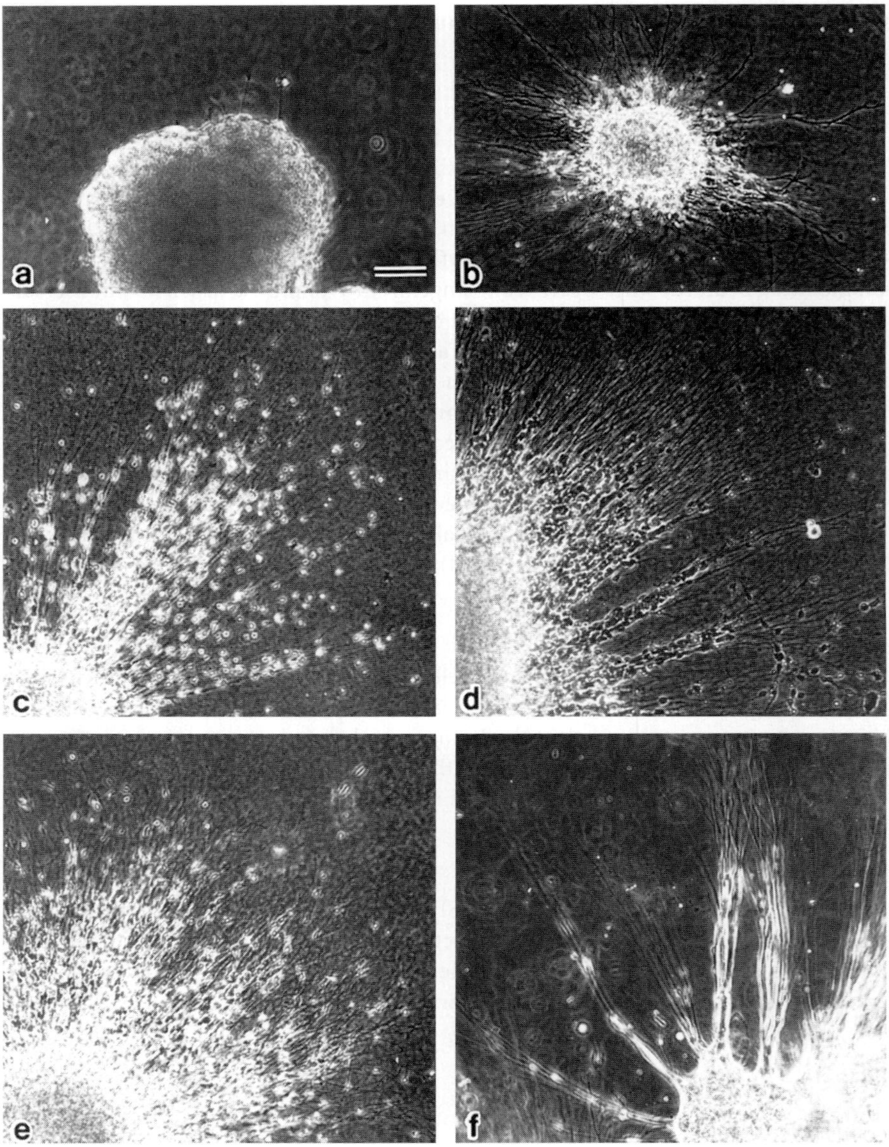

FIG. 1a–f. Neurite extension and migration of granule cells of cerebellar explant culture in 3-D gels. Concentrations of the components were as shown in Table 1. Laminin (LMN) was used at 400 μg/ml. **a** Methylcellulose (MET). **b** Fibrin (FIB). **c** FIB-LMNBT. **d** Collagen type 1 (COL)-LMNBT. **e** FIB-LMNCB. **f** Matrigel. *Bar* 100 μm

following 3-D experiments, therefore, two LMNs, LMNCB and LMNBT, were mainly used.

Migration of Granule Cells

As shown in Fig. 1, massive granule cells migrated out radially from the explants in FIB-LMN (c,e), COL-LMN (d), and Matrigel (f), all of which commonly contained LMN. However, neither FIB, COL, MET, and AGA alone nor FIB-HSPG and FIB-collagen type IV (Fig. 1, Table 1) promoted cell migration, although FIB and COL alone allowed the extension of neurite outgrowth. The migration distance of granule cells in FIB-LMN was longer than in COL-LMN (Fig. 2a) and depended on the concentration of LMN (Table 1). Mixing of FIB-LMN gel with HSPG or collagen type IV did not have an additional effect on cell migration (Fig. 2b). Furthermore, LMNBT, and LMNCB, both including entactin, were more effective for cell migration than ultrapure LMN without entactin in the 3-D gel (data not shown).

Discussion

These results indicate that both FIB and COL are suitable for constructing 3-D gel matrices and that LMN is essential for the promotion of granule cell migration in 3-D gels. It has been reported that the efficiency of LMN as a migratory substrate was enhanced when complexed with entactin and that the predominant motility-promoting site existed on LMN [5]. Entactin is colocalized and forms a very tight complex with LMN.

TABLE 1. Comparison of gel matrices and effects of laminin (LMN) and other components on neurite extension and granule cell migration in 3-D cultures

Gels[a]	Other components (μg/ml)			Neurite	Cell
	LMN[b]	HSPG	Collagen IV	extension	migration
Matrigel[c]	–	–	–	++++	++++
AGA	0	0	0	–	–
AGA-LMN	400	0	0	+	–
MET	0	0	0	–	–
MET-LMN	400	0	0	±	–
COL	0	0	0	+	–
COL-LMN	400	0	0	++++	+++
FIB	0	0	0	+	–
FIB-LMN_{50}	50	0	0	++	+
FIB-LMN_{200}	200	0	0	+++	+++
FIB-LMN_{400}	400	0	0	++++	+++−
FIB-HSPG	0	100	0	+	–
FIB-collagen IV	0	0	200	+	–

LMN, laminin; COL, collagen type I; FIB, fibrin; AGA, agarose; MET, methylcellulose; HSPG, heparin sulfate proteoglycan.

[a] Matrigel, 9.5 mg/ml; AGA, 0.4%; MET, 2.7%; COL, 0.1 mg/ml; FIB, 3 mg/ml.

[b] Two kinds of LMNs (LMN_{BT} or LMN_{CB}) were used (Fig. 2C).

[c] Matrigel contains approximately 60% LMN, 2% HSPG, and 30% collagen IV.

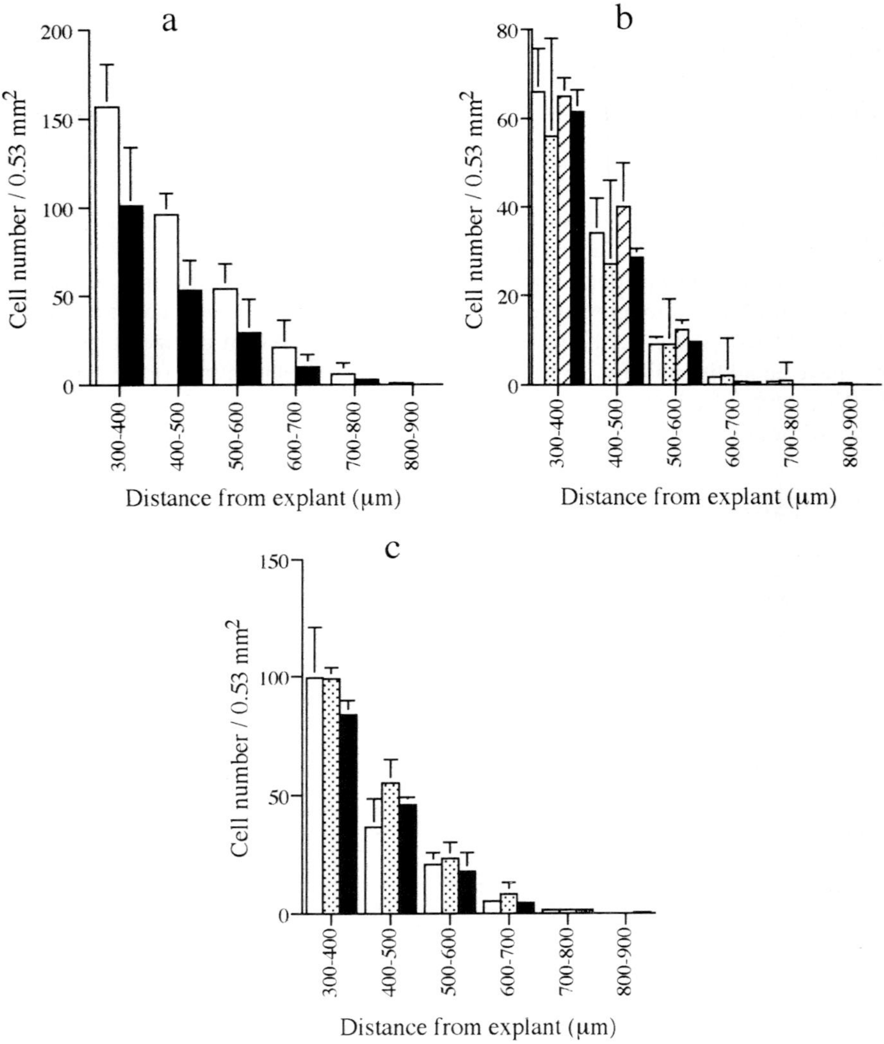

FIG. 2a–c. Quantitative comparison of each gel in terms of the migration distance of granule cells. Concentrations of FIB, COL, and Matrigel were as shown in Table 1. Error bars indicate standard deviation. **a** Comparison of FIB and COL gels in the presence of LMN. *Open column*, FIB-LMN; *filled column*, COL-LMN. LMNBT (400 μg/ml) was used. **b** Mixing effects of HSPG (heparin sulfate proteoglycan) and collagen type IV to FIB-LMN gel. *Open column*, FIB-LMNBT; *dotted column*, FIB-LMN-HSPG; *hatched column*, FIB-LMN-collagen IV; *filled column*, FIB-LMN-HSPG-collagen IV. LMNBT (200 μg/ml), HSPG (40 μg/ml), and collagen IV (200 μg/ml) were used. **c** Effects of LMNCB and LMNBT in FIB gel. *Open column*, FIB-LMNCB; *dotted column*, FIB-LMNBT; *filled column*, Matrigel. Both LMNCB and LMNBT were used at concentrations of 400 μg/ml

Immunohistochemistry by monoclonal anti-entactin antibody indicated the presence of entactin in parenchyma of the cerebellum of a 7-day-old mouse (data not shown). These results suggest that the binding of entactin may cause a conformational change of the LMN molecule, thereby exposing effective sites, such as the IKVAV site, that promotes neurite extension [4]. We are conducting a further investigation to test this hypothesis.

References

1. Nagata I, Nakatsuji N (1990) Granule cell behavior on laminin in cerebellar microexplant cultures. Dev Brain Res 53:86–95.
2. Wichterle H, Garcia-Verdugo JM, Alvarez-Buylla A (1997) Direct evidence for homotypic glia-independent neuronal migration. Neuron 18:779–791
3. Mosesson MW (1962) The preparation of human fibrinogen free of plasminogen. Biochim Biophys Acta 57:204–213
4. Tashiro K, Nagata I, Yamashita N, et al (1994) A synthetic peptide deduced from the sequence in the cross-region of laminin A chain mediates neurite outgrowth, cell attachment and heparin binding. Biochem J 302:73–79
5. Perris R, Paulsson M, Bronner-Fraser M (1989) Molecular mechanisms of avian neural crest cell migration on fibronectin and laminin. Dev Biol 136:222–238

Part 5
Glial Cell Lineage and Myelination

Mechanisms of Myelin Membrane Biogenesis by Oligodendrocytes

Winfried H. Krueger, Dana L. Madison, Taeyoon Kim, Yoshihide Yamaguchi, Brian Bouverat, Rashmi Bansal, and Steve Pfeiffer

Summary. Myelin membrane synthesis in the central nervous system by oligodendrocytes involves the coordinated intracellular transport and targeting of large amounts of specialized lipids and proteins over a relatively short span of time. The result is a radially polarized cell with vast sheets of myelin membrane composed of several biochemically distinct domains. We are currently investigating the utilization by oligodendrocytes of various elements of the vesicular trafficking pathway, in particular those mechanisms of membrane biogenesis predicted by the lipid raft and SNARE models. We hypothesize that specific components that we have identified, such as MVP17/rMAL, VAMP2/synaptobrevin 2, syntaxin 4, and Rab 3, are part of a plasma membrane-directed transport mechanism whereby nascent myelin vesicles are vectorially targeted to the forming myelin sheath.

Key words. Myelin—Oligodendrocytes—SNARE model—Lipid raft model—GTP-binding proteins

Introduction

The mechanisms by which cells achieve an asymmetric distribution of components into discreet membrane domains is a key issue in cell biology. Oligodendrocytes (OLs) form the myelin sheath in the central nervous system by producing extensive cell membrane expansions that form multilamellar wrappings around axons. This "myelination" process involves a coordinated assembly of specialized lipids and proteins into morphologically and biochemically distinct membrane domains [1]. Compacted myelin, formed through the extrusion of cytosol and apposition of opposing membrane faces, constitutes the major quantitative domain in which the myelin proteins proteolipid protein (PLP) and myelin basic protein (MBP) are concentrated. In contrast, the outer mesaxon and paranodal loops, and the inner mesaxon, retain cytoplasm and contain the myelin proteins cyclic nucleotide phosphohydrolase

Departments of Microbiology and Pharmacology, and Program in Neurological Sciences, University of Connecticut Medical School, 263 Farmington Avenue, Farmington, CT 06030-3205, USA

(CNP) [2] and connexin 32 [3,4], whereas myelin-associated glycoprotein (MAG) is exclusively localized to periaxonal membranes of the inner mesaxon [5,6]. The compartmentalization of the myelin membrane indicates differential sorting and translocation of myelin lipids and proteins.

Myelin sheath formation involves an impressive amount of membrane biogenesis. In the rat, the myelin membrane surface area expands by as much as 50 000 mm^2/cell/day [1], corresponding to the transport and insertion of as many as 5×10^5 50- to 100-nm-diameter vesicles/cell/day. At least three transport routes participate in myelin biogenesis. For example, myelin basic protein RNA is transported through the cytoplasm before its translation [7–9]; CNP is synthesized on free polyribosomes and subsequently transported to the plasma membrane [2]; PLP and MAG are processed through the vesicular (Golgi) pathway [10,11]. We have focused on the latter pathway, seeking to delineate trafficking mechanisms in OLs from the trans-Golgi network (TGN) to plasma membrane.

Transport to the plasma membrane consists of several steps [12]: (a) organization of cargo proteins and lipids into distinct domains of the sorting organelle, the Golgi network; (b) budding of the newly formed transport vesicle from the sorting compartment; (c) translocation of the transport vesicles along a selected translocation route to a specific plasma membrane microdomain; and (d) docking and fusion of the transport vesicle with the target membrane, resulting in the incorporation of the vesicle cargo into the plasma membrane, accompanied by the release of any luminal contents of the vesicle into the extracellular space (exocytosis).

Vesicular Transport to the Plasma Membrane and Myelin Sheath in Oligodendrocytes

Protein Sorting: The "Lipid Raft" Model

Membrane polarity requires the segregation of membrane components into distinct compartments, which are then budded off as vesicles and delivered to specific target membrane sites. It has been postulated that certain proteins are sorted in the TGN through association with detergent-insoluble, glycosphingolipid-enriched/cholesterol complexes (DIGs, or "lipid rafts") [13,14]. According to the model, association of a protein with lipid rafts results in detergent insolubility of the protein, as shown for GPI-anchored [15] and some apically targeted proteins [16]. Detergent insolubility has been used to isolate proteins associated with DIGs, leading to the identification of novel membrane proteins such as VIP21/cavaeolin [17,18] and the lectin-binding protein VIP36 [19].

The unusually large amount of glycosphingolipids in myelin, in particular, galactocerebroside and sulfatide, suggested a role for lipid rafts in myelin protein and lipid sorting [20,21]. Using detergent insolubility to identify myelin proteins associated with DIGs, we have isolated a novel developmentally regulated protein, MVP17/rMAL [20,22]. MVP17/rMAL is a member of the tetraspanin family that includes plasmolipin [23,24], a K^+ selective ion channel, PMP22 [25], a key protein in Charcot-Marie Tooth disease, and PLP, the major component of compact myelin. The function of MVP17/rMAL is under investigation.

Vesicular Docking and Fusion: The SNARE Model

The SNARE model proposes a mechanism for the docking and fusion of vesicles with their target membranes, whereby donor vesicles (v) interact with specific target (t) membranes through the specific recognition of so-called v- and t-SNAREs [12,26]. In synaptic vesicles, the v-SNARE VAMP-2/synaptobrevin 2 associates with the t-SNAREs SNAP-25 and syntaxin 1a, forming a 7S core docking complex [27–29]. The cytosolic protein α-SNAP associates with this complex [30], leading to enhanced binding of the soluble protein NSF and the subsequent formation of a 20S complex [31]. A number of accessory proteins, such as n-sec1/munc18(s), synaptophysin, synaptotagmin(s), Rab3, rabphilin-3a, Pra1, further fine tune the vesicular docking and fusion mechanism [32,33].

Docking and fusion of lipid raft-based vesicles may occur by a NSF-independent, non-SNARE mechanism [34,35]. However, SNARE-based docking and fusion nevertheless apparently does occur at the apical plasma membrane: apically directed vesicles carry v-SNAREs that are complexed with NSF [36]; syntaxins 2 or 4 are found at the apical surface of kidney cells [37,38]; synaptotagmin may substitute for VAMP proteins in docking [39]; and homologs for NSF exist, such as p97 [40].

The SNARE mechanism and components are conserved throughout cellular evolution, suggesting that this model describes a general membrane vesicle fusion mechanism adapted by all eukaryotic cells [30,36]. As a consequence, interorganelle transport is expected to be common among cell types, whereas plasma membrane-directed trafficking might demonstrate cell type-specific adaptations of this model. In fact, multiple VAMP (VAMP-1, -2, -3/cellubrevin; [41,42]), syntaxin (seven reported members; [36,43]), and SNAP 25 isoforms (SNAP25 and SNAP23 [44,45]) already have been identified, and genetic evolutionary analysis [46,47] suggests that there are more than 50 t-SNAREs (syntaxin and SNAP-25 homologs) and more than 8 VAMPs. Thus, a large number of combinations of v/t-SNARE interactions are available to confer specificity to target selection. We propose that for each plasma membrane/myelin subdomain in OLs, a specific v-and t-SNARE set may be employed for the targeting of the domain-specific membrane proteins.

We have analyzed OLs for key components of the SNARE model and have identified four groups on the basis of their expression pattern during OL lineage progression and terminal differentiation: (a) proteins that are downregulated, such as synapsin, synaptophysin, and syntaxin 2; (b) proteins showing little or no change in their expression levels, such as n-Sec1/Munc-18–1; (c) proteins that are upregulated during differentiation, such as Rab3a, syntaxin 4, and VAMP-2; and (d) proteins that are not expressed, such as syntaxins 1a and 3 and the regulatory proteins synaptotagmin 1 and rabphilin-3a (Madison et al., submitted).

Synapsin, synaptophysin, and syntaxin 2 are expressed at highest levels in OL progenitors. A common theme among these three proteins is their participation in cytoskeletal remodeling and reorganization of the plasma membrane. Syntaxin 2 has been implicated as a t-SNARE for "non-polarized" transport or transcytosis [48], whereas synaptophysin is thought to regulate VAMP-2 binding to its t-SNAREs [49,50], and synapsin functions in cytoskeletal remodeling during neuronal development [51]. Thus, these components may participate in the redistribution of plasma membrane from the trailing to the leading edge of the cell [52], thus facilitating directional migration of OLs toward the source of the chemotactic factors.

The developmental upregulation in maturing OLs of Syntaxin 4, Rab 3a, and VAMP-2 (and its deposition into the OL plasma membrane) (Madison et al., submitted) is intriguing because Syntaxin 4 and VAMP2 comprise a cognate set of v- and t-SNAREs for docking and fusion in plasma membrane-directed transport pathways [28,29,53,54]. We hypothesize that the VAMP-2/syntaxin 4 complex is important for myelin membrane assembly.

Rab Proteins in Oligodendrocytes

The Rab family of small GTPases has been implicated in the regulation of SNARE-based vesicular transport steps [55–58]. Rab proteins contain a core structure forming the catalytic domain, consisting of five highly conserved motifs [59]. GTP hydrolysis may provide a mechanism that, by inducing a high-affinity state, sterically activates v- and t-SNAREs [60] or other proteins that participate in budding, targeting, and fusion of transport vesicles. Rab proteins are posttranslationally lipidated at their C-terminus, allowing for their deposition into membranes of subcellular compartments [59,61,62]. The specificity of this association may be controlled by the interaction of individual Rab proteins with compartment-specific partners such as by the binding of Pra1 to both Rab 3a and VAMP-2, effectively linking Rab 3a to secretory vesicles [33]. Similar interactions may provide a molecular basis for the observation that individual Rab proteins act at specific transport steps [63–67].

We have analyzed the expression pattern of small GTPases during OL differentiation. In addition to the constitutively expressed GTPases Rap2, RalA, Rho, Rab5abc, Rab7, Rab11, Rab22, Rac, and RhoD, we have identified the expression of several developmentally regulated small GTP-binding proteins [68], including Rab 3a and 3c [69]. Using degenerative RT-PCR in combination with DNA probes and antibodies directed against known Rab proteins, we are now trying to establish the identity of the remaining unidentified Rab proteins in OLs.

Conclusion

The biochemical compartmentalization of the myelin sheath requires mechanisms for the sorting and translocation of myelin proteins and lipids to specific membrane domains, producing a polarized cell with an extensive, multidomain extension of the plasma membrane. We have demonstrated in OLs the presence of proteins involved in SNARE complex assembly leading to vesicular fusion, including VAMP-2 and Syntaxin 4, and small GTP-binding proteins that act as kinetic regulators for vesicle docking and fusion. Based on these observations, we hypothesized that OLs have adapted the general SNARE mechanism to assemble the myelin sheath (Fig. 1), and that prior sorting may involve elements of the lipid raft model as well. These data provide an initial mechanism from which to further study myelin membrane assembly.

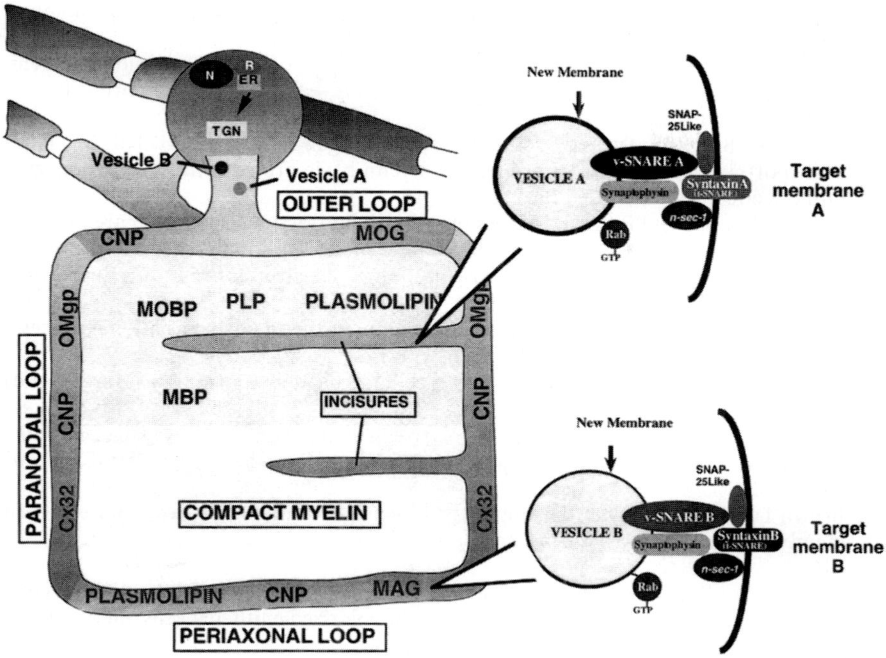

FIG. 1. Proposed mechanism for docking and fusion of myelin-containing vesicles that use SNARE-based target membrane recognition. Compartmentalization of the myelin membrane is achieved through docking and fusion of vesicles containing v-SNAREs and specific myelin components with myelin membrane subdomains specified through their t-SNAREs. *N*, nucleus; *R*, ribosome; *ER*, endoplasmic reticulum; *TGN*, trans-Golgi network; *CNP*, cyclic 2′,3′ nucleotide phosphodiesterase; *MOG*, myelin oligodendrocyte glycoprotein; *MOBP*, myelin oligodendrocyte basic protein; *PLP*, proteolipid protein; *MBP*, myelin basic protein; *MAG*, myelin-associated glycoprotein; *Cx32*, connexin 32; *OMgP*, oligodendrocyte myelin glycoprotein

References

1. Pfeiffer SE, Warrington AE, Bansal R (1993) The oligodendrocyte and its many cellular processes. Trends Cell Biol 3:191–197
2. Trapp BD, Bernier L, Andrews SB, Colman DR (1988) Cellular and subcellular distribution of 2′,3′-cyclic nucleotide phosphodiesterase and its mRNA in the rat central nervous system. J Neurochem 51:859–868
3. Miyazaki T, Shimazu T, Takeda Y, Toya S, Murakami Y, Uyemura K, Kawano H (1995) Distribution of PASII/PMP22 and connexin 32 proteins in the peripheral nervous system. Neurochem Int 27:377–383
4. Li J, Hertzberg EI, Nagy JJ (1997) Connexin 32 in oligodendrocytes and association with myelinated fibers in mouse and rat brain. J Comp Neurol 379:571–591
5. Trapp BD (1988) Distribution of the myelin-associated glycoprotein and P_0 protein during myelin compaction in quaking mouse peripheral nerve. J Cell Biol 107:675–685
6. Trapp BD, Kidd GJ, Hauer P, Mulrenin E, Haney CA, Andrews SB (1995) Polarization of myelinating Schwann cell surface membranes: role of microtubules and the trans-Golgi network. J Neurosci 15:1797–1807

7. Colman DR, Kreibich G, Frey AB, Sabatini DD (1982) Synthesis and incorporation of myelin polypeptide into CNS myelin. J Cell Biol 95:598–608
8. Kristensson K, Zeller NK, Dubois-Dalq ME, Lazzarini RA (1986) Expression of myelin basic protein gene in the developing rat brain as revealed by in situ hybridization. J Histochem Cytochem 34:467–473
9. Ainger K, Barry C, Avossa D, Barbarese E , Morgan F, Carson JH, Hill SJ (1993) Transport and localization of exogenous myelin basic protein mRNA microinjected into oligodendrocytes. J Cell Biol 123:431–441
10. Pereyra PM, Brown PE (1983) Studies on subcellular fractionation which are involved in myelin assembly: isolation from developing mouse brain and characterization by enzyme markers, electron microscopy and electrophoresis. J Neurochem 41:957–973
11. Townsend LE, Benjamins JA (1983) Effects of monensin on posttranslational processing of myelin proteins. J Neurochem 40:1333–1339
12. Rothman JE (1994) Mechanisms of intracellular protein transport. Nature (Lond) 372:55–63
13. Simons K, Wandinger-Ness A (1990) Polarized sorting in epithelia. Cell 62:207–210
14. Simons K, Ikonen E (1997) Functional rafts in cell membranes. Nature (Lond) 387:569–572
15. Brown DA, Rose JK (1992) Sorting of GPI-anchored proteins to glycolipid-enriched membrane subdomains during transport to the apical cell surface. Cell 68:533–544
16. Skibbens V, Roth M, Matlin K (1989) Differential extractability of influenza virus hemaglutinin during intracellular transport in polarized epithelial cells and nonpolar fibroblast. J Cell Biol 108:821–832
17. Kurzchalia T, Dupree P, Parton R, Kellner R, Virta H, Lehnert M, Simons K (1992) VIP21, a 21-kD membrane protein, is an integral component of trans-Golgi-network-derived transport vesicles. J Cell Biol 118:1003–1014
18. Dupree P, Parton R, Raposo G, Kurzchalia T, Simons K (1993) Caveolae and sorting in the trans-Golgi network of epithelial cells. EMBO J 12:1597–1605
19. Fiedler K, Simons K (1994) A putative novel class of animal lectins in the secretory pathway homologous to leguminous lectins. Cell 77:625–626
20. Kim T, Fiedler K, Madison DL, Krueger WH, Pfeiffer SE (1995) Cloning and characterization of MVP17: a developmentally regulated myelin protein in oligodendrocytes. J Neurosci Res 42:413–422
21. Kramer EM, Koch T, Niehaus A, Trotter J (1997) Oligodendrocytes direct glycosyl phosphatidylinositol-anchored proteins to the myelin sheath in glycosphingolipid-rich complexes. J Biol Chem 272:8937–8945
22. Schaeren-Wiemers N, Valenzuela DM, Frank M, Schwab ME (1995) Characterization of a rat gene, rMAL, encoding a protein with four hydrophobic domains in central and peripheral myelin. J Neurosci 15:5733–5764
23. Sapirstein V, Stadler I, Nolan C, Fischer I (1992) Expression of plasmolipin in the developing rat brain. J Neurosci Res 31:96–102
24. Fischer I, Sapirstein V, Durrie R (1994) Plasmolipin: the other myelin proteolipid. A review of studies on its structure, expression, and function. Neurochem Res 19:959–966
25. Suter U, Welcher A, Snipes G (1993) Progress in the molecular understanding of hereditary peripheral neuropathies reveals new insights into the biology of the peripheral nervous system. Trends Neurosci 16:50–56
26. Sollner T, Bennett MK, Whiteheart SW, Scheller RH, Rothman JE (1993) A protein assembly-disassembly pathway in vitro that may correspond to sequential steps of vesicle docking, activation, and fusion. Cell 75:409–418
27. Jacobsson G, Meister B (1996) Molecular components of the exocytic machinery in the rat pituitary gland. Endocrinology 137:5344–5356

28. Hayashi T, McMahon H, Yamasaki S, Binz T, Hata Y, Sudhoff TC, Niemann H (1994) Synaptic vesicle membrane fusion complex: action of clostridial neurotoxins on assembly. EMBO J 13:5051–5061
29. Hayashi T, Yamasaki S, Nauenburg S, Binz T, Niemann H (1995) Disassembly of the reconstituted synaptic vesicle membrane fusion complex in vitro. EMBO J 14:2317–2325
30. Bennett MK, Scheller RH (1993) The molecular machinery for secretion is conserved from yeast to neurons. Proc Natl Acad Sci USA 90:2559–2563
31. Bennett MK (1995) SNAREs and the specificity of transport vesicle targeting. Curr Opin Cell Biol 7:581–586
32. Calakos N, Scheller RH (1996) Synaptic vesicle biogenesis, docking, and fusion: a molecular description. Physiol Rev 76:1–29
33. Martincic I, Peralta ME, Ngsee JK (1997) Isolation and characterization of a dual prenylated Rab and VAMP2 receptor. J Biol Chem 43:26991–26998
34. Ikonen E, Tagaya M, Ullrich O, Montecucco C, Simons K (1995) Different requirements for NSF, SNAP and Rab proteins in apical and basolateral transport in MDCK cells. Cell 81:1–20
35. Wilson K (1995) NSF-independent fusion mechanisms. Cell 81:475–477
36. Hay JC, Scheller RH (1997) SNAREs and NSF in targeted membrane fusion. Curr Opin Cell Biol 9:505–512
37. Low S-H, Chapin SJ, Wiembs T, Komuves LG, Bennett MK, Mostov KE (1996) Differential localization of syntaxin isoforms in polarized Madin-Darby canine kidney cells. Mol Biol Cell 7:2007–2018
38. Mandon B, Nielsen S, Kishore BK, Knepper M (1997) Expressions of syntaxins in rat kidney. Am J Physiol 273:F718–F730
39. Schiavo G, Stenbeck G, Rothman JE, Söllner TH (1997) Binding of the synaptic vesicle v-SNARE, synaptotagmin, to the plasma membrane t-SNARE, SNAP-25, can explain docked vesicles at neurotoxin-treated synapses. Proc Natl Acad Sci USA 94:997–1001
40. Rabouille C, Levine TP, Peters JM, Warren G (1995) An NSF-like ATPase, p97, and NSF mediate cisternal regrowth from mitotic Golgi fragments. Cell 82:905–914
41. Elferink LA, Trimble WS, Scheller RH (1989) Two vesicle associated membrane proteins are differentially expressed in the rat central nervous system. J Biol Chem 264:11061–11064
42. McMahon HT, Sudhoff TC (1995) Synaptic core complex of synaptobrevin, syntaxin, and SNAP25 forms a high affinity a-SNAP binding site. J Biol Chem 270:2213–2217
43. Bennett MK, Garcia-Arraras JE, Elferink LA, Peterson K, Fleming A, Hazuka CD, Scheller RH (1993) The syntaxin family of vesicular transport receptors. Cell 74:863–873
44. Zhao N, Takahashi N, Hashida H, Sakaki Y (1994) Cloning and sequence analysis of the human SNAP25 cDNA. Gene 145:313–314
45. Ravichandran V, Chawla A, Roche PA (1996) Identification of a novel syntaxin- and synaptobrevin/VAMP-binding protein, SNAP-23, expressed in non-neuronal tissues. J Biol Chem 271:13300–13303
46. Weimbs T, Hui-Low S, Chapin SJ, Mostov KE (1997) Apical targeting in polarized epithelial cells: there's more afloat than rafts. Trends Cell Biol 7:393–399
47. Bock JB, Scheller RH (1997) Protein transport: a fusion of new ideas. Nature (Lond) 387:133–135
48. Apodaca G, Cardone MH, Whiteheart SW, DasGupta BR, Mostov KE (1996) Reconstitution of transcytosis in SLO-permeabilized MDCK cells: existence of an NSF-dependent fusion mechanism with the apical surface of MDCK cells. EMBO J 15:1471–1481
49. Edelmann L, Hanson PI, Chapman ER, Jahn R (1995) Synaptobrevin binding to synaptophysin: a potential mechanism for controlling the exocytic fusion machine. EMBO J 14:224–231

50. Washbourne P, Schiavo G, Montecucco C (1995) Vesicle-associated membrane protein 2 (synaptobrevin-2) forms a complex with synaptophysin. Biochem J 305:721–724
51. Han H-Q, Greengard P (1994) Remodeling of cytoskeletal architecture of non-neuronal cells induced by synapsin. Proc Natl Acad Sci USA 91:8557–8561
52. Bretscher MS (1996) Moving membrane up to the front of migrating cells. Cell 85:465–467
53. Calakos N, Bennett MK, Peterson KE, Scheller RH (1994) Protein-protein interactions contributing to the specificity of intracellular trafficking. Science 263:1146–1149
54. Gaisano HY, Ghai M, Malkus PN, Sheu L, Bouquillion A, Bennett MK, Trimble WT (1996) Distinct cellular localizations of the syntaxin family of proteins in rat pancreatic acinar cells. Mol Biol Cell 7:2019–2027
55. Zerial M, Sternmark H (1993) Rab GTPases in vesicular transport. Curr Opin Cell Biol 5:613–620
56. Simons K, Zerial M (1993) Rab proteins and the road maps for intracellular transport. Neuron 11:789–799
57. Novick P, Brennwald P (1993) Friends and family: the role of the Rab GTPases in vesicular trafficking. Cell 75:597–601
58. Nuoffer C, Balch WE (1994) GTPases: multifunctional molecular switches regulating vesicular traffic. Annu Rev Biochem 63:949–990
59. Takai Y, Kaibuchi K, Kikuchi A, Kawata T (1992) Small GTP-binding proteins. Int Rev Cytol 133:187–230
60. Brennwald P, Kearns B, Champion K, Keranen S, Bankaitis V, Novick P (1994) Sec 9 is a SNAP-25-like component of a yeast SNARE complex that may be the effector of SEC4 function in exocytosis. Cell 79:245–258
61. Farnsworth CC, Kawata M, Yoshida Y, Takai Y, Gelb MH, Glomset JA (1991) Carboxy terminus of the small GTP-binding protein smg-p25A contains two geranyl-geranylated residues and a methyl ester. Proc Natl Acad Sci USA 88:6196–6200
62. Yamane HK, Farnsworth CC, Xie H, Evans T, Howald WN, Gelb MH, Glomset JA (1991) Membrane binding domain of the small GTP-binding protein G25K contains an S-(all-*trans*-geranylgeranyl)-cysteine methyl ester at its carboxy terminus. Proc Natl Acad Sci USA 88:286–290
63. Orci L, Palmer DJ, Amherdt M, Rothman JE (1993) Budding from Golgi membranes requries coatamer complex of non-clatharin coat proteins. Nature (Lond) 362:648–652
64. Balch WE, McCaffery JM, Plutner H, Farquchar MG (1994) Vesicular stomatitis virus glycoprotein is sorted and concentrated during export from the endoplasmic reticulum. Cell 76:841–852
65. Goud B, Zahroui A, Tavitian A, Saraste J (1990) Small GTP-binding proteins associated with the Golgi cisternae. Nature (Lond) 345:553–556
66. Antony C, Cibert C, Geraud G, Santa-Maria A, Maro B, Maya V, Goud B (1992) The small GTP-binding protein rab6p is distributed from medial to the trans-Golgi network as determined by a confocal microscopy approach. J Cell Sci 103:785–796.
67. Lutcke A, Jansson S, Parton RG, Chavrier P, Valencia A, Huber LA, Lahtonen E, Zerial M (1993) Rab 17, a novel small GTPase, is specific for epithelial cells and is induced during cell polarization. J Cell Biol 121:553–564
68. Huber LA, Madison DL, Simons K, Pfeiffer SE (1994) Myelin membrane biogenesis by oligodendrocytes: developmental regulation of low molecular weight GTP-binding proteins. FEBS Lett 347:273–278
69. Madison DL, Krueger WK, Kim T, Pfeiffer SE (1996) Differential expression of rab3 isoforms in oligodendrocytes and astrocytes. J Neurosci Res 45:258–268

A Potential Supramolecular Assembly in Myelin

Greg Lemke

Summary. The structures of the integral membrane proteins of myelin, together with the molecular genetics of mutations in the genes encoding these proteins, suggest that they may be assembled into a single supramolecular complex within the myelin membrane. I summarize the data in favor of the existence of such a complex, and suggest the designation "myelisome."

Key words. Myelin—Genetics—Crystal structure—Mutation—Neuropathy

Myelin Proteins and Myelin Structure

As a result of research carried out over the past 15 years or so, we now have a fairly good understanding of the number, structure, and importance of the genes that encode the major protein constituents of both central and peripheral nervous system myelin. These proteins include (a) the integral membrane proteins protein zero (P_0) and peripheral myelin protein 22 (PMP-22) in the periphery, and Proteolipid Protein (PLP) in the CNS; (b) less abundant integral membrane proteins such as the myelin-associated glycoprotein (MAG); and (c) extrinsic myelin membrane proteins such as myelin basic protein (MBP) [1,2].

The integral membrane myelin proteins can be segregated into two sets based on the number and orientation of their transmembrane domains. P_0, a prototypical member of the first set, is a simple glycoprotein with a cleaved amino-terminal signal sequence and a single membrane-spanning domain. The extracellular portion of this protein consists entirely of an immunoglobulin-related domain that is known from both in vitro and in vivo work to mediate homophilic interactions that "zip up" each turn of the myelin membrane at apposed extracellular surfaces [3,4]. The P_0 cytoplasmic domain is similarly thought to mediate the tight association of apposed cytoplasmic membrane surfaces at each layer of the myelin sheath, but the experimental evidence for this hypothesis is more circumstantial, and the mechanism of action is largely conjectural. In contrast to P_0, PMP-22 and PLP, as prototypical members of the

Molecular Neurobiology Laboratory, The Salk Institute, 10010 N. Torrey Pines Rd., La Jolla, CA 92037-1099, USA second set of integral memebrane myelin proteins, both appear to either span or penetrate the myelin membrane multiple times. The most widely cited models predict four membrane-spanning domains, although the precise configuration of membrane domains is debated. Nonetheless, multiple membrane traverses are almost certainly the case, and in their hydrophobicity profiles, PMP-22 and related proteins resemble many gap junction subunits [5–7].

The level of our understanding of P_0 structure and function has recently been extended by the solving of a high-resolution X-ray crystal structure for the rat P_0 extracellular domain [8–10]. This crystal structure suggests several levels of organization, in which one extracellular domain first associates with three others, in a side-to-side fashion, to form a tetramer. Each P_0 extracellular domain tetramer resembles a square-shaped doughnut with a large central hole, and is hypothesized to further interact in a side-by-side fashion with the outside surfaces of four other tetrameric doughnuts tethered to the apposed membrane surface to form a very large, planar lattice of interacting P_0 tetramers. This large lattice is hypothesized to interlink with an equivalent lattice on an apposed extracellular membrane face, and to thereby form a large supramolecular network that locks facing membranes in place [8,10]. Several features of this model have important implications for the interpretation of the peripheral nerve phenotypes associated with human P_0 gene mutations (see following).

Molecular Genetics of Myelin Gene Mutations

Among the most compelling demonstrations of the role and importance of myelin proteins have come from the analysis of engineered and naturally occurring mutations in the genes encoding these proteins—in creatures as diverse as mice, humans, dogs, and rats. *Shiverer* mice, which as homozygotes express no MBP, experience increasingly debilitating tremors after about 2 weeks of age, and die prematurely from a failure to elaborate much myelin in the CNS [11]. The two most common alleles of the *Trembler* mouse carry point mutations in the *PMP-22* gene that lead to severe peripheral dysmyelination [12]. Also, a variety of point mutations in the *PLP* gene, e.g., the well-studied *jimpy* mutation in the mouse, lead to severe dysmyelination in the CNS [13]. At the same time, knock-outs of the mouse P_0 and *PMP-22* genes result in peripheral neuropathies and highly compromised myelination in the PNS [14,15].

Similarly, a variety of mutations in human populations affect the ability of Schwann cells and oligodendrocytes to normally myelinate their target axons. Mutations in the *PMP-22* gene lead to several forms of a relatively frequent inherited peripheral nerve disorder designated Charcot-Marie-Tooth (CMT) disease. Included in these mutations are a variety of point mutations [16–18]. The most common forms of CMT, however, result from either overdosage or underdosage of the *PMP-22* gene, rather than from a change in the structure of the protein. The dosage effects are caused by either the duplication or the deletion, on a single member of the human chromosome 17 pair, of the *PMP-22* gene, and lead to CMT1A and a disease called hereditary neuropathy with tendency to pressure palsies (HNPP), respectively. Interestingly, these dosage mutations in humans are dominant. In most instances, the chromosomal duplication that leads to over-expression of PMP-22 results in a markedly more severe disease than does the deletion that leads to underexpression. Similar observations have been made in transgenic mice, where heterozygous *PMP-22* gene knockouts exhibit only very mild deficits in peripheral myelination and nerve function, while homozygous knockouts (which express no PMP-22 at all) are worse, and mice or rats that overexpress the protein exhibit the strongest phenotype of all.

The genetics and effects of dosage and point mutation of the *PLP* gene are also particularly intriguing. Point mutations that prematurely truncate the PLP protein, as occur in the splice-site *jimpy* mutation in the mouse, or that result in certain missense mutations (e.g., the *myelin-deficient* mutation in the rat), are almost invariably debilitating [2,13]. Many of these mutations are thought to generate generate proteins that are not properly transported through the endoplasmic reticulum and Golgi apparatus. Also exhibiting a very strong phenotype are transgenic mice that overexpress PLP [19]. In marked contrast, heterozygous and even homozygous knockouts of the mouse *PLP* gene are viable and normally motile, and the axons of CNS neurons in these PLP-deficient mice exhibit only exceptionally subtle deficits in myelination and have normal conduction velocities [20]. It thus appears that altered structure or over-expression of PLP is a much more serious problem than loss of expression.

Mutations in the P_0 gene are similarly informative. Knockout mice that lack P_0 exhibit poor motor coordination, tremors, and occasional convulsions, and their Schwann cells are unable to elaborate normal myelin [14]. Heterozygous animals that contain half the normal amount of P_0 are initially normal in their behavior, but develop weak peripheral neuropathies and exhibit mild problems in myelin maintenance in old age [21]. Mutations in the human P_0 gene, many of them missense mutations, have also been identified, and these mutations result in inherited peripheral neuropathies that range from mild, e.g., CMT1B, to severe, e.g., Dejerine-Sottas syndrome (DSS) [9]. Both the position and genetics of many of these known mutations are of interest in light of the crystal structure. For example, several CMT1B-associated missense mutations cluster near the interface between individual monomers in the tetramer.

Most of the human missense mutations are dominant, as would be predicted for changes that perturb interactions in a multimeric assembly. At a few positions, differing mutations result in disease phenotypes of differing severity. Deletion of serine residue at position 34, for example, is predicted to yield an unstable protein that might never be transported into the myelin organelle. This mutation results in a mild CMT1B neuropathy that is not dissimilar to the phenotype of aged P_0 heterozygote mouse knockouts. Changing the same residue to cysteine, however, generates an outwardly pointing thiol that might be predicted to form disulfide aggregates in the extracellular space. This mutation leads to the more severe DSS [9].

The Myelisome: a Putative Supramolecular Assembly

The structure of integral membrane myelin proteins, together with the curious genetics of their mutations, suggests that these proteins structurally interact with each other in a defined way. If a particular protein multimerizes with itself alone, then it is entirely possible that point mutations that alter the folding or stability of the protein may in turn affect the properties of the multimer, and such mutations might very well be dominant as a consequence. Given self-multimerization alone, however, it is difficult to explain how mutations that simply alter the abundance of the protein might exhibit dominant effects. The dominance of such mutations, however, is readily explained if the proteins in question multimerize *with each other*, and if the stoichiometry of their multimerization must be maintained within set bounds.

The genetics of mutation of the *PMP-22*, *PLP*, and P_0 genes exhibit some similarities with those of the *Minute* mutations in *Drosophila*, which are also autosomal-dominant mutations that are capable of exhibiting both loss-of-function and gain-of-function effects [22,23]. There are in excess of 50 such loci in *Drosophila*, and many (perhaps most) of these correspond to genes encoding protein subunits of the ribosome. The ribosome is a complex macromolecular machine composed of a multiplicity of subunits, whose stoichiometry must be precisely fixed for the machine to be properly assembled and to efficiently function during protein translation.

Given these considerations, it is reasonable to hypothesize that each of the integral membrane proteins of the myelin sheath—P_0 and PMP-22 in the periphery, and PLP and perhaps a PMP-22-related protein in the CNS—are assembled into the same supramolecular assembly. It is possible that myelin basic protein is also associated with the aggregate cytoplasmic face of this assembly, but the genetics of MBP muations suggest that this protein does not play a role in formation of the assembly itself. A further expectation is that the stoichiometry of subunits within this assembly is relatively fixed, with P_0 in the PNS and PLP in the CNS predominating at least 10 fold over other assembly subunits (e.g., PMP-22 in the PNS). Because the supramolecular assembly being proposed has genetic features reminiscent of ribosomes and structural features reminiscent of intermembrane junctional complexes like desmosomes, I suggest that the name "myelisome" be applied.

The large lattice of multimerized tetrameric "doughnuts" that is observed in the crystal structure of the P_0 extracellular domain may reflect the backbone of the myelisome in the PNS, and PMP-22 or other proteins may occupy the "holes" that are evident in the center of each doughnut [10]. If either the relative abundance of the doughnuts or the holes is perturbed, then the organization of the entire assembly or its size and stability would be predicted to be alterred, which might lead to decreased myelin stability and peripheral nerve dysfunction.

The myelisome hypothesis makes one clear prediction that is probably testable given current anatomical methods; namely, that all the integral membrane proteins within the myelin sheath proper should colocalize. Because myelin is a protein-poor bilayer relative to most other biological membranes and as high-quality antibodies for both P_0 and PMP-22 are widely in use, it should be possible to use high-resolution immunoelectron microscopy to determine whether P_0 and PMP-22 are localized to the same space within the peripheral myelin sheath.

References

1. Lemke G (1992) In: Hall Z (ed) An introduction to molecular neurobiology. Sinauer Press, Sunderland, MA, pp 281–309
2. Lemke G (1993) The molecular genetics of myelination: an update. Glia 7:263–271

3. Lemke G, Axel R (1985) Isolation and analysis of a cDNA encoding the major structural protein of peripheral myelin. Cell 40:501–508
4. D'Urso D, Brophy PJ, Staugaitis SM, et al (1990) Protein zero of peripheral nerve myelin: biosynthesis, membrane insertion, and evidence for homotypic interaction. Neuron 4:449–460
5. Suter U, Snipes GJ (1995) Peripheral myelin protein 22: facts and hypotheses. J Neurosci Res 40:145–151
6. D'Urso D, Muller HW (1997) Ins and outs of peripheral myelin protein-22: mapping transmembrane topology and intracellular sorting. J Neurosci Res 49:551–562
7. Nave KA, Milner RJ (1989) Proteolipid proteins: structure and genetic expression in normal and myelin-deficient mutant mice. Crit Rev Neurobiol 5:65–91
8. Shapiro L, Doyle JP, Hensley P, et al (1996) Crystal structure of the extracellular domain from P_0, the major structural protein of peripheral nerve myelin. Neuron 17:435–449
9. Warner LE, Hilz MJ, Appel SH, et al (1996) Clinical phenotypes of different MPZ (P_0) mutations may include Charcot-Marie-Tooth type 1B, Dejerine-Sottas, and congenital hypomyelination. Neuron 17:451–460
10. Lemke G (1996) Unwrapping myelination. Nature (Lond) 383:395–396
11. Roach A, Takahashi N, Pravtcheva D, et al (1985) Chromosomal mapping of mouse myelin basic protein gene and structure and transcription of the partially deleted gene in shiverer mutant mice. Cell 42:149–155
12. Suter U, Welcher AA, Ozcelik T, et al (1992) Trembler mouse carries a point mutation in a myelin gene. Nature (Lond) 356(6366):241–244
13. Nave KA (1994) Neurological mouse mutants and the genes of myelin. J Neurosci Res 38:607–612
14. Giese KP, Martini R, Lemke G, et al (1992) Mouse P_0 gene disruption leads to hypomyelination, abnormal expression of recognition molecules, and degeneration of myelin and axons. Cell 71:565–576
15. Adlkofer K, Martini R, Aguzzi A, et al (1995) Hypermyelination and demyelinating peripheral neuropathy in Pmp22-deficient mice. Nat Genet 11:274–280
16. Scherer SS (1997) Molecular genetics of demyelination: new wrinkles on an old membrane. Neuron 18:13–16
17. Suter U, Snipes GJ (1995) Biology and genetics of hereditary motor and sensory neuropathies. Annu Rev Neurosci 18:45–75
18. Patel PI, Lupski JR (1994) Charcot-Marie-Tooth disease: a new paradigm for the mechanism of inherited disease. Trends Genet 10:128–133
19. Readhead C, Schneider A, Griffiths I, et al (1994) Premature arrest of myelin formation in transgenic mice with increased proteolipid protein gene dosage. Neuron 12:583–595
20. Klugmann M, Schwab MH, Puhlhofer A, et al (1997) Assembly of CNS myelin in the absence of proteolipid protein. Neuron 18:59–70
21. Martini R, Zielasek J, Toyka KV, et al (1995) Protein zero (P_0)-deficient mice show myelin degeneration in peripheral nerves characteristic of inherited human neuropathies. Nat Genet 11:281–286
22. Andersson S, Saeboe-Larssen S, Lambertsson A, et al (1994) A *Drosophila* third chromosome *Minute* locus encodes a ribosomal protein. Genetics 137:513–520
23. Saeboe-Larssen S, Urbanczyk Mohebi B, Lambertsson A (1997) The *Drosophila* ribosomal protein L14-encoding gene, identified by a novel *Minute* mutation in a dense cluster of previously undescribed genes in cytogenetic region 66D. Mol Gen Genet 255:141–151

Expression of the Myelin Basic Protein Gene in Cortical Pioneer Neurons During Embryonic Development

Anthony Campagnoni[1,2], Thomas Pribyl[1], Charles Landry[1], Julie Ellison[1], M. Irene Givogri[1], Kathy Kampf[1], and Celia Campagnoni[1]

Summary. The myelin basic protein (MBP) gene encodes two families of proteins that are structurally related to each other, the golli proteins and the "classic" MBPs. The golli–MBP transcription unit is unusual in many respects because it represents a rare arrangement of overlapping genes, or a gene within a gene. The MBP and golli transcription units also share multiple exons, which are alternatively spliced to produce the two families of mRNAs. This type of alternative splicing in overlapping transcription units of this size is very rare. The combined golli–mbp transcription unit contains three transcription start sites that are regulated independently to produce golli proteins or MBPs in neurons, oligodendrocytes, or immune cells with tissue, cell, and developmental specificity. Interestingly, golli proteins are expressed in cortical pioneer neurons and subplate cells and represent a new marker for these neurons. In an effort to identify regulatory elements responsible for the cellular specificity of the golli transcription unit, we performed in vitro and in vivo experiments with various upstream regions of the golli promoter. We identified a region that selectively targeted expression of a reporter gene in transgenic mice to the Cajal-Retzius and subplate neurons. The construct marked presubplate neurons at the earliest stages in their development and remained active throughout their developmental history.

Key words. Cerebral cortex—Cajal-Retzius cells—Cortical subplate—Transgene—Olfactory neurons

Historical Perspective on the Discovery of the Golli–mbp Gene

During the mid- to late 1980s, many laboratories were involved in identifying the multiple splice products of the myelin basic protein (MBP) gene. In our lab we isolated a number of MBP cDNA clones that were not predicted on the basis of the

[1] Mental Retardation Research Center, Neuropsychiatric Institute and [2] Brain Research Institute, UCLA Center for the Health Sciences, 760 Westwood Plaza, Los Angeles, CA 90024, USA

known MBP gene structure at that time. In addition to known MBP gene sequences, these clones contained unknown sequences at their 5′-ends. We mapped the genomic localizations of the 5′-end of one of these clones, M41-MBP, and found that it mapped to the MBP promoter region (i.e., directly upstream of the transcription start site [tss]) and to a new exon more than 25 kb upstream of the tss. We called these exons 1A and 0, respectively, and renamed the first exon of the MBP gene as exon 1B [1]. The M41-MBP mRNAs encode the 14 kDa MBP and appeared to be under the same developmental regulation as MBP mRNAs in the brain, but were much less abundant. Thus, these finding extended the length of the MBP gene by more than 25 kb and identified another, minor MBP tss.

We also had isolated several other MBP-related cDNAs that contained exon 1A and possessed the same 5′-ends, but these ends were different from the 5′-end of M41-MBP. From northern blots and RNase protection studies we knew these mRNAs were present in greater abundance than M41-MBP mRNAs in the brain, but their abundance was substantially lower than the MBP mRNAs. We assumed they would be products of a third MBP tss until we completed the cDNA sequencing studies and the genomic mapping work. It then became clear that these mRNAs (later called golli mRNAs by us) were the product of another transcription unit that overlapped and included the MBP gene. Furthermore, most, but not all, of the Golli mRNAs encoded unique proteins that contained MBP sequences, but the Golli proteins were not MBPs [2].

All our data had been gathered in the mouse, and we had some concern that these findings might be an oddity associated with the mouse, or even rodents. It was not until we had cloned analogous Golli mRNAs from human cell lines and established the structure of the human Golli gene that it was clear that the Golli–mbp transcription unit extended across mammalian species [3]. Interestingly, Golli mRNAs were expressed in the immune system as well as the nervous system. This established an important link between the expression of MBP epitopes in the immune system, which had important implications in our understanding of the mechanisms governing the autoimmune respose to MBP in experimental autoimmune encephalomyelitis (EAE). At about the same time that we published the structure, organization, and expression of the Golli gene in mouse and human, polymerase chain reaction (PCR) data were published by others confirming the expression of Golli mRNAs in the nervous system and the immune system [4,5]. Several groups have now identified Golli–mbp transcripts, firmly establishing their existence as products of the MBP gene locus in mice and humans.

In another approach, Mathisen et al. [6] published evidence for an embryonic form of MBP expressed in the CNS, PNS, thymus, spleen, and testes of the mouse. Essentially, these authors observed that the exon 1A region of the MBP gene was highly conserved between shark and mouse. They produced antibodies to a hypothetical peptide, assuming the region upstream of the tss was read in frame with the translation start site of MBP, which is exactly the arrangement in the Golli mRNAs. The tissue distribution of expression they found is remarkably similar to the distribution we have observed in embryos (Landry et al., in manuscript). The protein Mathisen et al. [6] identified on western blots does not correspond in size to any of the Golli proteins. The reason for this discrepancy remains unclear. Nonetheless, their observations on the expression of E-MBP in the embryo corresponds closely to our studies on Golli expression. Furthermore, their data suggest that at least the exon 1A region of the Golli gene has been well conserved during evolution.

Revised Structure of the Golli–MBP Gene

The structure of the golli–MBP transcription unit is shown in Fig. 1. In mice, the gene is 105 kb in length and in humans it is about 180 kb, placing it among the larger genes of neurobiological signifcance. It gives rise to two families of products. The Golli mRNAs are produced from the most upstream promoter at the first transcription start site (tss) at the beginning of exon 1. The MBP mRNAs are produced from two downstream promoters, tss2 at the beginning of exon 4 (which gives rise to M41-MBP mRNAs) and tss3 at exon 5B (which produces the vast majority of the MBP mRNAs). Both the Golli and MBP mRNA alternatively spliced families are more extensive than shown in the figure [7,8], and only the two major Golli and MBP mRNAs and proteins are shown in Fig. 1. In the mouse, BG21 is the most abundant Golli isoform, and it consists of 133 Golli-specific amino acids (encoded by exons 2,3, and 5A) and the first 56 amino acids of MBP followed by 5 Golli amino acids at its C-terminus. The principal Golli proteins share the same 133 aa as the Golli peptide plus varying amounts of the MBP sequence.

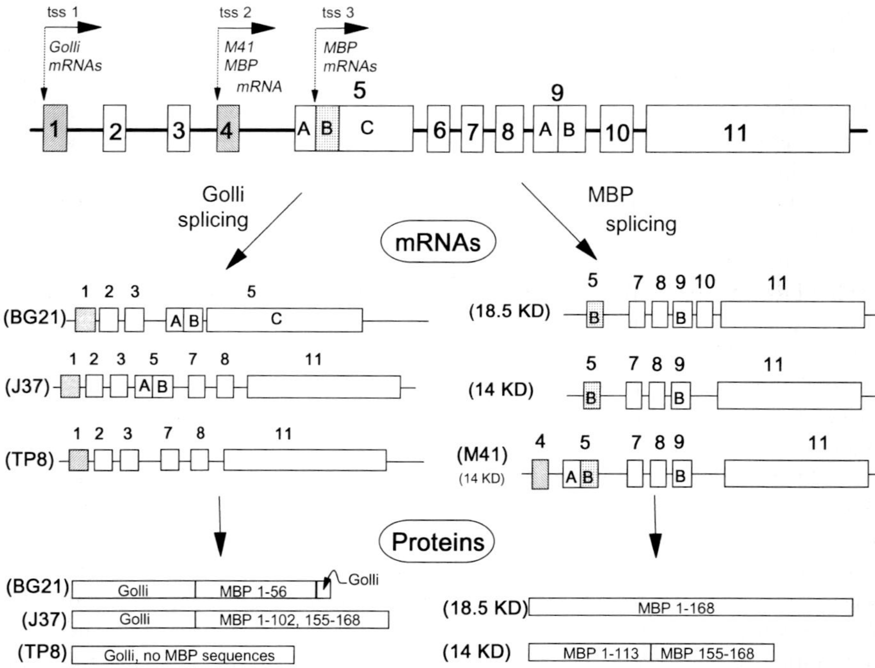

FIG. 1. Structure of the myelin basic protein (MBP) gene illustrating the three transcription start sites (*tss*) and the two sets of products produced from the gene, i.e., *MBPs* and *golli* proteins. Only the principal spliced mRNAs and the proteins they encode are shown in each of the two families

The Golli–mbp gene structure is very unusual and is unique in many respects. It numbers among the relatively small group of genes that have been found to be greater than 100kb in length. It also is an example of a very rare arrangement of overlapping genes, or a gene within a gene. For example, the transcription unit defining BG21, the most abundant Golli product, extends from exon 1 to exon 5C, overlapping both the M41-MBP and MBP transcription units. On the other hand, the transcription unit defining J37 extends from exon 1 to exon 11, completely encompassing the MBP transcription unit. In this respect the Golli–mbp gene is similar to the neurofibromatosis gene in that they are both large transcription units that encompass smaller genes [9,10]. However, they are distinctly different in that the smaller MBP and BG21 portions of the Golli–mbp gene are not simply included within an intron of the larger gene, but share alternatively spliced exons in common. This type of alternative splicing in overlapping transcription units of this size is very unusual, and for which there are very few examples in biology.

Another very unusual feature of the Golli–mbp gene is its regulation. The third transcription start site (tss3) is under very tight developmental control and is expressed only in myelin-forming cells. There is good evidence that this regulation resides within the promoter for tss3, which is exon 5A [11]. We have unpublished evidence that tss2 is active in oligodendrocytes and immune cells. There is now substantial evidence that tss1, from which Golli is expressed, is active in the immune system and in neurons and oligodendrocytes in the nervous system. It appears as if there is a gradient of increased cell and tissue specificity in moving from the upstream promoters to the downstream promoters in the gene. Nonetheless, the regulation of tss1 is particularly interesting because its expression is far from constitutive. While tss1 is active in a number of cells within the nervous and immune systems, it is quite selective about which cells it is active in and the developmental stage at which it is active.

The Golli Promoter and Neuron-Specific Promoters

Golli is expressed in neurons and in oligodendrocytes in the nervous system. To define the *cis*-acting elements that controlled expression of tss1, we created transgenic mice in which the *lac Z* gene was placed under the control of 1.3kb of genomic sequence directly upstream and including the golli tss1. Transgene expression was confined very specifically to a subset of neurons that normally express the gene. The onset of *lac Z* expression occurred around E11, demonstrating that this promoter region preserved the correct temporal pattern of gene expression within those neurons. Rather surprisingly, the promoter targeted developmentally accurate expression to the cortical pioneer neurons (i.e., Cajal-Retzius cells and presubplate neurons), among the earliest maturing neurons in the CNS. Of the neuron-specific promoter studies to date, almost none have exhibited the restricted expression of our Golli proximal promoter and none have targeted expression to the two sets of cortical pioneer neurons. It is particularly ironic that this neuronal-specific promoter is part of a quintessential glial gene, the myelin basic protein gene.

Pioneer Neurons and the Developing Cortical Plate

Cajal-Retzius neurons are first evident in the primordial plexiform layer (PPL) around E13.5 and assume their position in the most superficial level of the marginal layer when it begins to form the upper boundry of the cortical plate during the expansion of the plate between E15 and P0. In early postnatal life, most of the Cajal-Retzius cells in layer 1 disappear, primarily through cell death [12,13]. The PPL also contains the cells that will form the subplate, the lowest layer of the cortex immediately above the intermediate zone. A number of studies have suggested that subplate neurons (and, hence, the subcortical plate) largely disappear during early postnatal development either by cell death or morphological restructuring, analogous to the disappearance of Cajal-Retzius neurons in layer 1 [14–17]. However, other morphological and birth-dating studies indicate that neurons born on E12–E13 persist in significant numbers in a subplate as late as postnatal day 60 in rats and mice [18–21]. Thus, the fate of the cortical pioneer neurons, which are so important in the organization of the cortex, or of the subplate itself has not been completely clear.

Targeted Expression of a *lacZ* Transgene Under the Control of the Golli Promoter in Cortical Pioneer Neurons

We have found that the proximal promoter of the Golli tss1 gene targets expression to a subset of Cajal-Retzius cells and to presubplate neurons. One factor that has limited study on the function and fate of the Cajal-Retzius and presubplate neurons has been the lack of suitable markers of these cells. An important finding of our study was that the transgene could be used to mark the pioneer neurons and that it continued to be expressed throughout the developmental life of the subplate neurons. This afforded us the opportunity to follow the fate of and subplate neurons through development.

Throughout neuronal migration and expansion of the cortical plate, lac Z expression in the deep cortical layers was limited to, and consequently defined, these neurons as the subplate. Transgene expression persisted in cells that possessed the morphology of subplate neurons, and formed a distinct layer, perhaps 2–4 cells thick, between the white matter and the deepest portion of cortical layer 6 as late as P80. In a series of birthdating studies we found that these "blue" subplate cells that appeared to persist in the adult mouse were born between E10.5 and E12.5, the time when subplate cells are born in the mouse. These data were consistent with the notion that the cortical subplate exists as an intact and distinct structure in the mouse well after maturation of the cortical layers.

The presence of a distinct subplate in the postnatal mouse raises the issue as to whether the subplate truly persists through postnatal development in other mammals and what function(s) it might have in the adult animal. In the developing brain, the subplate neurons perform a wide variety of functions throughout corticogenesis [22]. They pioneer the pathway for the corticofugal projections that layers 5 and 6 neurons will ultimately follow [23]. Thalamic inputs pause in the subcortical plate until their intended neuronal targets in layers 2 and 4 have matured. Neurons in the subplate also establish ipsilateral and contralateral cortical connections [24]. The essential role(s) performed by the subplate has been illustrated by ablation of subplate neurons with kainic acid during development which then lead to the absence (misdirection) of thalamocortico- and corticofugal projections [25]. Goldman-Rakic and Galkin [26] have also suggested that the subplate zone may serve as a substrate for cortical repair following neonatal injury. Could the transgene mark a residual cohort of cells with such a function? Or, does the subplate persist in order to maintain critical, as yet undefined, connections?

These transgenic mice, in which these cells are marked, have provided some interesting insights into the ontological development of cells within the preplate, subplate, and a relatively unexplored deep cortical layer in postnatal animals. We believe that the regulatory element(s) and the transgenic animals themselves provide a useful set of tools for studying the formation, development, and function of pioneer neurons in the brain. These studies also have changed our perception of the function of the myelin basic protein gene and its products. It is clearly not only expressed in myelin-forming cells, but also appears to function in neurons and in many cells in the immune system.

References

1. Kitamura K, Newman SL, Campagnoni CW, Verdi JM, Mohandas T, Handley VW, Campagnoni AT (1990) Expression of a novel transcript of the myelin basic protein gene. J Neurochem 54:2032–2041
2. Campagnoni AT, Pribyl TM, Campagnoni CW, Kampf K, Amur-Umarjee S, Landry CF, Handley VW, Newman SL, Garbay B, Kitamura K (1993) Structure and developmental regulation of Golli-mbp, a 105-kb gene that encompasses the myelin basic protein gene and is expressed in cells in the oligodendrocyte lineage in the brain. J Biol Chem 268:4930–4938
3. Pribyl TM, Campagnoni CW, Kampf K, Kashima T, Handley VW, McMahon J, Campagnoni AT (1993) The 179 kilobase human golli-mbp gene: structure and expression in the immune and central nervous systems. Proc Natl Acad Sci USA 90:10695–10699
4. Zelenika D, Grima B, Pessac B (1993) A new family of transcripts of the myelin basic protein gene expression in brain and in immune system. J Neurochem 60:1574–1577
5. Fritz RB, Kalvakolanu I (1995) Thymic expression of the golli-myelin basic protein gene in the SJL/J mouse. J Neuroimmunol 57:93–99
6. Mathisen PM, Pease S, Garvey J, Hood L, Readhead C (1993) Identification of an embryonic isoform of myelin basic protein that is expressed widely in the mouse embryo. Proc Natl Acad Sci USA 90:10125–10129
7. Pribyl TM, Campagnoni CW, Kampf K, Ellison JA, Landry CF, Kashima T, McMahon J, Campagnoni AT (1996) Expression of the myelin basic protein gene locus in neurons and oligodendrocytes in the human fetal central nervous system. J Comp Neurol 374:342–353
8. Campagnoni AT (1995) Molecular biology of myelination. In: Kettenmann H, Ransom B (eds) Neuroglia. Oxford University Press, New York, pp 555–570
9. Xu G, O'Connell P, Viskochil D, Cawthon R, Robertson M, Culver M, Dunn D, Stevens J, Gesteland R, White R, Weiss R (1990) The neurofibromatosis type 1 gene encodes a protein related to GAP. Cell 62:599–608
10. Viskochil D, Cawthon R, O'Connell P, Xu G, Stevens J, Culver M, Carey J, White R (1991) The gene encoding the oligodendrocyte-myelin glycoprotein is embedded within the neurofibromatosis type 1 gene. Mol Cell Biol 11:906–912

11. Goujet-Zalc C, Babinet C, Monge M, Timsit S, Cabon F, Gansmueller A, Miura M, Sanchez M, Pournin S, Mikoshiba K, et al (1993) The proximal region of the MBP gene promoter is sufficient to induce oligodendroglial-specific expression in transgenic mice. Eur J Neurosci 5:624–632
12. Edmunds SM, Parnavelas JG (1982) Retzius-Cajal cells: an ultrastructural study in the developing visual cortex of the rat. J Neurocytol 11:427–446
13. Parnavelas JG, Edmunds SM (1983) Further evidence that Retzius-Cajal cells transform to nonpyramidal neurons in the developing rat visual cortex. J Neurocytol 12:863–871
14. Luskin MB, Schatz CJ (1985) Studies of the earliest generated cells of the cat's visual cortex: cogeneration of subplate and marginal zones. J Neurosci 5:1062–1075
15. Al-Ghoul WM, Miller MW (1989) Transient expression of Alz-50 immunoreactivity in developing rat neocortex: a marker for naturally occurring neuronal death? Brain Res 481:361–367
16. Ferrer I, Bernet E, Soriano E, Del Rio T, Fonseca M (1990) Naturally occurring cell death in the cerebral cortex of the rat and removal of dead cells by transitory phagocytes. Neuroscience 39:451–458
17. Woo TU, Beale JM, Finlay BL (1991) Dual fate of subplate neurons in a rodent. Cereb Cortex 1:433–443
18. Valverde F, Vacal-Valverde MV, Santacana M, Heredia M (1989) Development and differentiation of early generated cells of sublayer VIb in the somatosensory cortex of the rat: a correlated Golgi and autoradiographic study. J Comp Neurol 290:118–140
19. Valverde F, Lopez-Mascaraque L, Santacana M, De Carlos JA (1995) Persistence of early-generated neurons in the rodent subplate: assessment of cell death in neocortex during the early postnatal period. J Neurosci 15:5014–5024
20. Reep RL, Goodwin GS (1988) Layer VII of rodent cerebral cortex. Neurosci Lett 90:15–20
21. Bayer SA, Altman J (1990) Development of layer I and the subplate in the rat neocortex. Exp Neurol 107:48–62
22. Allendorfer KL, Schatz CJ (1994) The subplate, a transient neocortical structure: its role in the development of connections between the thalamus and cortex. Annu Rev Neurosci 17:185–218
23. Miller B, Chou L, Finlay BL (1993) The early development of thalamocortical and corticothalamic projections. J Comp Neurol 335:16–41
24. Goldman-Rakic PS (1982) Neuronal development and plasticity of association cortex in primates. Neurosci Res Prog Bull 20:520–532
25. Ghosh A, Antonini A, McConnell SK, Schatz CJ (1990) Requirement for subplate neurons in the formation of thalamocortical connections. Nature (Lond) 347:179–181
26. Goldman-Rakic PS, Galkin TW (1978) Prenatal removal of frontal association cortex in the rhesus monkey: anatomical and functional consequences in postnatal life. Brain Res 52:451–485

Function of Myelin Proteolipid Protein Gene Products During Oligodendrocyte Development

KAZUHIRO IKENAKA and ANNA IVANOVA

Summary. A gene for myelin proteolipid protein (PLP), a major structural protein in the CNS myelin, is activated at least a week before myelination. Mutations within the PLP gene cause abnormalities in myelination and affect the development of the oligodendrocytes in the premyelinating stage. We showed previously that PLP gene expression is directly associated with secretion of a factor that can increase the number of oligodendrocytes. Here we present the evidence that this activity is mediated by the C-terminal fragment of the PLP molecule and can be reproduced by the synthetic peptide including residues 215–232. The PLP gene product can be secreted and is able to affect the development of oligodendrocytes, at least in vitro.

Key words. PLP—DM-20—Development of oligodendrocytes

Introduction

Myelin proteolipid protein (PLP) and DM-20 are two alternatively spliced isoforms of the same gene transcript [1]. PLP is a structural protein in the CNS myelin. It appears at the early postnatal stage, when myelination starts [2]. PLP/DM-20 mRNA can be detected in the early embryonic stage, but the significance of this expression is not known [3,4]. It has been proposed that the presence of PLP gene products in the embryonic CNS might result from precocious expression of the gene in oligodendrocyte progenitor cells. Several reports indicate that early expression of the wild-type PLP gene is necessary for the normal development of premyelinating oligodendrocytes. The amino acid sequence of PLP is totally conserved among rat, mouse, and human, and almost any alteration from the wild-type PLP results in severe dysmyelination and the premature death of oligodendrocytes [5]. In the jimpy mouse, which splices out the fifth exon from the PLP mRNA, oligodendrocytes degenerate before the final maturation and rarely form myelin. Jimpy oligodendrocytes, however, can survive in vitro and produce myelin-like sheets in culture medium supplemented with the normal astrocyte-conditioned medium. However, conditioned medium from

National Institute for Physiological Sciences, Okazaki National Research Institutes, 38 Nishigonaka, Myodaiji-cho, Okazaki, Aichi 444-8585, Japan jimpy astrocytes cannot support the development of jimpy oligodendrocytes [6]. The partial recovery of jimpy oligodendrocyte function was also observed when the fragments of the embryonic jimpy brain were transplanted into host brains expressing the wild-type PLP gene [7].

Previously we showed that supernatants obtained from several cell lines that were shown to produce DM-20 can increase the number of oligodendrocytes in vitro [3]. Supernatants from the NIH3T3 fibroblasts forced to produce wild-type PLP or DM-20 increased the number of oligodendrocytes in the mixed glial cell culture. Supernatants from the nontransfected NIH3T3 cells or fibroblasts transfected with jimpy-PLP did not affect the number of oligodendrocytes [8]. We proposed that expression of the PLP gene products resulted in the secretion of an unknown factor affecting the development of the oligodendrocyte lineage. This factor was neither platelet-derived growth factor (PDGF) nor basic fibroblast growth factor (bFGF), but its ability to increase the number of oligodendrocytes in culture could be neutralized by the addition of the monoclonal antibody AA3 (Fig. 1a), recognizing the C-terminal of PLP or DM-20. Another monoclonal antibody, AH7–2a (Fig. 1a), recognizing a certain domain within the second extracellular loop of PLP or DM-20, could not neutralize the activity present in the supernatants. Therefore, the secreted PLP gene product should contain at least 60 amino acids of the PLP (DM-20) C-terminus.

To investigate the effect of purified PLP, DM-20, or its fragments on the development of oligodendrocytes, we performed a series of experiments using mixed glial cell culture. This chapter describes our findings and provides a brief discussion of the role of PLP gene products in the development of premyelinating oligodendrocytes.

Materials and Methods

ICR mice were obtained from Nihon SLC (Hamamatsu, Japan). Monoclonal anti-galactocerebroside (antibody, O1; mouse IgM) was obtained from Dr. Steven Pfeiffer (University of Connecticut, USA) [9]. The PLP peptides were kindly provided by Dr. M. Lees (Shriver Center, MA, USA) or obtained from Peptide Institute (Osaka, Japan). Low molecular weight PLP fractions were kindly supplied by Dr. Elizabeth Thrifilieff (Institut de Biologie Moleculaire et Cellulaire, Strasbourg, France) [10].

Primary cell cultures were prepared from E17 mouse embryos. Tissue from cerebral hemispheres was cut into pieces, incubated in 0.25% trypsin with DNaseI, and resuspended with a fire-polished glass pipette. Cells were seeded onto polyethylenimine-coated coverslips at a density of $3 \times 10^5/cm^2$ in Dulbecco's modified Eagle's medium containing 10% fetal bovine serum (FBS). After 3 days in culture, the medium was changed to a serum-free chemical-defined medium (N4:O3, 2:1). Control coverslips were maintained in the defined medium, while experimental slips were incubated with various factors. After further 4 days in vitro, the cells were processed for immunostaining, and the number of $O1^+$ cells was counted. In some cases primary cell cultures were prepared from P2 cerebral hemispheres and maintained according to the same protocol.

After 7 days in vitro, the coverslips were incubated with monoclonal mouse anti-GalC antibody O1 for 1 h at 37°C, washed three times, and fixed with 4% paraformaldehyde (PFA) for 10 min. After the nonspecific binding block (0.1% bovine serum

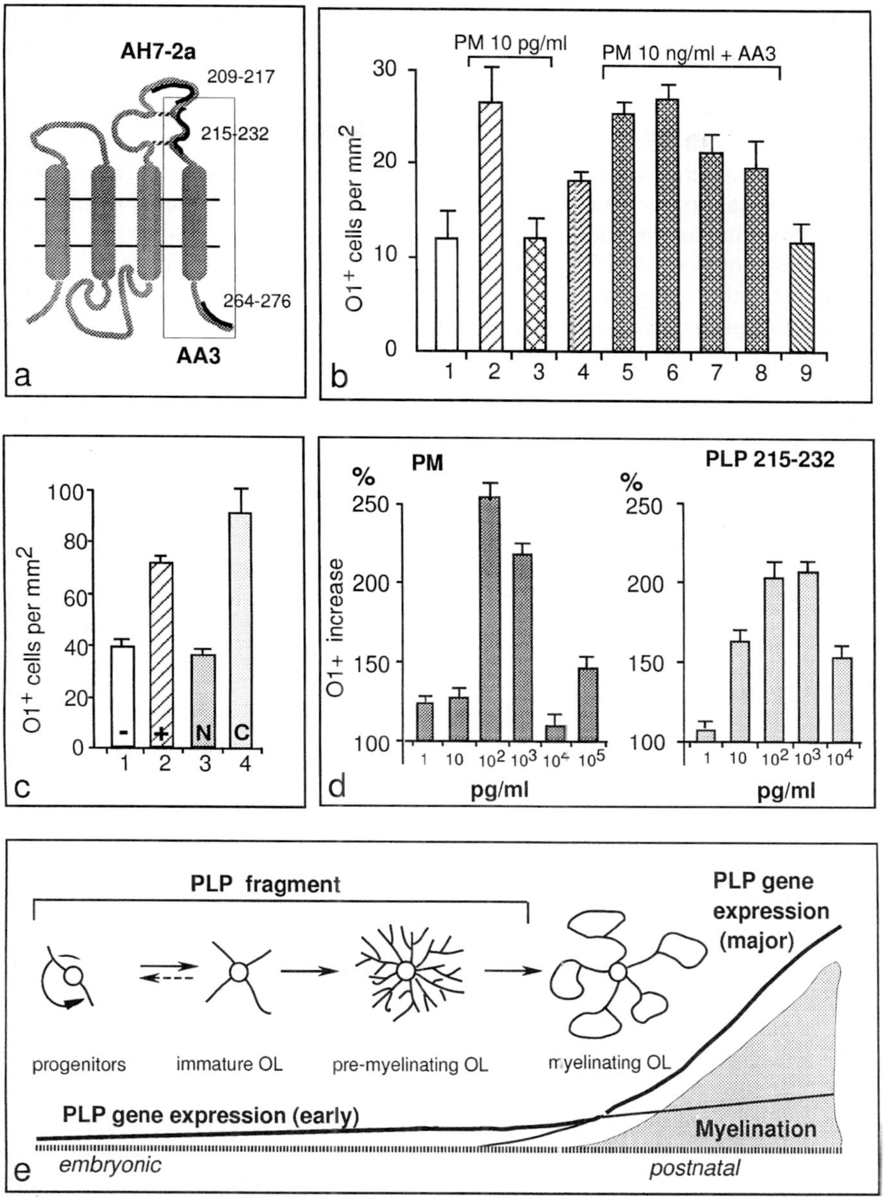

FIG. 1. **a** Schematic drawing of the proteolipid protein (PLP) protein molecule represents locations of the synthetic peptides and antibody-binding sites. **b** Effect of the myelin proteolipid mixture on the number of O1$^+$ cells in culture and neutralization experiments. **c** Effect of the low molecular weight PLP fractions on the number of oligodendrocytes. **d** *Left panel*, concentration curve of the proteolipid mixture (*PM*); *right panel*, concentration curve of the PLP peptide (PLP 215–232). **e** Schematic drawing represents the effect of the PLP fragment on the development of oligodendrocytes

albumin [BSA], 0.1% FBS in PBS), the cells were treated with fluorescein-conjugated goat antimouse IgM (Cappel; 1:200) for 1 h at room temperature, mounted onto glass slides, and examined under an Olympus Vanox-S fluorescence microscope.

In neutralization experiments, antibodies were incubated with a purified proteolipid mixture at 4°C for 60 min before addition to the culture medium. Culture was continued for another 4 days, and the number of $O1^+$ cells was counted. The anti-PLP monoclonal antibodies AA3 and AH7–2a were purified using a general ammonium sulfate precipitation method.

The proteolipid mixture was isolated from the young adult ICR mice by organic solvent extraction using the procedure of Montreal [11] with minor modifications. To transfer the proteolipid mixture from organic to aqueous solution, the samples were dissolved in 1% Triton X-100, 10 mM 2-mercaptoethanol, 20 mM Tris-HCl (pH 8.0), and stored at −80°C. Protein profile analysis of the proteolipid fraction was carried out using SDS-PAGE gels stained with Silver Stain II kit (Wako, Osaka, Japan). The proteolipid mixture contained mostly PLP, as well as a small amount of DM-20 and low molecular weight fragments. For the cell culture experiments, the proteolipid mixture was diluted to 10 ng/ml by PBS, and small aliquots were stored at −80°C.

Stock solutions of PLP peptides (10 mg/ml in 15 mM HCl) were stored at −80°C. They were diluted by the O3:N4-defined medium and added to the glial cell culture.

Results and Discussion

The Proteolipid Mixture from the Mouse Brain Affects the Number of $O1^+$ Cells in the Mixed-Glial Cultures

Our previous experiments have shown that PLP gene expression is linked to the secretion of a humoral factor(s) that affect development of oligodendrocytes in vitro [8]. This factor could be neutralized by the addition of anti-PLP antibody, AA3, to the supernatants obtained from the cells that express PLP or DM-20. Similar treatment with anti-PDGF or-bFGF antibodies had no effect, suggesting that the factor is neither PDGF nor bFGF [8].

To examine whether PLP itself can affect the development of oligodendrocytes in the mixed glial cultures, we purified the proteolipid mixture from adult mouse brain. The proteolipid mixture (mostly PLP) was solubilized with detergent and added to the cell culture as described in the previous section. Addition of 10 pg of the proteolipid mixture per 1 ml of medium caused a twofold increase in the number of $O1^+$ cells (Fig. 1b, lane 2). Incubation of the proteolipid mixture with AA3 antibody (5 μg/ml) completely disrupted the activity (Fig. 1b, lane 3). Therefore, the proteolipid mixture can increase the number of $O1^+$ cells in culture. Addition of 10 ng/ml of the proteolipid mixture caused a less notable effect in the number of $O1^+$ cells (Fig. 1b, lane 4).

Treatment of the proteolipid mixture (10 ng/ml) by increasing concentrations of AA3 antibody (1.25–10 μg/ml) resulted in a dose-dependent change in the number of $O1^+$ cells (Fig. 1b, lanes 5–8). AA3 antibody alone did not affect the number of $O1^+$ cells (Fig. 1b, lane 9) or its morphology (data not shown). Detergent (Triton X-100) used to solubilize the proteolipid mixture did not affect the cells at these dilutions (data not shown).

Our experiments showed that the proteolipid mixture exerts some biological effects at extremely low concentrations (10 pg/ml, 0.3 pM) (Fig. 1D, left panel). Neutralizing experiments demonstrated that PLP gene products affect the number of premyelinating oligodendrocytes in a dose-dependent manner. The loss of activity at high concentrations seems to be rather a feature of the PLP gene products than inhibition from other components of the proteolipid mixture.

Structural Analysis of the PLP Gene Product Secreted in the Medium

Our experiments demonstrated that the proteolipid mixture from adult mouse brain could increase the number of $O1^+$ cells because of the presence of PLP gene products. However, the PLP gene product should be less hydrophobic than the whole PLP molecule to be secreted into the aqueous solution under physiological conditions. Our previous experiments have shown that the activity present in the supernatants of the NIH3T3 cells forced to produce PLP or DM-20 can be neutralized by the monoclonal antibody AA3 against the C-terminal portion of the molecule. Treatment of the supernatants with monoclonal antibody AH7–2a, which recognizes a specific domain within the second extracellular loop of PLP, did not disrupt the activity [8]. This result suggests that the secreted PLP gene product contains the C-terminal portion of PLP or DM-20.

We started to look for the low-weight components of the proteolipid mixture that might represent PLP gene products. Lepage et al. [10] characterized at least two proteolipids with apparent molecular masses of 16 and 14 kDa; the 16-kDa proteolipid apparently corresponds to the carboxyl-terminal portion of PLP (residues 160–276), whereas the 14-kDa proteolipid contains the first 113 residues at the amino terminus, plus an additional N-terminal segment. In contrast to PLP and DM-20, both low molecular weight proteolipids have free N-terminal groups. The addition of these fractions to the mixed glial cell culture showed that the 16-kDa proteolipid could increase the number of $O1^+$ cells in a manner similar to the whole proteolipid mixture, while the 14-kDa proteolipid did not affect the number of $O1^-$ cells (Fig. 1c).

To identify and characterize the active region of the secreted fragment, three PLP peptides (209–217, 215–232, and 264–276) (Fig. 1a) were added to the culture system. PLP peptides containing residues 209–217 and 264–276 did not affect the number of $O1^+$ cells in the mixed glial culture (data not shown). However, PLP peptide corresponding to the residues 215–232 increased the number of $O1^+$ cells (Fig. 1d, right panel). The dose–response curve of PLP (215–232) was similar to the effect of the proteolipid mixture (Fig. 1d, left panel). The solvent used to solubilize the peptides (15 mM HCl) did not affect the number of $O1^+$ cells at these dilutions (data not shown). Thus, the active site of the PLP fragment includes residues 215–232. Addition of the proteolipid mixture and PLP peptide (215–232) to the mixed glial cultures prepared from different stages of development demonstrated that the PLP gene product can exert biological activity only when the PLP gene expression level is very low (data not shown).

These results suggest that the N-terminus of the PLP or DM-20 molecule is not required for the activity. We suppose that the secreted product of PLP gene contains residues 215–276 (Fig. 1a, boxed area) because the effect of the proteolipid mixture can be reproduced by addition of PLP peptide 215–232 to the culture system; the secreted product cannot be neutralized with AH7–2a antibody, which recognize residues 207–217; the secreted product can be neutralized with AA3 antibody, recognizing the carboxy-terminal of the molecule. Our findings suggest that PLP gene products are able to affect the development of oligodendrocytes, at least in vitro.

In situ hybridization of the paraffin sections with the probe against PLP/DM-20 mRNA demonstrated that the PLP gene transcript is expressed by a subset of oligodendrocytes in the embryonic mouse hindbrain. We performed in situ hybridization experiments to describe the source of the PLP gene product in the embryonic brain. Cells expressing PLP/DM-20 mRNA do not contain mRNA for the PDGF alpha receptor. They probably represent a subset of oligodendrocytes or even a different lineage. Our in vitro experiments presented here suggest that PLP gene products can be secreted and are able to regulate the development of immature oligodendrocytes that do not carry receptors to PDGF (Fig. 1e).

References

1. Nave KA, Lai C, Bloom FE, Milliner RJ (1987) Splice site selection in the proteolipid protein (PLP) gene transcript and primary structure of the DM-20 protein of central nervous system myelin. Proc Natl Acad Sci USA 83:5665–5669
2. LeVine SM, Wong D, Macklin WB (1990) Developmental expression of proteolipid protein and DM20 mRNAs and proteins in the rat brain. Dev Neurosci 12:235–250
3. Ikenaka K, Kagawa T, Mikoshiba K (1992) Selective expression of DM-20, an alternatively spliced myelin proteolipid gene product, in developing nervous system of the mouse. J Neurochem 58:2248–2253
4. Timsit SG, Bally-Cuif L, Colman DR, Zalc B (1992) DM-20 mRNA is expressed during early embryonic development of the nervous system of the mouse. J Neurochem 58:1172–1175
5. Knapp PE (1996) Proteolipid protein: is it more than just a structural component of myelin? Dev Neurosci 18:297–308
6. Knapp PE, Skoff RP, Benjamins JA, Studzinski DM (1992) Oligodendrocytes from hypomyelinated jimpy mice synthesize an altered PLP in vitro. Trans Am Soc Neurochem 23:134
7. Lachapelle F, Lapie P, Campagnoni AT, Gumpel M (1991) Oligodendrocytes of jimpy phenotype can be partially restored by environmental factors in vivo. J Neurosci Res 29:235–243
8. Nakao J, Yamada M, Kagawa T, et al (1995) Expression of proteolipid protein gene is directly associated with secretion of a factor influencing oligodendrocyte development. J Neurochem 64:2396–2403
9. Sommer I, Schachner M (1981) Monoclonal antibodies (O1 to O4) to oligodendrocyte cell surfaces: an immunological study in the central nervous system. Dev Biol 83:3111–327
10. Lepage PG, Helynck J, Chu B, et al (1986) Purification and characterisation of minor brain proteolipids: use of fast atom bombardment-mass spectrometry for sequencing. Biochimie (Paris) 68:669
11. Montreal J (1975) Chromatographic fraction of brain white matter proteolipid. J Neurochem 24:539–541

Effect of Substance P on a Glial Cell Lineage

Yasuo Takeda[1,3], Tomomi Notsu[1], Hitoshi Kawano[2], Koki Kawamura[2], Hiroaki Asou[4], and Keiichi Uyemura[1]

Summary. Accumulating evidence suggests that substance P plays a role not only as a neurotransmitter, but also as an active factor involved in neural development. By radioligand binding studies, substance P receptor is expressed in a certain type of glial cells such as reactive astrocytes and oligodendrocyte-type 2 astrocyte (O-2A) progenitor cells, suggesting its role in neuron–glia interaction. We focused on substance P as a new cue to regulate an oligodendrocyte cell lineage. The substance P receptor mRNA increased transiently at 1 week after birth. By antisubstance P receptor (anti-SPR) antibody, the marginal zone and cortical plate of rat cerebral cortex were stained densely at postnatal day 5. The staining pattern was similar to that of 1D11, which is expressed in radial glial cells. In vitro, the O-2A progenitor cells from newborn rat cerebral cortex were intensely positive to anti-SPR antibody. After differentiation of O-2A progenitor cells to mature oligodendrocytes, the expression of substance P receptor disappeared. Addition of substance P in culture inhibited differentiation of O-2A progenitor cells significantly. In conclusion, substance P may regulate the differentiation of O-2A progenitor cells that express substance P receptor.

Key words. Neuron–glia interaction—Neuropeptides—Substance P—Substance P receptor—Oligodendrocyte differentiation

Introduction

In gliogenesis, oligodendrocyte progenitor cells appear at embryonic day 16 after neuronal generation and continue through life in the ventricular zone [1]. Oligodendrocytes then differentiate and mature, mostly in the first 2 weeks after birth. Various types of neurotrophic factors and cell adhesion molecules are significantly involved in all processes of neuronal–glial cell interaction during neural development [2]. Recent reports have suggested the existence of several neuropeptides in an early stage of

Departments of [1]Physiology and [2]Anatomy, School of Medicine, Keio University, 35 Shinanomachi, Shinjuku-ku, Tokyo 160-8582, Japan
Departments of [3]Cell Recognition and [4]Neurobiology, Tokyo Metropolitan Institute of Gerontology, 35-2 Sakaecho, Itabashi-ku, Tokyo 173-0015, Japan

neural development [3]. In this chapter, we focus on neuropeptides, especially the interaction of substance P and its receptor, in the development of glial cells.

Nature of Substance P and Its Receptor

Substance P is one of five tachykinin peptides that have strong smooth muscle contractile activities. Substance P works as a neurotransmitter of primary sensory neurons, regulates blood pressure and salivary secretion, and is involved in immune and inflammatory disorders [4]. These biological responses are mediated by the NK-1 subtype receptor (also called substance P receptor), which is coupled with the Gq protein and stimulates IP_3 turnover and protein kinase C activation [5].

A Role for Substance P in the Guidance of Commisural Fibers

The possible function of substance P was suggested to release chemotrophic factors from floor plate cells in the embryonic spinal cord [6,7]. The substance P receptor mRNA and protein are specifically expressed in a subset of nonneuronal floor plate cells in embryonic spinal cord. On the other hand, substance P-positive neurons are identified in the ventral and medial region of the spinal cord, and its neurites elongate and across the floor plate just before crossing of the commisural axons. In in vitro experiments, by addition of substance P, the floor plate cells increase intracellular Ca^{2+} and release some chemotrophic factors, probably netrin. This factor may facilitate neurite extension of commisural neurons toward the floor plate. This is the first evidence that substance P could contribute to the fine tuning of developmental processes in the spinal cord. Substance P receptor was identified in floor plates not only in the spinal cord but also in hindbrain, in brainstem, and immediately underlying the fourth ventricle [7]. This result gives rise to the further possible contribution of substance P on other commisural axon guidance in the central nervous system.

Substance P Receptor Is Expressed by Reactive Astrocytes After Injury

By receptor binding assay, substance P-binding sites are present in the primary cultures of cortical astrocytes, and substance P stimulates phosphatidylinositol turnover in those cells [8]. While the binding site for substance P was not detected in normal optic nerve, substance P-binding sites were extensively upregulated in the glial scar composed of reactive astrocytes after optic nerve transection in vivo [9]. Concerning the source of ligand, the circulating leukocytes could produce several peptides including substance P [10]. Taken together, one of the possible roles of substance P is regulating glial response to injury in the central nervous system. On subtypes of astrocytes, type 1 and type 2 astrocytes were distinguished by morphology and antigenic phenotype in culture of the optic nerve [11]. However, it is difficult to identify type-2 astrocytes in the intact optic nerve. The type 2 astrocytes in culture may correspond to the reactive astrocytes that appear after neuronal injury in vivo. Interestingly, substance P receptor is also expressed in type 2 astrocytes in culture [12].

Substance P Receptor on O-2A Progenitor Cells

The developmental origin of each type of astrocytes is different. Type 1 astrocytes develop from the precursor cell of astrocytes, while type 2 astrocytes are generated from bipotential precursor cells, termed oligodendrocyte-type 2 astrocyte (O-2A) progenitors [11]. In vivo, O-2A progenitor cells differentiate exclusively to oligodendrocytes in normal development. However, in vitro, the O-2A progenitor cells can differentiate to either oligodendrocytes or type 2 astrocytes depending on culture conditions. By receptor binding studies, cultivated O-2A progenitor cells express substance P-binding sites [12]. Addition of substance P in culture induces phosphatidylinositol accumulation and Ca mobilization in the cells.

We examined the expression of substance P receptor in the O-2A progenitor cell lineage using a specific antibody against rat substance P receptor. The substance P receptor is strongly expressed in cultivated O-2A progenitor cells prepared from newborn rat cerebral cortex. After differentiation, expression of substance P receptor decreased. The immunoreactivity of substance P receptor was still detected in type 2 astrocytes, but practically disappeared in matured oligodendrocytes. Interestingly, addition of substance P significantly inhibited differentiation of O-2A progenitor cells to oligodendrocytes, and this effect was completely blocked by the presence of the substance P receptor antagonist CP-96,345. Additionally, expression of substance P receptor mRNA in the cerebral cortex was transiently increased at about 1 week after birth, which corresponds to the stage for migration and differentiation of O-2A progenitor cells. We also identified that substance P is exclusively expressed in developing neurons but not in glial cells in this developmental stage. These results suggested that a possible role of substance P is to regulate differentiation of O-2A progenitor cells via neuron–glia interaction in the developing cerebral cortex.

Conclusion

We have described several evidences of possible roles of substance P on neural development. It should be pointed out that neuropeptides coexist with classical neurotransmitters in a certain population of neurons [13]. For example, dopaminergic neurons originated in A9 and A10 areas contain cholecystokinin, while substance P is coexpressed in serotonergic neurons that form Raphe nuclei. It is known that these peptides appear in brain as early as the classical transmitters are expressed. If the subset of floor plate cells regionally corresponding to Raphe nuclei or A9 and A10 areas express the receptors for the corresponding peptides, a respective peptide-expressing neuron might be guided for target recognition by the trophic factors released from floor plate cells, as a result of the ligand–receptor interaction. Thus, substance P receptor activation might contribute to the formation of Raphe nuclei.

On the other hand, cholecystokinin might also contribute to the formation of A9 and A10 dopaminergic nuclei. The concept of peptide–receptor interaction, besides neurotrophic factors and cell adhesion molecules, may be a new tool for understanding neural development.

References

1. Cameron RS, Rakic P (1991) Glial cell lineage in the cerebral cortex: a review and synthesis. Glia 4:124–137
2. Pfeiffer SE, Warrington AE, Bansal R (1993) The oligodendrocyte and its many cellular processes. Trends Cell Biol 3:191–197
3. Schwartz JP, Taniwaki T (1994) Heterogeneity of expression of neuropeptide genes by astrocytes: functional implications. Dev Neurobiol 2:251–257
4. Krause JE, MacDonald MR, Takeda Y (1989) The polyprotein nature of substance P precursors. BioEssays 10:62–69
5. Krause JE, Bu JY, Takeda Y, et al (1993) Structure, expression and second messenger-mediated regulation of the human and rat substance P receptors and their genes. Regul Pept 46:59–66
6. De Felipe C, Pinnock RD, Hunt SP (1995) Modulation of chemotropism in the developing spinal cord by substance P. Science 267:899–902
7. Heath MJS, Lints TJ, Justin Lee C, et al (1995) Functional expression of the tachykinin NK1 receptor by floor plate cells in the embryonic rat spinal cord and brain stem. J Physiol (Camb) 486:139–148
8. Torrens Y, Daguet De Montety MC, El Etr M, et al (1989) Tachykinin receptors of the NK1 type (substance P) coupled positively to phospholipase C on cortical astrocytes from the newborn mouse in primary culture. J Neurochem 52:1913–1918
9. Mantyh PW, Johnson DJ, Boehmer CG, et al (1989) Substance P receptor binding sites are expressed by glia in vivo after neuronal injury. Proc Natl Acad Sci USA 86:5193–5197
10. Weinstock JV, Blum A, Walder J, et al (1988) Eosinophils from granulomas in murine schistosomiasis mansoni produce substance P. J Immunol 141:961–966
11. Raff MC (1989) Glial cell diversification in the rat optic nerve. Science 243:1450–1455
12. Marriott DR, Wilkin GP (1993) Substance P receptors on O-2A progenitor cells and type-2 astrocytes in vitro. J Neurochem 61:826–834
13. Lundberg JM, Hokfelt T (1983) Coexistence of peptides and classical neurotransmitters. Trends Neurosci 6:325–333

Multiple Restricted Origin of Oligodendrocytes

N. Spassky[1], C. Goujet-Zalc[1], E. Parmantier[2], S. Martinez[2], A. Ivanova[3], K. Ikenaka[3], B. Zalc[3], and J.-L. Thomas[3]

Dm-20, an alternative spliced product of the proteolipid protein (*plp*) gene, has been proposed, on the basis of its spatiotemporal pattern of expression, as an early embryonic marker of the oligodendrocyte lineage. To examine the possibility that the subset of ventricular cells expressing *dm-20* transcripts are oligodendrocyte precursors, we designed a transgenic mouse in which a fusion gene constituted by the positive selection marker *sh ble* (which confers resistance to the antibiotic zeomycine) and the *lacZ* reporter was placed under the control of the *plp* promoter and regulatory sequences. But for the exception of some dorsal territories, in three lines analyzed the expression of the transgene mirrored that of *dm-20* mRNA, i.e., emergence of β-galactosidase (β-gal) -expressing cells was restricted to the basal aspects of the neural tube and was discontinuous along the rostrocaudal axis: it started in the diencephalon at E.9.5, but was absent from the telencephalon and mesencephalon. Expression in the rhombencephalon and spinal cord was slightly delayed and occurred in rhombomeres r1–r2, was absent from rhombomeres r3–r5, and formed two continuous ventral and ventrolateral longitudinal columns starting from r6 and along the whole length of the spinal cord. Both in situ and in culture, β-gal$^+$ cells coexpressed markers of the oligodendrocyte lineage (NG-2, O4, RIP), but did not colocalize with markers of either radial glial cells (RC-2), or astrocytes (GFAP), or neurons (TuJ1, Phox-2b, *NeuroD*). In zeomycin-treated cultures, derived from E12.5 transgenic embryos, 80%–90% of the resistant cells differentiated into oligodendrocytes, compared to no more than 20% in nontreated control cultures. However, we show that the potentiality of the germinative neuroepithelium to generate oligodendrocytes is not restricted to *dm-20*-expressing territories, but is also shared by adjacent *dm-20*$^-$ but *pdgfrα*$^+$ territories, such as the basal plate of the mesencephalon or the r3–r5 region in the rhombencephalon. Altogether, our results suggest the existence of at least two distinct pools of oligodendrocyte precursors, defined by their expression of either *dm-20* or *pdgfrα*.

[1] Laboratoire de Neurobiologie Cellulaire, Moléculaire et Clinique, INSERM U-134, Université Pierre et Marie Curie, Hôpital de la Salpêtrière, 75651 Paris Cedex 13, France
[2] Departemento de Ciencas Morfologicas, Universitad de Murcia and Instituto de Neurociencias, Universidad de Alicante, 30071 Murcia, Spain
[3] National Institute for Physiological Sciences, Okazaki National Research Institutes, 38 Nishigonaka, Myodaiji-cho, Okazaki, Aichi 444-8585, Japan

Production and Survival of Oligodendrocytes in Developing Rodent Brain

Bruce D. Trapp, Akiko Nishiyama, and Wendy B. Macklin

Oligodendrocytes in vitro and in vivo are produced from progenitor cells that can be identified by the expression of platelet-derived growth factor receptor α and the proteoglycan NG2. We and others have provided evidence that NG2-positive cells give rise to premyelinating oligodendrocytes in developing rodent brain. Premyelinating oligodendrocytes express the myelin proteins DM20, MBP, MAG and CNP and extend multiple processes which occupy distinct neuropil domains. These cells have one of two fates: they myelinate axons or they undergo programmed cell death. Those that myelinate axons express and selectively target proteolipid protein (PLP) to compact myelin while DM20, a spliced variant of PLP, is detectable in the plasma membrane of myelinating oligodendrocyte cell bodies and processes and in compact myelin. Premyelinating oligodendrocytes which undergo programmed cell death have fragmented and condensed nuclear chromatin and their multiple processes have patchy myelin protein immunoreactivity. Most dying premyelinating oligodendrocytes are not labeled by the TUNEL method. We have investigated the effect of axotomy on the production and survival of premyelinating oligodendrocytes in rodent optic nerve. Rat optic nerves were transected at postnatal day (pd) 4 and analyzed at pd's 5, 6 and 7. Compared to non-transected nerves, the production of premyelinating oligodendrocytes during this period was not altered dramatically by nerve transection. There was no significant increase in the number of dying premyelinating oligodendrocytes and nerve transection did not induce apoptosis in oligodendrocytes. As described in vitro, therefore, the production survival of oligodendrocytes in vivo does not appear to depend upon the presence of viable axons.

Department of Neurosciences, Lerner Research Institute, The Cleveland Clinic Foundation, 9500 Euclid Avenne, Cleveland, OH 44195, USA

NG2+ Glial Progenitor Cells in Normal Adult Brain and in the Dysmyelinating Mutant *Jimpy*

Akiko Nishiyama[1,2] and Qian Wu[1,3]

Summary. We have been studying a population of glial cells in the rodent brain using antibodies against the NG2 proteoglycan and the alpha receptor for platelet-derived growth factor (PDGF α-receptor). In vitro, the two molecules are expressed on oligodendrocyte progenitor cells but disappear as they differentiate into mature oligodendrocytes. The distribution of NG2+/PDGF α-receptor− cells in the developing rat brain is similar to that previously reported for oligodendrocyte progenitor cells. Furthermore, immunoreactive cells exist in large numbers uniformly throughout the gray and white matter of the mature brain and coexpress NG2 and PDGF α-receptor, suggesting that they are antigenically related to oligodendrocyte progenitor cells. Double-labeling of adult rat brain sections with antibodies to NG2 and glial fibrillary acidic protein or with microglial markers revealed no double-labeled cells, indicating that NG2+ cells are distinct from mature astrocytes or microglia. NG2+ cells undergo increased proliferation in 18-day-old spinal cords of the dysmyelinating mutant *jimpy*, in which increased proliferation of "immature oligodendrocytes" has previously been observed by electron microscopy. These findings support the view that NG2+ cells in the developing and mature rodent central nervous system are oligodendrocyte progenitor cells.

Key words. Oligodendrocyte—NG2—PDGF alpha receptor—*Jimpy*—Glia

Introduction

We have been using antibodies to a cell surface chondroitin sulfate proteoglycan, NG2 [1,2] to identify and study oligodendrocyte progenitor cells in vitro and in vivo. In vitro, NG2 is expressed by oligodendrocyte progenitor cells and is downregulated as

[1]Department of Neurosciences, Lerner Research Institute, The Cleveland Clinic Foundation, 9500 Euclid Avenue, Cleveland, OH 44195, USA

[2]*Present address*: Department of Physiology and Neurobiology, University of Connecticut, 3107 Horsebarn Hill Rd, Storrs, CT 06269, USA

[3]*Present address*: Department of Neurosciences, Case Western Reserve University, Cleveland, OH 44106-4975, USA

they differentiate into mature oligodendrocytes [3–6]. In vivo, NG2+ cells begin to appear in the embryonic rat spinal cord, brainstem, and cerebrum around embryonic day 16 (E16) and continue to migrate and proliferate to occupy the entire central nervous system (CNS) by the end of the first postnatal week, when their maximal density is attained. Such a distribution of NG2+ cells is consistent with the previously reported distribution of cells of the oligodendrocyte lineage [7–9].

The alpha receptor for platelet-derived growth factor (PDGF α-receptor) has been previously shown to be expressed by oligodendrocyte progenitor cells by ligand binding studies and in situ hybridization [10–12]. We have further demonstrated by immunocytochemistry that NG2+ cells express PDGF-α both in vitro and in vivo [6,13]. These findings support the notion that NG2+ cells in vitro and in the developing brain are oligodendrocyte progenitor cells. Here we describe our observation that NG2+ cells are abundantly found throughout the mature CNS and are distinct from astrocytes and microglia.

A useful animal model to study the regulation of proliferation and differentiation of oligodendrocyte progenitor cells is the dysmyelinating mutant *jimpy*. *Jimpy* is an X-linked recessive mutation caused by a point mutation in the gene encoding proteolipid protein (PLP), which is a major constituent of the central myelin [14,15]. Affected males show typical tremor around postnatal day 10 (P10) and die during the fourth week after birth. Earlier electron microscopic analyses of *jimpy* CNS have revealed increased death of oligodendrocytes [16] and increased proliferation of "immature oligodendrocytes" [17]. We have found that NG2+ cells undergo increased proliferation in the *jimpy* spinal cord, providing further support that NG2+ cells are oligodendrocyte progenitor cells.

Materials and Methods

Animals

Sprague-Dawley rats were obtained from Charles River Laboratories (Raleigh, NC, USA). Breeding pairs for *jimpy* mice were obtained from the Jackson Laboratory (Bar Harbor, ME, USA). *Jimpy* and wild-type littermate males were used at postnatal day 18 (P18). *Jimpy* males were identified by their characteristic tremor. To label proliferating cells, 0.5 mg/g body weight of bromodeoxyuridine (BrdU; Boehringer-Mannheim, Indianapolis, IN, USA) was injected intraperitoneally every 2 h for a total of 6 h. Mice were anesthetized and perfused 2 h after the last injection.

Immunohistochemistry

Immunohistochemistry was carried out as previously described [18]. For detecting BrdU, 10-μm-thick sections were first pretreated with 2 N HCl for 10 min at room temperature, followed by neutralization in 0.1 M sodium borate (pH 8.6) for 10 min at room temperature. The sections were then processed for immunohistochemistry using a monoclonal antibody to BrdU (Amersham, Arlington Heights, IL, USA) and a rabbit antibody to NG2 (gifts from Drs. William Stallcup, The Burnham Institute, La Jolla, CA, and Joel Levine, SUNY, Stony Brook, NY). Immunolabeled sections were examined using a Zeiss Axiophot epifluorescence microscope. In some cases, *z*-series of optical slices were collected on a Leica TCS-NT confocal laser scanning microscope.

Results

Numerous NG2+ Cells Exist in the Mature Brain

Figure 1 demonstrates the morphology, number, and distribution of cells stained positively with monoclonal anti-rat NG2 antibodies in a 97-day (P97) rat cerebrum. Numerous immunoreactive cells with complex process arborization are present throughout the gray and white matter. In the cortex and hippocampus, the processes extend radially in all directions (Fig. 1a,b), whereas in the corpus callosum (Fig. 1c) the processes run primarily parallel to the axons. The processes of individual cells appear to cover the entire neuropil. Double-labeling of sections with monoclonal anti-NG2 antibodies and a rabbit antibody to PDGF α-receptor revealed that NG2+ cells in both gray and white matter also express PDGF α-receptor (not shown; see Fig. 2 in [18]), suggesting that these cells are antigenically related to oligodendrocyte progenitor cells.

To determine whether NG2+ cells are distinct from astrocytes, sections from P90 rat cerebrum were double-labeled with antibodies to NG2 and glial fibrillary acidic

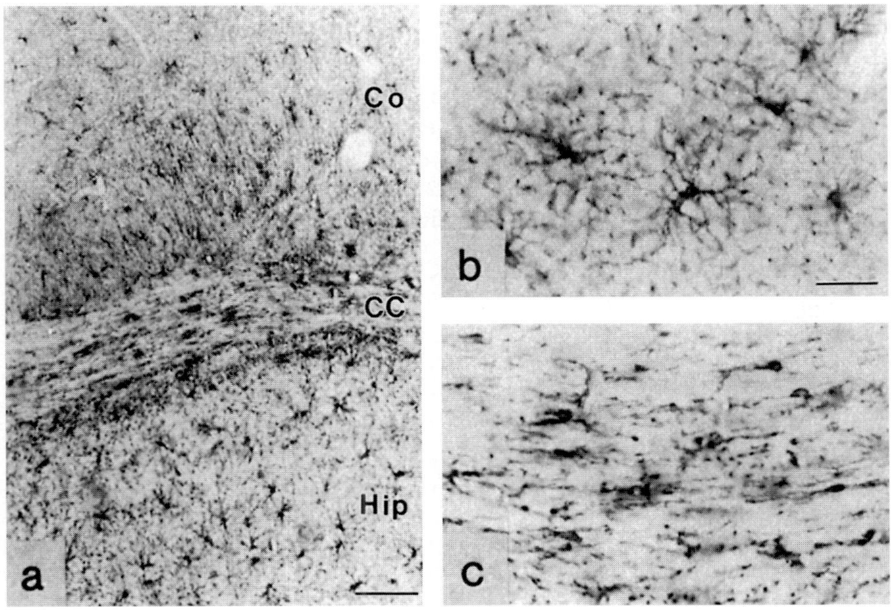

FIG. 1a–c. Immunoperoxidase labeling of a 30-μm section from a P97 rat brain using monoclonal anti-rat NG2 antibodies. Numerous NG2+ cells with multiple cell processes are uniformly distributed throughout gray and white matter. **a** Low-magnification photomicrograph of coronal section through the cerebral cortex (*Co*), corpus callosum (*CC*), and hippocampus (*Hip*). **b** Higher-magnification photomicrograph through the cerebral cortex. **c** Higher-magnification photomicrograph through the corpus callosum. *Bars:* **a**, 200 μm; **b**, 50 μm. (From [18], with permission)

protein (GFAP). Double-labeling did not reveal cells that bound both antibodies, indicating that NG2+ cells are distinct from GFAP+ astrocytes. Similarly, we did not detect cells that expressed both NG2+ and S100, which is a calcium-binding protein expressed by astrocytes [19] (data not shown). These findings suggest that NG2+/PDGF α-receptor + (PDGF α-receptor + cells) cells are distinct from mature astrocytes.

To examine the relationship between NG2+ cells and microglia, P90 rat sections were double-labeled with rabbit antibody to NG2 and microglial markers. Double-labeling did not reveal cells that bound both anti-NG2 antibody and GSA I-B_4 isolectin from *Griffonia simplicifolia* [20] or monoclonal antibody 4H1 [21], as previously described [18]. These findings suggest that NG2+ cells are distinct from microglia or mature astrocytes, consistent with previous reports by Levine and colleagues on cells in the rat cerebellum [5,22]. Our observation that NG2+ cells in the mature brain express PDGF α-receptor supports the view that they are oligodendrocyte progenitor cells.

Increased Proliferation of NG2+ cells in the Dysmyelinating Mutant Jimpy

We next examined changes in NG2+ cells in the *jimpy* mutant in which it was previously shown that "immature oligodendrocytes" identified by electron microscopic criteria undergo increased proliferation [17]. To identify proliferating cells, *jimpy* and wild-type mice were labeled with BrdU, a thymidine analog that is incorporated into cells in the S-phase of the cell cycle. These studies were carried out at P18, when most of the oligodendrocytes and myelin have already been generated, and little proliferation of progenitor cells is expected. Cells that incorporated BrdU were then visualized by antibody to BrdU. Figure 2 shows the total number of cells that were double-labeled for BrdU and NG2 in 10-μm-thick sections from P18 wild-type and *jimpy* spinal cords. There was a five- to tenfold increase in the number of

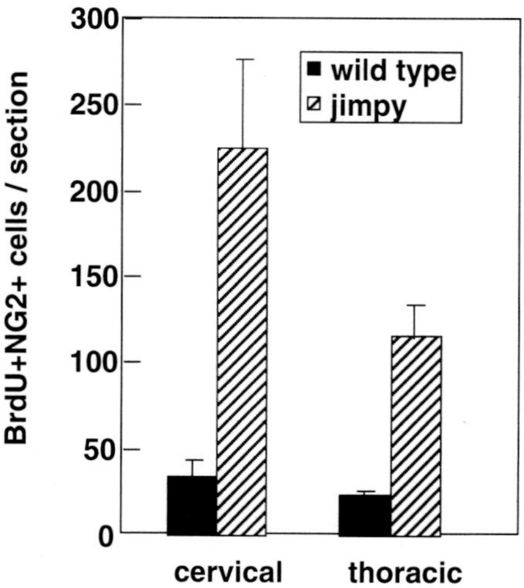

FIG. 2. Number of BrdU+/NG2+ cells in the cervical and thoracic spinal cord of wild-type (*filled bars*) and *jimpy* (*hatched bars*) mice. Numbers are expressed as the mean (n = 4) of the total number of BrdU+/NG2+ cells per cross section of the spinal cord. *Error bars*, SD

BrdU+/NG2+ cells in the *jimpy* spinal cord compared with that in the wild-type spinal cord. This result is consistent with the earlier electron microscopic studies that revealed a three- to tenfold increase in proliferating glial cells in the *jimpy* spinal cord [17]. In both *jimpy* and wild-type spinal cords, 70%–80% of the BrdU+ cells were NG2+.

Discussion

Our results demonstrate that NG2+ cells in the adult rat brain are distinct from mature astrocytes and microglia, consistent with previous observations by Levine and colleagues who demonstrated that NG2+ cells in the cerebellum do not express GFAP or the microglial marker OX42 [5,22]. Our observation that NG2+ cells in both gray and white matter of the mature CNS express PDGF α-receptor suggests that they are closely related to oligodendrocyte progenitor cells.

The numerous NG2+ cells in the mature CNS are relatively quiescent, as suggested by the low uptake of [^{3}H]thymidine [5] and BrdU (Nishiyama, unpublished observation). In the normal spinal cord, there is little proliferation of NG2+ cells at P18, when most of the oligodendrogliogenesis and myelinogenesis events have already occurred. However, in the *jimpy* spinal cord there is a significant increase in the number of proliferating NG2+ cells. This increase is consistent with the earlier electron microscopic studies [17] showing that immature oligodendrocytes proliferate in the *jimpy* spinal cord and also with our current hypothesis that the NG2+ cells are oligodendrocyte progenitor cells. Of the total proliferating cells in the *jimpy* and wild-type spinal cord, 70%–80% were NG2+. The remaining 20%–30% of the BrdU+ cells are likely to be astrocytes and microglial cells. We are currently carrying out experiments to identify growth factors and cytokines that induce proliferation of NG2+ cells in the *jimpy* spinal cord.

Acknowledgments. We thank Dr. William Stallcup (The Burnham Institute, La Jolla, CA), Joel Levine (SUNY, Stony Brook, NY), Carl-Henrik Heldin (Uppsala University), and William Hickey (Dartmouth Medical School, NJ) for the antibodies, and Drs. Bruce Trapp, Wendy Macklin (both at Cleveland Clinic), and Robert Miller (Case Western Reserve University, Cleveland, OH) for helpful discussions. This study was supported by the National Institutes of Health (NS35136 to A.N.) and the National Multiple Sclerosis Society (PP0453 to A.N.)

References

1. Nishiyama A, Dahlin KJ, Prince JT, Johnstone SR, Stallcup WB (1991) The primary structure of NG2, a novel membrane-spanning proteoglycan. J Cell Biol 114:359–371
2. Levine JM, Nishiyama A (1996) The NG2 chondroitin sulfate proteoglycan: a multifunctional proteoglycan associated with immature cells. Perspect Dev Neurobiol 3:245–259
3. Stallcup WB, Beasley L (1987) Bipotential glial precursor cells of the optic nerve express the NG2 proteoglycan. J Neurosci 7:2737–2744
4. Levine JM, Stallcup WB (1987) Plasticity of developing cerebellar cells in vitro studied with antibodies against the NG2 antigen. J Neurosci 7:2721–2731

5. Levine JM, Stincone F, Lee YS (1993) Development and differentiation of glial precursor cells in the rat cerebellum. Glia 7:307–321
6. Nishiyama A, Lin X-H, Giese N, et al (1996) Interaction between NG2 proteoglycan and PDGF α receptor on O2A progenitor cells is required for optimal response to PDGF. J Neurosci Res 43:315–330
7. LeVine SM, Goldman JE (1988) Spatial and temporal patterns of oligodendrocyte differentiation in rat cerebrum and cerebellum. J Comp Neurol 277:441–455
8. Reynolds R, Wilkin GP (1988) Development of macroglial cells in rat cerebellum. II. An in situ immunohistochemical study of oligodendroglial lineage from precursor to mature myelinating cell. Development (Camb) 102:409–425
9. Warrington AE, Pfeiffer SE (1992) Proliferation and differentiation of O4+ oligodendrocytes in postnatal rat cerebellum: analysis in unfixed tissue slices using anti-glycolipid antibodies. J Neurosci Res 33:338–353
10. Richardson WD, Pringle N, Mosley MJ, et al (1988) A role for platelet-derived growth factor in normal gliogenesis in the central nervous system. Cell 53:309–319
11. Pringle NP, Mudhar HS, Collarini EJ, et al (1992) PDGF receptors in the rat CNS: during late neurogenesis, PDGF alpha-receptor expression appears to be restricted to glial cells of the oligodendrocyte lineage. Development (Camb) 115:535–551
12. Pringle NP, Richardson WD (1993) A singularity of PDGF alpha-receptor expression in the dorsoventral axis of the neural tube may define the origin of the oligodendrocyte lineage. Development (Camb) 117:525–533
13. Nishiyama A, Lin X-H, Giese N, et al (1996) Co-localization of NG2 proteoglycan and PDGF α receptor on O2A progenitor cells in the developing rat brain. J Neurosci Res 43:299–314
14. Nave KA, Bloom FE, Milner RJ (1987) A single nucleotide difference in the gene for myelin proteolipid protein defines the *jimpy* mutation in mouse. J Neurochem 49:1873–1877
15. Ikenaka K, Furuichi T, Iwasaki Y, et al (1988) Myelin proteolipid protein gene structure and its regulation of expression normal and *jimpy* mutant mice. J Mol Biol 199:587–596
16. Knapp PE, Skoff RP, Redstone DW (1986) Oligodendroglial cell death in jimpy mice: an explanantion for the myelin deficit. J Neurosci 6:2813–2822
17. Skoff RP (1982) Increased proliferation of oligodendrocytes in the hypomyelinated mouse mutant-jimpy. Brain Res 248:19–31
18. Nishiyama A, Yu M, Drazba J, et al (1997) Normal and reactive NG2+ glial cells are distinct from resting and activated microglia. J Neurosci Res 48:299–312
19. Kligman D, Hilt DC (1988) The S100 protein family. TIBS (Trends Biochem Sci) 13:437–443
20. Streit WJ, Kreutzberg GW (1987) Lectin binding by resting and reactive microglia. J Neurocytol 16:249–260
21. Flaris NA, Densmore TL, Molleston MC, et al (1993) Characterization of microglia and macrophages in the central nervous system of rats: definition of the differential expression of molecules using standard and novel monoclonal antibodies in normal CNS and in four models of parenchymal reaction. Glia 7:34–40
22. Levine JM, Card JP (1987) Light and electron microscopic localization of a cell surface antigen (NG2) in the rat cerebellum: association with smooth protoplasmic astrocytes. J Neurosci 7:2711–2720

Glutamate Transporter (GLAST) as a Radial Glial Marker in the Mouse Cerebral Cortex

Yasuno Iwasaki[1], Yosuke Otake[1], Masahikc Watanabe[2], Kohichi Tanaka[3], and Kazuhiro Ikenaka[1]

Summary. Four types of glutamate transporter have been identified in the mammalian central nervous system (CNS). One of them, GLAST, is mainly localized in Bergmann glia in the adult cerebellar cortex and in the ventricular zone in the developing brain. A recent study with mouse spinal cord indicated that GLAST is expressed in a radial glia–astrocyte lineage in the spinal cord. In this study, we examined the expression pattern of GLAST gene in mouse cerebral cortex by in situ hybridization. At embryonic day 14 (E14), GLAST mRNA was expressed only in the ventricular zone adjacent to the lateral ventricle. Then, GLAST-expressing cells started to radially migrate from the ventricular zone toward the pial surface at E16. This migration became prominent at E18, and GLAST expression gradually decreased at the ventricular zone. Instead, GFAP mRNA was detected just above the ventricular zone at postnatal day 2 (P2). GLAST-expressing cells were also identified in the primary cultured cells from mouse cerebral cortex, some of which were overlapped with anti-GFAP immunoreactivity. These double-positive cells possibly represent a transient stage from radial glial cells to astrocytes. Our results show that GLAST is a useful marker of the astrocyte lineage also in the cerebral cortex.

Key words. Glutamate transporter—GLAST—Radial glia—Astrocyte—GFAP

Introduction

Glutamate is a major excitatory neurotransmitter in the mammalian central nervous system (CNS) and possibly plays key roles in CNS development [1]. Extracellular concentration of glutamate is regulated by transporters at the synaptic cleft to prevent the excessive stimulation of glutamate receptors. Four types of glutamate transporters have been identified in the mammalian CNS [2]. GLAST (GluT-1 or EAAT1) and GLT1

[1] National Institute for Physiological Sciences, Okagaki National Research Institute, 38 Nishigonaka, Myodaiji-cho, Okazaki, Aichi 444-8585, Japan
[2] Department of Anatomy, Hokkaido University School of Medicine, Sapporo 060-8638, Japan
[3] National Institute of Neuroscience, National Center of Neurology and Psychiatry, 4-1-1 Ogawa-higashi, Kodaira 187-8502, Japan

(EAAT2) are localized in astrocytes, whereas EAAC1 (EAAT3) and EAAT4 are localized in neurons. In situ hybridization studies revealed that these transporters are expressed in each distinct pattern in the fetal brain before synaptogenesis [3–5]. One of them, GLAST, exhibits prominent expression in Bergmann glial cells in the adult cerebellar cortex, which was also detected in the ventricular zone in the developing brain where most of the neurons and glia are generated. Analysis of mouse spinal cord by in situ hybridization and immunohistochemistry indicated that GLAST is first expressed in radial glial cells and then in astrocytes during spinal cord development, suggesting the generation of astrocytes from radial glial cells [6]. The lineage of astrocytes in the cerebral cortex, however, is still unclear, which might be because of the greater complexity of structural and functional organization in the cerebrum. It is now generally accepted that the astrocyte is derived from radial glial cells, which are generated from neuroepithelial cells [7–10]. It is, however, uncertain, whether all the astrocytes are derived from radial glial cells. Here, we examine GLAST expression in mouse cerebral cortex as a new marker in early gliogenesis and discuss astrocyte lineage in mouse cerebral cortex.

Materials and Methods

In Situ Hybridization of Mouse Brain

Embryos and neonates of ICR mice were fixed with 4% paraformaldehyde (PFA), and 8-μm paraffin sections were prepared according to the standard protocol. The sections were digested with 0.8% pepsin in 0.2 N HCl and then hybridized with digoxigenin- (DIG-) labeled GLAST cRNA probe [11]. After washing, the signal was detected with anti-DIG antibodies labeled with alkaline phosphatase.

Primary Culture and Double Labeling

Mixed glial culture was prepared from mouse cerebral cortex of embryonic day 13 (E13), E16, or neonate (P0), and cultured in DMEM–Ham's F-12 (1:1) medium with 10% fetal calf serum. After 24-h incubation, cells were fixed with 4% PFA and labeled with GLAST cRNA probe, followed by anti-glial fibrillary acidic protein (GFAP) (Sigma, St. Louis, MO, USA) staining.

Results

Mouse cerebral cortex was labeled with GLAST cRNA probes. Prominent expression was observed only in the the ventricular zone at E14 (Fig. 1a). Immunohistochemistry with anti-GLAST antibodies suggested that GLAST-expressing cells are radial glial cells with processes radially spanning from the ventricular zone to the brain surface (data not shown). At E16, GLAST-expressing cells migrated radially from the ventricular zone toward the brain surface throughout the cortex (data not shown). The migration of $GLAST^+$ cells became prominent at E18 (Fig. 1b); however, those cells were still negative for GFAP (Fig. 1c). Then, GLAST expression decreased at the ventricular zone at postnatal day 2 (P2) (Fig. 1d); instead, GFAP mRNA appeared, which was distributed just above the ventricular zone because of the short distance of migration (Fig. 1e).

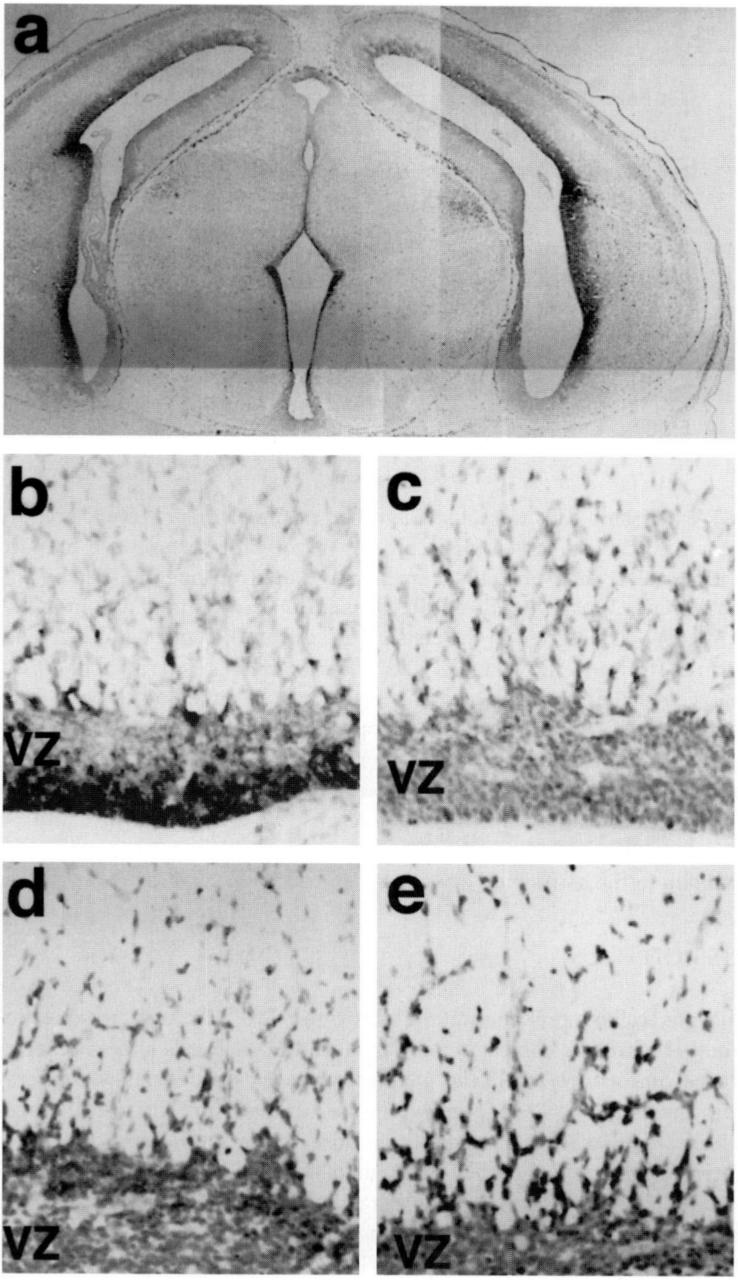

FIG. 1a–e. In situ hybridization analysis. **a** Expression pattern of glutamate transporter (GLAST) mRNA in the cerebral cortex of E14 mouse brain. Parahorizontal section of E14 mouse brain was hybridized with GLAST cRNA probe. GLAST is highly expressed in the ventricular zones adjacent to the lateral ventricle of cerebral cortex. **b–e** Comparative localization of GLAST and glial fibrillary acidic protein (GFAP) mRNA during development of mouse cerebral cortex. Parasagittal sections of E18 (**b**,**c**) and P2 (**d**,**e**) mouse brain were hybridized with GLAST (**b**,**d**) or GFAP (**c**,**e**) cRNA probes. Radial migration of GLAST$^+$ cells from the ventricular zone (*VZ*) toward the pial surface is observed at E18 in the cerebral cortex (**b**), when few GFAP$^+$ cells are there (**c**). At P2, the number of GLAST$^+$ cells decreased at the ventricular zone (**d**), whereas that of GFAP$^+$ cells increased just above the ventricular zone (**e**)

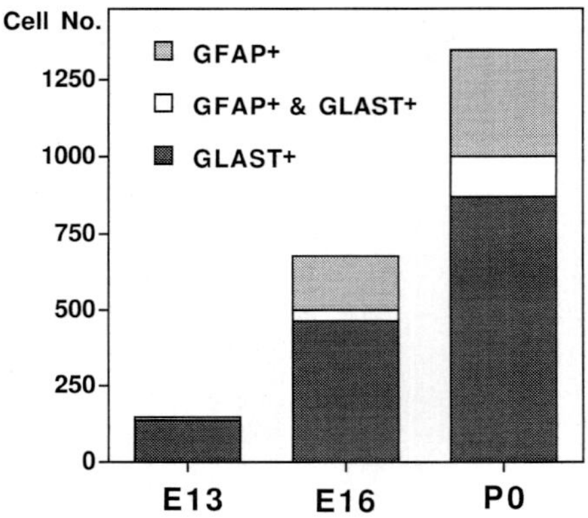

FIG. 2. Transforming of GLAST^{+} cells into GFAP^{+} astrocytes. Mixed glia culture of cerebral cortex was prepared from mouse embryos at the indicated age or from neonates and labeled with GLAST cRNA probe the next day, followed by anti-GFAP staining. Positive cells on the glass slide were counted and plotted. Double-positive cells for GLAST and GFAP increased with development

GLAST-expressing cells were also identified in the primary culture prepared from mouse cerebral cortex. Mixed glial cultures of cerebral cortex were prepared from mice (E13, E16, P0) and were labeled with GLAST cRNA probe followed by anti-GFAP stain the next day. Positive cells on the glass slide were counted and plotted (Fig. 2). GLAST mRNA was detected in the cells cultured from E13 mice and increased in the cells cultured from E16 and P0 mice. GLAST antibodies stained radial fibers extending from the ventricular zone toward the pial surface (data not shown). This result suggested that GLAST is expressed in radial glial cells, which was confirmed by staining with RC1, antibodies specific for radial glial cells (data not shown). In cells cultured from E16 mice, GFAP-immunoreactive cells appeared, some of which overlapped with GLAST cRNA labeling. Double-positive cells for GLAST and GFAP increased in P0 culture, which possibly represents a transient transforming stage from the radial glia to the astrocyte.

Discussion

The radial glial cell is a precursor of the astrocyte [7–10]. However, the question is still open whether astrocytes originate only from radial glial cells and whether all radial glial cells become astrocytes. In this chapter, we have provided a new marker, GLAST, to discuss the astrocyte lineage in the mouse cerebral cortex. In situ hybridization studies indicated that GLAST expression is observed in a radial glia–astrocyte lineage in the cerebral cortex. This was also confirmed by double-staining of primary cultured cells prepared from mouse cerebral cortex with GLAST cRNA probe followed by anti-GFAP staining. GLAST^{+} cells, localized in the ventricular zone at E14, extend radial fibers toward the pial surface. Most GLAST^{+} cells started to migrate around E18 and transformed to GFAP^{+} astrocytes. It is thought that neurons generated in the ventricular zone radially migrate along the fiber of the radial glia until about E18. Thus, the timing of transformation from radial glial cell to astrocyte is comparable with that of the completion of neuronal migration.

Our in situ hybridization studies also detected cells migrating from the ventricular zone toward the pial surface around E16. These cells seem to be different from the population of radial glial cells just described. Radial glial cells start to migrate or translocate around E18 and the distance of their migration is short, resulting in the formation of astrocytes near the ventricular zone, whereas the later population of GLAST$^+$ cells migrate earlier and further, resulting in wide distribution throughout the cortex. Immunohistochemical analysis indicated that the morphology is that of radially migrating cells extending short processes. Probably these cells also migrate along the processes of the radial glia. We speculate that these cells transform to GFAP$^+$ cells that appear along the pial surface and vascular vessels after E16.

We have provided here a new marker, GLAST, to study the astrocyte lineage in the mouse cerebral cortex. Analysis of GLAST mRNA by in situ hybiridization suggested the presence of multiple lineage for astrocytes in mouse cerebral cortex.

References

1. MacDonald JW, Johnston MV (1990) Physiological and pathophysiological roles and excitatory amino acids during central nervous system development. Brain Res Rev 15:41–70
2. Kanai Y, Trotti D, Nussberger S, et al (1997) The high affinity glutamate transporter family: structure, function, and physiological relevance. In: Reith MEA (ed) Neurotransmitter transporters. Structure, function, and regulation. Humana, Totowa, NJ, pp 171–213
3. Shibata T, Watanabe M, Tanaka K, et al (1996) Dynamic changes in expression of glutamate transporter mRNAs in developing brain. NeuroReport 7:705–709
4. Sutherland ML, Delaney TA, Noebels JL (1996) Glutamate transporter mRNA expression in proliferative zones of the developing and adult murine CNS. J Neurosci 16:2191–2207
5. Yamada K, Wada S, Watanabe M, et al (1997) Changes in expression and distribution of the glutamate transporter EAAT4 in developing mouse Purkinje cells. Neurosci Res 27:191–198
6. Shibata T, Yamada K, Watanabe M, et al (1997) Glutamate transporter GLAST is expressed in the radial glia-astrocyte lineage of developing mouse spinal cord. J Neurosci 17:9212–9219
7. Schmechel DE, Rakic P (1979) A Golgi study of radial glial cells in developing monkey telencephalon: morphogenesis and transformation into astrocytes. Anat Embryol 156:115–152
8. Misson J-P, Edwards MA, Yamamoto M, et al (1998) Identification of radial glial cells within the developing murine central nervous system: studies based upon a new immunohistochemical marker. Dev Brain Res 44:95–108
9. Voigt T (1989) Development of glial cells in the cerebral wall of ferrets: direct tracing of their transformation from radial glia into astrocytes. J Comp Neurol 289:74–88
10. Edwards MA, Yamamoto M, Caviness VS Jr (1990) Organization of radial glia and related cells in the developing murine CNS. Neuroscience 36:121–144
11. Tanaka K (1993) Cloning and expression of a glutamate transporter from mouse brain. Neurosci Lett 159:183–186

Involvement of Caspases in the TNF-α-Mediated Oligodendrocyte Cell Death

SHIN HISAHARA[1], SHIN-ICHI SHOJI[2], HIDEYUKI OKANO[1,3], and MASAYUKI MIURA[1,3]

Summary. Oligodendrocytes (OLGs) are selectively destroyed in chronically demyelinated multiple sclerosis (MS) lesions. Tumor necrosis factor (TNF) is thought to be one of the mediators responsible for the damage of OLGs in MS. The addition of TNF-α to primary cultures of mouse OLGs, which express caspase-1 and caspase-3 but not caspase-2, significantly decreased the number of live OLGs in 72 h. DNA fragmentation was detected in TNF-treated OLGs at 36 h with TUNEL staining. Z-Asp-CH_2-DCB, a broad inhibitor of the caspase family, prevented TNF-induced cell death of OLGs. Furthermore, Ac-YVAD-CHO (a specific inhibitor of caspase-1-like proteases) as well as Ac-DEVD-CHO (a specific inhibitor of caspase-3-like proteases) enhanced the survival of OLGs treated with TNF-α, indicating that the caspase-1- and the caspase-3-mediated cell-death pathways are involved in TNF-induced OLG cell death. Our results suggest that inhibition of caspases may be a novel approach to treat MS.

Key words. Oligodendrocytes—TNF—Apoptosis—Caspases—Demyelination

Multiple sclerosis (MS) is the most common autoimmune disease in the central nervous system (CNS). In MS patients, an induction of apoptosis of oligodendrocytes (OLGs) by immune cells results in a severe demyelination. A multifunctional cytokine, tumor necrosis factor-α (TNF-α), is produced by activated macrophages and T cells in CNS of MS patients. It is believed that TNF-α is one of the most important inflammatory cytokines associated with MS. TNF-α is increased during exacerbation of MS, and a downregulation of TNF-α with the phosphodiesterase type IV inhibitor rolipram is effective in reducing the progression of demyelination in experimental autoimmune encephalomyelitis (EAE) [1]. Transgenic mice that overexpress TNF-α in the CNS spontaneously develop a chronic inflammatory demy-

[1] Department of Neuroanatomy, Biomedical Research Center, Osaka University Medical School, 2-2 Yamadaoka, Suita, Osaka 565-0871, Japan
[2] Department of Neurology, Institute of Clinical Medical Sciences, University of Tsukuba, Tsukuba, 1-1-1 Tennoudai, Ibaraki 305-0006, Japan
[3] Core Research for Evolutional Science and Technology (CREST), Japan Science and Technology Corporation (JST), 2-6-15 Shibakoen, Minato-ku, Tokyo 105-0011, Japan

elinating disease [2]. TNF-α is also known as a pleiotrophic proinflammatory cytokine that elicits many biological functions including induction of apoptosis. Many tumor cells are sensitive to TNF and are effectively killed by it. In the CNS, it has been reported that OLGs are sensitive to TNF-mediated cytotoxicity [3,4]. Two TNF receptors (TNFR-1 and TNFR-2) are involved in transducing TNF signals into cytoplasm. TNFR-1 is mainly involved in transducing the cell-death signal, and OLGs selectively express TNFR-1 [5].

Recent studies by our laboratory and others have shown that TNF-induced cell death is executed by the activation of caspases, which are common mediators for programmed cell death [6,7]. Caspase (cysteine aspartate-specific proteinase) is the name for the family of proteases that includes mammalian caspase-1 and *Caenorhabditis elegans* CED-3 [8]. At least 12 family members have been found, and all the caspase members have been shown to be able to execute cell death when they are expressed in cultured cells.

We have shown that the activation of the members of the caspase family including caspase-1 itself is one of the most crucial events in TNF-induced apoptosis. HeLa cells treated with TNF-α and cycloheximide (CHX) undergo apoptosis [7]. A cowpox virus gene encoding a serpin named CrmA inhibits the enzymatic activity of the members of the caspase family including caspase-1 itself [9] and prevents cell death induced by these members [10,11]. Overexpression of *crmA* in HeLa cells prevents TNF-α/CHX-induced cell death and a secretion of mature IL-1β, suggesting that an enzymatic activation of the caspase family is involved in TNF-mediated apoptosis. More recently, it has been shown that recruitment of the signal transducer Fas-associating protein with a novel death domain (FADD) to the TNFR-1 complex mediates activation of caspase-8 [12,13].

We examined the expression of the members of the caspase family in OLGs that have been purified from mouse brain [14] and found that caspase-1, caspase-3, and caspase-11 are expressed in cultured mouse OLGs (Fig. 1). TNF-α-treated cells display apoptotic characteristics: degradation of cell processes and condensation of chromatin. We then examined the effects of TNF-α on the early stage of apoptosis [14]. We used in situ DNA fragmentation analysis (TUNEL assay) to detect early changes of DNA fragmentation in OLGs in 36 h of incubation. Control OLGs showed few signs of DNA fragmentation. We determined the percentage of nucleus-stained cells in the TUNEL assay from the total number of OLGs in the phase-contrast view. The basal level of apoptosis in this model was thus found to be approximately 8% (Fig. 2). After TNF-α treatment, we could detect 77% positively stained nuclei in OLGs in the TUNEL assay, whereas only 9.1% of the cells were TUNEL-positive with Z-Asp-CH_2-DCB, an aspartate-based inhibitor [15], respectively.

Administration of the caspase-1-like protease inhibitor (Ac-YVAD-CHO) or the caspase-3-like protease inhibitor (Ac-DEVD-CHO) into a primary culture of mouse OLGs could enhance the viability of TNF-treated OLGs. Most effectively, Z-Asp-CH_2-DCB, which prevents cell death induced by the caspase family with a specificity similar to that of p35, could prevent most of the TNF-α-induced OLG cell deaths (Fig. 3). N^{α}-*p*-tosyl-L-lysine choromethyl ketone (TLCK) and leupeptin could not inhibit OLG apoptosis induced by TNF-α. Our results suggest that activation of the caspase pathway is crucial in TNF-α-induced OLG apoptosis and that inhibitors of caspase protease may lead to therapeutic approaches for MS.

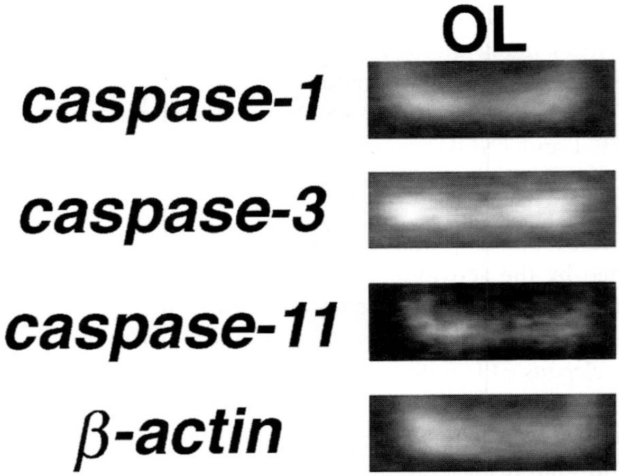

Fig. 1. Caspase-1, caspase-3, and caspase-11 are expressed in mouse oligodendrocytes (*OL*). Expression patterns of three caspase family members in oligodendrocytes as shown by RT-PCR analysis

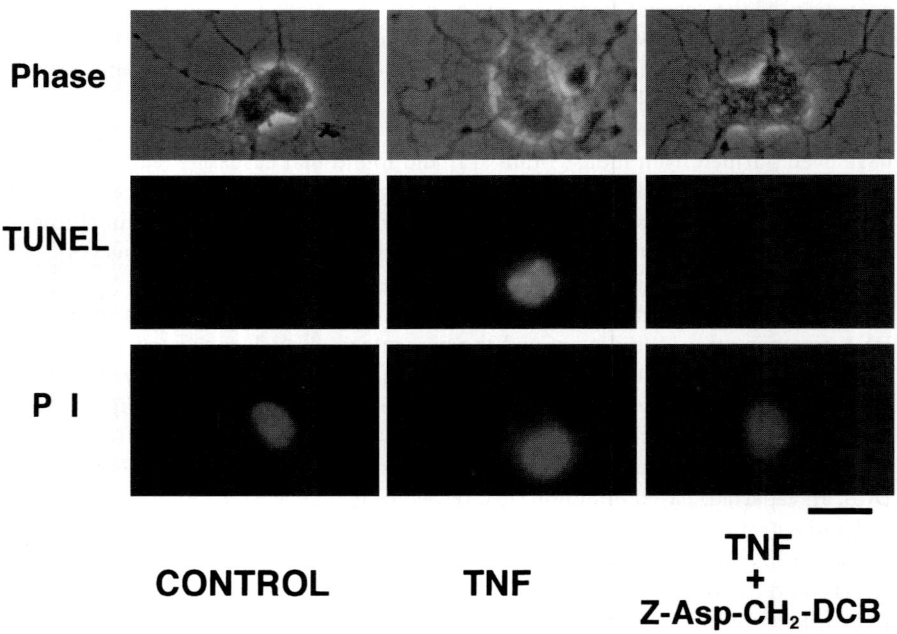

Fig. 2. DNA fragmentation observed in tumor necrosis factor-α- (TNF-α-) treated oligodendrocytes (OLGs) by TUNEL assay. TNF-α-treated OLGs showed DNA fragmentation after 36-h incubation, but control OLGs and TNF-α-treated OLGs together with 40 mg/ml of Z-Asp-CH_2-DCB did not. Nuclear morphology of TUNEL-positive/negative OLGs is still preserved after 36 h with propidium iodide (*PI*) staining

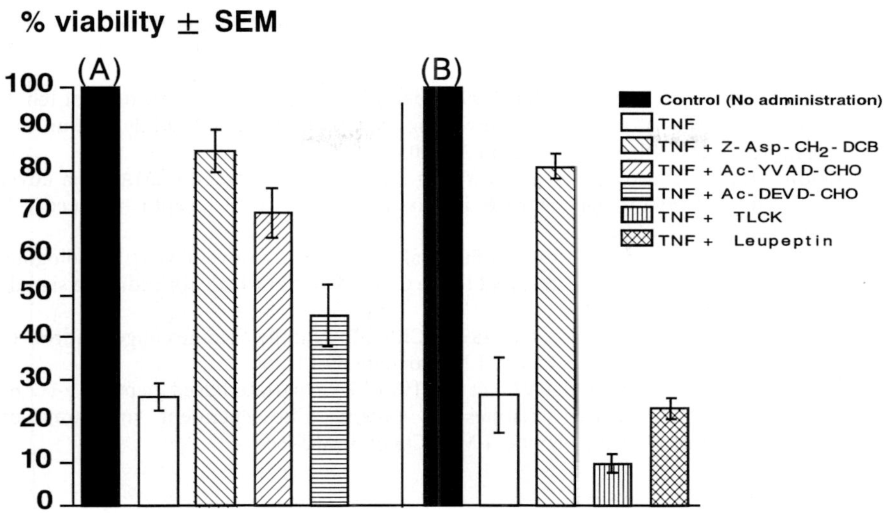

FIG. 3A,B. Caspase inhibitors prevent OLGs from undergoing TNF-α-induced apoptosis. **A** TNF-α (100 ng/ml) and Ac-YVAD-CHO (100 μM), Ac-DEVD-CHO (100 μM), or Z-Asp-CH_2-DCB (40 μg/ml) were administered to OLG culture simultaneously. **B** TNF-α and Z-Asp-CH_2-DCB, TLCK (100 μM), or leupeptin (100 μM) were administered to OLG culture simultaneously. The cells were fixed 72 h later and live mature OLGs were counted. In each experiment, the number of OLGs collected from each independent well was counted, and relative viability was expressed as mean percentage (±SEM)

References

1. Genain CP, Roberts T, Davis RL, et al (1995) Prevention of autoimmune demyelination in non-human primates by a cAMP-specific phosphodiesterase inhibitor. Proc Natl Acad Sci USA 92:3601–3605
2. Probert L, Akassoglou K, Pasparakis M, et al (1995) Spontaneous inflammatory demyelinating disease in transgenic mice showing central nervous system-specific expression of tumor necrosis factor α. Proc Natl Acad Sci USA 92:11294–11298
3. Selmaj KW, Raine CS (1988) Tumor necrosis factor mediates myelin and oligodendrocyte damage in vitro. Ann Neurol 23:339–346
4. D'Souza S, Alinauskas K, McCrea E, et al (1995) Differential susceptibility of human CNS-derived cell populations to TNF-dependent and independent immune-mediated injury. J Neurosci 15:7293–7300
5. Dopp JM, Makenzie-Graham A, Otero GC, et al (1997) Differential expression, cytokine modulation, and specific functions of type-1 and type-2 tumor necrosis factor receptors in rat glia. J Neuroimmunol 75:104–112
6. Tewari M, Dixit VM (1995) Fas- and tumor necrosis factor-induced apoptosis is inhibited by the poxvirus crmA gene product. J Biol Chem 270:3255–3260
7. Miura M, Friedlander RM, Yuan J (1995) Tumor necrosis factor-induced apoptosis is mediated by a crmA-sensitive cell death pathway. Proc Natl Acad Sci USA 92:8318–8322
8. Alnemri ES, Livingston DJ, Nicholson DW, et al (1996) Human ICE/CED-3 protease nomenclature. Cell 87:171
9. Ray CA, Black RA, Kronheim SR, et al (1992) Viral inhibition of inflammation: cowpox virus encodes an inhibitor of the interleukin-1β-converting enzyme. Cell 69:597–604

10. Miura M, Zhu H, Rotello R, et al (1993) Induction of apoptosis in fibroblasts by IL-1β-converting enzyme, a mammalian homolog of the *C. elegans* cell death gene ced-3. Cell 75:653–660
11. Wang S, Miura M, Jung Y-K, et al (1996) Identification and characterization of Ich-3, a member of the interleukin-1β-converting enzyme (ICE)/Ced-3 family and an upstream regulator of ICE. J Biol Chem 271:20580–20587
12. Boldin MP, Goncharov TM, Goltsev YV, et al (1996) Involvement of MACH, a novel MORT1/FADD-interacting protease, in Fas/APO-1- and TNF receptor-induced cell death. Cell 85:803–815
13. Muzio M, Chinnaiyan AM, Kischkel FC, et al (1996) Flice, a novel FADD-homologous ICE/CED-3-like protease, is recruited to the CD95 (Fas/APO-1) death-inducing signaling complex. Cell 85:817–827
14. Hisahara S, Shoji S, Okano H, et al (1997) ICE/CED-3 family executes oligodendrocyte apoptosis by tumor necrosis factor. J Neurochem 69:10–20
15. Dolle RE, Hoyer D, Prasad CVC, et al (1994) P_1 aspartate-based peptide a-((2,6-dichlorobenzoyl)oxy)methyl ketones as potent time-dependent inhibitors of interleukin-1β-converting enzyme. J Med Chem 37:563–564

Oligodendrocyte Differentiation Has a Strict Requirement for Astrocyte Contact In Vitro

Kazunori Yoshimura[1,3], Daisuke Nishimura[1,2], Yoko Sakurai[1], Thein Thein Aye[1], Harukazu Saito[1], Suguru Kawato[2], Masahiko Nomura[3], and Hiroaki Asou[1]

Summary. Cells at an intermediate stage of oligodendrocyte lineage, represented by the $O4^+$ and $GC/O1^-$ immature oligodendrocytes, are likely relate to the outcome of physiological events but they are not characterized well [1]. We have developed a method for generating homogeneous populations of immature oligodendrocytes (OLG) from primary culture of E18 rat brain, utilizing four types of intrinsic and extrinsic factors: cell proliferation, survival, differentiation, and omission of nonoligodendrocyte lineage cells. We have also demonstrated that a very large number of homogeneous populations of stage-specific OLGs can be supplied after 35 days of culture. In those cells, the total time for differentiation was extended and maturation occurred only with the direct contact of type 1 astrocytes.

Key words. Stage-specific oligodendrocytes—Oligodendrocytes (OLG)—Astrocytes—O1—O4

Oligodendrocytes (OLG) are myelin-forming cells in the central nervous system (CNS) that are generated from a neuroepithelial orgin [2]. During brain development, oligodendrocyte differentiation proceeds by a predictable schedule, but the timing mechanisms responsible are unknown. Oligodendrocyte-type 2 astrocyte progenitor cells (O-2A progenitor) from rat optic nerve or cerebrum grown in vitro in chemically defined medium, in the absence of mitogen, do not divide but instead differentiate rapidly into OLG [3]. OLG are dynamic cells that can respond to a variety of external environments in vitro [4]. Notably, OLG display a markedly homogeneous morphology in the optic nerve; however, there is abundant evidence that OLG are a heterogeneous population in the CNS overall [5]. Previously it has been extremely difficult to set a stage-specific homogeneous population of OLG in vitro; however, we have generated a mitogen and cell cycle-dependent long-term culture of oligodendrocyte

[1] Department of Neuro-Cell Biology, Tokyo Metropolitan Institute of Gerontology, 35-2 Sakaecho, Itabashi-ku, Tokyo 173, Japan
[2] Department of Biophysics and Lifescience, The University of Tokyo at Komaba, Meguro-ku, Tokyo 153, Japan
[3] Department of Physiology, Saitama Medical School, Iruma-gun, Saitama 350-04, Japan

lineage cells, and now large-scale homogeneous population of these cells have been isolated from embryonic rat brain. Those stage-specific OLG only have an ability to facilitate differentiation when in contact with type 1 astrocytes.

Figure 1 illustrates the principle of culture for production of stage-specific OLG. Our method for large-scale stage-specific pure oligodendrocyte production from embryonic rat cerebral hemispheres utilized four types of intrinsic or extrinsic OLG influence factors: cell proliferation and survival [supplemented with 10% fetal calf serum (FCS) into minimum essential medium (MEM); phase I, II in Fig. 1], differentiation [addition of basic fibroblast growth factor (bFGF) into serum-free medium (BS medium); phase III in Fig. 1], and the removal of nonoligodendrocyte lineage cells (treatment of cell passage with 0.25% trypsin; phase II, III, IV, V in Fig. 1). These stage-specific OLG yielded a final amount of 1×10^5 cells/dish after 35 days in culture. We postulate that O-2A progenitor cells have an intrinsic clock that operates by counting the number of cell division [6], or that immature OLG are also induced by bFGF to differentiate and proliferate [7]. These cells, however, become unresponsive to the growth factor, stop dividing, and spontaneously differentiate into mature OLG [8]. In culture, O-2A progenitor cells divided as many as eight times before differentiating into OLG [8,9].

To investigate the state of stage-specific OLG and whether a heterogeneous population of OLG gives rise in cultures, immunocytochemical studies were used to check every 2 days after cell passage (Fig. 2). In vitro, the ordered and partially overlapping expression of glycolipid antibodies, A2B5, O4, and O1/GC markers distinguishes a consecutive phenotypically defined stage of the oligodendrocyte lineage, represented by the biopolar A2B5$^+$, O4$^-$, and O1/GC$^-$ O-2A progenitor cells. At an early stage the OLG can be recognized by surface binding of salfateglycolipid antibody O4 and complex process-bearing O4$^-$, O1/GC$^-$ mature OLG [10]. Continued maturation of OLG results in elevated expression of major myelin protein, such as myelin basic protein (MBP) and proteolipid protein (PLP), and ultimately the elaboration of the myelin sheath [11]. Developmental expression of cell-specific markers in oligodendrocyte-lineage cells in cultures is demonstrated in Fig. 2. Stage-specific OLG were identified as immature OLG by their elaborate cell process and A2B5$^-$, O4$^+$, MBP$^-$, and O1/GC$^-$ immunoreactivity after 30 days of culture, suggesting differentiated OLG express myelin glycolipid galactocerebroside (GC) in their plasma membrane [12] and lose immunoreactivity for A2B5 [13] as migrating development continues.

To determine whether stage-specific (immature) OLG have an ability to differentiate into a mature type in vitro, we studied OLG cocultures with astrocytes or neurons. When the stage-specific OLG were cocultured with astrocytes, the expression of MBP (a major myelin component protein) on the western blotting and immnocytochemical level was increased, extending distribution of the multicomplex process of OLG in comparison with a single culture of OLG or coculture with neurons (Fig. 3). These results suggest that immature OLG are accelerated and differentiated to the mature OLG only when in contact with astrocyte plasma membranes, as OLG culture with astrocytes or neurons without direct cell–cell interactions has no effect on maturation (data not shown).

So far, one of the most important questions has been why O-2A progenitor cells in vivo/in vitro have a heterogenous capacity to proliferate and differentiate in response

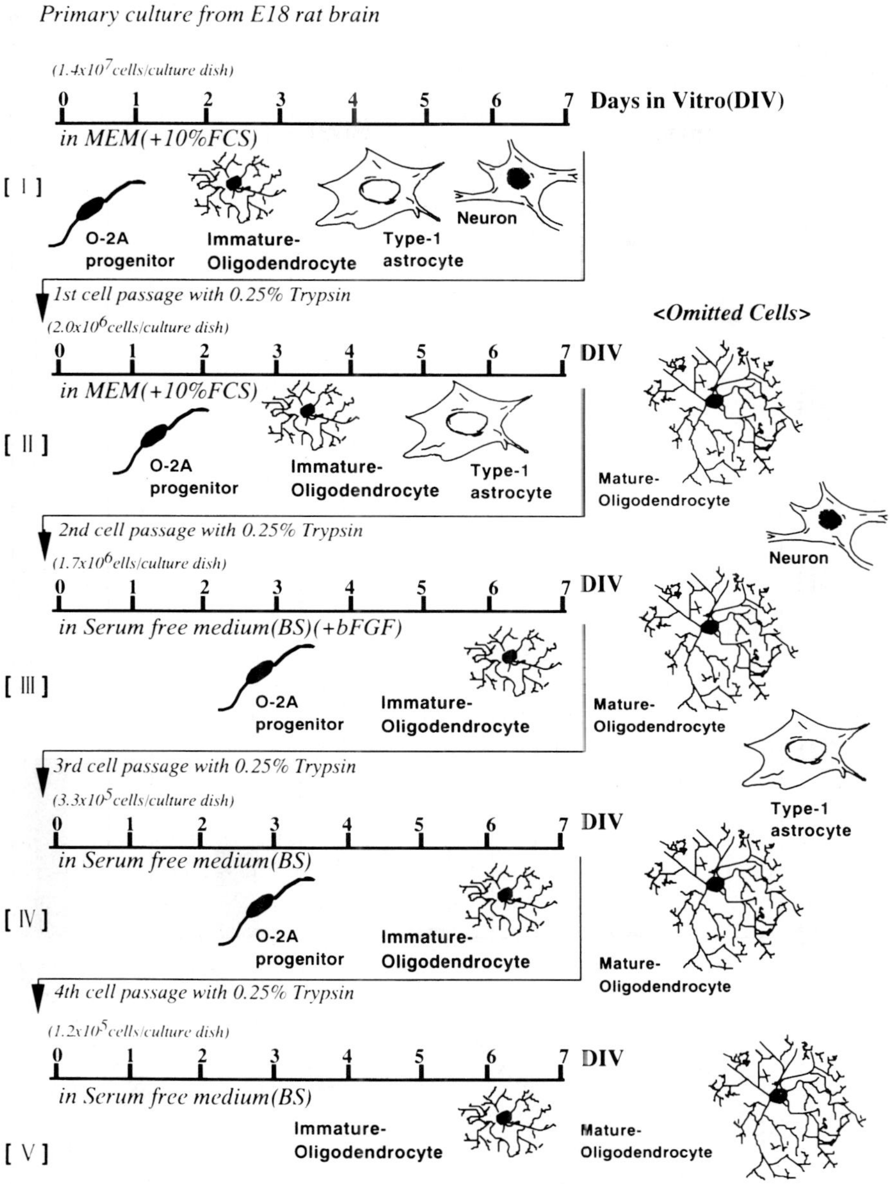

FIG. 1. Diagrammatic summary of stage-specific oligodendrocyte (OLG) cultures illustrates a flowchart of the procedure for isolation of stage-specific OLG. Oligodendrocyte differentiation could be controlled by the four stages of their development (for details, see Materials and Methods). *I, II, III, IV, V* represent the developmental stages of oligodendrocyte lineage

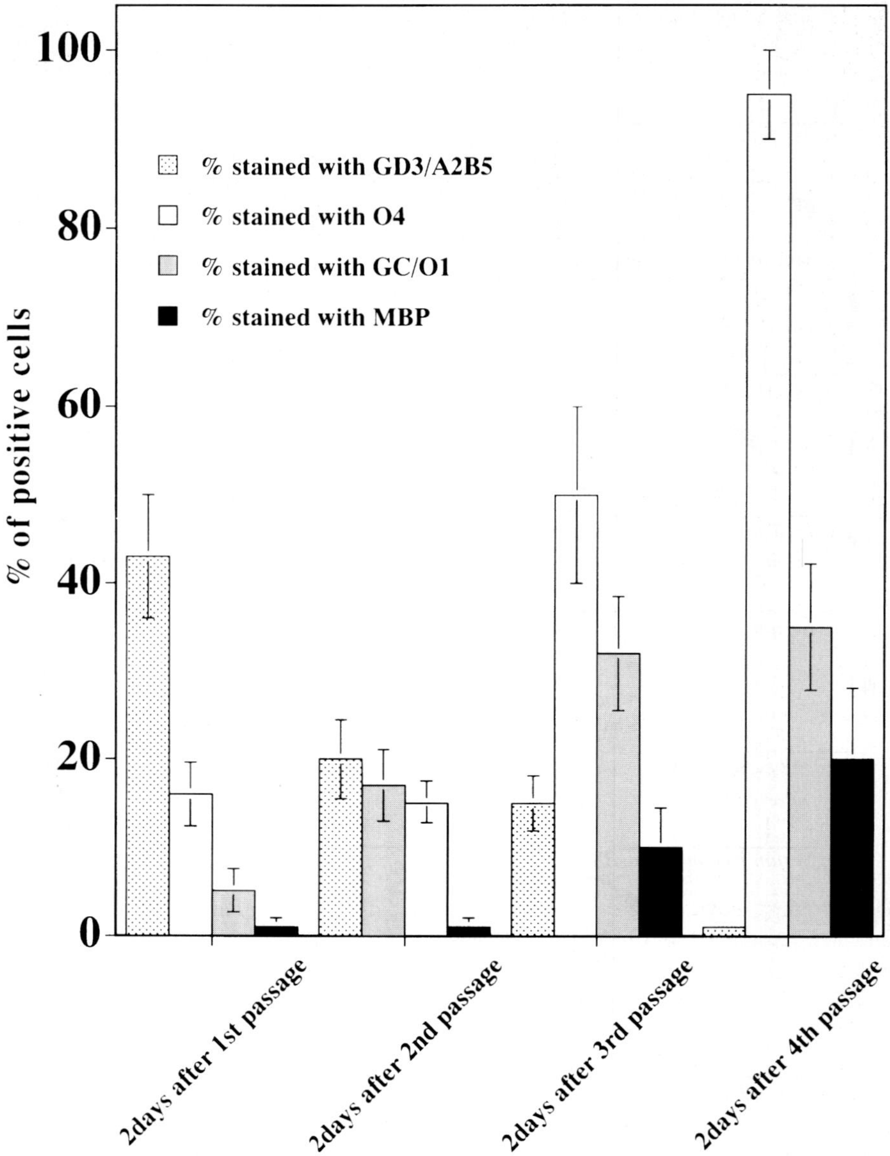

FIG. 2. Developmental expression of cell-specific marker in stage-specific OLG. The percentage of cells stained was calculated using the following protocol: the total number (1000) of cells was determined by counting 8–10 fields of cells following single/double label direct immunofluoresence using GD3/A2B5, O4, O1/GC, or MBP antibodies. Data from each stage (I, II, III, IV, V) of 2 days after cell passage represented the mean ± SE of the three separate experiments

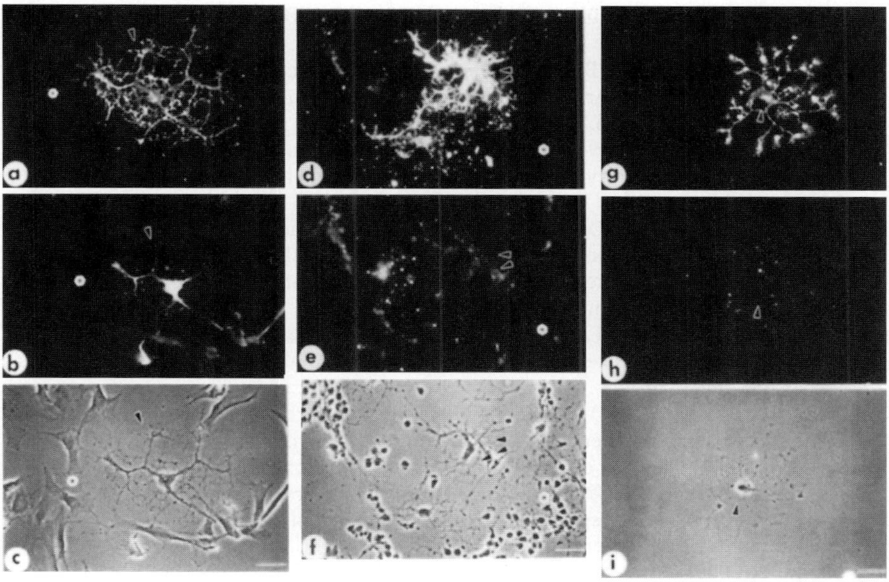

Fig. 3a–i. Double immunofluorescence staining of mixed cell cultures with anti-O4 and MBP antibodies. **a–c** Stage-specific OLG (*arrowhead*) + coculture with type 1 astrocytes (*asterisk*). **d–f** Stage-specific OLG (*arrowhead*) + cocultures with neurons (*asterisk*). **g–i** Stage-specific OLG alone (*arrowhead*) are the same microscopic field as 3-day cultures labeled with anti-O4 (**a,d,g;** fluorescein optics) and anti-MBP (**b,e,h;** rhodamine optics) antibodies with phase contrast (**c,f,i**). MBP immunoreactivity is accelerated in both cell body + processes only in cocultures with type 1 astrocytes (**b**). Data represent 1 of 8–10 separate experiments. *Bar* 31.2 μm

to growth factors. For this reason, homogeneous populations have not been used [14]. Our findings, however, clearly demonstrated that via the combination bFGF usage and cell passage procedure, oligodendrocyte precursor cells differentiate and become a homogeneous population of stage-specific OLG (immature OLG) in the serum-free medium after the fourth passage (Fig. 2). Those immature OLG also fail to differentiate into a mature type of OLG unless in contact with astrocytes.

Therefore, we concluded that cell-surface signaling molecules are required for the survival and differentiation of immature OLG. In our culture, at least, O-2A progenitor cells stopped dividing at the almost same time over the period of 3 weeks when cultured in the condition of serum-free medium after the third cell passage, suggesting that heterogeneity in differentiative capacity is also an intrinsic property of immature OLG. Finally, together with previous observations [15], our results indicate that stage-specific (immature) OLG can be provided for the role of remyelinating cells after demyelination in vivo.

References

1. Ludwin SK (1989) Evolving concepts and issues in remyelination. Dev Neurosci 11:140–148

2. Goldman JE (1992) Regulation of oligodendrocyte differentiation.Trends Neurosci 15:359–360
3. Gard AL, Williams WC, Burrell MR (1997) Oligodendroblasts distinguished from O-2A glial progenitors by surface phenotype ($O4^{+}GalC^{-}$) and response to cytokines using signal transducer LIFR-β. Dev Biol 19:197–203
4. Hardy R, Reynolds R (1993) Rat cerebral cortical neurons in primary culture release a mitogen specific for early ($GD3^{+}/O4^{-}$) oligodendroglial progenitors. J Neurosci Res 34:589–600
5. Bjartmar C, Hildebrand C, Loinder K (1994) Morphological heterogeneity of rat oligodendrocytes: electron microscopic studies on serial sections. Glia 11:235–244
6. Barres BA, Raff MC, Gaese F, et al (1994) A crucial role for neurotrophin-3 in oligodendrocyte development. Nature (Lond) 367:371–375
7. Grinspan JB, Reeves MF, Coulaloglou MJ, et al (1996) Re-entry into cell cycle is required for bFGF-induced oligodendroglial defferentiation and survival. J Neurosci Res 46:456–464
8. Temple S, Raff MC (1986) Clonal analysis of oligodendrocyte development in culture: evidence for a developmental clock that counts cell divisions. Cell 44:773–779
9. Raff MC (1989) Glial cell diversification in the rat optic nerve. Science 243:1450–1455
10. Gard AL, Pfeiffer SE (1989) Oligodendrocyte progenitors isolated directly from developing terencephalon at a specific phenotypic stage. Myelinogenic potential in a defined environment. Development (Camb) 106:119–132
11. Lemke G (1988) Unwrapping the genes of myelin. Neuron 1:535–543
12. Raff MC, Mirsky R, Fields KL, et al (1978) Galactocerebroside is a specific cell surface antigenic marker for oligodendrocytes in culture. Nature (Lond) 274:813–816
13. Raff MC, Williams BP, Millerm RH (1984) The in vitro differentiation of a bipotential glial progenitor cell. EMBO J 5:1857–1864
14. Prince J (1994) Glial cell lineage and development. Curr Opin Neurobiol 4:680–686
15. Gensert JM, Goldman JE (1997) Endogenous progenitors remyelinate demyelinated axons in the adult CNS. Neuron 19:197–203

Part 6
Circuit Formation

Guidance of Ventral Midline Decussating Axons in the Brain

Fujio Murakami, Ryuichi Shirasaki, and Ryuta Katsumata

Summary. The early pattern of the neuronal circuit in the vertebrate neural tube is composed of an orthogonal array of axonal tracts extending along circumferential and longitudinal axes. One major component that gives rise to the circumferentially as well as longitudinally growing axons are the commissural axons originating from the alar plate. Diffusible molecules secreted by the floor plate at the ventral midline of the neural tube appear to provide key signals for organizing neural patterning during development, by directing both ventrally growing circumferential axons: floor plate chemoattraction contributes to the guidance of ventrally growing axons toward the floor plate. Moreover, chemotropic guidance by the floor plate, initially discovered in the spinal cord, has been extended to all axial levels where floor plate cells are present. At the molecular level, netrin-1 appears to be a key molecule in guiding ventrally directed axons. The mechanisms that allow the growth of commissural axons across the floor plate remain less clear, but accumulating evidence suggests that commissural axons lose their responsiveness to floor plate-derived chemoattractant on their encounters with the floor plate. Loss of chemoattractant responsiveness of developing axons at intermediate targets may contribute to their leaving the intermediate targets.

Key words. Floor plate—Commissural axon—Chemotropism—Rat—Hindbrain

Introduction

A fundamental issue in neural development is to understand the mechanisms that give rise to neural patterning. The early pattern of the neuronal circuit in the vertebrate neural tube is composed of an orthogonal array of axonal tracts extending along circumferential and longitudinal axes. One major component that gives rise to the circumferentially as well as longitudinally growing axons are the commissural axons originating from the alar plate.

Axons of commissural neurons in the dorsal spinal cord, for example, follow stereotyped and characteristic trajectories: initial ventrally directed growth along the

---

Laboratory of Neuroscience, Division of Biophysical Engineering, Graduate School of Engineering Science, Osaka University, Machikaneyama 1-3, Toyonaka, Osaka 560-8531, Japan circumferential axis, midline decussation via the floor plate, and subsequent elongation along the longitudinal axis. Notably, the projection patterns of ventrally decussating commissural axons are conserved between the spinal cord and the caudal hindbrain in both chick and zebrafish embryos [1–4] as well as between the spinal cord and midbrain of chick embryos [5–6]. In the rat too, the axons originating from prospective deep cerebellar nuclei cells in the metencephalic alar plate follow growth patterns similar to those of commissural axons in the spinal cord [7].

Thus, it is possible that a common axon guidance system operates across axial levels as well as across species.

Mechanisms of Ventrally Directed Growth of Commissural Axons

Spinal commissural axons were, a decade ago, suggested to be guided toward the floor plate by a diffusible chemoattractant released from the floor plate [8; see also 9–11]. Similarity in the features of their trajectories leads to speculation that a common guidance mechanism operates among hindbrain commissural axons and spinal commissural axons. Indeed, cerebellar plate explants when cocultured with floor plate explants in collagen gels extended axons toward floor plate explants in the same fashion as spinal commissural axons [7]. This finding suggests that cerebellofugal axons are also guided toward the ventral midline by a floor plate-derived diffusible chemoattractant.

Consistent with these findings, commissural axons in the spinal cord and the brain express a common surface molecule, TAG-1, a glycoprotein that belongs to the immunoglobulin superfamily. TAG-1 is expressed on the spinal commissural axons in vivo and by axons elicited from the dorsal spinal cord in the presence of floor plate explant in vitro [8,9]. Moreover, antibody staining of flat whole-mount embryonic rat brain preparations has demonstrated that ventrally decussating commissural axons from the myelencephalon through to the level of the mesencephalon expressed TAG-1 in a "consecutive stripe" pattern and that its expression diminishes as they migrate through the floor plate [12]. Correspondingly, in vitro, TAG-1-positive axons originating from the alar plate of all these axial levels are attracted by the floor plate-derived chemoattractant [12]. Together these findings suggest that alar plate-derived ventrally decussating axons from the spinal cord up to the level of the mesencephalon are guided toward the ventral midline by a common cue, a diffusible factor(s) released from the floor plate.

Netrins (netrin-1 and netrin-2), diffusible chemotropic factors that can attract spinal commissural axons toward the floor plate in vitro, have recently been purified from embryonic chick brain [13,14]. In *netrin-1*-deficient mice, most spinal commissural axons were unable to reach the ventral midline, although initially directed ventrally [15]. This finding indicates that netrin-1 plays a crucial role in the guidance of ventrally growing commissural axons in vivo, although it is not required for initial ventral growth of the axons. However, floor plate explants from the netrin-1-knockout mouse can still attract spinal commissural axons in vitro [15], indicating that a molecule(s) other than netrin may also contribute to the guidance of the commissural axons.

Midline Crossing

The theory of chemoattractive guidance of axons by the floor plate predicts that the concentration of the chemotropic factor should peak at the ventral midline. According to this theory, commissural axons, having arrived at the floor plate, should require the action of an additional mechanism so as not to stall at the floor plate. One such mechanism would be cooperative interaction of the commissural axon growth cone with its contralateral counterpart at the midline and subsequent fasciculation with homologous axons. Indeed, in the grasshopper, contact and fasciculation with the contralateral commissural axon homolog seem to be involved in growth cone crossing of the midline [16,17]. However, a distinct mechanism seems to operate in vertebrates. In the rat, for example, cerebellofugal axons advance across the floor plate in whole-mount in vitro preparations with unilateral removal of the cerebellar primordium [18], indicating that these axons do not require the contralateral homolog to advance beyond the floor plate. Consistent with this observation, spinal commissural axons in the zebrafish embryo decussate in segments in which the contralateral commissural neuron is missing [19]. Furthermore, fasciculation of spinal commissural axons does not seem to occur in the chick embryo [20]. Thus, axon fasciculation with the contralateral counterpart may not contribute to the guidance of commissural axons across the midline in the vertebrate.

Change in Growth Cone Responsiveness to Floor Plate-Derived Chemoattractant

An alternative intriguing possibility is that growth cones alter their responsiveness to a floor plate-derived chemoattractant when they encounter floor plate cells, enabling them to navigate beyond the floor plate. The fact that commissural axons express TAG-1 until they reach the floor plate [12,21] but thereafter express L1 glycoprotein [21] might reflect changes in growth cone responsiveness.

This possibility has recently been tested directly, using an in vitro preparation that reproduces the crossing of the midline floor plate by cerebellar plate (CP) axons [22]. When a strip of the rostral metencephalon that included the entire circumferential trajectory of CP axons was cultured alone in collagen gel, axons originating from the CP grew across the midline floor plate to extend contralaterally. A floor plate explant was then juxtaposed to the metencephalic strip on one side, and the behavior of CP axons was examined by implanting the fluorescent tracer DiI into the CP contralateral to the explant, to test whether CP axons, after they have crossed the floor plate, are attracted by an ectopic floor plate explant. Under such conditions, CP axons that had crossed the midline floor plate did not show directed growth toward the ectopic floor plate explant.

Thus, floor plate chemoattraction of commissural axons is effective only until they arrive at the floor plate. Although it remains unknown how modification of growth cone responsiveness to the chemoattractant is achieved, one possible mechanism is that the interaction of commissural axons with floor plate cells directly changes the chemoattractant responsiveness of the axons by modifying chemoattractant-receptor mechanisms locally in the growth cones. Alternatively, the modification might be mediated by changes in transcription of the receptor or receptor-related molecule(s).

Although such interaction could be mediated either by a diffusible or membrane-associated factor(s), several lines of evidence suggest that commissural axons directly interact with floor plate cells. In the chick, axonin-1, a homolog of TAG-1, and Nr-CAM are expressed on spinal commissural axons and the floor plate cells, respectively. Application of these molecules themselves or antibodies against them leads to a failure of the axons to enter the floor plate both in vivo and in vitro [23,24]. Considering that axonin-1 binds to Nr-CAM, these results suggest that the events required for commissural axon growth cone advance across the floor plate may involve a direct interaction of commissural axons with floor plate cells. Commissural growth cone failure to enter the floor plate in the presence of anti-axonin-1 or anti-Nr-CAM antibody was taken to indicate that floor plate cells are inhibitory to commissural axon growth in the absence of axonin-1–Nr-CAM interaction [24].

In the wild-type zebrafish embryo, axons of VeLD neurons in the spinal cord and subsets of reticulospinal neurons in the hindbrain grow from the ventral spinal cord toward the ventral midline, but do not cross it. These axons, however, cross the midline in the *cyclops* mutant [25,26]. The trajectories of these axons in the mutant appear normal until they approach the floor plate, whereabouts axons in wild-type embryos abruptly change growth direction to course caudally. These observations also suggest that the midline decussation failure of axons in wild-type embryos is caused by a contact-mediated interaction with floor plate cells. A mechanism involving the interaction between commissural axons and floor plate cells is further supported by recent findings that floor plate cells in mice transfer large cytoplasmic proteins such as β-galactosidase to crossing axons [27]. In *Drosophila melanogaster*, formation of commissures appears to require the *commissureless* gene [28], which encodes a transmembrane protein and is expressed by CNS midline glia during formation of axon commissures [29]. An interesting finding is that the first commissural axons that contact the commissureless protein expressed in midline glial cells also accumulate commissureless protein along their axons, despite no *commissureless* transcript being observed in the cell bodies of these neurons, implying that the protein is transferred from midline glia to commissural axons [29].

Thus, it appears that an interaction at the ventral midline between commissural axons and midline cells plays an important role for midline crossing of the axons in a variety of species.

References

1. Trevarrow B, Marks DL, Kimmel CB (1990) Organization of hindbrain segments in the zebrafish embryo. Neuron 4:669–679
2. Clarke JDW, Lumsden AGS (1993) Segmental repetition of neuronal phenotype sets in the chick embryo hindbrain. Development (Camb) 118:151–162
3. Glover JC (1993) The development of brain stem projections to the spinal cord in the chicken embryo. Brain Res Bull 30:265–271
4. Kimmel CB (1993) Patterning the brain of the zebrafish embryo. Annu Rev Neurosci 16:707–732

5. Kröger S, Schwartz U (1990) The avian tectobulbar tract: development, explant culture, and effects of antibodies on the pattern of neurite outgrowth. J Neurosci 10:3118–3134
6. Shepherd IT, Taylor JS (1995) Early development of efferent projections from the chick tectum. J Comp Neurol 354:501–510
7. Shirasaki R, Tamada A, Katsumata R, et al (1995) Guidance of cerebellofugal axons in the rat embryo: directed growth toward the floor plate and subsequent elongation along the longitudinal axis. Neuron 14:961–972
8. Tessier-Lavigne M, Placzek M, Lumsden A, et al (1988) Chemotropic guidance of developing axons in the mammalian central nervous system. Nature (Lond) 336:775–778
9. Placzek M, Tessier-Lavigne M, Jessell T, et al (1990) Orientation of commissural axons *in vitro* in response to a floor plate-derived chemoattractant. Development (Camb) 110:19–30
10. Placzek M, Tessier-Lavigne M, Yamada T, et al (1990) Guidance of developing axons by diffusible chemoattractants. In: Cold Spring Harbor symposia on quantitative biology, vol LV. Cold Spring Harbor Laboratory Press, NY, pp 279–289
11. Yaginuma H, Oppenheim RW (1991) An experimental analysis of *in vivo* guidance cues used by axons of spinal interneurons in the chick embryo: evidence for chemotropism and related guidance mechanisms. J Neurosci 11:2598–2613
12. Shirasaki R, Mirzayan C, Tessier-Lavigne M, et al (1996) Guidance of circumferentially growing axons by netrin-dependent and -independent floor plate chemotropism in the vertebrate brain. Neuron 17:1079–1088
13. Kennedy TE, Serafini T, de la Torre JR, et al (1994) Netrins are diffusible chemotropic factors for commissural axons in the embryonic spinal cord. Cell 78:425–435
14. Serafini T, Kennedy TE, Galko MJ, et al (1994) The netrins define a family of axon outgrowth-promoting proteins homologous to *C. elegans* UNC-6. Cell 78:409–424
15. Serafini T, Colamarino SA, Leonardo ED, et al (1996) Netrin-1 is required for commissural axon guidance in the developing vertebrate nervous system. Cell 87:1001–1014
16. Myers PZ, Bastiani MJ (1993) Cell-cell interactions during the migration of an identified commissural growth cone in the embryonic grasshopper. J Neurosci 13:115–126
17. Myers PZ, Bastiani MJ (1993) Growth cone dynamics during the migration of an identified commissural growth cone. J Neurosci 13:127–143
18. Shirasaki R, Katsumata R, Murakami F (1994) Guidance of cerebellar plate axons at the floor plate: experimental manipulation using whole-mount in vitro preparation. Neurosci Res Suppl 19:S117
19. Kuwada JY, Bernhardt RR, Chitnis AB (1990) Pathfinding by identified growth cones in the spinal cord of zebrafish embryos. J Neurosci 10:1299–1308
20. Yaginuma H, Shiga T, Homma S, et al (1990) Identification of early developing axon projections from spinal interneurons in the chick embryo with a neuron specific β-tubulin antibody: evidence for a new "pioneer" pathway in the spinal cord. Development (Camb) 108:705–716
21. Dodd J, Morton SB, Karagogeos D, et al (1988) Spatial regulation of axonal glycoprotein expression on subsets of embryonic spinal neurons. Neuron 1:105–116
22. Shirasaki R, Katsumata R, Murakami F (1998) Change in chemoattractant responsiveness of developing axons at an intermediate target. Science 279:205–207
23. Stoeckli ET, Landmesser LT (1995) Axonin-1, Nr-CAM, and Ng-CAM play different roles in the in vivo guidance of chick commissural neurons. Neuron 14:1165–1179
24. Stoeckli ET, Sonderegger P, Pollerberg GE, et al (1997) Interference with axonin-1 and NrCAM interactions unmasks a floor-plate activity inhibitory for commissural axons. Neuron 18:209–221
25. Bernhardt RR, Nguyen N, Kuwada JY (1992) Growth cone guidance by floor plate cells in the spinal cord of zebrafish embryos. Neuron 8:869–882

26. Hatta K (1992) Role of the floor plate in axonal patterning in the zebrafish CNS. Neuron 9:629–642
27. Campbell RM, Peterson AC (1993) Expression of a *lacZ* transgene reveals floor plate cell morphology and macromolecular transfer to commissural axons. Development (Camb) 119:1217–1228
28. Seeger M, Tear G, Ferres-Marco D, et al (1993) Mutations affecting growth cone guidance in *Drosophila*: genes necessary for guidance toward or away from the midline. Neuron 10:409–426
29. Tear G, Harris R, Sutaria S, et al (1996) Commissureless controls growth cone guidance across the CNS midline in *Drosophila* and encodes a novel membrane protein. Neuron 16:501–514

Neuropilin May Be a Common Component of Receptors for Secreted Semaphorin Family Members

Jonathan A. Raper, Hiroaki Kobayashi, Leonard Feiner, and Adam M. Koppel

Collapsin-1, a member of the semaphorin family of signaling molecules, activates receptors on specific growth cones and inhibits their motility. Neuropilin, a previously cloned transmembrane protein, has recently been identified as a candidate receptor for collapsin-1. We show that the secreted collapsins-1, -2, -3, and -5 bind to overlapping but distinct axon tracts in situ. We infer that there are distinct receptors with different affinities for collapsin-1, -2, -3, and -5. In contrast, these four collapsins all bind recombinant neuropilin with similar affinities. Strong binding to neuropilin is mediated by the C-terminal third of the collapsins, while the N-terminal semaphorin domain confers their unique binding patterns in situ. Studies that relate the structures of collapsins-1, -2, and -3 to their ability to induce the collapse of sensory and sympathetic growth cones demonstrate that (1) the N-terminal semaphorin domain of collapsin-1 is both necessary and sufficient for biological activity, (2) the semaphorin domain contains a 70-amino-acid region that specifies the biological activity of the three family members, and (3) the positively charged C-terminal terminus that binds neuropilin potentiates activity without affecting specificity. From these studies we propose that neuropilin could be a common component of a semaphorin receptor complex, and that additional differentially expressed receptor components are likely to interact with the semaphorin domains of collapsins and mediate their functional specificities.

References

Koppel AM, Feiner L, Kobayashi H, Raper JA (1997) A 70 amino acid region within the semaphorin domain activates specific cellular response of semaphorin family members. Neuron 19:531–537

Feiner L, Koppel AM, Kobayashi H, Raper JA (1997) Secreted chick semaphorins bind recombinant neuropilin with similar affinities but bind different subsets of neurons in situ. Neuron 19:539–545

Kobayashi H, Koppel AM, Luo Y, Raper JA (1997) A role for collapsin-1 in olfactory and cranial sensory axon guidance. J Neurosci 17:8339–8352

University of Pennsylvania School of Medicine, Philadelphia, PA, USA

Neuropilin-1 Plays a Crucial Role in Peripheral Nerve Fiber Guidance

Hajime Fujisawa[1], Takashi Kitsukawa[2], Masayuki Shimizu[1], and Tatsumi Hirata[1]

Summary. Neuropilin-1 is a neuronal cell-surface protein that is highly conserved among vertebrate species and has been suggested to function as a receptor for a chemorepellent semaphorin-III/D. The roles of neuropilin-1-semaphorin-III/D-mediated chemorepulsive signals in vivo, however, is unknown. We produced neuropilin-1-deficient mutant mice by targeted disruption of the *neuropilin-1* gene, and showed that deficiency of neuropilin-1 caused severe abnormalities in the trajectory of efferent fibers of the peripheral nervous system (PNS) that express neuropilin-1 in normal development. We also showed that neuropilin-1-deprived dorsal root ganglion (DRG) neurons were perfectly protected from growth cone collapse elicited by semaphorin III/D. Our results indicate that neuropilin-1 is a functional receptor or an indispensable component of a receptor complex for the growth cone repellent semaphorin III/D, and that neuropilin-1-semaphorin III/D-mediated chemorepulsive signals are essential in directional guidance of PNS efferent fibers in vivo.

Key words. Neuropilin-1—Growth cone collapse—Chemorepellent—Axon guidance—Peripheral nerve

Introduction

During embryonic development, nerve fibers interact with various kinds of environmental cues of either attractive or repulsive properties to grow in a precise and orderly fashion, then arrive at appropriate targets [1]. Among these cues, collapsin-1/semaphorin III/D [2,3] are potent chemorepellents that can repel or collapse growth cones of dorsal root ganglion (DRG) neurons in vitro [2,4], and have been suggested to be repulsive guidance signals for DRG fibers.

Neuropilin-1 (formerly neuropilin or A5) is a type I membrane protein that is highly conserved among vertebrate species [5–8]. Neuropilin-1 is expressed in par-

[1] Group of Developmental Neurobiology, Division of Biological Science, Nagoya University Graduate School of Science, Chikusa-ku, Nagoya 464-8602, Japan

[2] Laboratory of Specification Mechanisms 1, National Institute for Basic Biology, 38 Nishigonaka, Myodaiji-cho, Okazaki, Aichi 444-8585, Japan

ticular classes of axons during the phase of active axon growth, and has been suggested to play roles in axon guidance and fasciculation [7–10]. Furthermore, we have shown that ectopic and excess expression of neuropilin-1 in mouse embryos induced defasciculation and abnormal sprouting of peripheral nervous system (PNS) efferent fibers [11]. More recently, expression cloning studies have shown that neuropilin-1 is a semaphorin-III/D-binding protein and mediates semaphorin-III/D-induced chemorepulsive signals to DRG neurons in vitro [12,13].

To clarify further the functions of neuropilin-1 in nerve fiber guidance, particularly in vivo, we produced *neuropilin-1* null mutant mice by targeted disruption of the *neuropilin-1* gene [14]. *Neuropilin-1* homozygous mutant embryos could still be observed as late as E12.5, but no viable homozygous mutant embryos were seen by E13.5, indicating that the *neuropilin-1* mutation is lethal during embryogenesis.

Pathways and Projections of PNS Efferent Fibers Are Disrupted in *Neuropilin-1* Mutant Embryos

Whole-mount immunostaining of homozygous *neuropilin-1* mutant embryos with antineurofilament antibody demonstrated severe abnormalities in the pathways and projections of cranial efferent fibers such as the ophthalmic, maxillary, mandibular, facial, glossopharyngeal, and vagus nerves that normally expressed neuropilin-1. These cranial nerves were often defasciculated and widely spread: the ophthalmic nerve of the homozygous mutant embryos at E12.5 overshot far beyond the growing front of the normal nerve within the front nasal part (Fig. 1). The maxillary nerves in the mutant embryos were also widely spread within the maxilla (Fig. 1).

The spinal nerves of both the sensory and motor nerves express neuropilin-1 in normal development [8,11]. Deprivation of neuropilin-1 disrupted patterned pathways and projections of the spinal nerves as found in the wild-type embryos (compare Fig. 2a and 2b). The spinal nerve trunks were defasciculated, and many nerve fibers left the main spinal nerve trunk at arbitrary points, then grew into the skin surface in an irregular manner. Several nerve fibers arrived at the opposite side of the embryos

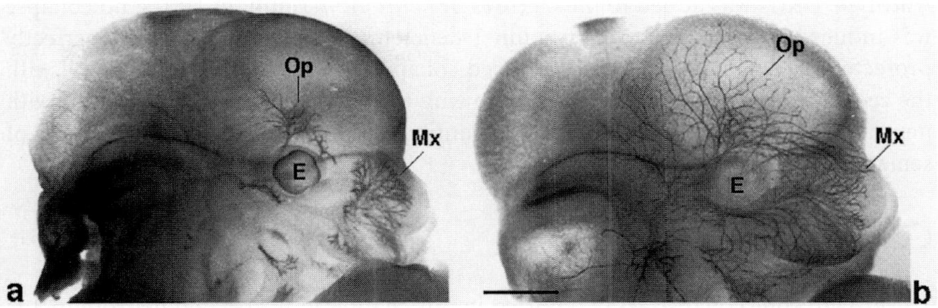

Fig. 1a,b. Whole-mount immunostaining with antineurofilament antibody of heads of embryos at E12.5. **a** Wild type. **b** Homozygous mutant. *Op*, ophthalmic nerve; *Mx*, maxillary nerve; *E*; eye. *Bar*, 1 mm

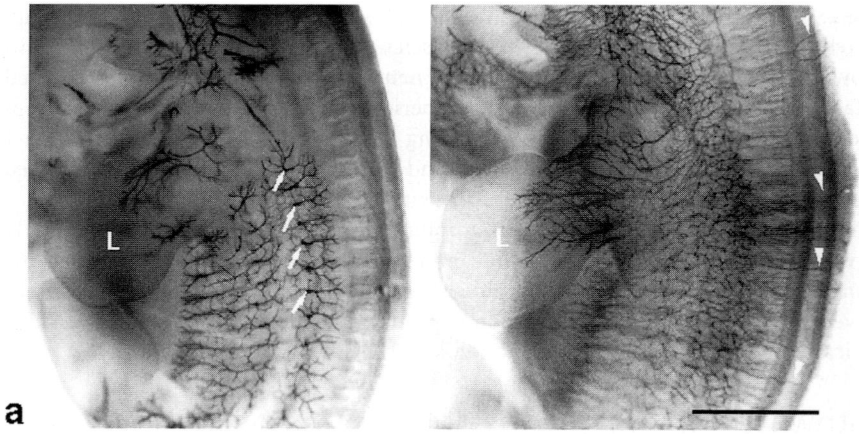

FIG. 2a,b. Whole-mount immunostaining with antineurofilament antibody of trunk and forelimb of embryos at E12.5. **a** wild type. **b** Homozygous mutant. *L*, forelimb. *Arrows*, penetration points of the lateral branches; *arrowheads*, fibers crossing the dorsal midline of the embryo. *Bar*, 1 mm

after crossing the dorsal midline, which was never observed in the normal or heterozygous mutant embryos. The pathways and projections of the spinal nerves into developing limbs were also abnormal. The spinal nerve fibers never formed thick nerve fiber bundles, then were distributed widely within the limb buds (Fig. 2).

Neuropilin-1 is Essential for Semaphorin III/D-Induced Growth Cone Collapse

To test whether neuropilin-1 is required for the semaphorin-III/D-induced growth cone collapse, we added semaphorin III/D recombinant protein to the cultures of DRGs of either the wild-type or homozygous mutant embryos at E12.5. An amount of semaphorin III/D recombinant protein that induces collapse of all growth cones of the wild-type DRGs was added to the cultures of *neuropilin-1* mutant DRGs; no collapse was induced, indicating that neuropilin-1-deficient DRG growth cones are perfectly protected from semaphorin III/D-induced collapse. These results coincide well with the recent expression cloning studies showing that semaphorin III/D can bind with neuropilin-1, and that anti-neuropilin-1 antibodies inhibit the repulsive action of semaphorin III/D on DRG fibers [12,13].

Conclusion

The growth cone collapse assay reported in this chapter indicates that neuropilin-1 is an indispensable cell-surface component which mediates semaphorin III/D-induced chemorepulsive signals to neurons. Moreover, the phenotype in the PNS efferent observed in the *neuropilin-1* mutant mouse embryos was similar to that of the *semaphorin III/D* mutant mouse embryos [15], suggesting that chemorepulsive signals which are mediated by neuropilin-1-semaphorin-III/D interaction provide a major mechanism in directional guidance of PNS efferent fibers in vivo.

Acknowledgments. This work was supported by grants from CREST (Core Research for Evolutional Science and Technology) of Japan Science and Technology Corporation (JST), and the Ministry of Education, Science and Culture (Japan).

References

1. Goodman CS (1996) Mechanisms and molecules that control growth cone guidance. Annu Rev Neurosci 19:341–377
2. Luo Y, Raible D, Raper JA (1993) Collapsin: a protein in brain that induces the collapse and paralysis of neuronal growth cones. Cell 75:217–227
3. Kolodkin AL, Matthes DJ, Goodman CS (1993) The semaphorin genes encode a family of transmembrane and secreted growth cone guidance molecules. Cell 75:1389–1399
4. Messersmith EK, Leonardo ED, Shatz CJ, et al (1995) Semaphorin III can function as a selective chemorepellent to pattern sensory projections in the spinal cord. Neuron 14:949–959
5. Takagi S, Tsuji T, Amagai T, et al (1987) Specific cell surface labels in the visual centers of *Xenopus laevis* tadpole identified using monoclonal antibodies. Dev Biol 122:90–100
6. Takagi S, Hirata T, Agata K, et al (1991) The A5 antigen, a candidate for the neuronal recognition molecule, has homologies to complement components and coagulation factors. Neuron 7:295–307
7. Takagi S, Kasuya Y, Shimizu M, et al (1995) Expression of a cell adhesion molecule, neuropilin, in the developing chick nervous system. Dev Biol 170:207–222
8. Kawakami A, Kitsukawa T, Takagi S, et al (1996) Developmentally regulated expression of a cell surface protein, neuropilin, in the mouse nervous system. J Neurobiol 29:1–17
9. Fujisawa H, Takagi S, Hirata T (1995) Growth-associated expression of a membrane protein, neuropilin, in *Xenopus* optic nerve fibers. Dev Neurosci 17:343–349
10. Satoda M, Takagi S, Ohta K, et al (1994) Differential expression of two cell surface proteins, neuropilin and plexin, in *Xenopus* olfactory axon subclasses. J Neurosci 15:942–955
11. Kitsukawa T, Shimono A, Kawakami A, et al (1995) Overexpression of a membrane protein, neuropilin, in chimeric mice causes anomalies in the cardiovascular system, nervous system and limb. Development (Camb) 121:4309–4318
12. He Z, Tessier-Lavigne M (1997) Neuropilin is a receptor for the axonal chemorepellent semaphorin III. Cell 90:739–751
13. Kolodkin AL, Levengood DV, Rowe EG, et al (1997) Neuropilin is a semaphorin III receptor. Cell 90:753–762
14. Kitsukawa T, Shimizu M., Sanbo M, et al (1997) Neuropilin-semaphorin III/D-mediated chemorepulsive signals play a crucial role in peripheral nerve projection in mice. Neuron 19:995–1005
15. Taniguchi M, Yuasa S, Fujisawa H, et al (1997) Disruption of *semaphorin III/D* gene causes severe abnormality in peripheral nerve projection. Neuron 19:519–530

Role of Intrinsic Neuronal Determinants for Nerve Sprouting and Synaptic Plasticity

T. Laux, O. Tarabal, D. Frey, C. Schneider, and P. Caroni

Growth of neuritic processes can be subdivided into long distance and local. Long-distance elongation is restricted to early development and regeneration after fiber tract lesions, and appears to involve the activation of a program of growth-associated gene expression in neurons. Local growth involves alterations in the spatial arrangement of preexisting and the formation of the new synaptic connections. In addition to mediating the formation and repair of synaptic connections, it may provide a substrate for the long-term changes of synaptic circuitry involved in learning and memory.

The differential expression of intrinsic neuronal components may affect competence, responsiveness, and mode of local growth by neurons. According to this possibility, plasticity and stability may be affected by the particular complement of competence-conferring components in neurons and their subcellular compartments.

To test some of these concepts, we overexpressed candidate growth-associated proteins specifically in adult neurons of transgenic mice. We initially focused on GAP-43, CAP-23, and MARCKS, three abundant acylated protein kinase C substrates that associate with the plasmalemma and the cortical cytoskeleton, where they promote actin cytoskeleton and cell-surface dynamics. These proteins are expressed at high levels during axonal elongation and target innervation, and their expression is retained in partially overlapping populations of neurons in the adult. GAP-43 specifically targets to axons, whereas CAP-23 and MARCKS accumulate in axons, in dendrites, and at the soma.

Transgenic mice overexpressing GAP-43, CAP-23, or MARCKS exhibited greatly elevated local structural plasticity in the adult peripheral and central nervous system. These effects of the transgenes were cell autonomous. In agreement with their subcellular targeting, the effects of GAP-43 were restricted to axons, whereas CAP-23 and MARCKS also induced growth of filopodia and spines at the soma and dendrites on certain types of neurons. Ultrastructural analysis of sprouting neuromuscular junctions revealed that the presence of transgene in the nerve terminal induced characteristic alterations in the arrangement and association of terminal Schwann cell processes with nerve. A similar type of alteration was detected when sprouting was induced by local paralysis in nontransgenic mice. Contact between nerve sprouts and

Friedrich Miescher-Institute, P. O. Box 2543, CH-4002 Basel, Switzerland muscle required paralysis. The results suggest a role for intrinsic neuronal components in the local control of structural plasticity, and point to the importance of selective cell adhesion in controlling local growth.

Relevant Publications

Aigner L, Arber S, Kapfhammer JP, Laux T, Schneider C, Botteri F, Brenner HR, Caroni P (1995) Overexpression of the neural growth-associated protein GAP-43 induces nerve sprouting in the adult nervous system of transgenic mice. Cell 83:269–278

Bailey CH, Alberini C, Ghirardi M, Kandel ER (1994) Molecular and structural changes underlying long-term memory storage in *Aplysia*. Adv Second Messenger Phosphoprotein Res 29:529–544

Caroni P (1997) Intrinsic neuronal determinants that promote axonal sprouting and elongation. BioEssays 19:767–775

Caroni P, Aigner L, Schneider C (1997) Intrinsic neuronal determinants locally regulate extrasynaptic and synaptic growth at the adult neuromuscular junction. J Cell Biol 136:679–692

Darian-Smith C, Gilbert CD (1994) Axonal sprouting accompanies functional reorganization in adult cat striate cortex. Nature (Lond) 368:737–740

Skene JHP (1989) Axonal growth-associated proteins. Annu Rev Neurosci 12:127–156

Son Y-J, Thompson WJ (1995) Nerve sprouting in muscle is induced and guided by processes extended by Schwann cells. Neuron 14:133–141

Development and Function of Glutamic Acid Decarboxylase-Deficient Mice

Kunihiko Obata, Hideo Asada, and Kei Maruyama

Summary. γ-Aminobutyric acid (GABA) is an inhibitory neurotransmitter and probably a neurotrophic substance. It is synthesized by glutamic acid decarboxylase (GAD), but the respective roles of the two isoforms (GAD65 and GAD67) have not been determined. The selective elimination of each GAD isoform by gene targeting is expected to clarify these issues. We have produced both GAD65−/− and GAD67−/− mice. The lack of GAD65 did not change brain GABA contents or animal behavior, except for a slight increase in susceptibility to seizures. GAD67−/− mice were born at the expected frequency but died of severe cleft palate soon after birth. GAD activities and GABA contents were reduced to 20% and 7%, respectively, in the cerebral cortex of the newborn GAD67−/− mice. Their brain, however, did not show any discernible defects. Involvement of GABA in palate formation have been suggested by previous pharmacological and genetic investigations and was confirmed by the present study.

Key words. GABA—Glutamic acid decarboxylase—Mutant mouse—Cleft palate

Introduction

γ-Aminobutyric acid (GABA), the principal inhibitory neurotransmitter [1,2], is synthesized from glutamic acid by glutamic acid decarboxylase (GAD) in GABAergic neurons of the mammalian central nervous system [3]. GAD exists as two isoforms with molecular masses of 65 and 67 kDa (GAD65 and GAD67, respectively), which are encoded by independent genes [3]. Several different properties of these isoforms [3,4] have suggested that they have distinct roles in neural functions. It has also been proposed that GAD65 participates more actively in synaptic function than GAD67, because GAD65 appears to be concentrated in the nerve terminals while GAD67 is not [5]. Such separate roles, however, have not been substantiated. The use of gene targeting for the selective elimination of each GAD isoform is expected to address this question. Therefore, we produced both GAD65- and GAD67-knockout mice [6,7] and

Laboratory of Neurochemistry, National Institute for Physiological Sciences, 38 Nishigonaka, Myodaiji-cho, Okazaki, Aichi 444-8585, Japan analyzed their phenotypes. In this chapter, our results are summarized and discussed. After publication of our papers, two papers on the production of GAD67- [8] and GAD65-knockout mice [9] have appeared, largely confirming our observations.

Production of GAD Mutant Mice

Exon 3 of GAD65 and exon 1 of GAD67 gene in TT2 ES cells were disrupted by insertion of a Neo casette. The TT2 cell line is derived from blastocysts of C57BL/6 x CBA F1 mice [10]. After confirmation of homologous recombination, the cloned ES cells were injected into eight-cell embryos from ICR mice. The mutant mice were obtained by mating chimeric mice with C57BL/6 mice. Deletion of the GAD genes was confirmed by southern blotting of tail tissue DNA.

GAD and GABA of GAD65 Mutant Brain

Mice were obtained from heterozygous parents at the expected Mendelian frequency. GAD65−/− mice appeared healthy, were of similar size and weight to wild-type and GAD65+/− littermates, and had no gross defects. No morphological abnormalities were observed in brain sections.

GAD65 mRNA and protein were absent in the GAD65−/− brain and were reduced to one-half of wild-type brain levels in the GAD65+/− brain. However, compensatory upregulation of GAD67 was not observed in GAD65 mutant mice. Although total GAD activity in the GAD65−/− cerebral cortex was 44% of the wild-type one, GABA content did not differ among three genotypes of mice. Immunohistochemistry of GABA also demonstrated normal number and distribution of GABA-positive neurons in GAD65−/− brain sections. These results indicate that there is some homeostatic regulation of GABA synthesis, and the remaining GAD67 activity might be enough to maintain the level of GABA in GAD65-deficient mice.

Seizure Susceptibility of GAD65 Mutant Mice

No behavioral abnormalities in GAD65−/− mice were observed in locomotor activity, passive avoidance, Rota-Rod, and Morris water maze tests. Sensitivity to convulsants, however, was higher in GAD65−/− mice. Intraperitoneal administration of picrotoxin (5 mg/kg) induced a generalized convulsive fit after 14.5 ± 0.9 min in control mice. Latency to seizure was significantly shortened to 8.6 ± 1.1 min for GAD65−/− mice. Continuous infusion of pentylenetetrazol induced twitch, clonic convulsion, and tonic extension, sequentially. Doses required for induction of convulsion were significantly lower for homozygous mutants (54.6 ± 7.9 mg/kg) than for wild-type mice (86.7 ± 8.8 mg/kg).

Picrotoxin blocks the GABA receptor complex, and one of the targets of pentylenetetrazol is the GABA receptor. Therefore, higher sensitivity of GAD65−/− mice to these convulsants may reflect any impairment of GABAergic transmission. Recently, Kash et al. [9] have produced GAD65 mutant mice using 129/SvJ-derived ES cells. When the genetic background was changed by backcross generation with the

NOD/LtJ strain of mice, seizure susceptibility of GAD65−/− mice was greatly increased [9].

GAD67 Mutant Mice

Mutant mice were obtained from heterozygous parents at the expected Mendelian frequency. At birth, the littermates were indistinguishable from each other in appearance. Afterward, however, GAD67−/− mice displayed gradual swelling of the bowels and died during the first morning after birth (day P0.5). Severe cleft palate was found in all of them. Their stomach and duodenum were filled with swallowed air instead of milk.

GAD and GABA in the GAD67 Mutant Cerebral Cortex

As shown in Fig. 1, total GAD activity and GABA content were greatly reduced in E13.5–14.5 and P0.5 (newborn) brains of GAD67−/− mice. In P0.5 GAD67−/− mice, GAD activity was less than 20% and GABA content was 7% of the control. The marked reductions of GABA in GAD67 mutants are in sharp contrast with GAD65-knockout mice, which maintain normal levels of GABA. These results indicate that the active component of GAD in developing mice is GAD67.

The brain of the GAD67−/− newborn appeared histologically normal. If GABA is required during the early stage of neurogenesis [11,12], the small amount of GABA synthesized by GAD65 may be sufficient to allow normal or nearly normal develop-

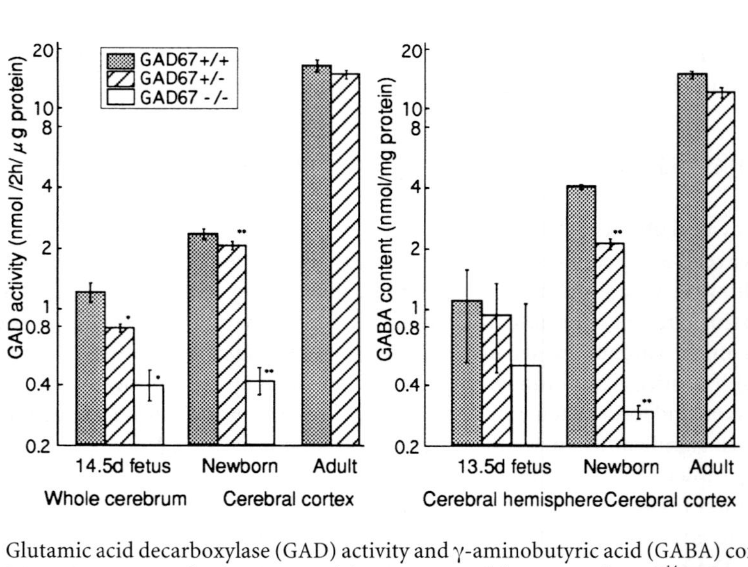

FIG. 1. Glutamic acid decarboxylase (GAD) activity and γ-aminobutyric acid (GABA) content in the GAD67 mutant mouse brain. GAD activity was assayed by conversion of ^{14}C-glutamate to $^{14}CO_2$. GABA was measured by HPLC. Values are mean ± SD. Statistical differences from GAD67+/+ brain are marked with *dots*

ment of the GAD67−/− brain. This hypothesis will be explored using GAD65- and GAD67-double-knockout mice that completely lack GABA.

Neural Functions of GAD67-Deficient Mice

Rescue of the cleft palate is essential for investigation of neural function and behavior of GAD67-deficient mice. Closure of the cleft palate with several materials and pharmacological manipulation during pregnancy have been unsuccessful so far. We are now trying the alteration of genetic background to cleft palate-resistant strain and the production of conditional knockout effective only after birth.

Cleft Palate in GAD67-Deficient Mice

The occurrence of cleft palate in the absence of apparent neural defects in GAD67-knockout mice was unexpected. We could not demonstrate GAD- or GABA-positive cells in the palate shelf, a primordium of the palate, by in situ hybridization or immunohistochemistry, and it is not known at present whether GABA functions directly or indirectly in palate formation.

Previous pharmacological studies have suggested the involvement of GABA in the palate [13]. Cleft palate and neurological disorders are observed in pink-eyed cleft palate (*p*) mutant mice, which are defective in the β3 subunit of $GABA_A$ receptor [14]. A recent paper [15] demonstrated that targeted disruption of the $GABA_A$ receptor β3 subunit gene actually induced cleft palates, required for palate formation. Growth factors, intracellular signal transduction pathways, and intercellular interactions have been investigated in connection with palate and craniofacial development [16]. How GABA is involved in the signaling cascades that function during palatal formation remains to be investigated.

Conclusion

Previous investigations have suggested that GAD65 participates more actively in synaptic function than GAD67 [3–5] and that GABA is intimately involved in the early stage of neurogenesis by regulating proliferation, migration, and process extension of neurons [11,12], The present studies on GAD mutant mice, however, indicate that neural functions are not much impaired by GAD65 deficiency and that substantial reduction of GABA by GAD67 deletion apparently does not affect neurogenesis. Furthermore, it has been disclosed that GABA synthesized by GAD67 is involved in the development of the palate.

Acknowledgment. This work was supported by grants-in-aid 07279107 and 08458263 from the Ministry of Education, Science, Sports and Culture of Japan.

References

1. Obata K (1972) The inhibitory action of γ-aminobutyric acid, a probable synaptic transmitter. Int Rev Neurobiol 15:167–187

2. Roberts E, Chase TN, Tower DB (eds) (1976) GABA in nervous system function. Raven, New York
3. Erlander MG, Tillakaratne NJK, Feldblum S, et al (1991) Two genes encode distinct glutamate decarboxylases. Neuron 7:91–100
4. Dirkx R Jr, Thomas A, Li L, et al (1995) Targeting of the 67-kDa isoform of glutamic acid decarboxylase to intracellular organelles is mediated by its interaction with the NH_2-terminal region of the 65-kDa isoform of glutamic acid decarboxylase. J Biol Chem 270:2241–2246
5. Esclapez M, Tillakaratne NJK, Kaufman DL, et al (1994) Comparative localization of two forms of glutamic acid decarboxylase and their mRNAs in rat brain supports the concept of functional differences between the forms. J Neurosci 14:1934–1855
6. Asada H, Kawamura Y, Maruyama K, et al (1996) Mice lacking the 65 kDa isoform of glutamic acid decarboxylase (GAD65) maintain normal levels of GAD67 and GABA in their brain but are susceptible to seizures. Biochem Biophys Res Commun 229:891–895
7. Asada H, Kawamura Y, Maruyama K, et al (1997) Cleft palate and decreased brain γ-aminobutyric acid in mice lacking the 67-kDa isoform of glutamic acid decarboxylase. Proc Natl Acad Sci USA 94:6496–6499
8. Condie BG, Bain G, Gottlieb DL, et al (1997) Cleft palate in mice with a targeted mutation in the γ-aminobutyric acid-producing enzyme glutamic acid decarboxylase 67. Proc Natl Acad Sci USA 94:11451–11455
9. Kash SF, Johnson RS, Tecott LH, et al (1997) Epilepsy in mice deficient in the 65-kDa isoform of glutamic acid decarboxylase. Proc Natl Acad Sci USA 94:14060–14065
10. Yagi T, Aizawa S, Tokunaga T, et al (1993) A role for Fyn tyrosine kinase in the suckling behaviour of neonatal mice. Nature (Lond) 366:742–745
11. Barker JL, Behar T, Li Y-X, et al (1998) GABAergic cells and signals in CNS development. Perspect Dev Neurobiol 4:1–16
12. LoTurco JJ, Owens DF, Heath MJS, et al (1995) GABA and glutamate depolarize cortical progenitor cells and inhibit DNA synthesis. Neuron 15:1287–1298
13. Wee EL, Norman EJ, Zimmerman EF (1986) Presence of γ-aminobutyric acid in embryonic palates of AJ and SWV mouse strains. J Craniofac Genet Dev Biol 6:53–61
14. Culiat CT, Stubbs L, Montgomery CS, et al (1994) Phenotypic consequences of deletion of the γ_3, α_5, or β_3 subunit of the type A γ-aminobutyric acid receptor in mice. Proc Natl Acad Sci USA 91:2815–2818
15. Homanics GE, DeLorey TM, Firestone LL (1997) Mice devoid of γ-aminobutyrate type A receptor β3 subunit have epilepsy, cleft palate, and hypertensive behavior. Proc Natl Acad Sci USA 94:4143–4148
16. Greene RM (1989) Signal transduction during craniofacial development. Crit Rev Toxicol 20:137–152

Time-Lapse Video Microscopic Analysis of Cell Process Formation from Drebrin A-Expressing Fibroblast

Tomoaki Shirao, Ren Yong, Kensuke Hayashi, and Yuko Sekino

Summary. The cytoskeleton has been suggested to be one of the important endogenous factors that control the elaborate morphology of neuronal processes. In a previous paper, we showed that overexpression of neuronal actin-binding protein, drebrin A, in fibroblasts results in the appearance of thick, curving bundles of actin filaments. In this chapter, we have analyzed, using a time-lapse recording microscopy system, how drebrin-expressing fibroblasts formed highly branching cell processes. Because drebrin is a cytoskeletal protein, drebrin in living cells could not be immunostained. Therefore, L cells were cotransfected with cDNAs of both drebrin A and green fluorescent protein (GFP), and drebrin-expressing cells were identified as GFP-positive cells. Their cell shape was at first thick and round, seldom bearing any filopodia. Then, outgrowth of filopodia-like cell processes suddenly increased, apparently because of accumulation of drebrin A. The cell processes of drebrin-expressing cells were more stable than normal filopodia and seldom retract. Finally, these cell processes changed into long cell processes.

Key words. Drebrin—Actin-binding protein—Cell morphology—Dendrite—Gene transfer

Introduction

Axon and dendrite formation are the most characteristic features of nerve cells. Many proteins that are specifically expressed in association with outgrowth of axons and dendrites have been reported [1]. The cytoskeleton has been suggested to be one of the important endogenous factors that control the formation of neuronal processes [2–7]. Neuronal actin binding protein, drebrin A, has been suggested to be one of the endogenous factors [8–10]. Drebrins were first identified from chicken brain as 95- and 100-kDa proteins on SDS-polyacrylamide gel electrophoresis [11]. It is classified into two types of isoforms, embryonic type (drebrin E), which exists in large amounts in whole-cell bodies of developing neurons, and adult-type (drebrin A), which exists

Department of Neurobiology and Behavior, Gunma University School of Medicine, 3-39-22 Showa-machi, Maebashi, Gunma 371, Japan mainly in dendritic spines in the adult brain [12–14]. Drebrin strongly inhibits the actin binding of tropomyosin and α-actinin [15,16]. Drebrin reduces the sliding velocity of actin filaments on immobilized myosin and inhibits the actin-activated ATPase activity of myosin, suggesting that drebrin may be an actin-linked regulator of actomyosin interaction in neurons [14].

In previous papers [15,17], we have shown that overexpression of drebrin A in fibroblasts results in the appearance of thick, curving bundles of actin filaments by excluding tropomyosin from actin filaments. Additionally, the formation of highly branching cell processes that resemble neuronal dendrites resulted from this overexpression of drebrin A. However, it has not yet been clarified whether the neurite-like cell processes of these transfected fibroblasts are a real outgrowth of cell processes or just retraction cell processes. In this chapter, we have analyzed how drebrin-expressing cells form these cell processes using time-lapse video microscopy, and found that they first formed many filopodia-like extensions from round cell bodies and that these extensions gradually changed into long cell processes.

Materials and Methods

Mouse L cells were maintained in Dulbecco's modified Eagle's medium (DMEM) supplemented with 10% fetal bovine serum (FBS) in an atmosphere of 5% CO_2 in air at 37°C. For time-lapse video recording, L cells were plated onto poly-L-lysine (PLL)-coated coverslips (24 mm in diameter) that were placed between the stainless steel securing ring and the Teflon dish, which had a silicone rubber ring on the bottom. The cells were transfected with two plasmids, MIW-DA [17] and pEGFP-c1 (Clontech, Palo Alto, CA, USA), at a ratio of 10:1. Plasmids MIW-DA contained a full-length cDNA for rat drebrin A and pEGFP-c1 contained a cytomegalovirus (CMV) promoter and green fluorescent protein (GFP) cDNA by the standard calcium phosphate precipitation method. After 18 h, the cells were rinsed with DMEM plus 10% FBS, and the dish was inserted into a heated culture chamber on an inverted microscope (Olympus, Tokyo, Japan). Images were acquired at 5-min intervals by video camera (C2741; Hamamatsu Photonics, Hamamatsu, Japan) with Nomarski differential interference contrast optics, and were recorded on laser videodisc recorder (Sony, Tokyo, Japan). Fields of view were chosen in which GFP-positive cells were identified.

Results and Discussion

Cotransfection of drebrin and GFP cDNAs enabled us to identify drebrin-expressing cells as GFP-positive cells (Fig. 1). The morphological changes of these cells were then observed with an inverted microscope for the following 32 h by Nomarski differential interference contrast optics. Because transfection with GFP cDNA did not affect L-cell morphology (data not shown), the effect of drebrin A on the formation of cell processes could be analyzed in the present experiment.

The GFP-positive cells (see arrows in Fig. 2a) were round without filopodia, although the control GFP-negative cells were flat and polygonal with a few filopodia. Control cells had the highest rate of filopodia formation at the beginning, and the activity had gradually decreased. In contrast, filopodia formation in transfected cells

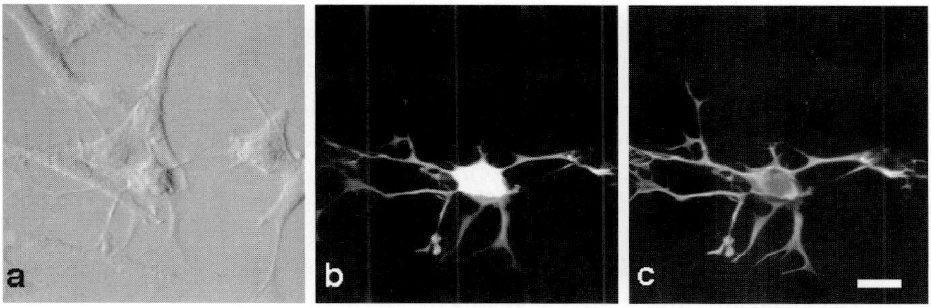

FIG. 1a–c. Green fluorescent protein (GFP) and drebrin cDNA were cotransfected into L cells. After 48 h, they were immunostained with antidrebrin monoclonal antibody as a first antibody and with rhodamine-labeled goat antimouse IgG as a second antibody. **a**, **b**, and **c** are Nomarski, GFP, and rhodamine images, respectively. Note that GFP-expressing cells were concurrently expressing drebrin. *Bar* 20 μm

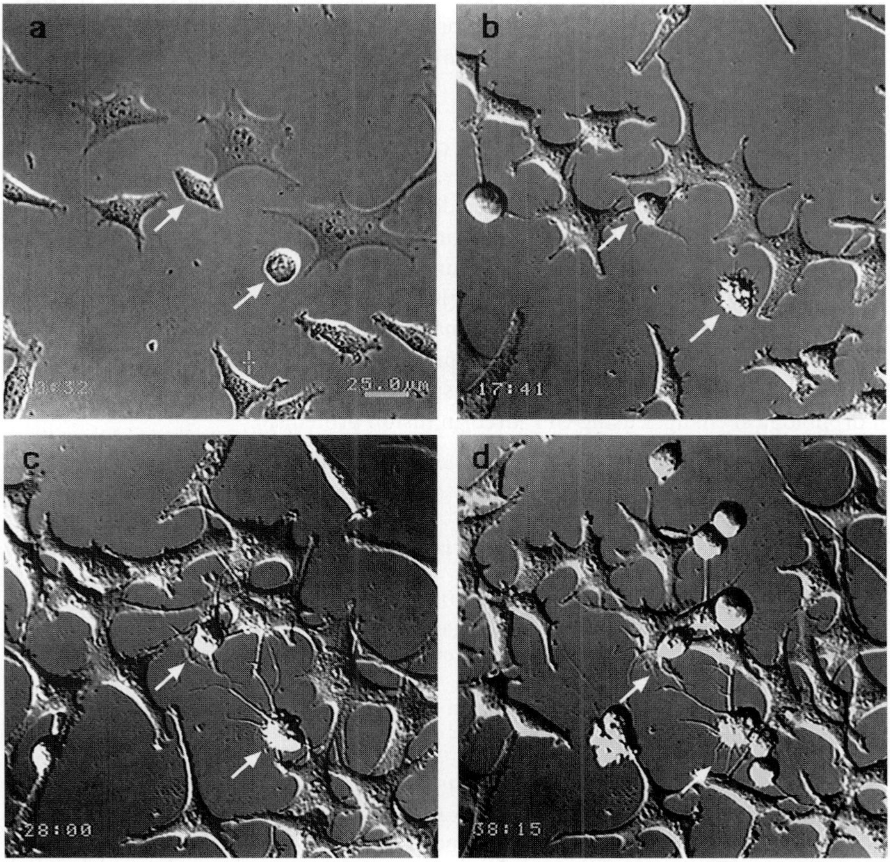

FIG. 2a–d. Time-lapse observation of GFP and drebrin coexpressing cells (*arrows*). Note that these cells have thick round cell bodies and there is a gradual outgrowth of many long cell processes. Incubation times are indicated at the *left-bottom corner* of each panel: **a**, 32 min; **b**, 17 h, 41 min; **c**, 28 h, **d**, 38 h, 15 min, after start of time-lapse recording. *Bar* 25 μm

was absent at the beginning of the experiment. At 17 h the transfected cells suddenly increased formation of filopodia-like cell processes to attain a similar level of activity as control cells, although these processes were stable and seldom retracted (Fig. 2b) and became long thick processes (Fig. 2c).

We have already reported that actin filaments bound to drebrin are stable against cytochalasin treatment [18] and that drebrin-expressing cells have strong adhesiveness to substrata [19,20]. It will be of interest to investigate whether the strong cytoskeletal stability within these filopodia or the strong adhesiveness to substrata may inhibit the retraction of their filopodia. At 38 h, the length of cell processes of GFP-positive cells were similar to that at 28 h, but they seemed to have more complex branching than those at 28 h (Fig. 2d). This indicates the possibility that cell processes from drebrin-expressing cells have high branching activity, which is comparable to that of in vivo developing dendrites. However, occasionally, two independent filopodia from drebrin-expressing cells juxtaposed on each other and became impossible to be discriminated as two independent processes by light microscopy. Therefore, it was not clarified whether a single process actually bifurcated or one of the two adjacent processes that had been recognized as a single cell process changed its growing direction.

The results clearly demonstrate that dendrite-like cell processes of drebrin-expressing fibroblasts are not retraction cell processes but a real outgrowth of cell processes from cell bodies. Original observations showed the drebrin-expressing cells to be thick and of a round cell shape, seldom bearing any filopodia. We have previously reported that a small amount of drebrin-expression results in the disappearance of stress fibers [15,17]. Therefore, the round cell shape and low activity of filopodia formation may be related to the disorganization of actin filaments in the fibroblast. The sudden increase of cell process formation of drebrin-expressing cells suggests that a critical concentration of drebrin expression is necessary to form cell processes. It will be of future interest to monitor subcellular distribution of drebrin during morphological changes using GFP-drebrin fusion proteins.

References

1. Skene JH (1989) Axonal growth-associated proteins. Annu Rev Neurosci 12:127–156
2. Baas PW, Pienkowski TP, Kosik KS (1991) Processes induced by tau expression in Sf9 cells have an axon-like microtubule organization. J Cell Biol 115:1333–1344
3. Solomon F (1981) Specification of cell morphology by endogenous determinants. J Cell Biol 90:547–553
4. Hirokawa N (1982) Cross-linker system between neurofilaments, microtubules, and membranous organelles in frog axons revealed by the quick-freeze, deep-etching method. J Cell Biol 94:129–142
5. Drubin DG, Feinstein SC, Shooter EM, et al (1985) Nerve growth factor-induced neurite outgrowth in PC12 cells involves the coordinate induction of microtubule assembly and assembly-promoting factors. J Cell Biol 101:1799–1807
6. Letourneau PC, Ressler AH (1984) Inhibition of neurite initiation and growth by taxol. J Cell Biol 98:1355–1362
7. Drubin D, Kobayashi S, Kellogg D, et al (1988) Regulation of microtubule protein levels during cellular morphogenesis in nerve growth factor-treated PC12 cells. J Cell Biol 106:1583–1591

8. Shirao T, Kojima N, Nabeta Y, et al (1989) Two forms of drebrins, developmentally regulated brain proteins, in rat. Proc Jpn Acad 65:169–172
9. Shirao T (1995) The roles of microfilament-associated proteins, drebrins, in brain morphogenesis: a review. J Biochem (Tokyo) 117:231–236
10. Sasaki Y, Hayashi K, Shirao T, et al (1996) Inhibition by drebrin of the actin-bundling activity of brain fascin, a protein localized infilopodia of growth cones. J Neurochem 66:980–988
11. Shirao T, Obata K (1985) Two acidic proteins associated with brain development in chick embryo. J Neurochem 44:1210–1216
12. Shirao T, Obata K (1986) Immunochemical homology of 3 developmentally regulated brain proteins and their developmental change in neuronal distribution. Brain Res 394:233–244
13. Shirao T, Inoue HK, Kano Y, et al (1987) Localization of a developmentally regulated neuron-specific protein S54 in dendrites as revealed by immunoelectron microscopy. Brain Res 413:374–378
14. Hayashi K, Ishikawa R, Ye L-H, et al (1996) Modulatory role of drebrin on the cytoskeleton within dendritic spines in the rat cerebral cortex. J Neurosci 16:7161–7170
15. Shirao T, Hayashi K, Ishikawa R, et al (1994) Formation of thick curving bundles of actin by drebrin A expressed in fibroblasts. Exp Cell Res 215:145–153
16. Ishikawa R, Hayashi K, Shirao T, et al (1994) Drebrin, a development-associated brain protein from rat embryo, causes the dissociation of tropomyosin from actin filaments. J Biol Chem 269:29928–29933
17. Shirao T, Kojima N, Obata K (1992) Cloning of drebrin A and induction of neurite-like processes in drebrin-transfected cells. Neuroreport 3:109–112
18. Asada H, Uyemura K, Shirao T (1994) Actin-binding protein, drebrin accumulates in submembranous regions in parallel with neuronal differentiation. J Neurosci Res 38:149–159
19. Ikeda K, Shirao T, Toda M, et al (1995) Effect of a neuron-specific actin-binding protein, drebrin A, on cell-substratum adhesion. Neurosci Lett 194:197–200
20. Ikeda K, Kaub P, Asada H, et al (1996) Stabilization of adhesion plaques by the expression of drebrin A in fibroblasts. Dev Brain Res 91:227–236

Development of Neurites, Dendritic Filopodia, and Synaptic Spines in Hippocampal Low-Density Culture

TAKASHI NAGANO[1], YURIKO IWAKURA[1], TATSUO USHIKI[2], and HIROYUKI NAWA[1]

Summary. Low-density cultures of hippocampal neurons are useful for studies of synaptic development. The system of culture with a glial cup we have developed can be maintained for more than a month. Dendritic spine structures were identified after 2 weeks of culture, whereas presynaptic vesicle accumulation was apparent 1 week after plating. The precedence of presynaptic maturation may reflect a process that presynaptic factors induce spine development. However, some neurons began to extend many dendritic filopodia, irrespective of presynaptic contacts, about a week after plating. The filopodial processes that show no presynaptic contacts are suggested to differentiate into spines and therefore are also called free spines. Most of these neurons displayed the dendritic morphology typical of intrinsic GABAergic cells. Previous studies on dendritic spines have shown that morphological spine development in hippocampal pyramidal cells requires presynaptic influences, while that in cerebellar Purkinje cells does not. Taken together, these results suggest that morphological development of dendritic spines in some neuronal types requires presynaptic components and that in the other neuronal types does not.

Key words. Dendritic spine—Dendritic filopodia—GABA—Hippocampal neuron—Plasticity

Introduction

Dr. Ramon y Cajal first described the postsynaptic structure called the "dendritic spine" more than 100 years ago. This structure is commonly observed at postsynaptic sites that receive excitatory inputs in the brain. Dendritic spines generally include the structure called the post-synaptic density (PSD), which consists of numerous transmitter receptors and protein kinases. These spines change their shapes, as well as their molecular organization, in response to various stimuli including hormones and learn-

[1] Department of Molecular Neurobiology, Brain Research Institute, Niigata University, 1-757 Asahimachi-dori, Niigata 951-8585, Japan
[2] Department of Anatomy III, Niigata University School of Medicine, 1-757 Asahimachi-dori, Niigata 951-8510, Japan

ing experiences. Woolley and McEwen [1] reported that 30% of the dendritic spines on cornu ammonis (CA) 1 pyramidal cells were gained and lost during the estrous cycle. The spines and PSDs enlarge after tetanic stimulation that induces long-term potentiation [2,3], although these observations are controversial [4]. In contrast, mentally retarded patients tend to show poor spine development and to retain many thin spines [5,6]. All these findings indicate that there is a significant correlation between synaptic plasticity and spine structures, but the precise mechanism of spine development is poorly understood. In this study, we attempted to explore the mechanism using a low-density hippocampal culture system.

Materials and Methods

Primary cultures of hippocampal neurons from embryonic Sprague-Dawley rats were prepared as described by Goslin and Banker [7]. Briefly, whole hippocampi of embryonic day 18 rats were dissociated enzymatically and cells were plated at a density of 1–2 $\times$ 10^2 cells/mm^2 on poly-D-lysine-coated dishes. First, the dissociated cells were cultured for 24 h in Dulbecco's modified Eagle's medium containing 1 mM purified glutamine (Ajinomoto, Tokyo, Japan), 2% fetal bovine serum, nutrient mixture N_2 (100 μg/ml transferrin, 5 μg/ml bovine insulin, 100 nM putrescine, 30 nM sodium selenite, and 20 nM progesterone), 10 mM HEPES, 100 U/ml penicillin, and 100 μg/ml streptomycin [8]. Then, each culture was grown in glia-conditioned N_2 medium and treated with 5 μM cytosine arabinoside for a few days to prevent nonneuronal cell

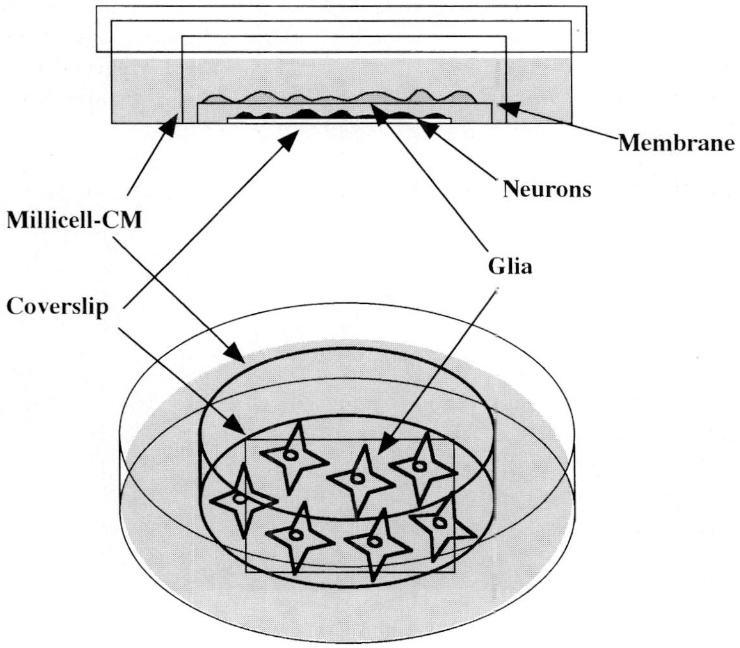

FIG. 1. A novel low-density neuronal culture system using a Millicell-CM cup with glial cells

proliferation. Astrogilal cells from forebrains of neonatal rats were cultured separately in this medium containing 5% rat serum. Glial cells at the third passage were plated on Millicell-CM cups (Millipore, Bedford, MA, USA) (Fig. 1), cultured in serum-free N_2 medium, and then transferred to the neuronal cultures just described. Ultrastructural analysis was done with the conventional procedure described previously [9].

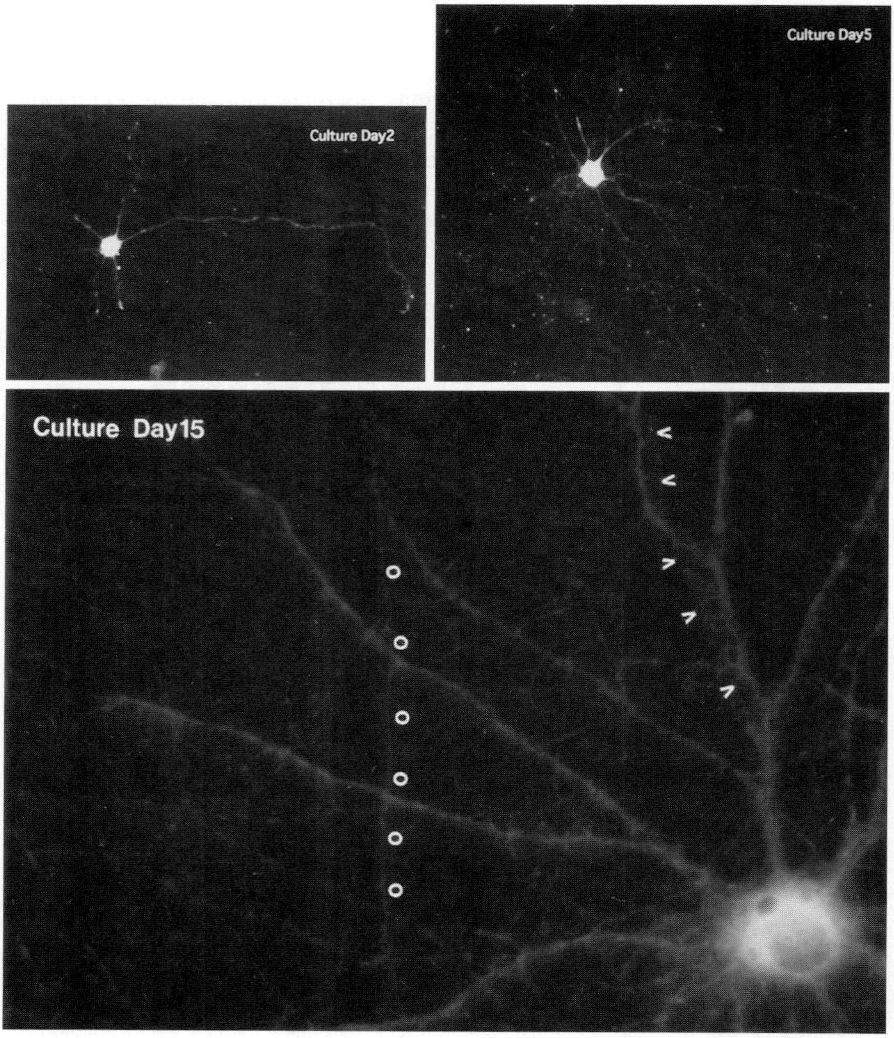

FIG. 2. Typical neuronal growth visualized with DiO at days 2, 5 and 15 of culture, respectively. An axon running along a dendrite is marked with *arrowheads* and one crossing dendrites with *circles*

Results

New Method for Low-Density Culture of Hippocampal Neurons

In the low-density culture system developed previously [7], the neurons grow facing the glial cells on a cover glass. After several weeks, the neurons, particularly those at the central area of the glass, tend to degenerate presumably because the glass limits their oxygen supply. To overcome this problem, we used an air-permeable transparent membrane filter on which glial cells can be maintained. Now, culture devices with such filters on their bottoms are commercially available, e.g., the Millicell system (Millipore) (Fig.1) that we used. This modification has improved the reproducibility and stability of the original low-density culture system.

Classification of Cultured Hippocampal Neurons

We monitored neuronal growth by staining cells with a hydrophobic dye, DiO. Extensive growth of dendrites and axons was seen 2 days after plating (Fig. 2), when these two processes were distinguishable. Dendritic growth had virtually ceased by around day 10 and dendritic spines appeared around day 15. Neurons in hippocampal culture exhibited various types of cell morphology: pyramidal, bipolar, and radial multipolar (Fig. 3). The multipolar cells could be subclassified into two groups: cells with numerous straight and thin dendrites elongating radially (radial cells) and others with giant triangular soma with thick and curved dendrites (giant cells). The former group

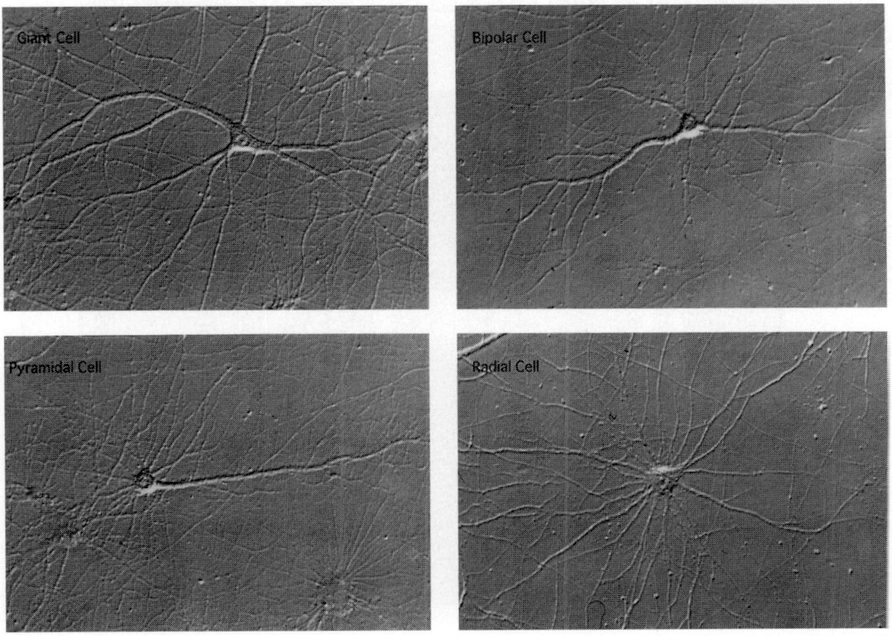

Fig. 3. Various morphological types of neurons found in culture

presumably represented CA pyramidal neurons, whereas the other appeared to be GABAergic interneurons as they were GABA immunoreactive (data not shown).

Axonal Connectivity in Low-Density Culture

We found two types of axonal arbors and synaptic formation in culture: some axons ran along dendritic processes and formed many synapses with a single postsynaptic

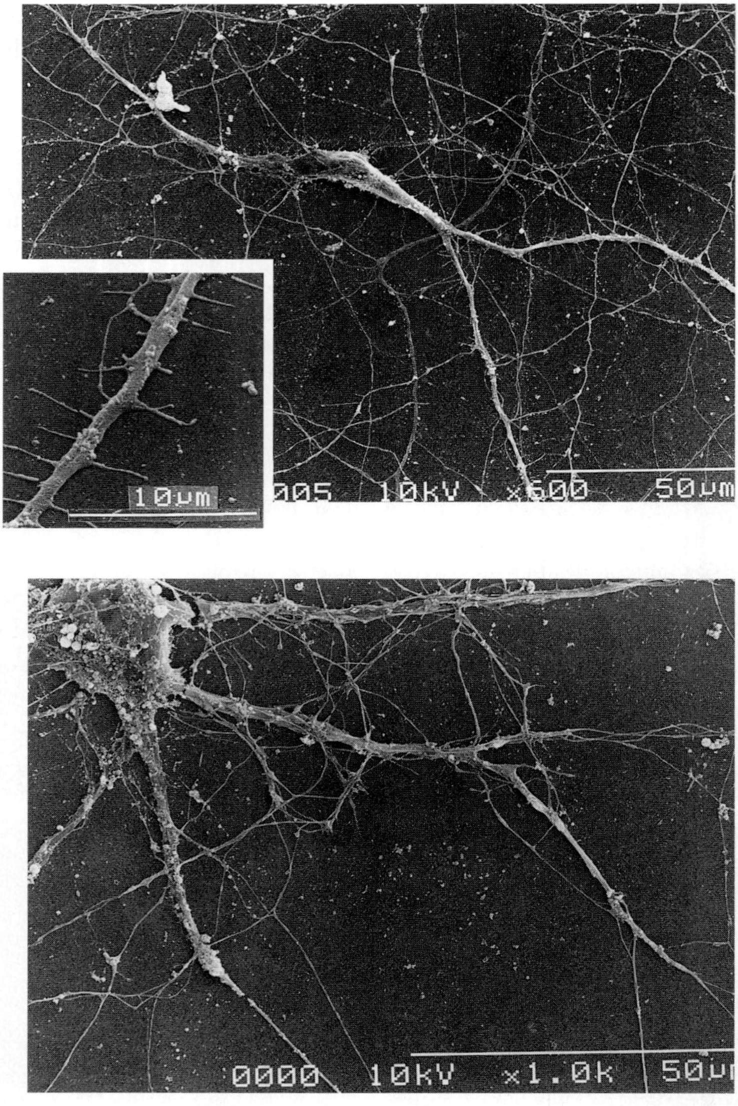

Fig. 4. Dendritic filopodia or free spines were often associated with bipolar cells (*upper panel*) but less often with radial cells (*lower panel*)

neuron, and others tended to cross the dendrites and formed synapses with several separate neurons (Fig. 2, day 15).

Association of Dendritic Filopodia with GABAergic Neurons

The ultrastructures of the dendrites in culture were characterized at day 15 by scanning electron microscopy. Typically, dendritic filopodia were often associated with bipolar or giant cells, as shown in Fig. 4 (upper panel). In contrast, radial cells rarely have such dendritic processes (Fig. 4, lower panel). In some cases, axons running on the dendrite extended branches that resembled dendritic filopodia. Thus, care must be taken to distinguish these structures.

Discussion

The precise mechanisms underlying synaptic differentiation and maturation in the brain are poorly understood, although such mechanisms in the neuromuscular junction have been studied extensively. Molecular and morphological specialization of neuromuscular synapses requires various diffusible molecules, including neurotransmitters and growth factors, that are produced and released by both pre- and postsynaptic cells [10]. However, we know little about the signals regulating synaptic development in the central nervous system (CNS). The complexity of the cell types and methodological difficulties have prevented neuroscientists exploring individual processes involved in synaptic differentiation and maturation in the CNS. Culture techniques can provide a useful approach for studying the precise processes during synaptogenesis of brain neurons.

Dendritic spine formation in the brain often follows extensive elongation of filopodial processes from the dendrites. This observation led to the hypothesis that the dendritic filopodia (free spines) can change into spine structures in response to maturation signals. Alternatively, the filopodial structure may be a bud of the dendritic branch that is transformed into the dendritic shaft. Results of recent microscopic studies on dendritic dynamics support the former hypothesis. Dailey and Smith [11] found that the dendritic filopodia were most dynamic and abundant in the hippocampus during early developmental stages and that they disappeared later in development as more spines grew. Moreover, the filopodia could make contact with nearby axons and sometimes became stabilized as the postsynaptic structure developed [12].

The hippocampus has relatively simple anatomical organization and fewer cell types than other brain regions. The low-density hippocampal culture system is therefore popularly used to investigate synaptic development [7]. Brain neurons in culture appear to form mature synapses that accumulate vesicles at their presynaptic sites and show neurotransmitter receptor clustering at their postsynaptic sites [13,14]. In previous reports, however, the cell types of cultured hippocampal neurons were rarely distinguished. Generally, our cultures contained four types of cell, which we classified according to their dendritic morphology and cell polarity as bipolar cells, giant cells, radial cells, and pyramidal cells. The former two populations were GABAergic neurons as revealed by GABA immunostaining. Both types of neurons had a few thick dendritic processes and elaborated numerous dendritic filopodia. As these cells showed good survival and maintained their viability even in old cultures, we cannot deny the possibility that their dendritic filopodia may have resulted from presynaptic neuronal loss and subsequent presynaptic degeneration.

The association of dendritic filopodia with GABAergic neurons in culture is consistent with the in vivo observation that GABAergic Purkinje cells of reeler mutant mice have free spines containing PSD without synaptic contacts as the result of granule cell loss. However, Gaiarsa et al. [15] reported a different possibility for spine formation on CA3 pyramidal neurons: degeneration of mossy fibers resulted in smooth aspiny dendritic regions on the neurons, suggesting presynaptic influences on the spine formation. Our observations indicated that Smith's hypothesis (the dendritic filopodia can change into spine structures in response to maturation signals) [11,12] appears to be applicable to GABAergic neurons but may not apply to other cell types. Further study to characterize the types of neurons possessing the dendritic filopodia or free spines is required, as multiple cell types were observed in the present study.

With respect to neural connectivity in culture, some axons tended to form synapses with several separate neurons by crossing their dendrites. Formation of such neural networks appears to occur in vivo: in the hippocampal CA1 region, Schaffer collateral fibers tended to form synapses with many dendrites [16]. Therefore, it would be significant to identify the origins of the afferent fibers in culture and elucidate factors that regulate such patterns of synapse formation. Our low-density neuronal culture system appears to be a powerful tool for analyzing synaptic development. However, this system has limitations because all the cell types and networks formed in culture are complex and need to be identified. Therefore, its use in conjunction with slice cultures and in vivo manipulation techniques seems essential to enable drawing definitive conclusions about the cell-specific synaptic development in the CNS.

This work was supported by the Japan Society for the Promotion of Science (RFTF-96L00203).

References

1. Woolley CS, McEwen BS (1990) Estradiol regulates synapse density in the CA1 region of the hippocampus in the adult female rat. Soc Neurosci Abstr 16:328
2. Desmond NL, Levy WB (1986) Changes in the postsynaptic density with long-term potentiation in the dentate gyrus. J Comp Neurol 253:476–482
3. Fifkova E, Van Harreveld A (1977) Long-lasting morphological changes in dendritic spines of dentate granular cells following stimulation of the entorhinal area. J Neurocytol 6:211–230
4. Lee KS, Schottler F, Oliver M, et al (1980) Brief bursts of high-frequency stimulation produce two steps of structural change in rat hippocampus. J Neurophysiol (Bethesda) 44:247–258
5. Marin-Padilla M (1972) Structural abnormalities of the cerebral cortex in human chromosomal aberrations. a Golgi study. Brain Res 44:625–629
6. Purpura DP (1974) Dendritic spine "dysgenesis" and mental retardation. Science 186:1126–1128
7. Goslin K, Banker G (1991) Rat hippocampal neurons in low-density culture. In: Goslin K, Banker G (eds) Culturing nerve cells. MIT Press, Cambridge, pp 251–281
8. Nawa H, Bessho Y, Carnahan J, et al (1993) Regulation of neuropeptide expression in cultured cerebral cortical neurons by brain-derived neurotrophic factor. J Neurochem 60:772–775

9. Ushiki T, Takeda M (1997) Three-dimensional ultrastructure of the perivascular space in the rat thymus. Arch Histol Cytol 60:89–99
10. Hall ZW, Sanes JR (1993) Synaptic structure and development: the neuromuscular junction. Cell 72 (suppl):99–121
11. Dailey ME, Smith SJ (1996) The dynamics of dendritic structure in developing hippocampal slices. J Neurosci 16:2983–2994
12. Ziv NE, Smith SJ (1996) Evidence for a role of dendritic filopodia in synaptogenesis and spine formation. Neuron 17:91–102
13. Craig AM, Blackstone CD, Huganir RL, et al (1993) The distribution of glutamate receptors in cultured rat hippocampal neurons: postsynaptic clustering of AMPA-selective subunits. Neuron 10:1055–1068
14. Fletcher TL, De Camilli P, Banker G (1994) Synaptogenesis in hippocampal cultures: evidence indicating that axons and dendrites become competent to form synapses at different stages of neuronal development. J Neurosci 14:6695–6706
15. Gaiarsa JL, Beaudoin M, Ben-Ari Y (1992) Effect of neuronal degeneration on the morphological development of rat CA3 pyramidal neurons: inductive role of mossy fibers on the formation of thorny excrescences. J Comp Neurol 321:612–625
16. Bayer SA (1985) Hippocampal region. In: Paxinos G (ed) The rat nervous system. Academic, Sydney, pp 335–352

Electrophysiological Evidence That $G\alpha_q$ Is Required for Climbing Fiber Synapse Elimination During Postnatal Cerebellar Development

KOUICHI HASHIMOTO[1,2], STEFAN OFFERMANNS[3], MELVIN I. SIMON[4], and MASANOBU KANO[1,5]

Summary. Each cerebellar Purkinje cell (PC) in adult mice is supplied with strong excitatory synapses from a single climbing fiber (CF). This one-to-one relationship is preceded, at early developmental stages, by a multiple innervation of PCs by CFs. Elimination of supernumerary CF synapses occurs during the first 3 postnatal weeks. We report that mutant mice lacking the α-subunit of a GTP-binding protein G_q ($G\alpha_q$), have persistent multiple CF innervation. The $G\alpha_q$ mutant mice are viable but ataxic, and exhibit clear sign of motor discoordination. The morphology of the cerebellum and the morphology and density of parallel fibers (PFs) to PC synapses appear normal. Excitatory synaptic transmission from CF to PC and that from PF to PC is functional. However, similar to the previously reported two strains of mutant mice lacking the type 1 metabotropic glutamate receptor (mGluR1) and the γ-isoform of protein kinase C (PKCγ), about 40% of adult $G\alpha_q$ mutant PCs remain multiply innervated by CFs. This abnormality results from a defect in regression of supernumerary CFs in the third postnatal week. These results strongly suggest that $G\alpha_q$ mediates signals from mGluR1 to PKCγ that are involved in CF synapse elimination during the third postnatal week.

Key words. Cerebellar Purkinje cell—Synapse elimination—Climbing fiber—G protein—Mouse

[1] Laboratory for Cellular Neurophysiology, Brain Science Institute, RIKEN, Wako, Saitama 351-0198, Japan
[2] Department of Physiology, Jichi Medical School, Minamikawachi-machi, Tochigi 329-0498, Japan
[3] Institut für Pharmakologie, Freie Universität Berlin, Thielallee 69-73, 14195 Berlin, Germany
[4] Division of Biology 147-75, California Institute of Technology, Pasadena, CA 91125, USA
[5] President address: Department of Physiology, Kanazawa University School of Medicine, 13-1 Takaramachi, Kanazawa 920-8640, Japan

Introduction

The climbing fiber (CF) system of the cerebellum has been a good model to study how mature and functional synaptic connections are formed during postnatal development [1]. In newborn animals, cerebellar Purkinje cells (PCs) are innervated by multiple CFs. Elimination of supernumerary CFs occurs during postnatal development, and the mature state is attained around postnatal day 20, at which time most PCs are innervated by single CFs. Previous studies have demonstrated that multiple CF innervation persists in the cerebella of adult animals that are devoid of parallel fibers (PFs) [1], the other major excitatory inputs to PCs. These results suggest that the developmental change from multiple- to mono-CF innervation is dependent on the formation of PF-to-PC synapses. However, molecular mechanisms of this developmental change have been largely unknown. We have recently reported that mutant mice lacking the type 1 metabotropic glutamate receptor (mGluR1) and the γ-isoform of protein kinase C (PKCγ) are impaired in CF synapse elimination, although their PF–PC synapses are functional and morphologically normal [2,3]. mGluR1 is known to couple to two similar members of the heterotrimeric G_q family, $G\alpha_q$ and $G\alpha_{11}$. To explore which of the two G_q members is involved in CF synapse elimination, we examined the cerebellum of mice deficient in $G\alpha_q$. We found that the $G\alpha_q$ knockout mice were viable and ataxic and had persistent multiple CF innervation to adulthood [4]. Our results strongly suggest that $G\alpha_q$ mediates signals from mGluR1 to PKCγ that are involved in the elimination of multiple CF innervation.

Results and Discussion

$G\alpha_q$ Mutant Mice Are Viable but Ataxic

Homozygous mutant mice showed increased mortality during the first postnatal day, presumably from internal bleeding during birth trauma that results from a defect in platelet aggregation [5]. Of homozygous mutant newborns, 60%–70% survived the postnatal period and developed clear signs of motor coordination deficit that became obvious about 3 weeks after birth. The animals displayed awkward and jerky movements and loss of balance during rearing, and occasionally fell on their side [4].

No obvious morphological defects could be observed on examination of the peripheral and central nervous system of $G\alpha_q$ homozygous mutant mice. Extensive morphological examination of the cerebellar cortex by histological, immunohistochemical, and electron microscopic techniques indicated that deletion of the $G\alpha_q$ gene did not affect gross development of the cerebellar anatomy, cell production, cytodifferentiation, and formation of PF–PC synapses [4].

$G\alpha_q$ Mutant PCs Are Multiply Innervated by CFs

In sagittal cerebellar slices prepared from wild-type mice or age-matched $G\alpha_q$ mutant animals (P22–P57), single PCs were recorded using patch-clamp techniques in the whole-cell configuration [6], and CFs were stimulated in the granule cell layer. As the stimulus intensity was gradually increased (pulse width, 0.1 ms; strength, 0–100 V) a large excitatory postsynaptic current (EPSC) was elicited in an all-or-none fashion in

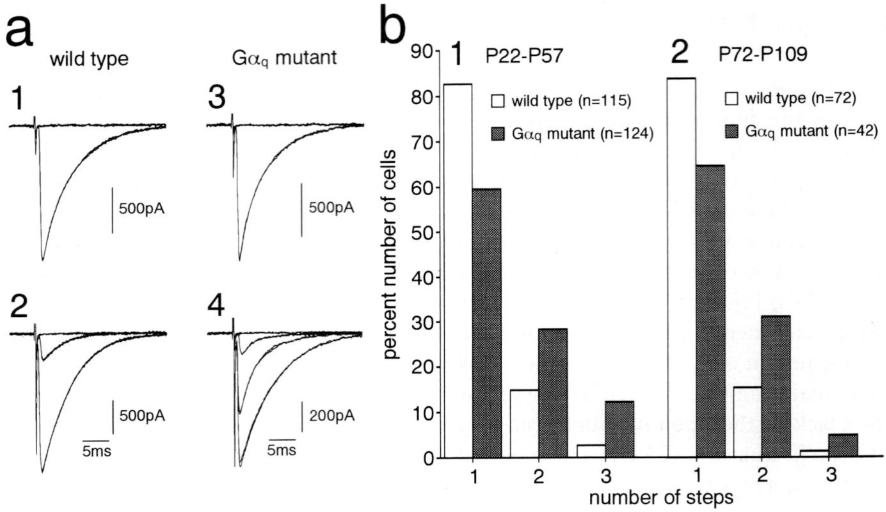

FIG. 1. **a** Excitatory postsynaptic currents (EPSCs) elicited by climbing fiber (CF) stimulation in a wild-type (*1,2*) and a Gα_q mutant (*3,4*) Purkinje cells (PCs). One or two traces are superimposed at each threshold intensity. Stimuli were applied at 0.2 Hz. Holding potentials were −10 mV for *1–3* and 0 mV for *4*. **b** Summary histograms show the number of discrete steps of CF-EPSCs of the wild type (*open columns*) and mutant (*hatched columns*) PCs. Data were obtained from mice at P22–P57 (*1*) and at P72–P109 (*2*), respectively. All cells were studied blind to the mouse genotype. (Modified from [4])

the majority of wild-type PCs (Fig. 1a, 1). In other wild-type PCs, CF-EPSCs showed two or three discrete steps (Fig. 1a, 2). The number of discrete steps of CF-EPSCs represents the number of CFs innervating the recorded PC [1–3]. Thus as summarized in the frequency distribution histogram (Fig. 1b, left), 82.6%, 14.8%, and 2.6% of wild-type PCs recorded in the present study (P22–P57) are innervated by one, two, and three CFs, respectively. On the other hand, in Gα_q mutant PCs, 59.6% had single CF-EPSC steps (Fig. 1a, 3), whereas 28.2% and 12.1% had two and three discrete steps, respectively (Fig. 1a, 4).

The difference in CF innervation between the wild-type and Gα_q mutant mice during P22–P57 could result from the retardation of general cerebellar maturation in the Gα_q mutant mice. To assess this possibility, we examined CF-EPSCs in older mice (P72–P109) (Fig. 1b, right). The differences between the frequency distributions of PCs during P22–P59 and those during P70–P109 were not significant for either the wild-type or the mutant mice ($P > 0.05$, Chi-square test). We did not find any abnormality in the kinetics, pharmacology, and short-term synaptic plasticity of Gα_q mutant CF-EPSCs, suggesting that CF synapses are functional in Gα_q mutant mice.

Next, we followed the developmental course of CF innervation. During the first postnatal week (P1–P7), the majority of PCs were multiply innervated by CFs in both wild-type and mutant mice (Fig. 2a). However, the frequency distribution of the number of CF-EPSC steps per PC in the mutant mice was not significantly different from that in the wild-type mice ($P > 0.05$, Chi-square test). During the second

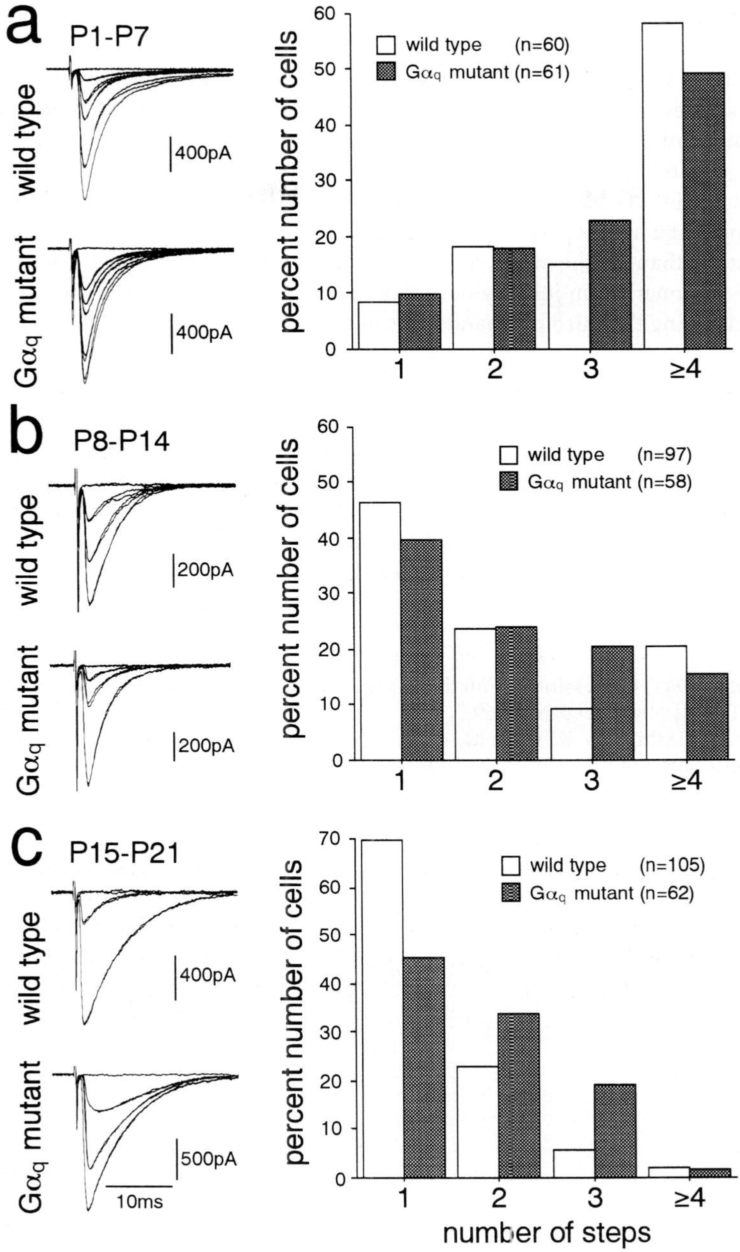

FIG. 2. **a** *Left*: CF-EPSCs of the wild type (P7; holding potential, −30 mV) and Gα_q mutant (P6; holding potential, −60 mV) PCs. *Right*: Summary graph shows numbers of discrete steps of CF-EPSCs of the wild-type (*open columns*) and mutant (*hatched columns*) PCs from mice at P1–P7. **b** Examples of CF-EPSCs of the wild-type (P10; holding potential, −20 mV) and Gα_q mutant (P10; holding potential, −20 mV) PCs and summary graph of the data from mice at P8–P14. **c** Examples of CF-EPSCs of the wild-type (P21; holding potential, −10 mV) and mutant (P18; holding potential, −10 mV) PCs and summary graph of data from mice at P15–P21. (Modified from [4])

postnatal week (P8–P14), the percentage of PCs with multiple CF-EPSC steps decreased markedly in both wild-type and mutant mice (Fig. 2b). The distributions were not significantly different between the two strains ($P > 0.05$, Chi-square test). In contrast, the frequency distributions between the two strains during the third postnatal week (P15–P21) were markedly different (Fig. 2c) ($P < 0.01$, Chi-square test), indicating that a higher percentage of the PCs were multiply innervated by CFs in the $G\alpha_q$ mutant than in the wild-type mice. These results suggest that regression of multiple CF innervation initially occurs normally but that the process is specifically impaired during the third postnatal week in the Gq mutant mice.

Acknowledgments. We thank T. Kitamura, C. Wakabayashi, S. Pease, V. Mancino, Y.-H. Hu, and J. Silva for expert technical help. S.O. was a recipient of a fellowship from the Deutsche Forschungsgemeinschaft and the Guenther Foundation. This work has been supported by grants to M.K. from the Japanese Ministry of Education, Science, Sports and Culture, and by CREST (Core Research for Evolutional Science and Technology) of Japan Science and Technology Corporation (JST), and by a grant from the National Institutes of Health to M.I.S.

References

1. Crépel F (1982) Regression of functional synapses in the immature mammalian cerebellum. Trends Neurosci 5:266–269
2. Kano M, Hashimoto K, Chen C, et al (1995) Impaired synapse elimination during cerebellar development in PKCγ mutant mice. Cell 83:1223–1232
3. Kano M, Hashimoto K, Kurihara H, et al (1997) Persistent multiple climbing fiber innervation of cerebellar Purkinje cells in mice lacking mGluR1. Neuron 18:71–79
4. Offermanns S, Hashimoto K, Watanabe M, et al (1997) Impaired motor coordination and persistent multiple climbing fiber innervation of cerebellar Purkinje cells in mice lacking $G\alpha_q$. Proc Natl Acad Sci USA 94:14089–14094
5. Offermanns S, Toombs CF, Hu Y-H, et al (1997) Defective platelet activation in $G\alpha_q$ deficient mice. Nature (Lond) 389:183–186
6. Edwards FA, Konnerth A, Sakmann B, et al (1989) A thin slice preparation for patchclamp recordings from neurons of the mammalian central nervous system. Pflügers Archiv 414:600–612

Visualization of Dendritic Spine Formation with GFP-Fusion Proteins

Kensuke Hayashi and Tomoaki Shirao

Summary. We have previously reported that an actin-binding protein, drebrin, accumulates at spines and modulates actin–myosin interaction within spines. Drebrin may be one of the regulators of actin filament reorganization during the shape change of spines. We constructed chimeric cDNAs carrying a green fluorescent protein (GFP) cDNA and various fragments of drebrin cDNA and introduced them into CHO cells to test their actin-binding activity. We found that GFP-drebrin bound to actin filaments in CHO cells and caused remodeling of actin filaments and cell shape. A central 28-amino-acid sequence was identified as the core for actin binding. Next, we introduced the GFP-drebrin fusion cDNA into primary cultured cortical neurons. Three weeks after plating, spines branching off the neurons were visible under a fluorescence microscope. The fusion protein was localized at the head of spines. Furthermore, spines of neurons transfected with a GFP-drebrin cDNA were longer than those of GFP-transfected neurons. These results support our hypothesis that drebrin regulates the morphology of spines.

Key words. GFP—Drebrin—Synapse formation—Spine—Primary culture

Introduction

Spines change their shape during neuronal development, as well as in response to environmental cues [1]. Actin filaments are likely to play an important role in the shape change of spines [2]. We have reported that an actin-binding protein, drebrin, accumulates at spines and modulates actin–myosin interaction within spines [3,4]. We have hypothesized that drebrin is one of the regulators of actin filament reorganization during formation and shape change of spines. The purpose of this study was to confirm our hypothesis by overexpressing drebrin tagged with green fluorescent protein (GFP) in neurons.

First, we constructed chimeric cDNAs carrying GFP cDNA and various fragments of drebrin cDNA. To test the ability of these constructs to bind to actin filaments, we

Department of Neurobiology and Behavior, Gunma University School of Medicine, 3-39-22 Showamachi, Maebashi, Gunma 371, Japan introduced them into CHO cells and analyzed the association of the fusion proteins with actin filaments in cells. Next, we introduced the GFP-drebrin fusion cDNA into primary cultured cortical neurons to verify the ability of drebrin to localize at spines and the involvement of drebrin in mechanisms of spine morphogenesis.

Materials and Methods

cDNA Constructs

EGFP-C1 vector (Clontech, Palo Alto, CA, USA) was used to construct GFP-drebrin fusion cDNAs.

CHO Cells

Transfection was performed with Tfx-20 (Promega, Madison, WI, USA). Four days after transfection, cells were fixed and stained with rhodamine-phalloidin to visualize actin filaments.

Primary Culture of Cortical Neurons

The cerebral cortices of ED20 rat embryos were dissociated with papain and plated on polyethylenimine-coated cover glasses. Cells were cultured in MEM, 6 g/l glucose, 1 mM pyruvate, 5% horse serum, and 5% fetal bovine serum. On day 5, AraC was added to reduce proliferation of glial cells. On days 5–7, cells were transfected with cDNAs and Transfectum (Biosepra, Marlborough, MA, USA). Three weeks after plating, cells were fixed and observed.

Confocal Microscopy and Measurements of Spine Length

Confocal images of 0.8-μm thickness were projected onto one plane. Spine length was measured on the projected images with NIH Image software, and the simple average of 25–200 measurements was taken from each neuron and ascribed to that neuron as its spine length. Twelve GFP-transfected and 16 GFP-drebrin-transfected neurons were analyzed and statistically compared.

Results and Discussion

In all CHO cells transfected with GFP-drebrin full-length cDNA (1–707 a.a.), the fusion protein was associated with actin filaments (Table 1). Of the cells, 60% exhibited severe change in shape (Fig. 1, *Type IV*), and 30 percent exhibited remodeling of actin filaments (Fig. 1, *Type III*). The association of the fusion protein and actin filaments was also observed in some cells transfected with a shorter drebrin cDNA, although the change in cell shape and remodeling of actin filaments were less frequent. We found that a central 28-amino-acid sequence (273–370 a.a.) would allow the fusion protein to be associated with actin filaments (Fig. 1, *Type II*). This observation indicates that this short sequence is a core for actin binding. The frequency of the association was, however, relatively small (60%) in the case of the 28-amino-acid

TABLE 1. Cell-type ratios of transfected CHO cells

Drebrin sequence[a] (amino acids)	Type I[b] (%)	Type II	Type III	Type IV
1–707[c]	0	16	24	60
1–319, 364–707	0	20	32	48
319–707	100	0	0	0
1–366	0	68	18	14
169–366	0	74	24	2
233–366	0	64	16	10
273–366	0	72	22	6
301–366	72	28	0	0
319–366	66	34	0	0
233–317	0	60	40	0
233–300	50	50	0	0
233–279	86	14	0	0
273–300	36	64	0	0

Fifty cells were analyzed for each sequence.
[a] Sequences were fused with green fluorescent protein and introduced into CHO cells.
[b] Refer to Fig. 1 for the definition of cell types.
[c] Full-length cDNA for drebrin.

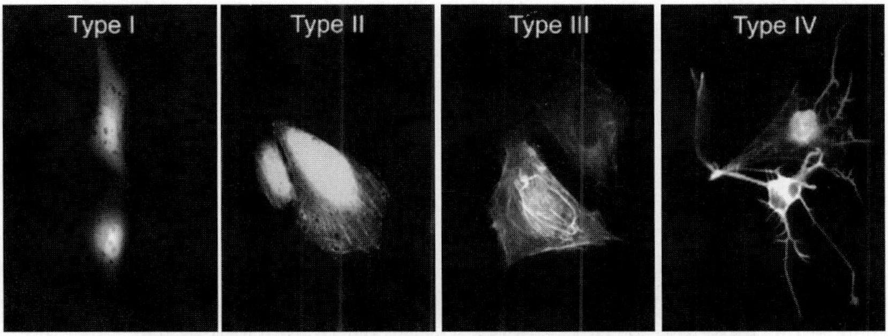

FIG. 1. Fluorescence microscopic observation of CHO cells transfected with green fluorescent protein (GFP) -drebrin cDNA. The transfected cells were classified into four types: *Type I*, the fusion protein was not associated with actin filaments; *Type II*, the fusion protein was associated with actin filaments; *Type III*, the fusion protein was associated with actin filaments and the filaments were remodeled; *Type IV*, the fusion protein was associated with actin filaments and cells shape was changed

sequence (Table 1), indicating that the larger region around this core sequence may be necessary for strong binding.

Next, we introduced the GFP-drebrin full-length cDNA into primary cultured cortical neurons. Three weeks after plating, when synapses were formed between neurons, spines were visible branching off the neurons (Fig. 2a). GFP was distributed diffusely in cell bodies and dendrites, whereas GFP-drebrin fusion protein was localized at spines but only weakly at dendrites (Fig. 2b).

Confocal observation revealed that GFP-drebrin was localized at spine heads (Fig. 3b). Statistic analysis of spine length revealed that spines of neurons transfected with

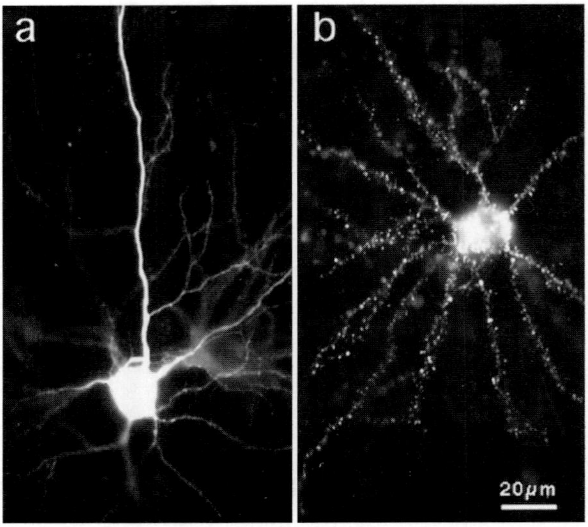

FIG. 2a,b. Fluorescence microscopic observation of primary cultured neurons transfected with GFP-drebrin cDNA. **a** A neuron transfected with GFP only. **b** A neuron transfected with GFP-drebrin (full length)

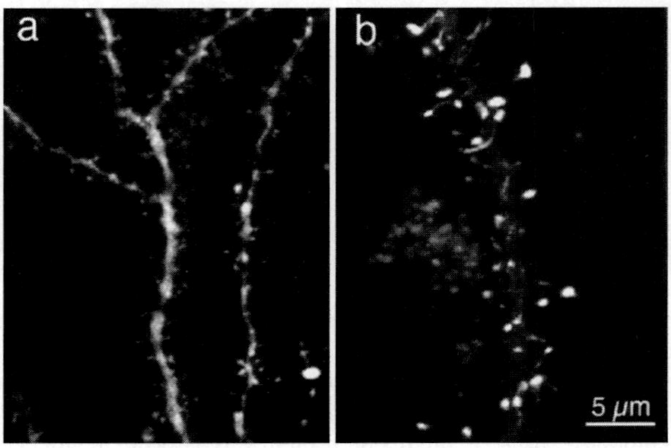

FIG. 3a,b. Confocal microscopic observation of neurons transfected with GFP-drebrin cDNAs. **a** A neuron transfected with GFP only. **b** A neuron transfected with GFP-drebrin (full length)

a GFP-drebrin cDNA were significantly longer than those of GFP-transfected neurons (1.4 ± 0.2 and 2.2 ± 0.5, respectively; average $\pm$ SD, $P = 1.6 \times 10^{-6}$). This observation confirms our hypothesis that drebrin is involved in the regulation of spine morphology.

We have also found that this spine-visualization system using GFP-cDNA is effective to examine the effect of molecules on dendrite and spine morphogenesis. In particular, it is possible to keep cells alive while monitoring the formation of dendrites and spines using this method.

References

1. Koch C, Zador A (1993) The function of dendritic spines: devices subserving biochemical rather than electrical compartmentalization. J Neurosci 13:413–422
2. Fifkova E, Delay RJ (1982) Cytoplasmic actin in neuronal processes as a possible mediator of synaptic plasticity. J Cell Biol 95:345–350
3. Ishikawa R, Hayashi K, Shirao T, et al (1994) Drebrin, a development-associated brain protein from rat embryo, causes the dissociation of tropomyosin from actin filaments. J Biol Chem 47:29928–29933
4. Hayashi K, Ishikawa R, Ye L-H, et al (1996) Modulatory role of drebrin on the cytoskeleton within dendritic spines in the rat cerebral cortex. J Neurosci 16:7161–7170

Optical Mapping of Neural Responses in the Embryonic Rat Brainstem: Early Functional Organization of Vagal Nuclei

KATSUSHIGE SATO, YOKO MOMOSE-SATO, YOSHIYASU ARAI, and KOHTARO KAMINO

Summary. We examined the functional organization of the vagal nuclei of the rat embryo during morphogenesis. Electrical activity in response to vagal stimuli was recorded optically from many sites of the slice preparation stained with a voltage-sensitive dye. In the E13–E14 preparations, two types of spike-like optical signals were detected: one was a narrow signal, and the other was a broader signal. The narrow-signal response area corresponded to the nucleus of the tractus solitarius (NTS), and the broader-signal response area to the dorsal motor nucleus of the vagus nerve (DMNV). In the E15–E16 preparations, the narrow signals were followed by a slow signal related to glutamate-mediated excitatory postsynaptic potentials, suggesting that synaptic function is organized in the NTS by the 15-day-old embryonic stage. In the E15 and E16 preparations, although the nucleus ambiguus is identified morphologically, no neural response was observed, indicating that the embryonic organization of morphology and physiological function is not necessarily temporally coincident.

Key words. Optical mapping—Vagal nuclei—Embryonic development—Synaptic activity—Voltage-sensitive dye

Introduction

Ontogenetic approaches to the emergence of spatiotemporal patterns of electrical activity in neural circuits could be a useful and rational strategy to study the functional organization and architecture of the extremely complex central nervous systems of mammals. In the present study, we have monitored electrical activity evoked by vagal stimulus in the early embryonic rat brainstem with the goal of exploring the spatiotemporal patterning of the neural responses, and then traced the embryonic functional organization of the vagus-related nuclei during morphogenesis.

Department of Physiology, Tokyo Medical and Dental University School of Medicine, 1-5-45 Yushima, Bunkyo-ku, Tokyo 113-8519, Japan

Materials and Methods

Brainstem slices were prepared by sectioning 13- to 16-day-old embryonic rat brainstems transversely at the level of the vagus nerve. The slice preparations were stained with a voltage-sensitive merocyanine-rhodanine dye NK2761. For preparations in which the vagus nerve was stimulated, the cut end of the nerve was drawn into a microsuction electrode, and positive (depolarizing) square current pulses were applied to the vagus nerve. Light from a tungsten-halogen lamp was collimated, rendered quasimonochromatic with interference filters with a transmission maximum at 703 $\pm$ 15 nm, and focused on the preparation. The objective and photographic eyepiece projected a real image of the preparation (magnification, $\times$25) onto a 1020- or 128-element silicon photodiode matrix array mounted on an Olympus Vanox microscope. The output of each detector was passed to an amplifier and was recorded simultaneously. The optical recording was carried out at room temperature, 26°–28°C.

Results and Discussion

Optical Signals in Response to Vagal Stimulation

Neural activity-related optical signals in response to vagal stimulation were recorded simultaneously from multiple sites of 13- to 16-day-old embryonic (E13–E16) rat brainstem slice preparations, which were stained with a voltage-sensitive dye (NK2761). Two examples are shown in Fig. 1.

In the recordings obtained from E13 and E14 preparations, two response areas that were spatially separated were discriminated in the dorsal portion of the preparation: one (area I) was located on the dorsolateral portion, and the other (area II) was located on the dorsomedial portion on the stimulated side of the preparation. Although the signals detected from the two areas were simple in shape, a definite difference was observed in the waveforms of the signals from area I and area II: area I exhibited shorter duration signals (type I) and area II displayed longer duration signals (type II). Type I and II signals were completely blocked by tetrodotoxin (20 μM), indicating that these signals were the result of action potentials or terminal depolarization.

Slow Signals

In E15 and E16 preparations, another type of signal, composed of two components, a first, fast component that appeared to correspond to a type I or type II signal, and a second, delayed slow component, was detected. The amplitudes of the slow signals were decreased exponentially with repetitive stimulation applied to the vagus nerve. This effect seems to reflect synaptic fatigue, and if so, it argued that the slow signal is related to the postsynaptic potential. Further, the later phase of the slow signals was reduced by APV (200 μM), and the initial phase was reduced by CNQX (5 μM). When APV was added together with CNQX, the slow signals were eliminated. In addition, the later phase was enhanced in a Mg^{2+}-free bathing solution. These results suggest that the slow signal reflects the glutamate-mediated excitatory postsynaptic potential and that the initial phase of the slow signal was mainly the result of non-NMDA

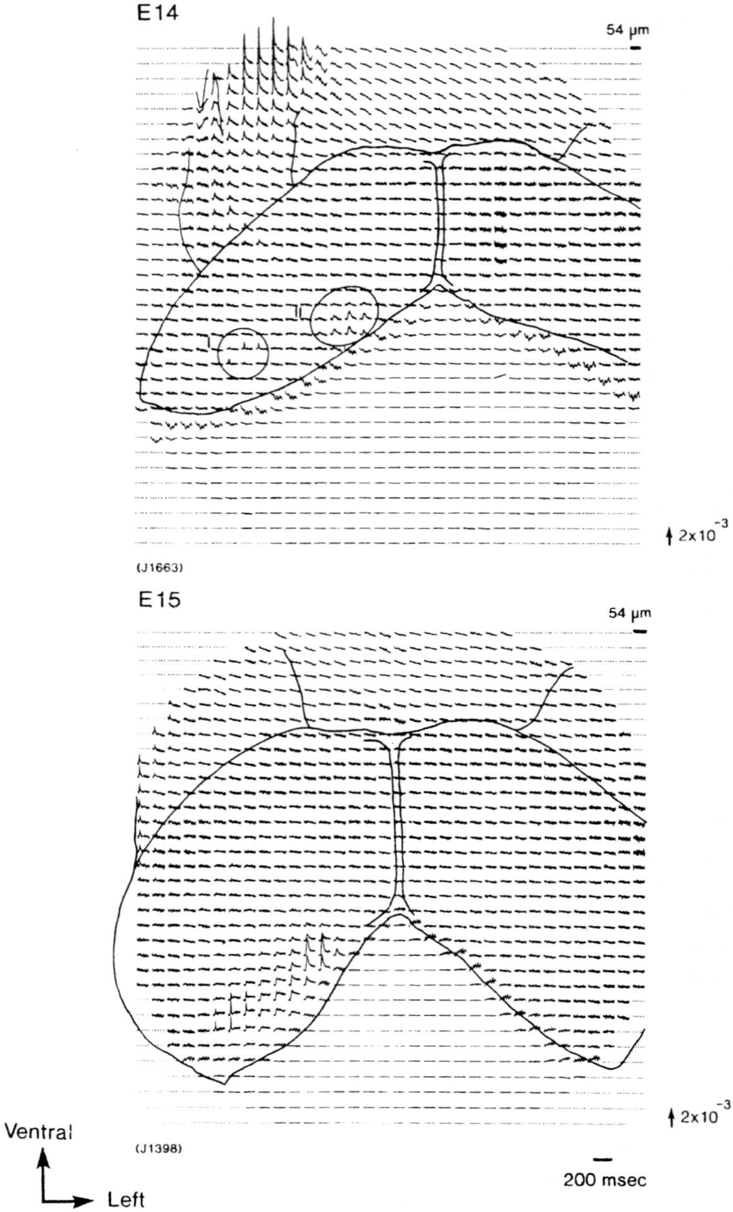

FIG. 1. Simultaneous multisite optical recordings of neural responses in E14 (*top*) and E15 (*bottom*) preparations

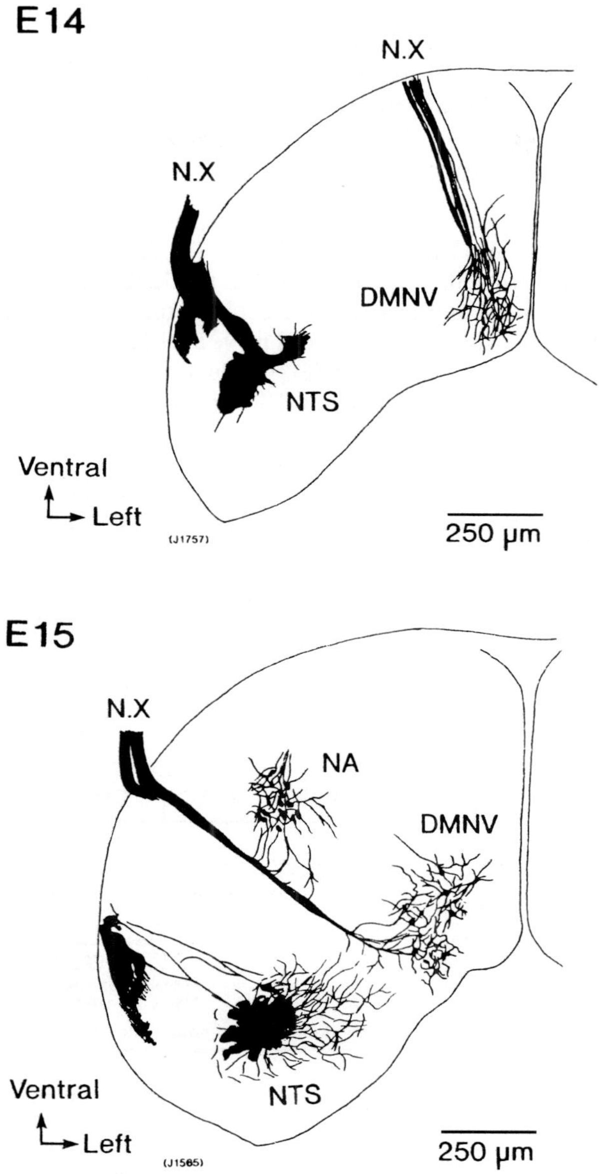

FIG. 2. Morphological views of the vagus nerve pathways into the brainstem

receptors and the later phase of NMDA receptors. In the E14 preparations, the slow signals were only elicited in Mg^{2+}-free Ringer's solution, suggesting that postsynaptic function related to NMDA receptor emerges, in latent form, within the vagus-related nucleus at the 14-day-old embryonic stage.

Comparison with Morphogenesis

Figure 2 shows the pathways of the vagus nerve into the E14 and E15 rat brainstems. These traces were obtained from photographic views of the preparation in which the vagus nerve was labeled with a carbocyanine dye, DiI. In the E15 preparation, three neural populations linked to the vagus nerve were identified, and it is likely that they indicate the relative positions of three cranial nerve nuclei related to the vagus nerve: the nucleus of the tractus solitarius (NTS), the dorsal motor nucleus of the vagus nerve (DMNV), and the nucleus ambiguus (NA). In contrast, in the E14 preparation two neural populations corresponding to the NTS and the DMNV were observed, but the population corresponding to the NA was missing.

In comparison of the optical response areas with the morphological profile, it is likely that areas I and II are correlated to the NTS and the DMNV, respectively. In E15 to E16 preparations, although the NA was observed morphologically, no optical response has been detected from the area corresponding to the NA, suggesting that the NA is not yet functionally organized at these embryonic stages. In Fig. 3, the sequence of embryonic expression of neural responses within the vagal-related nuclei is summarized, compared with morphogenesis.

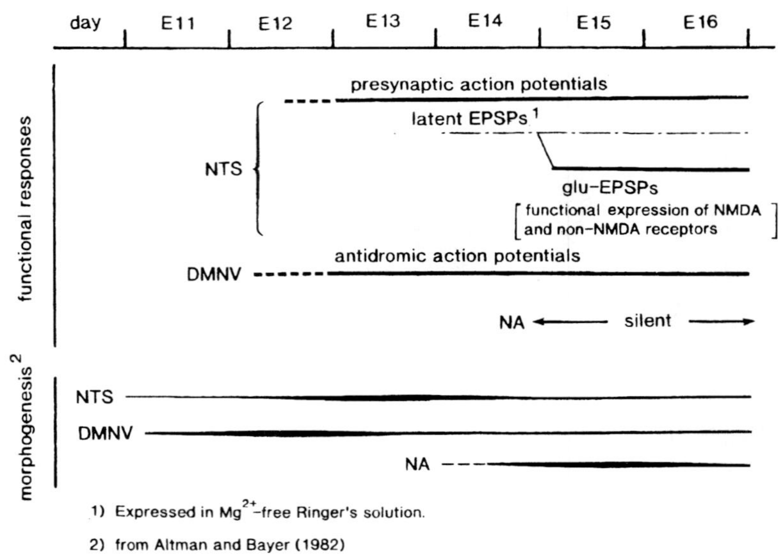

FIG. 3. Summary of sequence of events in the initial embryonic emergence of the neural responses in the rat vagus-related nuclei. The functional events are compared with morphogenesis of the nuclei [2]

References

1. Sato K, Momose-Sato Y, Hirota A, et al (1998) Optical mapping of neural responses in the embryonic rat brainstem with reference to the early functional organization of vagal nuclei. J Neurosci 18:1345–1362
2. Altman J, Bayer SA (1982) Development of the cranial nerve ganglia and related nuclei in the rat. Springer, Berlin Heidelberg New York

Central Respiratory Chemoreception in the Developing Rat Brainstem: Dynamic Imaging Analysis with Fluorescent Voltage-Sensitive Dye

Yasumasa Okada[1], Zibin Chen[2,3], and Frederic L. Eldridge[2]

Summary. We applied an optical recording technique to analyze dynamic processes of central respiratory chemoreception in developing rat brainstems. The brainstem was isolated, stained with fluorescent voltage-sensitive dye, and continuously superfused in vitro with artificial cerebrospinal fluid (aCSF). Responses of cellular membrane potentials to CO_2 changes were analyzed as dynamic images with a high-sensitivity optical recording system. Regions with cellular depolarization were found on the ventral (midline, parapyramidal, and ventrolateral) medullary surface, and the distribution of these CO_2-depolarized regions corresponded almost exactly with that of surface blood vessels. Propagation of excitatory signals from the ventral medullary surface to deep regions was dynamically visualized on a transverse section. Utilizing a synaptic blockade, cellular depolarization was not found except in the surface layer, and the ventrodorsal excitation propagation was inhibited. This result suggests that intrinsic CO_2-excitatory regions are located in the surface layer of the ventral medulla, and that the excitation is synaptically transmitted to deep medullary regions. We would like to propose a model of cellular architecture in central chemoreception whereby primary chemoreceptor cells are located in the medullary marginal glial layer adjacent to surface vessels, and interneurons in the subsurface layer amplify and transmit the chemical information to the respiratory neuronal network.

Key words. Optical recording—Rat—Brainstem—Central chemosensitivity—Control of breathing

[1] Department of Medicine, Keio University Tsukigase Rehabilitation Center, Amagiyugashima, Shizuoka 410-3293, Japan
[2] Department of Physiology, University of North Carolina, Chapel Hill, NC 27599, USA
[3] BioMet-Science Development Group, GlaxoWellcome Inc., Research Triangle Park, NC 27709, USA

Introduction

The regional and cellular specification of the central respiratory chemosensing mechanism in the mammalian brainstem has not been defined. So far, this issue has been studied mainly by microelectrode techniques [1]. However, with microelectrodes, activities of only a limited number of points in the brainstem can be simultaneously analyzed. To overcome this limitation and analyze regional activities of a physiologically important brainstem area, we have applied an optical recording technique in the developing rat brainstem.

Materials and Methods

The brainstem was isolated from the neonatal rat (Sprague-Dawley; 1–5 days old), stained with fluorescent voltage-sensitive dye (RH-795), and continuously superfused with artificial cerebrospinal fluid (aCSF) in a recording chamber as described elsewhere [2]. Responses of cellular membrane potentials to an increase of aCSF CO_2 fraction (from 2% to 8%) on the ventral medullary and transversely sectioned surfaces were analyzed as dynamic images using a high-sensitivity optical recording system (HR Deltaron 1700; Fuji Film, Tokyo, Japan) [3,4]. The HR Deltaron 1700 is a high-resolution (128 × 128 pixels), rapid (maximum 1700 frames per second) image acquisition system.

Results

Regions with cellular depolarization were found in the ventral (midline, parapyramidal, and ventrolateral) medulla, and the distribution of the CO_2-depolarized regions almost exactly corresponded with that of surface blood vessels (Fig. 1a,b). Propagation of the excitatory signals from the ventral medullary surface to a deeper region was dynamically visualized on a transverse section (Fig. 2). With synaptic blockade (low Ca^{2+}-high Mg^{2+}, tetrodotoxin [TTX], or Cd^{2+}) aCSFs, cellular depolarization was found only in the surface layer (marginal glial layer), and the ventrodorsal excitation propagation ceased to occur (Fig. 3).

Discussion

In this study we have shown that the intrinsically CO_2-excitatory regions are located in the surface layer of the medial and lateral ventral medulla. Interestingly, we found that the CO_2-depolarized regions corresponded with that of surface vessels (arteries and veins). This anatomical arrangement seems to be physiologically reasonable; chemoreceptor cells are located close to vessels so that chemoreceptor cells can rapidly and sensitively detect the CO_2 change of the blood that enters or leaves the brainstem. The CO_2-excitatory regions identified in the present study were in relatively good accordance with previous reports [1,5,6]. However, our data indicated that the lateral border of the ventrolateral chemosensitive area in the medulla is more lateral than previously reported. This lateral border of the ventrolateral

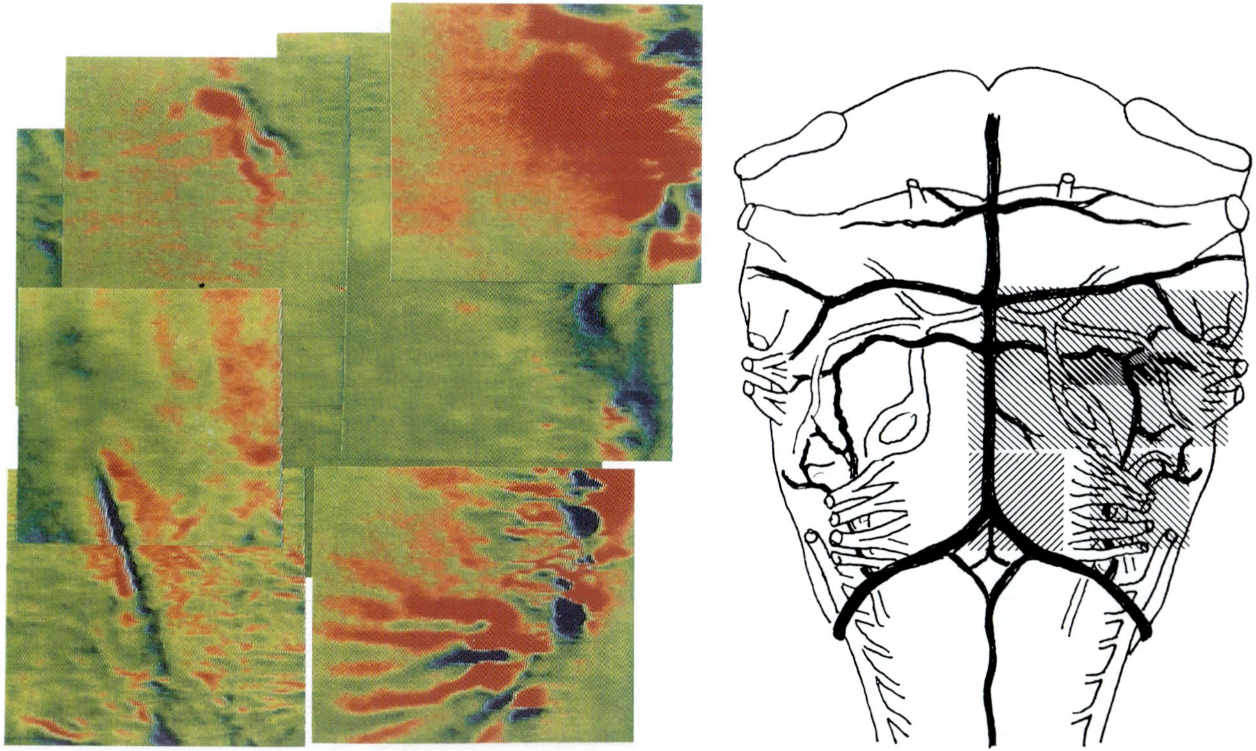

a b

FIG. 1. **a** Regions with cellular depolarization were found in the ventral (midline, parapyramidal, and ventrolateral) medullary surface. Excitatory regions are depicted as *red* and inhibitory as *blue*. **b** Schema of recorded area (*hatched area*). The distribution of the CO_2-depolarized regions corresponded well with that of surface vessels

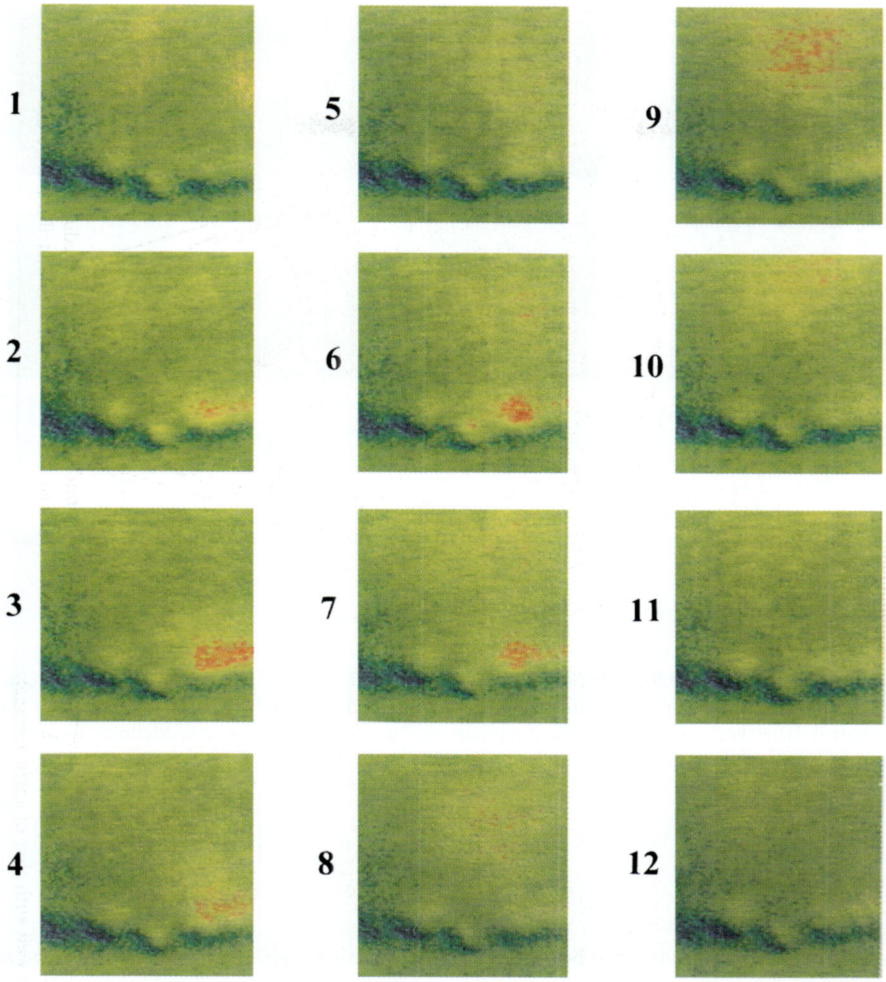

FIG. 2. Propagation of the excitatory signals from the ventrolateral medullary surface (*down* in each panel) to a deeper region (*upper* in each panel) is dynamically visualized in a transverse section of the medulla. Excitation (shown as *red*) initiated on the surface layer (see frame *6*) and propagated to the ventral respiratory group (VRG) region (see frame *9*). Interval between frames is 10 msec

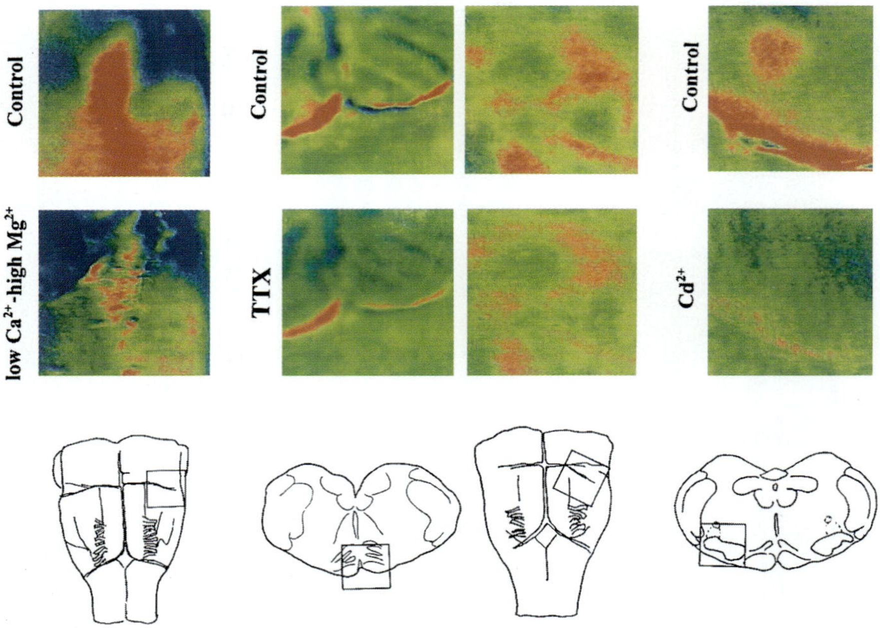

FIG. 3. With synaptic blockade (low Ca^{2+}-high Mg^{2+}, TTX, or Cd^{2+}) aCSFs, CO_2-induced excitation (shown as *red*) on the medullary surface remained, although it was weaker than in the control. However, excitations in deep regions were not found in transverse sections. Recorded areas are shown in schemata in the *lower panels*

chemosensitive area has not been thoroughly investigated in the past because surgical exposure of this lateral medullary region was technically impossible in conventional in vivo preparations. In dynamic images we have also shown that the excitation is synaptically transmitted from the medullary surface to deep regions. This finding is consistent with our model (see following), but is not in agreement with the hypothesis of Kawai et al. [7], who claimed that respiratory neurons themselves in the deep medulla were intrinsically chemosensitive. Obviously, this disagreement requires further investigation.

We propose a model of cellular architecture in central chemoreception on the basis of present and previous experiments [8,9] whereby primary chemoreceptor cells are located in the marginal glial layer adjacent to surface blood vessels and interneurons in the subsurface layer amplify and relay the chemical information to the respiratory neuronal network (Fig. 4).

Acknowledgment. The authors thank Fuji Photo Film Co. Ltd. (Tokyo, Japan) for giving us the opportunity to use their optical recording system, the HR Deltaron 1700.

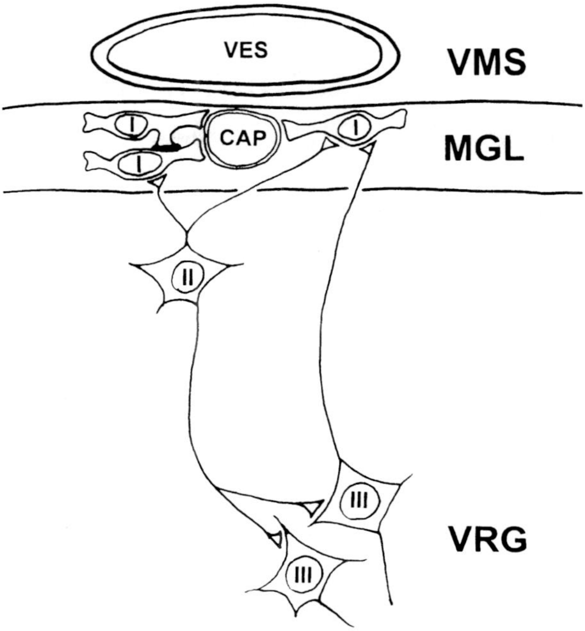

FIG. 4. Model of cellular architecture in central chemoreception. Primary chemoreceptor cells (type I cells, *I*) are located in the marginal glial layer (*MGL*) adjacent to surface vessels (*VES*) of the ventral medullary surface (*VMS*). Type I cells also surround capillaries (*CAP*). Interneurons (type II cells, *II*) in the subsurface layer relay the chemical information from type I cells to the respiratory neurons (*III*) in the ventral respiratory group (*VRG*)

References

1. Okada Y, Mückenhoff K, Scheid P (1993) Hypercania and medullary neurons in the isolated brain stem–spinal cord of the rat. Respir Physiol 93:327–336
2. Okada Y, Mückenhoff K, Holtermann G, et al (1993) Depth profiles of pH and P_{O_2} in the isolated brain stem–spinal cord of the neonatal rat. Respir Physiol 93:315–326
3. Obaid AL, Salzberg BM (1996) Micromolar 4-aminopyridine enhances invasion of a vertebrate neurosecretory terminal arborization. J Gen Physiol 107:353–368
4. Iijima T, Witter MP, Ichikawa M, et al (1996) Entorhinal–hippocampal interactions revealed by real-time imaging. Science 272:1176–1179
5. Loeschcke HH (1982) Central chemosensitivity and the reaction theory. J Physiol (Lond) 332:1–24
6. Bernard DG, Li A, Nattie EE (1996) Evidence for central chemoreception in the midline raphé. J Appl Physiol 80:105–115
7. Kawai A, Ballantyne D, Mückenhoff K, et al (1996) Chemosensitive medullary neurons in the brainstem–spinal cord preparation of the neonatal rat. J Physiol (Lond) 492:277–292
8. Okada Y, Chen Z , Eldridge FL (1996) Sites of the central respiratory chemoreception in the rat medulla investigated by microinjection of CO_2-enriched mock cerebrospinal fluid (CSF). FASEB J 10:A643
9. Chen Z, Okada Y, Eldridge FL (1997) Hypercapnia-induced *c-fos* expression in cholinergic cells within the marginal glial layer of the rat ventral medulla oblongata. FASEB J 11:A131

Regulation of Microtubule-Associated Protein MAP1B–Tubulin Interaction by Acidic Phospholipids

EMIKO YAMAUCHI[1,2] and HISAAKI TANIGUCHI[2]

Summary. Microtubule-associated protein MAP1B, a major neuronal cytoskeletal protein, is highly expressed during an early stage of brain development and has been shown to be important for normal brain development. Because MAP1B is involved in neurite outgrowth and is concentrated in growth cone membranes, the protein may interact directly with the membrane components. In this study, we showed that MAP1B purified from young rat brain can bind to acidic phospholipids such as phosphatidylserine and phosphatidylinositol, but not to a neutral phospholipid, phosphatidylcholine. Furthermore, the binding of MAP1B to microtubules was inhibited by acidic phospholipids. Phosphatidylcholine showed no effect on the binding of MAP1B to the microtubules. Using synthetic peptides derived from the tubulin-binding domain of MAP1B, the phospholipid-binding site of MAP1B was located to the C-terminal half of the tubulin-binding domain. These results suggest that MAP1B binds to biological membranes through its tubulin-binding site and that the binding may play an important role in the dynamic regulation of cytoskeletons in the growth cone.

Key words. MAP1B—Phospholipid—Microtubule-associated protein—Cytoskeleton—Growth cone

Introduction

MAP1B (microtubule-associated protein 1B) is a major MAP in developing brain [1]. Although the physiological function of MAP1B has yet to be elucidated, its phosphorylation has been suggested to be important for synapse formation. MAP1B is strongly phosphorylated in cultured cortical neurons undergoing synaptogenesis, and inhibition of the phosphorylation results in the suppression of synapse formation [2]. Recent data from gene-targeting experiments in the mouse indicate that MAP1B is

[1] Department of Biological Chemistry, Faculty of Pharmaceutical Sciences, Nagoya City University, Mizuho-ku, Nagoya 467, Japan

[2] Division of Biomedical Polymer Science, Institute for Comprehensive Medical Science, Fujita Health University, Toyoake, Aichi 470-11, Japan

necessary for normal development of the central nervous system [3]. MAP1B is localized in the growth cone, especially in its membrane fractions. However, no molecular basis for the membrane localization of MAP1B has been presented. In this study, we showed that purified MAP1B can bind specifically to acidic phospholipids. The binding site of acidic phospholipids was located to the C-terminal half of the tubulin-binding domain, and phospholipid binding induced MAP1B dissociation from microtubules. These results suggest acidic phospholipids may play an important role in the MAP1B-dependent regulation of microtubules in the growth cone.

Materials and Methods

Materials

MAP1B was purified from young rat brain as described, and tubulin was prepared from bovine brain [4]. Synthetic peptides were obtained from Research Genetics (Huntsville, AL).

Binding of MAP1B to Phospholipids

Binding of MAP1B and that of the MAP1B-derived peptides to phospholipid membranes were analyzed by sedimentation analysis [4].

Results

Binding of MAP1B to Phospholipid Membranes

Binding of MAP1B to phospholipid vesicles was directly studied by sedimentation assay. Purified MAP1B was incubated with large unilamellar vesicles of phosphatidylserine (PS), phosphatidylinositol (PI), phosphatidic acid (PA), or phosphatidylcholine (PC), and subjected to centrifugation. The phospholipid vesicles sedimented under the conditions employed, while the unbound protein remained in the supernatants. With all three acidic phospholipids studied, PS, PI, and PA, MAP1B was found cosedimented with the lipids. On the contrary, no significant binding was observed when PC was used. Therefore, MAP1B has the intrinsic ability to bind to membrane phospholipids.

Effects of Phospholipids on MAP1B–Tubulin Interaction

Because the main function of MAP1B is its interaction with tubulin, the effects of the phospholipid binding to MAP1B on the protein–protein interaction were then studied. When polymerized tubulin and MAP1B were mixed and sedimented by centrifugation, MAP1B was found in the pellets with tubulin. The purified MAP1B, therefore, still retained its activity. The addition of PC to the mixture did not show any effect on the MAP1B–tubulin interaction. However, when acidic phospholipids were included in the mixture, MAP1B did not cosediment with tubulin, but remained in the supernatants (Fig. 1a). Therefore, addition of the acidic phospholipids induced dissociation of MAP1B from the polymerized tubulin. The formation of a ternary complex between

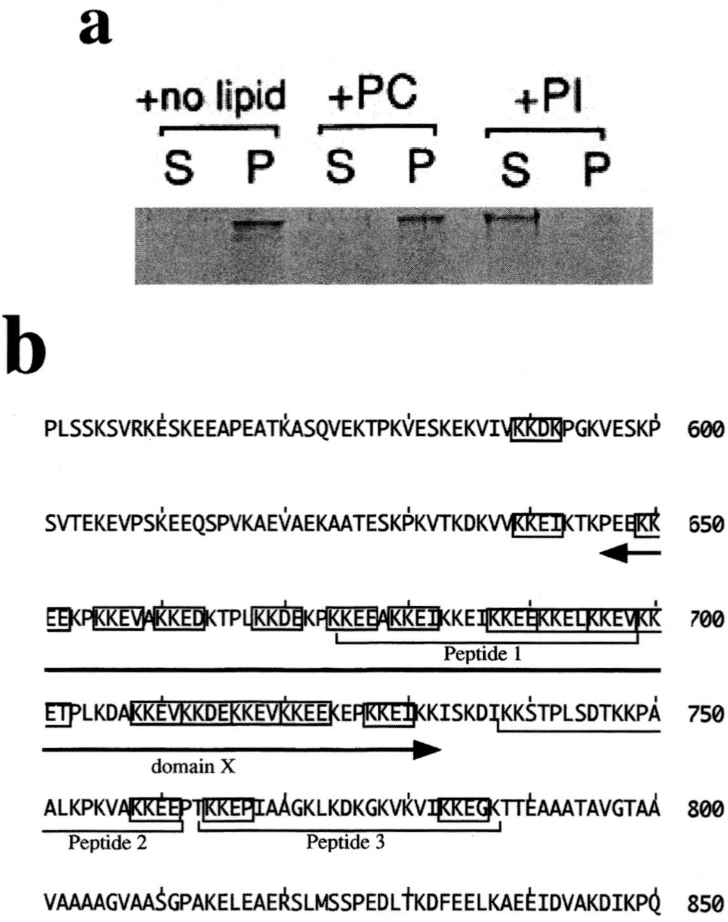

FIG. 1. **a** Effects of phospholipids on MAP1B–microtubule interaction. MAP1B (0.1 mg/ml) was incubated with polymerized tubulin (2.4 mg/ml) in the presence or absence of phospholipids (0.2 mg/ml). MAP1B remaining in the supernatants (*S*) and that cosedimented with tubulin (*P*) were analyzed. **b** Tublin-binding domain of rat MAP1B. The typical KKE(E/I/V) motifs are *boxed*. Positions of domain X and three peptides synthesized in the present study are also indicated

the phospholipid membranes, MAP1B, and tubulin was not observed, suggesting either that the binding of acidic phospholipids to MAP1B induce a conformational change that results in the loss of binding to tubulin or that the acidic phospholipids share the same binding site(s) in MAP1B with tubulin. Interestingly, MAP1B did not bind to phosphatidylinositol 4-monophosphate or phosphatidylinositol 4,5-diphosphate to any significant extent, and the addition of these acidic phospholipids did not affect the MAP1B–microtubule interaction significantly. These results demonstrate that the interaction between MAP1B and phospholipids involves not only ionic interactions but also specific structural interactions.

Identification of the Phospholipid-Binding Site of MAP1B

To identify the phospholipid-binding site, three peptides corresponding to the different parts of the microtubule-binding domain were designed. Previous studies [5] have demonstrated that in addition to the "domain X" between amino acids 646–732, another domain either N- or C-terminal to the domain X constitutes a second tubulin-binding site (Fig. 1b). The typical KKE(E/I/V) tubulin-binding motif is clustered in the domain X. Peptide 1 (KEEAKKEIKKEIKKEEKKELKKEV, residues 675–698) was chosen from the central part of domain X, and peptide 3 (KKEPIAAGKLKDKGKVKVIKKEGK, residues 764–786) from the C-terminal half of the tubulin-binding domain. Peptide 2 (KKSTPLSDTKKPAALKPKVAKKEE, residues 738–761) is adjacent to peptide 3 and similar to the latter, but still retains one typical KKEE motif. Peptide 1 did not show any significant binding either to PC or PS. On the contrary, peptides 2 and 3 bound to PS vesicles specifically. Thus the C-terminal half of the microtubule-binding domain of MAP1B, but not the N-terminal half containing typical KKE(E/I/V) repeats, has the ability to bind to acidic phospholipids specifically.

Discussion

This study established that MAP1B has an intrinsic ability to bind to acidic phospholipids and that the C-terminal subdomain of the microtubule-binding domain is responsible for the binding. Furthermore, we found that binding of acidic phospholipids to the protein affects its interaction with tubulin.

A hydropathy plot of MAP1B sequence shows a single hydrophobic domain between residues 791 and 817 that is located just C-terminal to the tubulin-binding domain [2]. Because MAP1B binds to acidic phospholipids rather than to neutral phospholipids, such a neutral hydrophobic domain cannot be the primary binding site. On the contrary, a closer look of the adjacent tubulin-binding site revealed an interesting variation of the character within the domain. The N-terminal half of the domain consists mainly of the typical KKE(E/I/V) motif, while the C-terminal half contains a derivative of the motif that lacks the acidic residues. This gives the latter domain a basic amphiphilic character in which basic residues and hydrophobic residues appear alternatingly.

We have previously shown that such basic amphiphilic domains serve as reversible phospholipid-binding sites in various proteins such as MARCKS, GAP-43, and nitric oxide synthase [6]. It is of interest to note that the membrane interactions of these domains are often regulated by protein phosphorylation. Because MAP1B is known to be phosphorylated at multiple sites and the interactions of other MAPs such as tau and MAP2 with tubulin are regulated by protein phosphorylation of the binding domains, this raises an intriguing possibility that such a regulatory mechanism is operational in the MAP1B–membrane interactions. The interaction of MAP1B with membrane phospholipids, in turn, may regulate the dynamics of the neuronal cytoskeleton during the axonal neurite elongation.

References

1. Bloom GS, Luca FC, Vallee RB (1985) Microtubule-associated protein 1B: identification of a major component of the neuronal cytoskeleton. Proc Natl Acad Sci USA 82:5404–5408
2. Muramoto K, Taniguchi H, Kawahara M, et al (1994) A substrate of ecto-protein kinase is microtubule-associated protein 1B in cortical cell cultures undergoing synaptogenesis. Biochem Biophys Res Commun 205:1467–1473
3. Edelmann W, Zervas M, Costello P, et al (1996) Neuronal abnormalities in microtubule-associated protein 1B mutant mice. Proc Natl Acad Sci USA 93:1270–1275
4. Yamauchi E, Titani K, Taniguchi H (1997) Specific binding of acidic phospholipids to microtubule-associated protein MAP1B regulates its interaction with tubulin. J Biol Chem 272:22948–22953
5. Noble M, Lewis SA, Cowan NJ (1989) The microtubule binding domain of microtubule-associated protein MAP1B contains a repeated sequence motif unrelated to that of MAP2 and tau. J Cell Biol 109:3367–3376
6. Matsubara M, Titani K, Taniguchi H (1996) Interaction of calmodulin-binding domain peptides of nitric oxide synthase with membrane phospholipids: regulation by protein phosphorylation and Ca^{2+}-calmodulin. Biochemistry 35:14651–14658

Site-Specific Phosphorylation and Dephosphorylation of Microtubule-Associated Protein MAP1B During Neuronal Development

EMIKO YAMAUCHI[1,2], KAZUYO MURAMOTO[3], YOICHIRO KURODA[3], and HISAAKI TANIGUCHI[2]

Summary. Microtubule-associated protein MAP1B is a major neuronal cytoskeletal protein essential for brain development. Its function has been suggested to be regulated by protein phosphorylation. To characterize the in vivo phosphorylation sites of MAP1B, the protein was purified from young rat brain and subjected to liquid chromatography/electrospray mass spectrometry. The results obtained revealed that MAP1B is highly phosphorylated; 30 phosphorylation sites were identified. Most of the phosphorylation sites were immediately followed by proline, suggesting that MAP1B is an in vivo substrate of proline-directed protein kinases such as MAP kinase and cdk 5. When similar analyses were carried out with MAP1B purified from different stages of rat brain, only the 2 phosphorylation sites were found to undergo dephosphorylation between embryonic day 18 and postnatal day 15. This specific dephosphorylation during the early stage of brain development suggests that dephosphorylation plays an important role in the regulation of MAP1B function.

Key words. MAP1B—Mass spectrometry—Microtubule-associated protein—Phosphorylation—Neuronal development

Introduction

Microtubule-associated protein MAP1B is a major component of the neuronal cytoskeleton in the developing brain [1]. Gene-targeting experiments in the mouse indicate that MAP1B is essential for normal development of the central nervous system [2]. MAP1B has been shown to be phosphorylated and the phosphorylation is regulated during development. Immunostaining of brain homogenate using antibodies that recognize different epitopes of phosphorylated MAP1B changes drasti-

[1] Department of Biological Chemistry, Faculty of Pharmaceutical Sciences, Nagoya City University, Mizuhoku, Nagoya 467, Japan
[2] Division of Biomedical Polymer Science, Institute for Comprehensive Medical Science, Fujita Health University, Toyoake, Aichi 470-11, Japan
[3] Department of Molecular & Cellular Neurobiology, Tokyo Metropolitan Institute for Neuroscience, Fuchu, Tokyo 183, Japan

cally during brain development [3]. We have previously shown that its phosphorylation is essential for synaptogenesis in cultured cortical neurons; MAP1B is strongly phosphorylated in the cultured cortical neurons undergoing synaptogenesis, and the inhibition of the phosphorylation results in the suppression of the synapse formation [4]. Therefore, the phosphorylation of MAP1B plays an important role in the regulation of the physiological function of the protein in neurite elongation/synapse formation. Phosphorylated MAP1B is also found in paired-helical filaments in Alzheimer's disease [5].

In the present study, in vivo phosphorylation of MAP1B was studied by capillary high performance liquid chromatography/electrospray mass spectrometry (LC/MS). The result obtained revealed that MAP1B was hyperphosphorylated and contained more than 30 phosphorylation sites. Furthermore, of the 30 phosphorylation sites, only a few phosphorylation sites underwent specific dephosphorylation between embryonic day 18 (E18) and postnatal day 15 (P15).

Materials and Methods

Materials

MAP1B was purified from fetal and young rat brains as described [6]. All other chemicals and biochemicals used were of analytical grade.

Protease Digestion and LC/MS Analysis of the Digests

MAP1B was digested with *Achromobacter* lysyl endoprotease and analyzed directly by LC/MS as described [7]. The peptide mixture were also separated off-line by reversed-phase high performance liquid chromatography, and fractions obtained were subjected to the tandem mass (MS/MS) analysis.

Results

LC/MS Analysis of MAP1B Lysyl Endoprotease Digests

To study the in vivo phosphorylation sites of MAP1B, the protein was purified from brains of 1-week-old rats. The protein was digested with lysyl endoprotease, and the resulting peptide mixtures were analyzed directly by LC/MS. Peptides were identified by comparing the molecular masses observed with those predicted from the amino acid sequence deduced from the cDNA sequences [8,9]. Phosphopeptides were identified based on the 80-Da mass difference between the predicted and observed molecular masses [7]. The LC-MS analysis revealed that MAP1B is hyperphosphorylated; 20 phosphopeptides containing 30 phosphorylation sites were identified. The extent of the phosphorylation varied from about 5% to almost 90% depending on the peptide. Ten phosphopeptides were found clustered in the region C-terminal to the microtubule-binding domain (Fig. 1). The characteristic heptadeca-repeat domain near the C-terminus contained 6 phosphorylation sites. To identify the phosphorylation sites in the individual phosphopeptide, the phosphopeptides were separated by reversed-phase high performance liquid chromatography and subjected to tandem

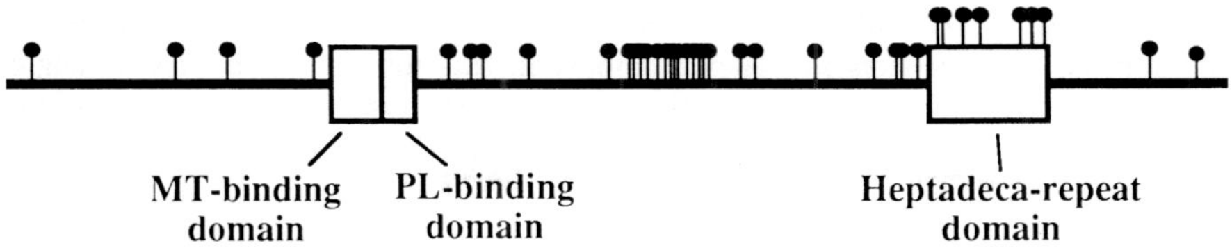

FIG. 1. Schematic representation of the in vivo phosphorylation sites of MAP1B. Phosphorylation sites identified in this study are indicated with *filled circles*. The microtubule (MT) binding domain, phospholipid (PL) binding domain [6], and heptadeca-repeat domain are indicated with *boxes*

mass spectrometry. From the results obtained, it became evident that the majority of the phosphorylation site are followed by Pro, suggesting that the so-called proline-directed protein kinases such as MAP kinase are involved.

Change in Phosphorylation During Development

Because the expression of MAP1B and its phosphorylation state are considered to be closely related to brain development, the phosphorylation state of MAP1B was followed during development. MAP1B was isolated from rat brains at varying stages, i.e., embryonic day E18 and postnatal day P7, and subjected to LC/MS analysis. Most of the phosphopeptides identified in MAP1B isolated from P7 rat brain were also observed with MAP1B preparations isolated from E18 rat brain. However, a significant decrease in phosphorylation was observed with a peptide derived from the heptadeca-repeat domain near the C-terminus. These results indicate that only the protein kinase(s) responsible for the phosphorylation at the heptadeca repeat domain are turned off or the protein phosphatase(s) specific for the sites activated.

Discussion

MAP1B is a major microtubule-associated protein in the neuronal cells undergoing neurite outgrowth and is concentrated in the growth cone, which is a key organization guiding the extending neurite to form correct synapses. It has been recognized as a phosphoprotein that was distinguished to two species by the two bands on the SDS-PAGE gel [3]. A few monoclonal antibodies were raised which can distinguish various phosphorylation sites, although the epitopes of these antibodies were not established [3]. Thus, the present study was the first systematic study on the MAP1B in vivo phosphorylation sites.

It should be noted that many of the phosphorylation sites identified were serine/threonine residues followed immediately by proline, suggesting that MAP1B is the in vivo substrate of proline-directed protein kinases. In addition to CKII, which was suggested to phosphorylate essential sites of MAP1B for neural development, proline-directed protein kinases such as MAP kinase may also function as MAP1B kinase. The "hyperphosphorylation" at Ser(Thr)-Pro sites has also been observed with other MAPs such as MAP2 and tau. The physiological function of MAP1B may be regulated by multiple kinases and phosphatases, and the complex regulation of the phosphorylation at the multiple sites should play important roles in the dynamic cytoskeletal reorganization during synaptogenesis.

References

1. Bloom GS, Luca FC, Vallee RB (1985) Microtubule-associated protein 1B: identification of a major component of the neuronal cytoskeleton. Proc Natl Acad Sci USA 82:5404–5408
2. Edelmann W, Zervas M, Costello P, et al (1996) Neuronal abnormalities in microtubule-associated protein 1B mutant mice. Proc Natl Acad Sci USA 93:1270–1275
3. Ulloa L, Avila J, Diaz-Nido J (1993) Heterogeneity in the phosphorylation of microtubule-associated protein MAP1B during rat brain development. J Neurochem 61:961–972
4. Muramoto K, Taniguchi H, Kawahara M, et al (1994) A substrate of ecto-protein kinase is microtubule-associated protein 1B in cortical cell cultures undergoing synaptogenesis. Biochem Biophys Res Commun 205:1467–1473
5. Hasegawa M, Arai T, Ihara Y (1990) Immunochemical evidence that fragments of phosphorylated MAP5 (MAP1B) are bound to neurofibrillary tangles in Alzheimer's disease. Neuron 4:909–918
6. Yamauchi E, Titani K, Taniguchi H (1997) Specific binding of acidic phospholipids to microtubule-associated protein MAP1B regulates its interaction with tubulin. J Biol Chem 272:22948–22953
7. Taniguchi H, Manenti S, Suzuki M, et al (1994) Myristoylated alanine-rich C kinase substrate (MARCKS), a major protein kinase C substrate, is an in vivo substrate of proline-directed protein kinase(s). A mass spectroscopic analysis of the post-translational modifications. J Biol Chem 269:18299–18302
8. Zauner W, Kratz J, Staunton J, et al (1992) Identification of two distinct microtubule binding domains on recombinant rat MAP 1B. Eur J Cell Biol 57:66–74
9. Liu D, Fischer I (1996) Isolation and sequencing of the 5′ end of the rat microtubule-associated protein (MAP1B)-encoding cDNA. Gene 171:307–308

The Connection Between Granule Cells and Mossy Fibers Changes Dynamically During Postnatal Cerebellar Development

Miwako Ozaki[1], Tsutomu Hashikawa[2], and Ryoji Yano[1]

Summary. We have examined the postnatal developmental process of axonal elongation of mossy fibers and their target cells in the mouse pontocerebellar system by DiI tracing and immunostaining with antibodies against marker proteins for neurons, Bergmann glia, and granule cells. The results showed that mossy fibers from the pontine nuclei had already entered the cerebellar cortex as early as postnatal day 1 (P1). DiI-labeled postmigratory granule cells rarely appeared until P3, even in the internal granular layer (IGL). The number of labeled granule cells increased gradually from P5. A large number of granule cells established contacts with mossy fibers on P7–P9. A portion of mossy fibers were found to pass through the molecular layer to reach regions just beneath the external granular layer (EGL); however, they hardly entered the EGL. Labeled granule cells had a tendency to be assembled in clusters from premigratory through postmigratory stages. No labeled granule cells were observed in a later stage. A definite glomerular structure of mossy fiber terminals appeared in IGL from P14. Purkinje cells and Bergmann glia were also found to be tentative targets of mossy fibers at a certain period.

Key words. Cerebellum—Synaptogenesis—Granule cell—Mossy fiber—DiI

Introduction

During the development of the central nervous system, neurons are communicating with each other to form specific neural networks. Understanding how each neuron selectively contacts and communicates with another particular neuron is significant in studying the mechanism of synaptogenesis and synaptic organization. In dorsal root ganglion (DRG) neurons, neuronal communication through neuronal activities or growth factors was suggested to be essential to form mature synapses [1]. In the cerebellum, granule cells migrate from the external granular layer (EGL) to the internal granular layer (IGL) and make synapses with mossy fibers originating from the brainstem and the spinal cord. An NMDA (*N*-methyl-D-aspartate) type glutamate

[1]Laboratory for Cellular Information Processing and [2]Laboratory for Neural Architecture, Brain Science Institute, RIKEN, 2-1 Hirosawa, Wako, Saitama 351-0198, Japan

receptor composed of various subunits (NR2A–2D) has been suggested to play a key role in cerebellar development. A subunit constituting NMDA receptors changes dramatically from NR2B to NR2C in postmigratory granule cells. Interestingly, NR2C subunit gene expression expands slowly from the posterior to anterior lobules in the first 2 weeks after birth.

We recently found that neuregulin-β, a ligand for ErbB receptor tyrosine kinases, could induce NR2C subunit expression in combination with neuronal activities in organotypic cultured granule cells [2]. The coincidence of gene expression and the start of synaptic formation between mossy fibers and granule cells may suggest that mossy fiber innervation promotes synaptic maturation via the neuregulin-ErbB receptor system. The molecular mechanism of functions of the neuregulin-ErbB receptor system has been extensively studied in neuromuscular formation, in which neuregulins accumulated in neuromuscular synapses induce expression of the acetylcholine receptor gene in muscle cells [3,4]. Our recent study revealed that neuregulins function at the central synapse as well, suggesting that mossy fiber innervation controls gene expression in a similar manner to the peripheral nervous system.

In this study, we tried to reveal relationships between mossy fiber innervation and cerebellar cortical development using a tracer, DiI (1,1′-dioctadecyl-3,3,3′,3′-tetramethylindocarbocyanine perchlorate), a lipophilic carbocyanine dye, in the pontocerebellar system. DiI labeling is a useful method in not only a study of process of axoplasmic transport but also in studying neural connections in fixed brain [5]. DiI experiments exhibited developmental process of mossy fibers and dramatic changes of mossy fiber target cells, suggesting that mossy fibers might be involved in cerebellar morphogenesis and synaptic organization.

Materials and Methods

Tissues and DiI Application

Mice (postnatal day 1–14) were perfused with saline followed by a fixative containing 4% paraformaldehyde in 0.1 M phosphate buffer (pH 7.3). The brains were immersed in phosphate-buffered 4% paraformaldehyde solution for about 1 week. DiI crystals were placed at the dorsal portion of the brachium pontis to label mossy fibers arising from the pontine nuclei. The brains were stored in the dark at room temperature for 3 months before sectioning. The vibratome sections were examined with a laser conforcal microscope using a YHS filter system (Bio Rad, Hercules, CA, USA).

Preparation of Tissue Sections and Immunoreaction

Brain slices (approximately 500–700 μm thick) were incubated in PBS containing 1% Triton X-100 (PBST) with gentle agitation for at least 30 min and then in normal horse serum (HS) (diluted 1:10 in PBST) for 2 h. They were treated with primary antibody for 3 h and washed at least six times for 30 min each in PBST containing 1% HS. Subsequently, the slices were incubated with Cy3 conjugated antimouse IgG as a secondary antibody for 3 h, washed six times for 30 min, and then examined by laser confocal microscopy.

Results

The sequential process of postnatal cerebellar cortical development demonstrated by the results of our DiI labeling experiments is summarized in Fig. 1. It has been generally accepted that cerebellar granule cells develop in a strict sequential manner, with major processes of neural development such as proliferation, axonal elongation, migration, and formation of synaptic connections. In the first stage, granule cell precursors proliferate in the superficial aspect of the EGL, from which postmitotic cells descend into the deeper aspect of EGL. Second, the cells migrate from the EGL to the IGL, extending parallel fiber axons and passing through the Purkinje cell layer. The granule cell migration is complete at P14 in mice. In the final stage, granule cell dendrites establish connections with mossy fibers in the IGL, forming synapses with the mature glomerular structure.

In our resent study, however, more complicated developmental processes were observed. Distinctive DiI-positive cells are seen as a red color in Fig. 1. Mossy fibers reached the cerebellar cortex of postnatal day 1 (P1), although most of them did not seem to make contact even with postmigratory granule cells in the IGL until P3. A few mossy fibers contacted individual granule cells in the IGL at the earlier stage of cerebellar development. The number of granule cells stained by contact with mossy fibers increased gradually from P5, and their number reached a peak around P7–P9.

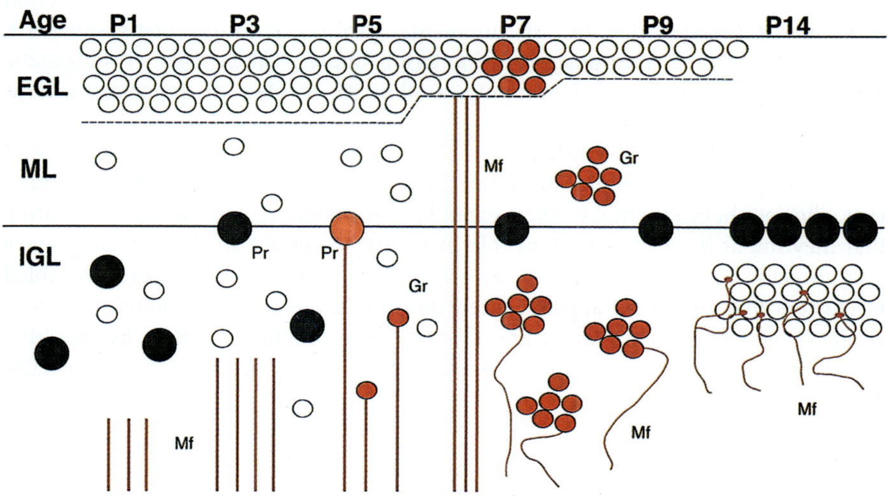

Fig. 1. Summary diagram of mossy fiber target. During postnatal development, granule cells (*small open circles*) migrated from the EGL to the IGL and were localized in the IGL by P14. Purkinje cells (*large gray circles*), which have settled as immature cells in the Purkinje cell dense zone, spread into a single layer during development. *Red color* shows DiI-labeled fibers and neurons that were characteristically observed at each age; *P1–P14* indicates days after birth. *EGL*, external granular layer; *ML*, molecular layer; *IGL*, internal granular layer; *Mf*, mossy fiber; *Gr*, granule cell; *Pr*, Purkinje cell

During these stages, the majority of mossy fibers remained in the IGL. A portion of them seemed to keep extending and go through the molecular layer (ML), although they scarcely entered the EGL. Thereafter, DiI-positive granule cells were observed in cluster-like form in the EGL, ML, and IGL at the middle stage of development. Only mossy fibers were stained by DiI at P14 when cerebellar development entered the final stage. From P14, mossy fiber terminals started swelling and making glomerular structures in the IGL. Overall orientation of development in the cerebellar cortex was from posterior to anterior lobules.

Mossy fiber target cells (summarized in Table 1) were identified by their morphological features and immunostaining patterns with antibodies against neurofilament 68, tenascin, and adhesion molecule L1, as marker proteins for neurons, Bergmann glia, and granule cells, respectively. Mossy fibers changed their target cells during development. Bergmann glia were stained with DiI from the early postnatal stage until granule cell migration was complete. A contact between mossy fibers and Purkinje cells was observed during a short period, from P3 to P5.

Discussion

In the cerebellum, NR2C gene expression proceeds from posterior to anterior lobules during the first postnatal 2 weeks. The role of mossy fibers in NR2C gene expression as suggested from the results of in vitro experiments is further supported by concentrated localization of neuregulins in mossy fibers and granule cells and the existence of their receptor tyrosine kinase, ErbB, in granule cells [2,6]. In addition, DiI injection experiments revealed that the developmental process of mossy fibers showed a tendency to proceed from posterior to anterior lobules. Taken together, epigenetic gene regulation by mossy fiber input was thought to play an important role in synaptogenesis and synaptic plasticity.

In the previous cerebellar developmental model, mossy fibers stayed inside the IGL, waiting for granule cell migration to form synapses between them. This study, however, showed that some mossy fibers extend to the inside edge of the EGL, indicating that they might have an important role in sending a certain signal to premigratory granule cells. Furthermore, DiI-labeled granule cells in EGL and ML were assembled in clusters that contained approximately 30–50 granule cells. The staining pattern might suggest that premigratory granule cells form function modules by receiving signals from a certain group of mossy fibers directly or indirectly during cerebellar development.

TABLE 1. Changes in DiI staining pattern depending on age

Target cells of mossy fibers	Period	Method of identification
Bergmann glia	P1 ~ P14	Morphology and whole-mount immunostaining
Purkinje cells	P3 ~ P5	Morphology and whole-mount immunostaining
Clustered granule cells	P7 ~ P9	Morphology

The stained cells were identified by morphological features and whole-mount immunostaining pattern. Which DiI-stained cells appeared in a certain period may suggest that mossy fiber change target cells through cerebellar development. *P* indicates postnatal day.

Mossy fiber target cells other than granule cells were observed in this study. Bergmann glial cells were stained continuously before P14, and Purkinje cells stained transiently in a certain period. These findings may suggest much more efficient functions of mossy fibers in not only granule cell development but also in the entire development of cerebellar neuronal organization.

Acknowledgments. This work was supported by RIKEN Foundation and a grant-in-aid for encouragement of young scientists from the Ministry of Education, Science, Sports and Culture of Japan (to M.O.).

References

1. Itoh K, Ozaki M, Stevens B, et al (1997) Activity-dependent regulation of *N*-cadherin in DRG neurons: differential regulation of *N*-cadherin, NCAM, and L1 by distinct patterns of action potentials. J Neurobiol 33:735–748
2. Ozaki M, Sasner M, Yano R, et al (1997) Neuregulin-β induces expression of an NMDA receptor subunit. Nature (Lond) 390:691–694
3. Fischbach GD, Rosen KM (1997) ARIA: a neuromuscular junction neuregulin. Annu Rev Neurosci 20:429–458
4. Sandrock AW Jr, Dryer SE, Rosen KM, et al (1997) Maintenance of acetylcholine receptor number by neuregulins at the neuromuscular junction in vivo. Science 276:599–603
5. Honig MG, Hume RI (1989) DiI and DiO: versatile fluorescent dyes for neuronal labeling and pathway tracing. Trends Neurosci 12:333–335
6. Ozaki M, Kishigami S, Yano R (1998) Expression of receptors for neuregulins ErbB2, ErbB3 and ErbB4 in developing mouse cerebellum. Neurosci Res 30:351–354

Developmental Expression of Voltage-Gated Potassium Channels in Mouse Cerebellar Granule Cells In Vivo and in Microexplant Culture

KENSUKE NAKAHIRA[1], YOSHIHIKO WAKAZONO[1,3], RIICHI SHIBATA[1], TAKASHI KURAHASHI[1,3], AKIMICHI KANEKO[1,2], and KAZUHIRO IKENAKA[1]

Summary. We examined the expression of K^+ channel genes during the development of mouse cerebellar granule cells by in situ hybridization. The expression of Kv4.2 mRNA was observed in P3 cerebellum, mainly in the internal granule cell layer. Positive signals were also found in the inner side of the external granule cell layer, suggesting that the Kv4.2 gene starts to be expressed in the postmitotic granule cells. Because the Kv4.2 subunit is known to form a fast-inactivating K^+ channel, we then examined K^+ currents during the maturation of granule cells using an microexplant culture system. First, we identified cerebellar granule cells in this microexplant culture system by immunostaining with the granule cell-specific marker, Zic. This staining revealed that granule cells in this culture system migrated radially out from explants and then formed small clusters. Simultaneous recording of their morphology and channel activity was performed by a recording patch-pipette containing 0.2% Lucifer yellow. Both bipolar and T-shaped/dendrite-bearing cells generated voltage-gated fast-activating K^+ currents; however, inactivation of their currents was different. T-shaped cells, which were considered to be mature, showed a fast-inactivating current component. These results suggested that regulation of specific subunit expression might lead to electrophysiological maturation.

Key words. Patch-clamp—Zic—Kv4.2—Kv3.1

Introduction

Neurons originated from neuroepithelial cells migrate to their final positions and maturate to generate action potential, which is the most important characteristic of the neuron. However, the molecular mechanism that regulates maturation of the

[1] National Institute for Physiological Sciences, Okazaki National Institute, 38 Nishigonaka, Myodaiji-cho, Okazaki, Aichi 444-8585, Japan
[2] Department of Physiology, School of Medicine, Keio University, 35 Shinamomachi, Shinjuku-ku, Tokyo 160-8582, Japan
[3] Department of Biology, Osaka University, Toyonaka, Osaka 560-0043, Japan

action potential is still unknown. We have been trying to understand the maturation process of action potentials by studying the developmental profile of expression of voltage-gated ion channel genes.

In the amphibian spinal cord neuron, it is well known that the properties of the action potential, such as ionic dependency, amplitude, and duration, change during the course of development. The maturation of action potential is controlled by developmental changes of voltage-gated K^+ currents. Barish (1986) observed that voltage-gated K^+ currents changed from slowly activating current to a large fast-activating current, and this change occurred in parallel with the developmental changes in the action potential [1]. Ribera and Spitzer (1990) found that a fast-inactivating voltage-gated K^+ current, known as A-type current, appeared at the final stage of neuronal maturation [2]. This current is believed to control the frequency of repetitive firing. These findings suggested that the expression of voltage-gated K^+ current plays a key role in the maturation of the action potential in amphibian neurons.

It is reasonable to assume that K^+ currents play a similar role in the mammalian neuron; however, few studies have been reported so far, partly because of the lack of a convenient model system like the amphibian motoneuron. To clarify the molecular mechanism that regulates the maturation of action potential in mammals, we examined the expression of voltage-gated K^+ channels in developing granule cells by in situ hybridization and electrophysiological recording using a mouse cerebellar microexplant culture.

Materials and Methods

In Situ Hybridization of Mouse Cerebellum

cDNA fragments of Kv3.1 (nucleotide 94–283 of rat NGK2 [3]) and Kv4.2 (nucleotide 1248–1893 of rat shal1 [4]) were amplified by RT-PCR from adult mouse brain. Probes for both channels were prepared by labeling these cDNA fragments with digoxigenin (DIG). Neonatal mice (P3, P5, P7, and P10) were fixed with 4% paraformaldehyde (PFA) and 8-μm paraffin sections were prepared. The sections were treated with 0.8% pepsin in 0.2 N HCl and then hybridized with DIG-labeled cRNA probes. Signals were detected with anti-DIG antibody conjugated with AP.

Electrophysiological Recording

Microexplant culture of early postnatal ICR mice (P2–P4) cerebellum was performed according to the method of Nagata and Nakatsuji [5]. The electrical recording was performed with a solution containing 150 mM NaCl, 2.5 mM KCl, 20 mM glucose, 2.0 mM HEPES, pH 7.4, 1 mM tetrodotoxin (TTX), and 4 mM $CoCl_2$. The pipette was filled with a solution containing 150 mM KCl, 4.0 mM NaCl, 5 mM EGTA, 10 mM HEPES (pH 7.2), and 0.2% Lucifer yellow. Currents were recorded in a whole-cell voltage-clamp mode by using a patch-clamp amplifier (CEZ-2000; Nihon Khoden, Tokyo, Japan). All recordings were performed at room temperature (23°–25°C).

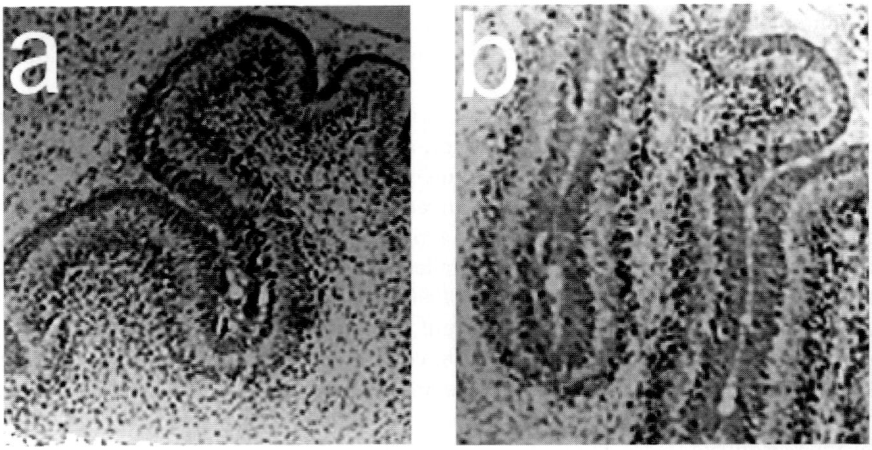

FIG. 1a,b. The expression of Kv3.1 and Kv4.2 mRNA in P3 mouse cerebellum was examined by in situ hybridization. **a** Kv3.1 mRNA was found in external granule layer (EGL) and internal granule layer (IGL). **b** Kv4.2 mRNA was found in the inner side of EGL and IGL but not in the outer zone of EGL

Results and Discussion

To elucidate the expression pattern of voltage-gated K^+ channel genes, the developmental profile of the mRNA expression was examined by in situ hybridization. We focused on Kv3.1 and Kv4.2 because their expression was reported to be specific to granule cells in adult cerebellum. Kv3.1 expression was observed in neonatal mouse cerebellum at all stages (Fig. 1a). The signal was restricted mainly in granule cells as reported and found in the external granule layer (EGL) and inner granule layer (IGL). Kv3.1 is reported to form a delayed rectifier-type channel when expressed in oocytes [6]; thus, this observation suggests that mitotic granule cells in the EGL already express delayed rectifier-type K^+ currents. On the other hand, Kv4.2 mRNA was not detected in the cells in the mitotic zone of EGL. The mRNA was found in the inner side of EGL, as well as in the cells in IGL, suggesting that granule cells start to express the Kv4.2 after their last mitosis. Because the Kv4.2 channel conducts fast-inactivating (A-type) current when expressed in oocytes [7], it is possible that these postmitotic granule cells start to express A-type currents.

Because developing granule cells change their morphology at the inner side of EGL from bipolar to T-shaped, we then examined morphological change and K^+ current expression during development using the microexplant culture system. This microexplant culture system was developed to investigate the process of neuronal differentiation in vitro by Nagata and Nakatsuji [5]. Unlike widely used primary culture methods, it is not necessary to culture either in high-K^+ medium or in contact with glia cells in this culture system. Immunostaining with an anti-Zic antibody (kindly provided by Dr. Jun Aruga) [8], which is specific to cerebellar granule cells, revealed that about 50% of migrating cells were Zic immunopositive, and that most of the Zic-immunopositive cells were localized in small clusters.

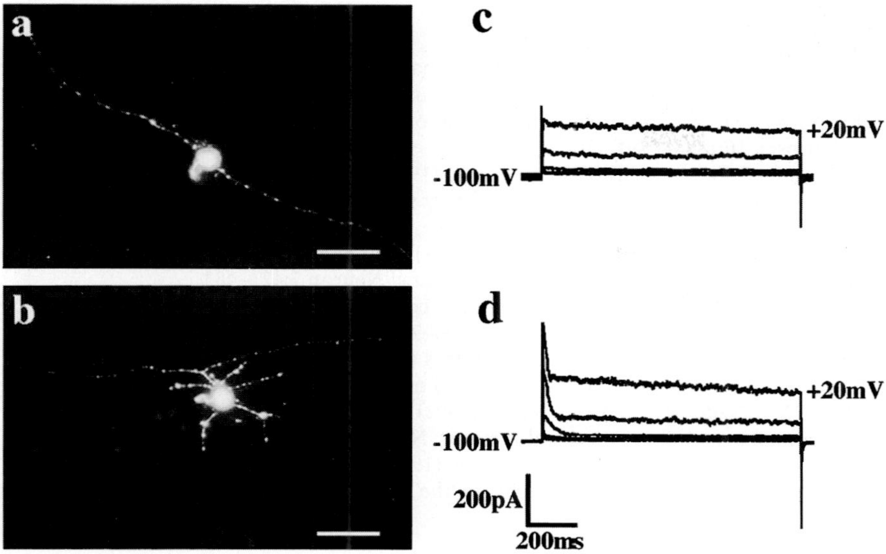

FIG. 2a–d. Simultaneous recording of K^+ currents and morphology of the cells in cerebellar microexplant culture. K^+ currents were recorded from bipolar granule cells (**a** and **c**) and T-shaped/dendrite-bearing cells (**b** and **d**). T-shaped cells showed fast-inactivating component. *Bar* 50 μm

By simultaneous recording of morphology and electrophysiological activity, bipolar and T-shaped cells in this culture were observed (Fig. 2a,b), and we considered that bipolar and T-shaped cells corresponded to immature and mature granule cells, respectively. Both bipolar and T-shaped cells exhibited a fast-activating K^+ current; however, the inactivation kinetics of their currents were remarkably different (Fig. 2c,d). T-shaped cells showed a fast-inactivating component in addition to a slow-inactivating component; in contrast, bipolar cells exhibited a very small or no fast-inactivating component. Activation and inactivation of this component were voltage dependent and it was 4-AP sensitive (data not shown). These data indicate that this current is a typical A-type current. These findings demonstrated that voltage-gated K^+ currents of developing granule cells change from delayed rectifier to A-type current in parallel with their morphological change from bipolar to T-shaped cell.

In the present study, developmental changes of voltage-gated K^+ currents were similar to that found in amphibian spinal cord neurons. It is suggested that the A current, which appeared in the maturation process of cerebellar granule cells, also plays a similar role in maturation of the action potential. Interestingly, this A current appeared in parallel with the morphological change of developing granule cells. Therefore, we also speculate that this A current plays an important role not only in the maturation of the action potential but also in neuronal differentiation. Further examination of the membrane excitability and morphological maturation using this culture system will provide the basis for understanding the role of the voltage-gated K^+ current in neuronal development.

References

1. Barish ME (1986) Differentiation of voltage-gated potassium current and modulation of excitability in cultured amphibian spinal neurons. J Physiol (Camb) 375:229–250
2. Ribera AB, Spitzer NC (1990) Differentiation of IKA in amphibian spinal neurons. J Neurosci, 10:1886–1891
3. Yokoyama S, Imoto K, Kawamura T, et al (1989) Potassium channels from NG108-15 neuroblastoma-glioma hybrid cells. Primary structure and functional expression from cDNAs. FEBS Lett 259:37–42
4. Baldwin TJ, Tsaur ML, Lopez GA, et al (1991) Characterization of a mammalian cDNA for an inactivating voltage-sensitive K^+ channel Neuron 7:471–483
5. Nagata I, Nakatsuji N (1990) Granule cell behavior on laminin in cerebellar microexplant cultures. Dev Brain Res 52:63–73
6. Weiser M, Vega-Saenz de Miera E, Kentros C, et al (1994) Differential expression of Shaw-related K^+ channels in the rat central nervous system. J Neurosci 14:949–972
7. Sheng M, Tsaur ML, Jan YN, et al (1992) Subcellular segregation of two A-type K^+ channel proteins in rat central neurons. Neuron 9:271–284
8. Aruga J, Yokota N, Hashimoto M, et al (1994) A novel zinc finger protein, Zic, is involved in neurogenesis, especially in the cell lineage of cerebellar granule cells. J Neurochem 63:1880–1890

Regulation of GABAergic Cell-Specific Gene Transcription

Yuchio Yanagawa[1], Takashi Kobayashi[1], Kenji Ishii[1], Hiroyuki Sakagami[2], Hisatake Kondo[2], Fumi Tashiro[3], Jun-ichi Miyazaki[3], and Shinri Tamura[1]

Summary. γ-Aminobutyric acid (GABA) is synthesized by glutamic acid decarboxylase (GAD), which is the most specific marker for GABAergic neurons. GAD has been shown to exist in two forms, GAD65 and GAD67, encoded by separate genes in mammals. To elucidate the regulatory mechanisms of mouse GAD67 (mGAD67) gene expression and to define the promoter region essential for its GABAergic cell-specific expression, we have isolated and characterized the 5′-flanking region and noncoding exons of the mGAD67 gene. Sequence analysis of the mGAD67 gene revealed a number of putative regulatory elements, including the neuron-restrictive silencer element, the CCAGGAG neuronal motif, and Krox-24 binding site. In transgenic mice, we have examined the expression pattern conferred by a 10.2-kilobase (kb) fragment of the mGAD67 gene upstream region fused to the *lacZ* reporter gene. The transgene expression was observed in neurons of particular brain regions such as olfactory tubercle, cerebral cortex, and basal ganglia, where expression of GAD67 gene has been reported to be abundant. These results suggest that the 10.2-kb DNA fragment of the mGAD67 gene may be essential for its GABAergic cell-specific expression in vivo.

Key words. GABAergic neuron—GAD67—Promoter—Transcription—Transgenic mouse

Introduction

γ-Aminobutyric acid (GABA) is considered to be a principal inhibitory neurotransmitter, and GABAergic neurons comprise more than 30% of all neurons in the adult mammalian central nervous system (CNS). GABA is synthesized by glutamic acid decarboxylase (EC 4.1.1.15; GAD), currently the most specific marker for GABAergic neurons. GABAergic neurons in the CNS are widely distributed, including cells in the

[1] Departments of Biochemistry, Institute of Development, Aging and Cancer and [2] Department of Anatomy, School of Medicine, Tohoku University, Seiryomachi, Aoba-ku, Sendai 980-8575, Japan

[3] Department of Nutrition and Physiological Chemistry, Osaka University Medical School, Suita, Osaka 565-0871, Japan

olfactory bulb, cerebral cortex, basal ganglia, hypothalamus, and cerebellum. These nuerons acquire their GABAergic phenotype progressively as the expression of GAD gene increases during development.

Two different genes encode two forms of glutamic acid decarboxylase (GAD65 and GAD67). They are distinguished by their molecular masses as well as their cofactor interactions and subcellular distributions [1]. The characterization of mouse GAD67 (mGAD67) cDNA has been shown to encode 515 amino acids [2]. The characterization of the mGAD67 gene has been shown to span more than 45 kilobases (kb) and to contain 17 coding exons and 2 alternatively spliced 5′-noncoding exons in addition to a fetal-specific exon [3].

To understand the molecular mechanisms for the transcriptional regulation of a GABAergic gene, we have isolated the 5′-flanking region of the mGAD67 gene. We report here the structural analysis of the mGAD67 gene promoter and the functional study of the upstream region with transgenic mice.

Materials and Methods

A 0.2-kb cloned fragment (corresponding to exon 0A and part of intron 0A) was used to screen a mouse genomic library. A recombinant phage clone extending from the 5′-flanking region to exon 1 was isolated, and then the 10.2-kb DNA fragment of the mGAD67 gene (comprising the 5′-flanking region, the first noncoding exons <exons 0A and 0B>, introns 0A and 0B, and part of the third exon named exon 1) was subcloned into pBluescript KS (−). This fragment was fused to the *lacZ* reporter gene in frame followed by a polyadenylation signal (Fig. 1).

Transgenic mice were created using standard microinjection techniques. Offspring and subsequent generations were genotyped using polymerase chain reaction and tail

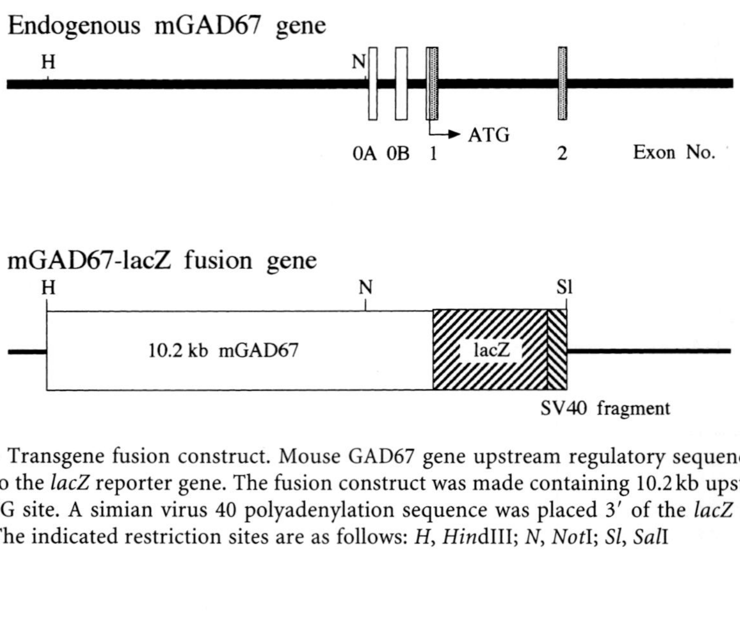

Fig. 1. Transgene fusion construct. Mouse GAD67 gene upstream regulatory sequences were fused to the *lacZ* reporter gene. The fusion construct was made containing 10.2 kb upstream of the ATG site. A simian virus 40 polyadenylation sequence was placed 3′ of the *lacZ* reporter gene. The indicated restriction sites are as follows: *H*, *Hin*dIII; *N*, *Not*I; *Sl*, *Sal*I

DNA Southern blots. Distribution of the transgene β-galactosidase expression driven by the mGAD67 promoter was determined using X-gal histochemical staining techniques [4].

DNA was sequenced by automated methods using PRISM Ready Reaction Dye Deoxy Termination Cycle Sequencing Kit (Perkin Elmer, Vaterstetten, Germany) according to the manufacturer's instructions.

Results and Discussion

To determine if mGAD67-lacZ expression within the brain was occurring in a fashion similar to that which is known for GAD67, β-galactosidase expression was analyzed in the brain of transgenic mice with X-gal histochemical staining. Table 1 lists a number of characteristic β-galactosidase staining areas of the nervous systems in the transgenic mice. The LacZ expression pattern in the forebrain and brainstem was largely in accord with GAD67 expression reported previously by immunohistochemistry and in situ hybridization: e.g., olfactory tubercle, piriform cortex, amygdaloid nuclei, cerebral neocortex, and basal ganglia. Moreover, *lacZ* expression was not observed in regions that are described to contain few or no GAD67-positive cells (e.g., dorsal thalamus and hanenula).

These results demonstrate that 10.2 kb of the mGAD67 upstream region may contain the necessary *cis*-regulatory elements to direct targeted expression of the transgene in the GABAergic neurons. However, little or no expression was observed in several regions containing GABAergic neurons such as hypothalamus and cerebellar cortex, suggesting the possibility that additional regulatory elements associated with these region-specific expression might be embedded in the far upstream sequences or intragenic sequences or in the 3′-flanking sequences.

TABLE 1. Transgene expression in the CNS of mGAD67-lacZ transgenic mice

Brain region	Transgene expression[a]	GAD67 mRNA expression[b]
Olfactory bulb	++	++
Cerebral cortex	++	++
Olfactory tubercle	+++	+++
Piriform cortex	+++	+++
Amygdaloid nuclei	++	++
Hippocampal formation	−/+	+
Basal ganglia	++/+++	++
Dorsal thalamus	−	−
Ventral thalamus	−	+++
Hypothalamus	−	++
Habenula	−	−
Midbrain	++	++
Pons	+	+
Medulla oblongata	+	+
Cerebellar cortex	−/+	++

[a] Relative intensities of the LacZ-staining in the cell bodies were determined by visual comparison as highest (+++) and undetectable (−).
[b] According to Benson et al. [8].

To aid the characterization of the promoter, we determined the nucleotide sequence of the 5′-flanking regions and noncoding exons (exons 0A and 0B) (Fig. 2). Previous studies demonstrated that each of the noncoding exons was expressed in mouse brain, suggesting the presence of regulatory elements in the promoter regions associated with neuron-specific expression [3]. As shown in Fig. 2, neuron-restrictive silencer element- (NRSE-) like sequences were located at position +24 to +38 (in exon 0A) and +811 to +797 (in exon 0B), and a neuronal motif-like element was

```
-535  CATACTGGCT CAGTAGAGCG CTCCCTAGGA CACCACCCAG AGAAAACCTT CTGGAATCGC CAATCCAGGA CAACCACAGG
                                                                                  Neuronal
-455  AGATTGATCA AGGCAAAGCA AATATTCTAG TCTTCTCCAA GATACGGGAT GGAGGGCTAA GAAGAGGGGA AGGAAGAATC
      motif
-375  ACCCAGTGAG ACAAAATTCT TCGTAGGAAT TATATTTTCC CTTGCCCTCA CCCAACATCG CCTATCTCAA AAATAATTTA

-295  AAAACAAAAA ACAAACAAAC AAACAAAAAA CAGAGCGTGC TGAGTGCATT CTGGATTACT CATAGGACTT TGTCACACAC

-215  ACCCTCCTTT CTGGTCGCAA ACCCGTGAGC TGGATTTATA ATCGCCCTAC AAAGCTCCAG AGGCAGTCAG ACACCTGCAA

-135  AGGAGCCCCA GCGCTCCGCG GACGAGCTGC CCCCGCGAGC AACGGCCTCG TGATTCCCCC CGCCGAGCGG GTCCCCGCCT
                                                             Sp1
 -55  CCCCACTCCG CCCCCGCCTC CCCCAAGCCC AGCGGCCGCC TCTCCGGATC TCTCCCTTCT TCAGGCTCTC CCGTGCCGGA
         EGR-1/Krox-24                                       (+1)  Exon 0A
 +26  CCAGGGATCG TGCAAGCAAG GAAGCAGCCC TGGGGTGACA CCCAGCACGT ACTCCTGTGA CAGAGCCGAG CCCAGCCCAG
        NRSE
+106  CCCCGGGACG CTTCGCAGAG GAGTCGCGGG AGGGTCCAGC TCGCTGTCGC TGAACCGAGG TGGGTAAATA CCGGATCACC

+186  CGGACTTGCC AGAGCCCCGC CTCGGTTATT CAACCCGGTG CCCTCAGCCT GCGGCTTCTT GTCCAGCCCT TTGCCTCACC

+266  CCACCAGCAC GCGCCTGTCG CCCACCTTCC ACAGCCGAGT GCTTCCCGCC AGTCCTGTCG CAGCTGTGGT TTGGCGCACC
                                                   Sp1
+346  CATCCCCTAC CCTCCTCGCT CTTGAAACCC TTATCAATTT CATTCGGGAG GACACATAAG GGCTGAAGTG ACCCCAGAGC

+426  CATTTTTTGC CTGCGCTTGG GGGATCGGAG CTCACGGGCT GTCGGATACG CGCCTGGGTA GCAGCGGCAG AAGGTGCGAG

+506  GAGCGATAAC CTGTTACTTC CCCAGCACCC GCGTCCCACC GCCGCCCGCG GAGCTGCGAG AGCGCGCGGA GGTTGCGGGG
                                          Sp1
+586  GGGGGGGGGA CAGATGCTCG GCCCGAGGGA CGACACGCAA AAGGCGGCTG GTGGGCTGCG CTCTGCGGGG TCTTAGATTT
                 E-box
+666  ACCCAGATTA CCGTTCTCCT CCCTCTCCTC TCATTCTTTC GCCCGCTCCT TCTCCCCATC TCTTCCTCTT CTCTCGCCCT

+746  CTCTCCCTCA CTCTGTCCAC GAACAAACAC CAAATAGGGG AACCGCGCAC TCTTCTATCC GAGGTCTCTT GCGCCCTAGG
                                         Exon 0B                NRSE
+826  CGCAAATTCG CCACCGGGTG ACGGTCTCCT TTTTACCCTC TGCCACCTTC TCTGGCTCAG TGAAACGAAC TCTGGGGACT
          F-ACT1
+906  AGGTAGAGTA GAAACAGTGA ATCTGGGACG GAGCCTCGCT CTAGGGTTGA GAAAGCTAGA AACGAGGTGG GACAAGAAAC

+986  TGATCAGCAA GGCTTCGACA CGCTGCCTTT TGGCTGTCTG GAGTGCAAGG TGACTGATTC TTAGCGGGTC AAGTCCTGAG
```

FIG. 2. Nucleotide sequence of the 5′-flanking region of the mouse GAD67 gene. The *numbers* along the sequence refer to nucleotide position relative to the transcription start site (+1) in exon 0A. The sequence of the noncoding exons (exon 0A and exon 0B) is *underlined*. Potential consensus sequences for regulatory elements and transcriptional factor binding sites are *double underlined*. The consensus elements are abbreviated as follows: *E-box*, E-box sequence; *NRSE*, neuron-restrictive silencer element; *Sp1*, Sp1-binding site; *EGR-1/Krox-24*, EGR-1/Krox-24 binding site; *F-ACT-1*, F-ACT-1 binding site present at position −460 to −450. NRSE, which represses transcription by association with a repressor protein known as NRSF, has been shown to downregulate the transcription of the type II sodium channel and SCG10 genes in nonneuronal cells [5].

Schoenherr et al. identified about 20 genes carrying NRSE-like sequences, and among these genes, in 10 of them NRSE was located in the 5′-noncoding region [6]. It is of interest to note that NRSE-like sequences in the mGAD67 gene are located in the 5′-noncoding regions. The neuronal motif CCAGGAG was previously described to be common to the rat type II sodium channel gene and the rat GAP-43 gene [5,7]. Mutational analysis of the NRSE- or neuronal motif-like elements in the mGAD67 gene will establish their specific contribution to the mGAD67 gene expression in the nervous system.

In conclusion, we have demonstrated that the 10.2-kb DNA fragment of the mGAD67 gene directed expression of the *lacZ* reporter gene to the restricted brain region containing GABAergic neurons. These studies provide insight into the molecular mechanisms regulating mGAD67 gene expression in brain, which ultimately account for the phenotype selection of GABAergic synthesis and the maintenance of GABAergic function.

Acknowledgments. This study was supported by Grants-in-Aid for Scientific Research from the Ministry of Education, Science, Sports, and Culture of Japan, ONO Medical Research Foundation, Smoking Research Foundation, Nakayama Foundation of Human Science, and 1st Toyota High-Tech Research Grant Program.

References

1. Erlander MG, Tobin AJ (1991) The structure and function of glutamic acid decarboxylase. Neurochem Res 16:215–226
2. Katarova Z, Szabo G, Mugnaini E, et al (1990) Molecular identification of the 62-kD form of glutamic acid decarboxylase from mouse. Eur J Neurosci 2:190–202
3. Yanagawa Y, Kobayashi T, Kamei T, et al (1997) Structure and alternative promoters of the mouse glutamic acid decarboxylase 67 gene. Biochem J 326:573–578
4. Langford KG, Shai SY, Howard TE, et al (1991) Transgenic mice demonstrate a testis-specific promoter for angiotensin-converting enzyme. J Biol Chem 266:15559–15562
5. Maue RA, Kraner SD, Goodman RH, et al (1990) Neuron-specific expression of the rat type II sodium channel gene is directed by upstream regulatory elements. Neuron 4:223–231
6. Schoenherr CJ, Anderson DJ (1995) The neuron-restrictive silencer factor (NRSF): a coordinate repressor of multiple neuron-specific genes. Science 267:1360–1363
7. Nedivi E, Basi GS, Akey IV, et al (1992) A neural-specific GAP-43 core promoter located between unusual DNA elements that interact to regulate its activity. J Neurosci 12:691–704
8. Benson DL, Isackson PJ, Gall CM, et al (1992) Contrasting patterns in the localization of glutamic acid decarboxylase and Ca^{2+}/calmodulin protein kinase gene expression in the rat central nervous system. Neuroscience 46:825–849

Part 7
Synaptic Refinement and Neurotransmitter Release

Membrane Trafficking in Neurons

Richard H. Scheller

In an attempt to define the molecular mechanisms that regulate membrane flow in the nerve terminal, we have characterized the proteins associated with the critical organelle in this process, the synaptic vesicle.

Studies of these molecules have led to a working hypothesis for synaptic vesicle docking and fusion. They propose that the plasma membrane protein syntaxin is associated with a soluble factor, n-secl, on the phasma membrane. As the vesicle docks, a 7S particle is formed as n-secl is displaced from the complex. The 7S particle is composed of two vesicle-associated proteins, VAMP and synaptotagmin, and two components of the plasma membrane, SNAP-25 and syntaxin. At least part of the specificity of vesicle docking may be determined by interactions of the components of this 7S particle. Formation of the 7S particle is proposed to drive membrane fusion. Only after α-SNAP is associated with the complex can NSF bind, forming an approximately 20S particle. On ATP hydrolysis by NSF, the particle dissociates, allowing recycling of the vesicle components.

Many of the molecules discussed in this model are structurally and functionally homologous to proteins that have been genetically identified in yeast to function in the secretory process. In yeast, 15 genes have been demonstrated to be required for the fusion of vesicles derived from the Golgi with the plasma membrane. Of these 15 genes, only 7 have known homologs in mammalian systems. Recently we have identified homologs of the yeast proteins sec6 and sec8; the rat homologs are referred to as rsec6 and rsec8. Both rsec6 and rsec8 are present in a 750-kDa complex composed of eight proteins. Amino acid sequencing of peptides derived from each of the bands reveals that each of the molecules is a novel, previously undescribed protein. The rsec6/8 complex is present in nerve terminals and is largely associated with the plasma membrane. Further biochemical and structural studies are aimed at understanding the specific roles of this large complex of proteins in secretion.

Howard Hughes Medical Institute, Department of Molecular and Cellular Physiology, Stanford University Medical School, Stanford, CA 94305-5323, USA

Doc2 as a Presynaptic Modulator for Neurotransmitter Release

Satoshi Orita[1], Takuya Sasaki[2], and Yoshimi Takai[2]

Summary. Doc2α is a protein having two C2 domains interacting with Ca^{2+} and phospholipid. Evidence is accumulating that Doc2α is implicated in Ca^{2+}-dependent neurotransmitter release. The mode of action of Doc2α is not fully understood, but we have recently found that it is not essential for the Ca^{2+}-dependent neurotransmitter release process but increases the efficacy of recruitment of synaptic vesicles in the reserve pool to the readily releasable pool in cooperation with Munc13.

Key words. Doc2—Munc13—C2 domain—Neurotransmitter release

Introduction

Synaptic vesicles are transported to the presynaptic plasma membrane where Ca^{2+} channels are located. Depolarization-induced Ca^{2+} influx through the Ca^{2+} channels into the cytosol of nerve terminals initiates the fusion of the vesicles with the plasma membrane, causing neurotransmitter release. Recent studies have revealed that this synaptic vesicle exocytosis process consists of many complicated steps, which are regulated by many components, including at least recruitment of vesicles from the reserve pool to the readily releasable pool and the fusion of the vesicles in the readily releasable pool.

Three systems that regulate these processes have thus far been proposed: the SNARE, Rab3A small G protein (Rab3A)-Rab GDI, and Doc2-Munc13 systems. In the SNARE system [1], syntaxin and SNAP-25 localized on the presynaptic plasma membrane (t-SNAREs) form a 7S complex with VAMP and synaptotagmin localized on synaptic vesicles (v-SNAREs). Synaptotagmin in this 7S complex is replaced by NSF and α-SNAP to form a 20S complex. The formation of these complexes causes the docking and/or fusion of the vesicles with the plasma membrane. In the Rab3A-Rab

[1] Shionogi Institute for Medical Science, Settsu, Osaka 566-0022, Japan
[2] Department of Molecular Biology and Biochemistry, Osaka University Medical School, Suita, Osaka 565-0871, Japan

GDI system [2], the GDP-bound form of Rab3A is complexed with Rab GDI and stays in the cytosol of nerve terminals. This complex is recruited to synaptic vesicles, where the GDP-bound form is converted to the GTP-bound form and interacts with rabphilin3, causing the translocation of the vesicles to the presynaptic plasma membrane. Before or after the fusion of the vesicles, the GTP-bound form is converted to the GDP-bound form and dissociates from rabphilin3. The GDP-bound form is then complexed with Rab GDI and the complex returns to the cytosol. In the Doc2-Munc13 system [3–6], Doc2 is localized on the vesicles whereas Munc13 is localized on the presynaptic plasma membrane. Binding of diacylglycerol to Munc13 causes the interaction of Munc13 with Doc2, and this interaction induces the docking of the vesicles to the plasma membrane.

In this chapter, we focus on the Doc2-Munc13 system in neurotransmitter release and induction of long-term potentiation (LTP).

Discovery of Doc2 as a Protein Having Two C2 Domains

During the course of the studies on rabphilin3, we isolated a novel protein having two C2 domains and named it Doc2 (Double C2) [3]. The C2 domain was first found in protein kinase C, which is activated by diacylglycerol or phorbol ester in the presence of Ca^{2+} and phospholipid [7]. Protein kinase C has one C1 domain and one C2 domain. Synaptotagmin and rabphilin3 had been shown to have two C2 domains, C2A and C2B [2,8]. Munc13, a mammalian homolog of unc-13, also has two C2 domains [9]. Munc13 has another atypical C2 domain and one C1 domain. Figure 1 schematically shows the domain organization of Doc2 in comparison with those of other proteins having the C2 domain. Doc2 has two members, Doc2α and Doc2β [3,4]. Both the members have two

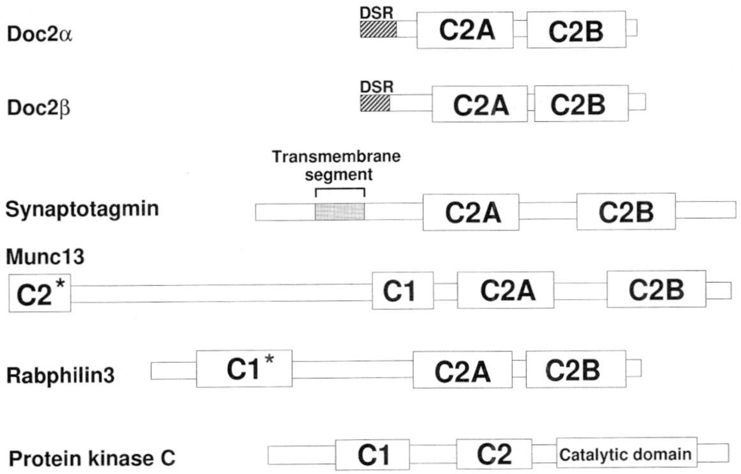

FIG. 1. The structure of Doc2 in comparison with the structures of other proteins having the C2 domain. The shows the relative locations of the Doc2-specific region (*DSR*), C2A domain (*C2A*), C2B domain (*C2B*), C1 domain (*C1*), C2 domain (*C2*), atypical C2 domain (*C2**), and Rab3A-binding C1-like domain (*C1**)

C2 domains and one highly homologous region, named a Doc2-specific region (DSR). Doc2α and -β have neither a transmembrane segment nor a Rab3A-binding domain. Doc2α is specifically expressed in neural tissue, where it is highly concentrated on synaptic vesicles, whereas Doc2β is ubiquitously expressed.

Implication of Doc2α in Ca^{2+}-Dependent Exocytosis

The properties of Doc2α, which are similar to those of synaptotagmin and rabphilin3, suggested that it might also be implicated in Ca^{2+}-dependent neurotransmitter release. To prove this assumption, we took advantage of the growth hormone (GH) coexpression assay system of PC12 cells, in which GH is stored in dense core vesicles and released in response to high K^+ in an extracellular Ca^{2+}-dependent manner. Overexpression of Doc2α stimulated the Ca^{2+}-dependent, high K^+-induced release of coexpressed GH without affecting the basal release [5]. In the PC12 cells transfected with a plasmid with the coding sequence of Doc2α in the antisense orientation, the high K^+-induced release was inversely inhibited. The Doc2α mutant expressing an N-terminal fragment or a C-terminal fragment inhibited the high K^+-induced release. These results suggest that Doc2α is involved in Ca^{2+}-dependent exocytosis from PC12 cells and that Doc2α interacts with another component of this exocytotic machinery.

Interaction of Doc2α with Munc13

To clarify the mode of action of Doc2α in Ca^{2+}-dependent exocytosis, it is important to isolate its interacting protein(s). We attempted to isolate a Doc2α-interacting protein from a rat brain cDNA library by use of the yeast two-hybrid system and isolated Munc13 as a Doc2α-interacting protein [6]. Doc2α and Munc13 interacted with each other through the N-terminal region of Doc2α (the Mid domain localized in DSR: amino acid residues 13–37) and the C-terminal region of Munc13 (the Did domain between the C2A and C2B domains: amino acid residues 851–1461). We confirmed that Doc2α interacted with Munc13 both in a cell-free system and in intact PC12 cells during the high K^+-induced Ca^{2+}-dependent exocytosis. The Doc2α–Munc13 interactions were stimulated by phorbol ester through the C1 domain of Munc13. Overexpression of the Did domain of Munc13 reduced the Ca^{2+}-dependent exocytosis from PC12 cells, and coexpression with Doc2α suppressed this reduction. These results, together with the earlier findings that secretagogues produce diacylglycerol and elevate cytoplasmic Ca^{2+} [10], suggest that diacylglycerol-induced Doc2α–Munc13 interactions play an important role in Ca^{2+}-dependent exocytotic machinery, as schematically shown in Fig. 2.

Implication of Doc2α–Munc13 Interactions in Ca^{2+}-Dependent Neurotransmitter Release

We next attempted to clarify whether the Doc2α–Munc13 interactions are indeed implicated in Ca^{2+}-dependent neurotransmitter release and at which stage in the neurotransmitter release process these interactions regulate in the neuron. For this

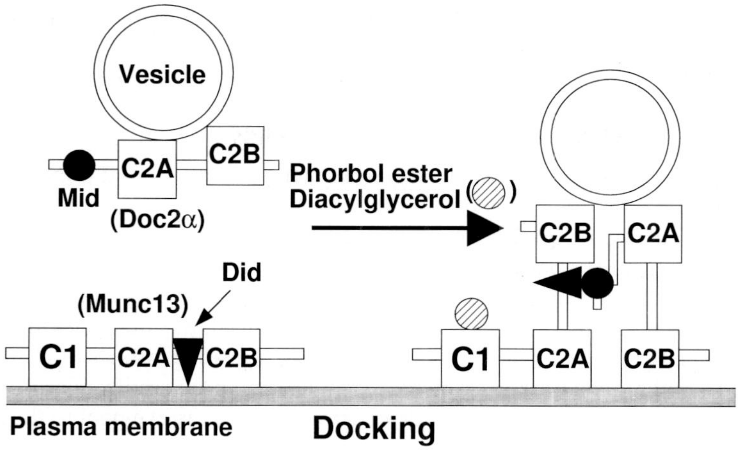

FIG. 2. A model for the mode of action of the Doc2α-Munc13 system in Ca^{2+}-dependent exocytosis

purpose, we analyzed nerve-impulse-evoked transmitter release between pairs of cultured superior cervical ganglion neurons (SCGNs) in which the synthetic peptide of the Mid domain (the Mid peptide) is introduced into the presynaptic partner [11]. Cultures of SCGNs are favorable for these experiments because peptides can be introduced into the relatively large presynaptic cell bodies by microinjection, the injected peptides can rapidly diffuse to nerve terminals forming synapse with adjacent neurons, and the effects of stimulated release of acetylcho-line can be accurately monitored by recording the excitatory postsynaptic potentials evoked by action potentials in the presynaptic neurons. The Mid peptide inhibited in vitro Doc2α–Munc13 interactions. Introduction of the Mid peptide into presynaptic neurons of cholinergic synapses, formed between pairs of SCGNs, reversibly inhibited synaptic transmission. This inhibitory effect was dependent on presynaptic activity and on extracellular Ca^{2+} concentrations. The onset of the Mid peptide effect was shortened when stimulated at a higher frequency, and the inhibition was more potent at 1 mM Ca^{2+} than at 5.1 mM Ca^{2+}. These results suggest that Doc2α–Munc13 interactions play a Ca^{2+}-dependent role in a step before the final fusion step of synaptic vesicles with the presynaptic plasma membrane in the neurotransmitter release process.

Implication of Doc2α in Postsynaptic LTP Formation

LTP is a candidate cellular mechanism for learning and memory. LTP in the CA1 region of the hippocampus is induced postsynaptically by activation of the *N*-methyl-D-aspartate receptor (NMDAR) [12] and it is proposed that the metabotropic glutamate receptor (mGluR) may also be involved in LTP induction [13,14]. Ca^{2+} influx through NMDAR and the components of the signaling pathway from mGluR are implicated in the LTP induction. These receptors are activated by glutamate released from nerve terminals, and a continuous supply of glutamate to its receptors during conditioning is essential for the postsynaptic induction of LTP [15]. Therefore, the components implicated in recruitment of the vesicles from the reserve pool to the readily releasable pool, which would ensure the continuous supply of vesicles, are also likely to regulate the postsynaptic mechanism for the LTP induction. The mode of action of Doc2α, that is, it regulates this recruitment, suggests that it is implicated in LTP formation.

To study this role of Doc2α, we generated homozygous null Doc2α mutant mice (Doc2α −/− mice) (G. Sakaguchi et al., in manuscript). In Doc2α −/− mice, frequency facilitation, a form of short-term plasticity induced by 5-Hz stimulation in the CA1 region of the hippocampus, was markedly enhanced, but the following decline was significantly steeper than in wild-type mice, suggesting two distinct roles for Doc2α: (1) inhibition of the release of docked vesicles from the readily releasable pool; and (2) efficient recruitment of synaptic vesicles from the reserve pool to the readily releasable pool. The latter mode of action of Doc2α is consistent with the observations that Doc2α interacts with Munc13 and Munc18 [6,16], because Munc18 interacts with syntaxin and inhibits initiation of the SNARE complex formation [17]. It is possible that the Doc2α–Munc13 interactions increase the efficacy in displacing Munc18 from syntaxin, initiating the SNARE complex formation. As to the role of Doc2α in the fusion step, it could be speculated that synaptotagmin I, Doc2α, and rabphilin3 may form a scaffold-like structure and serve as Ca^{2+} sensors at the fusion step, because all these proteins have two C2 domains interacting with Ca^{2+} and phospholipid, and electrophysiological studies have revealed that there are Ca^{2+} sensors other than synaptotagmin [18].

In addition to altered short-term plasticity, LTP was impaired in Doc2α −/− mice, but the impairment was overcome by a low concentration of mGluR agonist. These results suggest that the activation of mGluR is essential for LTP induction and that the reduced LTP in mutant mice is likely to be caused by rapid depletion of vesicles during tetanic stimulation, which suppresses the continuous supply of glutamate for activation of mGluR.

Conclusion

Doc2α is not essential for the Ca^{2+}-dependent neurotransmitter release process but increases the efficacy of the recruitment of synaptic vesicles in the reserve pool to the readily releasable pool in cooperation with Munc13. This Doc2α-Munc13 system may regulate the SNARE system, which is essential for neurotransmitter release. It may be noted that, of the many components of the neurotransmitter release machinery, Doc2α, which is not essential but increases efficacy for neurotransmitter release, is involved in the LTP formation. In addition to this role, Doc2α appears to regulate the fusion step, but further study is necessary for the mode of action of Doc2α in the fusion step.

Acknowledgment. Work concerning the analysis of the physiological function of Doc2 was performed in collaboration with Dr. Mochida (Tokyo Medical College) and Drs. Manabe and Takahashi (Tokyo University).

References

1. Rothman JE (1994) Mechanisms of intracellular protein transport. Nature (Lond) 372:55–63
2. Takai Y, Sasaki T, Shirataki H, et al (1996) Rab3A small GTP-binding protein in Ca^{2+}-dependent exocytosis. Genes Cells 1:615–632
3. Orita S, Sasaki T, Naito A, et al (1995) Doc2: a novel brain protein having two repeated C2-like domains. Biochem Biophys Res Commun 206:439–448
4. Sakaguchi G, Orita S, Maeda M, et al (1995) Molecular cloning of an isoform of Doc2 having two C2-like domains. Biochem Biophys Res Commun 217:1053–1061
5. Orita S, Sasaki T, Komuro R, et al (1996) Doc2 enhances Ca^{2+}-dependent exocytosis from PC12 cells. J Biol Chem 271:7257–7260
6. Orita S, Naito A, Sakaguchi G, et al (1997) Physical and functional interactions of Doc2 and Munc13 in Ca^{2+}-dependent exocytotic machinery. J Biol Chem 272:16081–16084
7. Nishizuka Y (1988) The molecular heterogeneity of protein kinase C and its implications for cellular regulation. Nature (Lond) 334:661–665
8. Südhof TC, Rizo J (1996) Synaptotagmins: C2-domain proteins that regulate membrane traffic. Neuron 17:379–388
9. Brose N, Hofmann K, Hata Y, et al (1995) Mammalian homologues of *Caenorhabditis elegans unc-13* gene define novel family of C2-domain proteins. J Biol Chem 270:25273–25280
10. Wakade TD, Bhave SV, Bhave AS, et al (1991) Depolarizing stimuli and neurotransmitters utilize separate pathways to activate protein kinase C in sympathetic neurons. J Biol Chem 266:6424–6428
11. Mochida S, Orita S, Sakaguchi G, et al (1998) Role of the Doc2α-Munc13-1 interaction in the neurotransmitter release process. Proc Natl Acad Sci USA (in press)
12. Malenka RC, Nicoll RA (1993) NMDA-receptor-dependent synaptic plasticity: multiple forms and mechanisms. Trends Neurosci 16:521–527
13. Bashir ZI, Bortolotto ZA, Davies CH, et al (1993) Induction of LTP in the hippocampus needs synaptic activation of glutamate metabotropic receptors. Nature (Lond) 363:347–350
14. Lu YM, Jia Z, Janus C, et al (1997) Mice lacking metabotropic glutamate receptor 5 show impaired learning and reduced CA1 long-term potentiation (LTP) but normal CA3 LTP. J Neurosci 17:5196–5205
15. Malenka RC (1991) Postsynaptic factors control the duration of synaptic enhancement in area CA1 of the hippocampus. Neuron 6:53–60
16. Verhage M, de-Vries KJ, Roshol H, et al (1997) DOC2 proteins in rat brain: complementary distribution and proposed function as vesicular adapter proteins in early stages of secretion. Neuron 18:453–461
17. Pevsner J, Hsu SC, Braun JE, et al (1994) Specificity and regulation of a synaptic vesicle docking complex. Neuron 13:353–361
18. Zucker RS (1996) Exocytosis: a molecular and physiological perspective. Neuron 17:1049–1055

Roles of Synaptotagmin and Inositol Polyphosphates in the Mechanism of Exocytosis: The Clamp Hypothesis

KONOSUKE KUMAKURA[1], MICA OHARA-IMAIZUMI[1], NOBUYUKI SASAKAWA[2], MITSUNORI FUKUDA[3], MICHIO NIINOBE[4], and KATSUHIKO MIKOSHIBA[3]

Summary. We have studied the effects of antibodies against the C2A and C2B domains of synaptotagmin (Syt) on catecholamine release from digitonin-permeabilized chromaffin cells to elucidate the roles of the C2A and C2B domains in the regulation of exocytosis [1]. Anti-C2A antibody reduced Ca^{2+}-evoked ATP-independent release, while anti-C2B antibody increased spontaneous release with little effect on Ca^{2+}-evoked release. Inositol polyphosphates (IPS) that bind the C2B domain [2] inhibited both Ca^{2+}-evoked and spontaneous release. The inhibition was reversed by anti-C2B antibody and by high Ca^{2+}. Therefore, IPS binding to the C2B domain may keep the primed vesicles away from spontaneous fusion at resting stage. Binding of Ca^{2+} to the C2A domain may facilitate or trigger vesicular fusion by releasing this suppression by IPS. In these cells, a depolarizing stimulation produces a rapid and transient accumulation of IP_5 [3]. Stimulation of the permeabilized cells with Ca^{2+} also resulted in a rapid and transient increase of IPS in the medium. The effects of anti-C2A and anti-C2B antibodies suggested that the accumulation is partly caused by the liberation of IPS from the C2B domain of Syt on binding of Ca^{2+} to the C2A domain, thus supporting our clamp hypothesis.

Key words. Chromaffin cells—Exocytosis—Synaptotagmin—C2 domains—Inositol polyphosphates

Introduction

Synaptotagmin (Syt) may function as a Ca^{2+} sensor in synaptic vesicle exocytosis [4,5]. The cytoplasmic carboxyl terminus of Syt has two repeats, known as the C2A and C2B domains, which are homologous to the C2 regulatory domain of protein

[1] Life Science Institute, Sophia University, 7-1 Kioi-cho, Chiyoda-ku, Tokyo 102-8554, Japan
[2] Department of Pharmacology, School of Medicine, Keio University, 35 Shinanomachi, Shinjuku-ku, Tokyo 160-8582, Japan
[3] Department of Molecular Neurobiology, Institute of Medical Science, The University of Tokyo, 4-6-1 Shirokanedai, Minato-ku, Tokyo 108-8639, Japan
[4] Institute for Protein Research, Osaka University, 3-2 Yamadaoka, Suita, Osaka 565, Japan

kinase C [6]. It has recently been shown that the C2A domain of Syt binds phospholipid in a Ca^{2+}-dependent manner, while the C2B domain binds phospholipids irrespective of Ca^{2+} and binds also inositol polyphosphates (IPS) [2]. Microinjection studies with IPS [7], anti-C2A antibody [8], or anti-C2B antibody [9] in the squid giant synapse suggested a bidirectional role of Syt in fast transmitter release in the neuron. However, the role of Syt in slow release in secretory cells as well as the functional relationship between the C2A and C2B domains remained obscure. We examined the effects of antibodies against C2A and C2B domains and of IPS on the release of catecholamines (CA) in digitonin-permeabilized chromaffin cells to address these questions [1]. In addition, we examined the effects of these antibodies on the Ca^{2+}-induced rapid accumulation of IPS in the medium for the permeabilized cells.

Materials and Methods

Chromaffin cells were isolated from fresh bovine adrenal medulla and cultured as described [10]. The cells were permeabilized by incubating for 4 min at 37°C with digitonin buffer (DG buffer), and CA release from the permeabilized cells was detected as previously described [1]. Polyclonal antibodies against the C2A and C2B domains of the mouse Syt II (referred to as anti-C2A antibody and anti-C2B antibody, respectively) were prepared and characterized as previously described [11].

To examine the accumulation of IPS in the medium, the cells were labeled with [^{3}H]inositol for 3 days in culture [3]. The cells were permeabilized with DG buffer for 4 min, and then stimulated with Ca^{2+} (0 or 100 μM) buffer for 1 min in the absence of MgATP. An aliquot of the medium was taken at 15 s and 1 min after the stimulation, and the accumulated isoforms of IPS were separated by HPLC, and the effluent radioactivity was monitored by an on-line flow detector [3]. Inositol 1,3,4,5-tetrakisphosphate (IP_4) was obtained from Dojindo (Kumamoto, Japan) and inositol 1,3,4,5,6-pentakisphosphate (IP_5) and inositol 1,2,3,4,5,6-hexakisphosphate (IP_6) from Calbiochem (SanDiego, CA, USA).

Results and Discussion

Both anti-C2A and anti-C2B antibodies used in this study recognized Syt I or II in bovine adrenal chromaffin cells as a 65-kDa immunoreactive band [1]. The anti-C2A antibody at 40 μg/ml inhibited CA release evoked by 10 μM Ca^{2+} in the presence of MgATP to about 60% of that in the presence of control IgG. The anti-C2A antibody did not affect the spontaneous Ca^{2+}-independent release (Fig. 1a). Ca^{2+}-evoked CA release from the permeabilized chromaffin cells has been resolved into an ATP-dependent priming step and ATP-independent Ca^{2+}-triggered step [12]. Inhibition by anti-C2A antibody was more evident when the release was evoked by 100 μM Ca^{2+} in the absence of MgATP, where the release was reduced to 40% of that in control or in the presence of control IgG [1]. Anti-C2A antibody clearly reduced the rate of initial release for 1–1.5 min from 3.6%/min to 1.5%/min, while the rate of the release in the second phase was less affected [1]. Because the same anti-C2A antibody inhibited Ca^{2+}/phospholipid binding to the C2A domain [11], these results suggested that the C2A domain functions as a Ca^{2+} sensor that directly facilitates vesicular fusion.

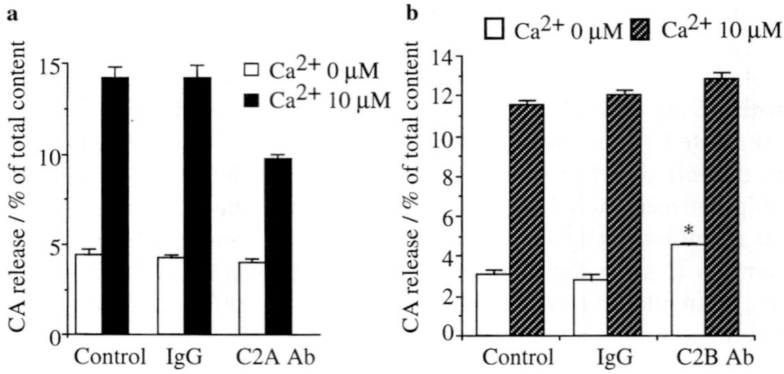

FIG. 1a,b. Effects of (**a**) anti-C2A antibody (40 μg/ml) and (**b**) anti-C2B antibody (100 μg/ml) on the spontaneous and evoked release of CA from digitonin-permeabilized chromaffin cells. The cells were permeabilized with digitonm (DG) buffer for 4 min, and then stimulated with Ca^{2+} (0 or 10 μM) buffer for 1.5 min in the presence of MgATP. The antibodies or control IgG (40 or 100 μg/ml) was added during permeabilization and stimulation. Data are mean ± SEM of at least eight determinations from individual wells. *, $P < 0.001$

In contrast, anti-C2B antibody did not affect Ca^{2+}-evoked release by itself, even at the higher concentration (100 μg/ml), but consistently augmented the spontaneous release (Fig. 1b). This increase in the spontaneous release by anti-C2B antibody presumably results because it displaces the bound IPS at C2B domain of Syt. In fact, IPS such as IP_4, IP_5, and IP_6 reduced the initial rate of CA release evoked by 10 μM Ca^{2+} as well as the spontaneous release (Fig. 2a). The inhibitory action of IPS was overcome by anti-C2B antibody. The Ca^{2+}-evoked release during the initial 1.5 min was reduced to 10%–20% of control by 3×10^{-5} M IP_6 and was reversed to 50% of control by anti-C2B antibody. The inhibition by IP_6 of the spontaneous release was also reversed by anti-C2B antibody [1]. Taken together with the facts that anti-C2B antibody inhibits IPS binding to the C2B domain [11], IPS binding to the C2B domain of Syt appears to prevent its interaction with phospholipids or with proteins of plasma membrane and thus suppresses Ca^{2+}-triggered vesicular fusion.

The binding of IPS to the C2B domain may be important as an inhibitory clamp to keep primed or docked vesicles away from spontaneous fusion at the resting level of intracellular Ca^{2+}. IP_6 at 10^{-4} M strongly inhibited the ATP-independent release evoked by Ca^{2+} at less than 10 μM, but the inhibitory action of IP_6 was much less when the release was evoked by Ca^{2+} greater than 50 μM. Moreover, when the release was evoked by Ca^{2+} greater than 200 μM, IP_6 could not produce any significant inhibition (Fig. 2b). These results suggest the physiological significance of the inhibitory action of IPS and a role of Ca^{2+} to counterbalance the inhibitory clamp.

On the basis of these observations, we assume that Ca^{2+} binding to the C2A domain of Syt, which results in release of IPS from the C2B domain, may facilitate an interaction of Syt with other proteins such as syntaxin [13,14], thus facilitating Ca^{2+}-triggered vesicular fusion. Rapid accumulation of IP_5 at a depolarizing stimulation has been reported in adrenal chromaffin cells [3], as well as in N1E-115 neuroblastoma cells and in cultured rat granule cells [15,16]. A rapid accumulation of IPS in the cytosolic

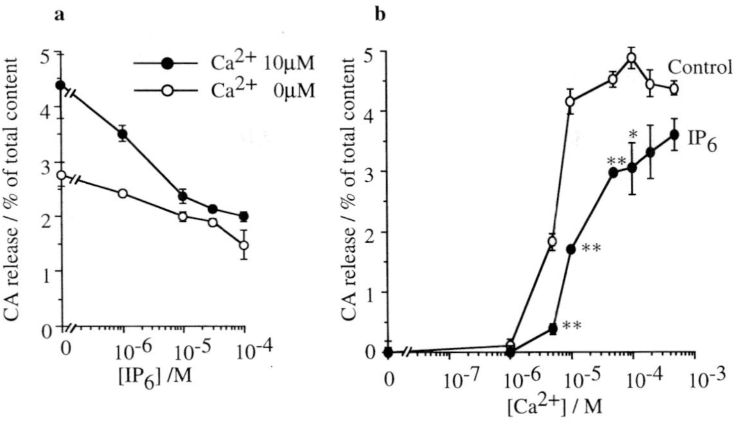

FIG. 2. **a** Dose–response curve for inositol polyphosphate (IPS) inhibition of evoked and spontaneous catecholamine (CA) release from digitonin-permeabilized cells during the initial phase. The cells were permeabilized with DG buffer for 4 min, incubated two times for 2 min with the buffer without digitonin, and then stimulated with Ca^{2+} (0 or 10 μM) buffer for 1.5 min. IPS (IP_6) was added during incubation for the last 2 min and the subsequent stimulation. **b** Effects of Ca^{2+} on inhibition by IP_6 of ATP-independent release from digitonin-permeabilized cells. The cells were permeabilized as described in **a**, then stimulated with Ca^{2+} (0–500 μM) buffer (Ca^{2+}-HEEDTA buffer) for 1.5 min in the absence of MgATP; 5 mM MgATP was present during permeabilization and incubation for the first 2 min. IP_6 (10^{-4} M) was added during incubation for the last 2 min and the subsequent stimulation. The amount of CA released in the absence of Ca^{2+} was measured for each group, and the mean value for each group was subtracted from each evoked release. Data are mean ± SEM of at least eight determinations from individual wells. *, $P < 0.01$, **, $P < 0.001$, compared to control

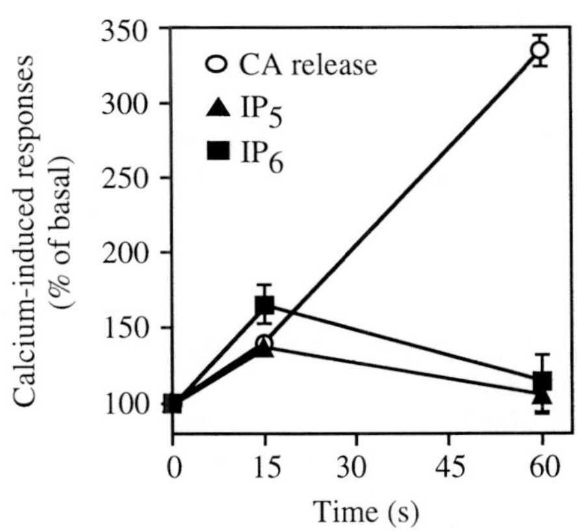

FIG. 3. Rapid accumulation of IPS in the medium after stimulation of digitonin-permeabilized cells with Ca^{2+} (100 μM). Inositol polyphosphates and CA in the same aliquots were measured as described in Materials and Methods. Data are mean ± SEM of at least six determinations from individual wells

fraction also occurred in the permeabilized chromaffin cells (Fig. 3). The concentration of the cytosolic IPS reached the maximum at 15 s following Ca^{2+} challenge preceding the release of CA, and declined to the basal level at 1 min while CA release was persisting. These accumulations may be partly the result of liberation of IPS from the C2B domain of Syt on the binding of Ca^{2+} to the C2A domain. Indeed, anti-C2A antibody abolished Ca^{2+}-induced rapid increase of cytosolic IPS, and anti-C2B

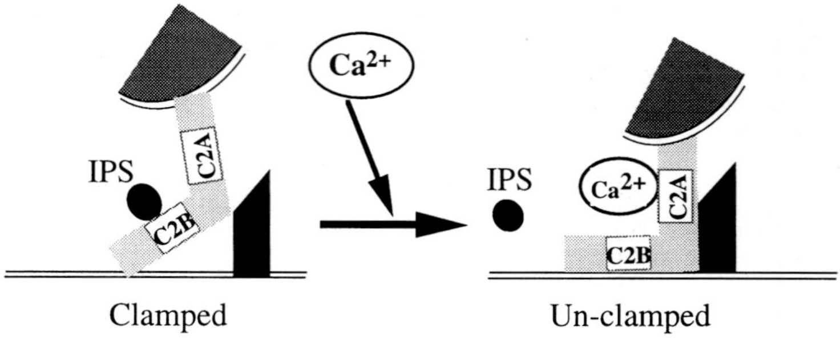

FIG. 4. Clamp by synaptotogmin (Syt) through binding of IPS to the C2B domain, and the role of Ca^{2+} to counterbalance the inhibitory clamp through binding to the C2A domain

antibody by itself resulted in an accumulation of IPS in DG buffer during the permeabilization (Sasakawa et al., in preparation).

In conclusion, all the results suggested that IPS binding to C2B domain arrests membrane fusion by presumably preventing interaction of Syt with phospholipids or with proteins of the plasma membrane. Thus, IPS binding to the C2B domain may keep the docked or primed vesicles away from spontaneous fusion at the resting level of intracellular Ca^{2+}. Binding of the increased intracellular Ca^{2+} to the C2A domain may facilitate or trigger vesicular fusion by counterbalancing this inhibitory clamp. The hypothesis was supported by the effects of anti-C2A and anti-C2B antibodies on the Ca^{2+}-induced accumulation of IPS in the cytosolic fraction. The hypothesis is schematically summarized in Fig. 4.

References

1. Ohara-Imaizumi M, Fukuda M, Niinobe M, et al (1997) Distinct roles of C2A and C2B domains of synaptotagmin in the regulation of exocytosis in adrenal chromaffin cells. Proc Natl Acad Sci USA 94:287–291
2. Fukuda M, Aruga J, Niinobe M, et al (1994) Inositol-1,3,4,5-tetrakisphosphate binding to C2B domain of IP4BP/synaptotagmin II. J Biol Chem 269:29206–29211
3. Sasakawa N, Nakaki T, Kato R (1990) Stimulus-responsive and rapid formation of inositol pentakisphosphate in the cultured adrenal chromaffin cells. J Biol Chem 265:17700–17705
4. Scheller RH (1995) Membrane trafficking in the presynaptic nerve terminal. Neuron 14:893–897
5. Südhof TC (1995) The synaptic vesicle cycle—a cascade of protein-protein interactions. Nature (Lond) 375:645–653
6. Perin MS, Fried VA, Mignery GA, et al (1990) Phospholipid binding by a synaptic vesicle protein homologous to the regulatory region of protein kinase C. Nature (Lond) 345:260–263
7. Llinás R, Sugimori M, Lang EJ, et al (1994) The inositol high-polyphosphate series blocks synaptic transmission by preventing vesicular fusion. Proc Natl Acad Sci USA 91:12990–12993
8. Mikoshiba K, Fukuda M, Moreira JE, et al (1995) Role of the C2A domain of synaptotagmin in transmitter release as determined by specific antibody injection into the squid giant synapse preterminal. Proc Natl Acad Sci USA 92:10703–10707

9. Fukuda M, Moreira JE, Lewis FMT, et al (1995) Role of the C2B domain of synaptotagmin in vesicular release and recycling as determined by specific antibody injection into the squid giant synapse preterminal. Proc Natl Acad Sci USA 92:10708–10712
10. Kumakura K, Sasaki S, Sakurai T, et al (1994) Essential role of myosin light chain kinase in the mechanism for MgATP-dependent priming of exocytosis in adrenal chromaffin cells. J Neurosci 14:7695–7703
11. Fukuda M, Kojima T, Aruga J, et al (1995) Functional diversity of C2 domains of synaptotagmin family: mutational analysis of inositol high polyphosphate binding domain. J Biol Chem 270:26523–26527
12. Bittner MA, Holz RW (1992) Kinetic analysis of secretion from permeabilized adrenal chromaffin cells reveals distinct components. J Biol Chem 267:16219–16225
13. Chapman ER, Hanson PI, An S, et al (1995) Ca^{2+} regulates the interaction between synaptotagmin and syntaxin. J Biol Chem 270:23667–23671
14. Li C, Ullrich B, Zhang JZ, et al (1995) Ca^{2+} dependent and independent activities of neural and non-neural synaptotagmins. Nature (Lond) 375:594–599
15. Sasakawa N, Nakaki T, Kashima R, et al (1992) Stimulus-induced accumulation of inositol tetrakis-, pentakis-, and hexakisphosphates in N1E-115 neuroblastoma cells. J Neurochem 58:2116–2123
16. Sasakawa N, Nakaki T, Kakinuma E, et al (1993) Increase in inositol tris-, pentakis- and hexakisphosphates by high K^+ stimulation in cultured rat cerebellar granule cells. Brain Res 623:155–160

Functional Analysis of HPC-1/Syntaxin 1A and Its Relation to Williams Syndrome

Kimio Akagawa

Summary. The function of HPC-1/syntaxin 1A in exocytosis was studied using various types of cells that had exocytotic processes. Suppression of HPC-1 expression in neurons or PC12 cells increased transmitter release, while overexpression of this protein in the β-cells resulted in decrease of insulin secretion. These results strongly suggested that HPC-1 might play a suppressive role in regulated exocytosis. In addition, suppression of HPC-1 in neurons enhanced neuronal sprouting, implying the involvement of HPC-1 in the sprouting mechanism of neurons. Fluorescent in situ hybridization studies revealed that the HPC-1 gene was hemizygously deleted in Williams syndrome patients (WS), implying that HPC-1 might be a candidate gene responsible for the neuropsychological symptoms in WS.

Key words. HPC-1—Exocytosis—Neuronal sprouting—Gene deletion—Williams syndrome

Introduction

HPC-1/syntaxin1A was originally found as a neuronal membrane protein [1,2]. HPC-1 belongs to the syntaxin protein family, which mediates intracellular membrane transport in many types of cells [3]. It has been believed that HPC-1 played an important role in exocytosis [4]. However, the precise physiological function of this protein is not as clear because this protein is distributed widely on the axonal membrane in addition to the synaptic regions. In this context, we have been performing cell biological studies on the function of HPC-1. This paper reviews our recent work on HPC-1 and shows a possible relevancy between its gene and a human congenital disease.

Department of Physiology, Kyorin University School of Medicine, 6-20-2 Shinkawa, Mitaka, Tokyo 181, Japan

Suppressive Role of HPC-1 in Exocytosis

To study the physiological meaning of HPC-1/syntaxin 1A in exocytosis, several distinct types of experiments were attempted. First of all, intracellularly applied anti-HPC-1 IgG enhanced the amplitude of an excitatory postsynaptic current by increasing transmitter release from the presynaptic terminals of self-recurrent synapses in vitro [5]. Also, application of the anti-IgG into the permeabilized PC12 cells increased calcium-dependent catecholamine release [6]. In a cell line of the β cell, overexpression of HPC-1 decreased insulin release by high glucose stimulation [7]. Ectopic expression of HPC-1 protein in an amphibian early embryo caused accumulation of secretory vesicles containing extracellular matrix (ECM) in the vicinity of the plasma membrane of the ectodermal cells, resulting in reduction of ECM secretion (Fig. 1) [8]. In the transformed PC12 cells in which HPC-1 expression was partially

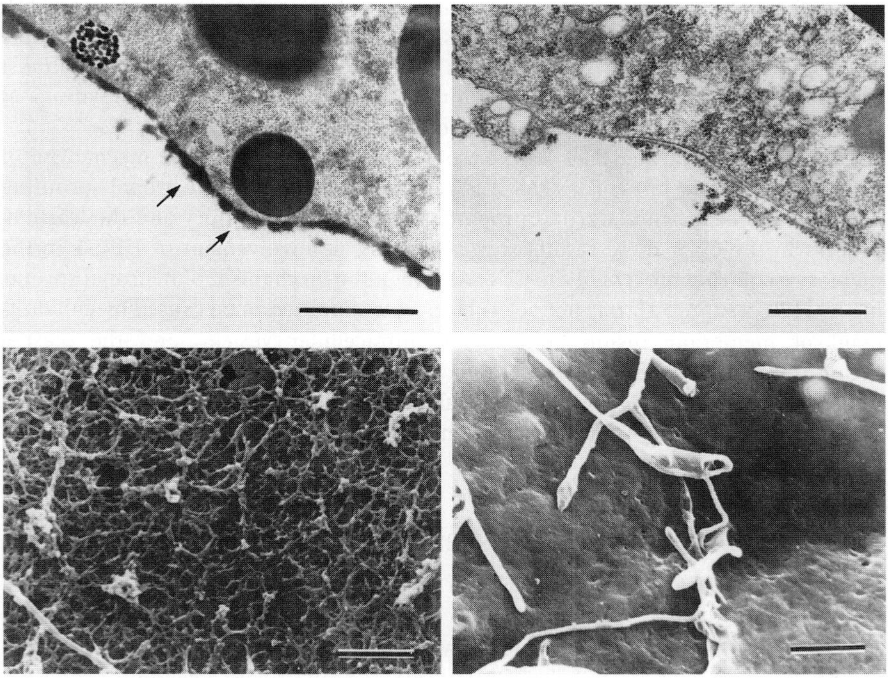

FIG. 1. Reduction of extracellular matrix (ECM) secretion by ectopic HPC-1 expression in early amphibian embryos. HPC-1cRNA was injected into 2-cell-stage embryos of the Japanese newt to express ectopic HPC-1 protein; the surfaces of their ectodermal cells were studied 48 h after injection. *Upper figures*: electron photomicrographs show the ECM on the cell surface of the control (*left*) and HPC-1-expressing embryo (HP, *right*). The control embryo does not express HPC-1 at this developmental stage [8]. Deposit of electron-dense ECM (*arrows*) is visible in the control. In HP, no ECM deposit is present and accumulation of secretory vesicles containing ECM is observed near the plasma membrane. *Lower figures*: scanning electron microscopic observations of the cell surfaces of the same regions as the upper figures show ECM as meshlike materials in the control (*left*); the cell surface of HP (*right*) is devoid of ECM. *Bars* 0.5 μm

suppressed by the antisense RNA transfection method, catecholamine secretion stimulated by high potassium or secretagogues increased [9]. These results indicated that HPC-1 would play an inhibitory role in the exocytotic process, possibly by suppressing intracellular membrane fusion between the secretory vesicles and the plasma membrane.

Enhancement of Neurite Sprouting by Suppression of HPC-1 in Neurons

On the other hand, it has been postulated that HPC-1 might have another function than exocytosis because distribution of HPC-1 protein was not restricted to the synaptic region of neurons, but also appeared on the axonal plasma membrane where no exocytosis occurred [10]. Interestingly, it was demonstrated that suppression of HPC-1 expression by applying the antisense oligonucleotides against HPC-1 mRNA elicited enhancement of axonal sprouting in cultured dorsal root ganglion neurons [11]. Neuronal sprouting was also elicited in a similar experiment using cerebellar granule cells in vitro. In the PC12 cells in which expression of HPC-1 protein was almost completely blocked, abundant microprocess formation and degranulation of catecholamine-containing vesicles were observed (Fig. 2).

These results indicated that HPC-1 would also be involved in the mechanism of neuronal sprouting in neurons. As it is generally accepted that neuronal sprouting could be one of the fundamental properties in long-term memory and the learning process in the CNS, these results suggested possible involvement of HPC-1 in the memory-acquiring process of the CNS. Although the mechanism of neuronal sprouting by HPC-1 suppression is not yet as clear, it is supposed to be caused by enhancement of membrane fusion between the intracellular vesicles and the plasma membrane and by rearrangement of the cytoskeletal structures, because it had been found that although HPC-1 was an integral membrane protein it possessed a tubulin-binding domain in the molecule and could affect microtubule reassembly in vitro [12].

Hemizygous Deletion of the HPC-1 Gene in Williams Syndrome Patients

To examine a possible relationship between the HPC-1 gene and human genetic disorders, fluorescent in situ hybridization (FISH) analysis with a genomic probe was performed. The result of FISH revealed that the locus of the human HPC-1 gene (STX1A) was mapped to the chromosome 7 band q11.23 [13]. Because this locus corresponded to the region of microdeletion in the chromosomes of patients with Williams syndrome (WS) [14], the lymphocytes of WS were studied as well. The FISH study clearly demonstrated that all 46 patients with WS studied so far had hemizygous deletion of the HPC-1 gene (Fig. 3) [15]. These findings suggested that HPC-1 was likely to be one of the genes responsible for the peculiar neuropsychological symptoms of WS, especially that of having good memory ability despite mental retardation.

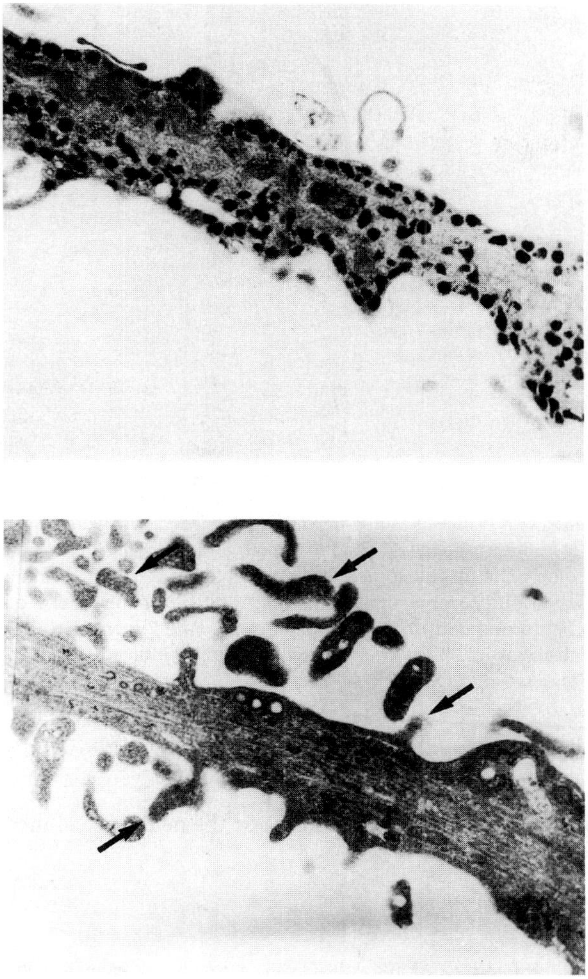

FIG. 2. Enhancement of microprocess formation in HPC-1-deficient PC12 cells. PC12 cells were transfected with the vector expressing the antisense RNA against HPC-1 mRNA to reduce HPC-1 protein expression [9] and were studied under an electron microscope. In the control parent cells (*upper figure*), the cell membrane of the main process is smooth and only a few microprocesses are seen. Many secretory granules containing catecholamine are also observed as electron-dense substance. In the transfected PC12 cell in which HPC-1 protein is completely abolished (*lower figure*), filopodia-like processes (*arrows*) have formed extensively and almost no secretory granules are found

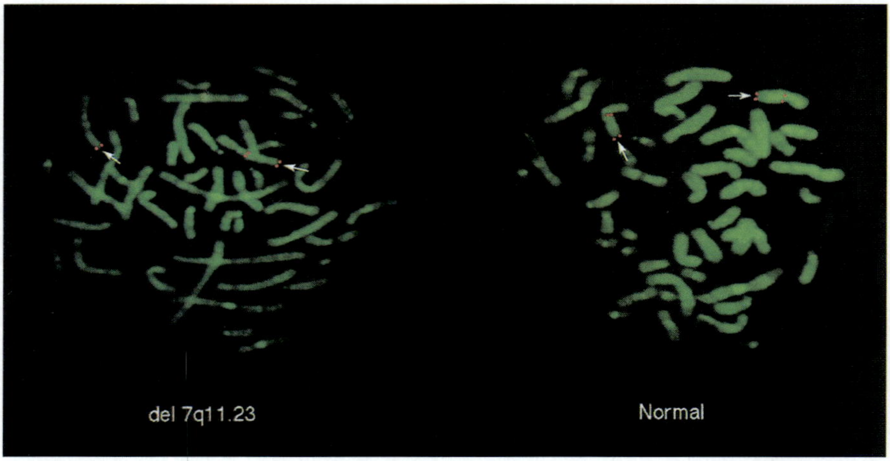

FIG. 3. Fluorescent in situ hybridization (FISH) analysis on the lymphocytes from the normal control and patients with Williams Syndrome (WS) [15]. The hybridization signals were viewed using a confocal laser microscope (Bio-Rad MRC600) and the constructed images were printed out with pseudocolors. The chromosomes are shown in *green*. *Red points* with or without *arrows* indicate the signals of the chromosome 7 marker (D7S742) or the HPC-1 probe (BAC27H2), respectively. In the normal control (*right*), all chromosome 7 have HPC-1 signal. In the WS patient (*left*), a chromosome 7 lacking the HPC-1 signal, indicates hemizygous deletion of HPC-1 gene in patients with WS

However, further studies are necessary to assess the neurological mechanism underlying the unique symptoms in WS.

References

1. Barnstable CJ, Hofstein R, Akagawa K (1985) A marker of early amacrine cell development in rat retina. Dev Brain Res 20:286–290
2. Inoue A, Obata K, Akagawa K (1992) Cloning and sequence analysis of cDNA for a neuronal cell membrane antigen, HPC-1. J Biol Chem 267:10613–10619
3. Bennett MK, Garcia-Arraras JE, Elferink LA, et al (1993) The syntaxin family of vesicular transport receptors. Cell 74:863–873
4. Söllner T, Rothman JE (1994) Neurotransmission: harnessing fusion machinery at the synapse. Trends Neurosci 17:344–348
5. Yamaguchi K, Takada M, Fujimori K, et al (1997) Enhancement of synaptic transmission by HPC-1 antibody in cultured hippocampal neuron. NeuroReport 8:3641–3644
6. Kushima Y, Fujiwara T, Morimoto T, et al (1995) Involvement of HPC-1/syntaxin 1A in transmitter release PC12h cells. Biochem Biophy Res Commun 212:97–103
7. Nagamatsu S, Fujiwara T, Nakamichi Y, et al (1996) Expression and functional role of syntaxin 1/HPC-1 in pancreatic β cells: syntaxin 1A but not 1B plays a negative role in regulatory insulin release pathway. J Biol Chem 271:1160–1165
8. Komazaki S, Fujiwara T, Takada M, et al (1995) Rat HPC-1/syntaxin 1A and syntaxin 1B interrupt intracellular membrane transport and inhibit secretion of the extracellular matrix in embryonic cells of an amphibian. Exp Cell Res 221:11–18

9. Akagawa K, Watanabe T, Fujiwara T, et al (1977) Activation of dopamine release from HPC-1/syntaxin 1A reduced PC12h cells. Neurosci Res (Suppl) 21:S88
10. Koh S, Yamamoto A, Inoue A, et al (1993) Immunoelectron microscopic localization of the HPC-1 antigen in the rat cerebellum. J Neurocytol 22:995–1005
11. Yamaguchi K, Nakayama T, Fujiwara T, et al (1996) Enhancement of neuronal sprouting by suppression of HPC-1/syntaxin 1A activity in the cultured vertebrate nerve cells. Brain Res 740:185–192
12. Fujiwara T, Yamaguchi K, Yamamori T, et al (1997) Interaction of HPC-1/syntaxin 1A with the cytoskeletal protein, tubulin. Biochem Biophys Res Commun 231:352–355
13. Nakayama T, Fujiwara T, Miyazawa A, et al (1997) Mapping of the human HPC-1/syntaxin 1A gene (STX1A) to chromosome 7 band q11.2. Genomics 42:173–176
14. Osborn LR, Martindale D, Scherer SW, et al (1996) Identification of genes from a 500-kb region at 7q11.23 that is commonly deleted in Williams syndrome patients. Genomics 36:328–336
15. Nakayama T, Matsuoka R, Kimura M, et al (in press) Hemizygous deletion of the HPC-1/syntaxin 1A gene (STX1A) in patients with Williams syndrome (WS). Cytogenet Cell Genet

Regulatory Roles of Complexins in Neurotransmitter Release from Mature Presynaptic Nerve Terminals

Shoichiro Ono[1], Mariko Sekiguchi[1], Itsuko Nihonmatsu[1], Kanako Hirata[1], Takeo Awaji[2], Seiichi Takahashi[3], and Masami Takahashi[1]

Summary. Complexins are presynaptic proteins whose functional roles in synaptic transmission are still unclear. In cultured rat hippocampal neurons, complexins distributed throughout the cell bodies, dendrites, and axons, whereas synaptotagmin I and synaptobrevin/VAMP-2, essential proteins for neurotransmitter release, accumulated in the synaptic releasing sites as early as 1 week in culture. Complexins also accumulated in the synaptic release sites and colocalized with synaptotagmin I and synaptobrevin/VAMP-2 after 3–4 weeks in culture. Complexins I and II were expressed in more than 90% and 70% of the neurons, respectively; however, they distributed largely in different populations of synaptic terminals. These results suggest that complexins are not essential but have some regulatory roles in neurotransmitter release from mature neurons.

Key words. Complexin—Synaptic maturation—Synapse—Neurotransmitter release—SNARE proteins

Introduction

Synaptic transmission is mediated by neurotransmitters released from presynaptic nerve terminals by Ca^{2+}-triggered exocytosis of synaptic vesicles; however, the precise mechanism is still not clarified. Complexin (also called 921-L, 921-S, and synaphin) is a 18- to 20-kDa protein that was originally discovered by three groups independently either as a brain-specific protein [1] or as a SNARE complex-associated protein [2,3]. So far, two complexin isoforms have been discovered in mammalian brain [1,3]. The amino acid sequences of complexins are highly rich in charged amino acids, having no apparent homology with other known proteins [1–3]. Recombinant complexin binds to SNARE complex containing syntaxin, SNAP-25, and synaptobrevin/VAMP-2, and

[1] Mitsubishi Kasei Institute of Life Sciences, Machida, Tokyo 194, Japan
[2] Department of Molecular and Cell Pharmacology, National Children's Medical Research Center, 3-35-31 Taishido, Setagaya-ku, Tokyo 154, Japan
[3] Department of Physiology, Kochi Medical School, Nankoku, Kochi 783, Japan

competes with αSNAP for binding to the SNARE complex [3]. Injection of anticomplexin II antibody into presynaptic neurons of *Aplysia* buccal ganglia caused a stimulation of neurotransmitter release [4]. Injection of recombinant complexin II and αSNAP caused depression and facilitation, respectively, of neurotransmitter release from nerve terminals. The effect of complexin was reversed by a subsequent injection of αSNAP, and vice versa. All these results suggest that complexins have some regulatory roles in neurotransmitter release from neurons. In this study, we investigated the intracellular localization and distribution of complexin isoforms in developing rat brain and cultured hippocampal neurons.

Results and Discussion

Rat brain was homogenized in 0.32 M sucrose to pinch off nerve terminals and fractionated into a soluble and a particulate fraction that included the syaptosomes by single-step ultracentrifugation. Distribution of complexin isoforms was examined by immunoblotting. Although most of complexin I and II were recovered in the soluble fraction on P1, the proportion recovered in the particulate fraction containing synaptosomes increased markedly during postnatal development (Fig. 1), suggesting that the distribution of complexin I and II is changed during development.

The intracellular localization and distribution of complexin isoforms are further studied in cultured rat hippocampal neurons. Complexin I was expressed in most of the neurons in culture, and 93%–98% of the β-tubulin-positive cells were labeled with anticomplexin I antibody throughout the culture period up to 4 weeks. In contrast, not all neurons were stained with anticomplexin II antibodies, and the percentage of complexin II-positive neurons increased from 55% to 75% when the culture period was prolonged from 1 to 4 weeks (Fig. 2A). After grown in cell culture for 2 weeks, most hippocampal neurons form functional synapses [5] and show a synchronous oscillation of firing and resulting Ca^{2+} transients [6,7], and the synaptic proteins

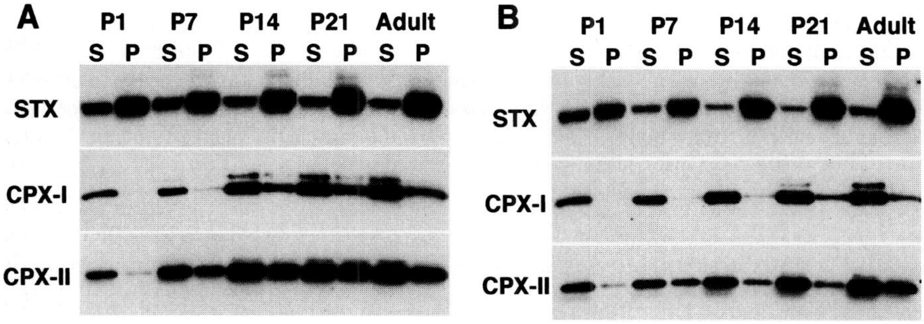

FIG. 1A,B. Subcellular distribution of complexin isoforms in developing rat brain. Hippocampus (**A**) and cerebellum (**B**) from rat brain of indicated ages were homogenized and fractionated into soluble (*S*) and particulate fractions (*P*). In each fraction, 10 μg of proteins were separated by SDS-PAGE and analyzed by immunoblotting with either antisyntaxin I (*STX*), anticomplexin I (*CPX-I*), or anticomplexin II antibodies (*CPX-II*). Note that the percentage of complexin I and II recovered in the particulate fractions increased during development

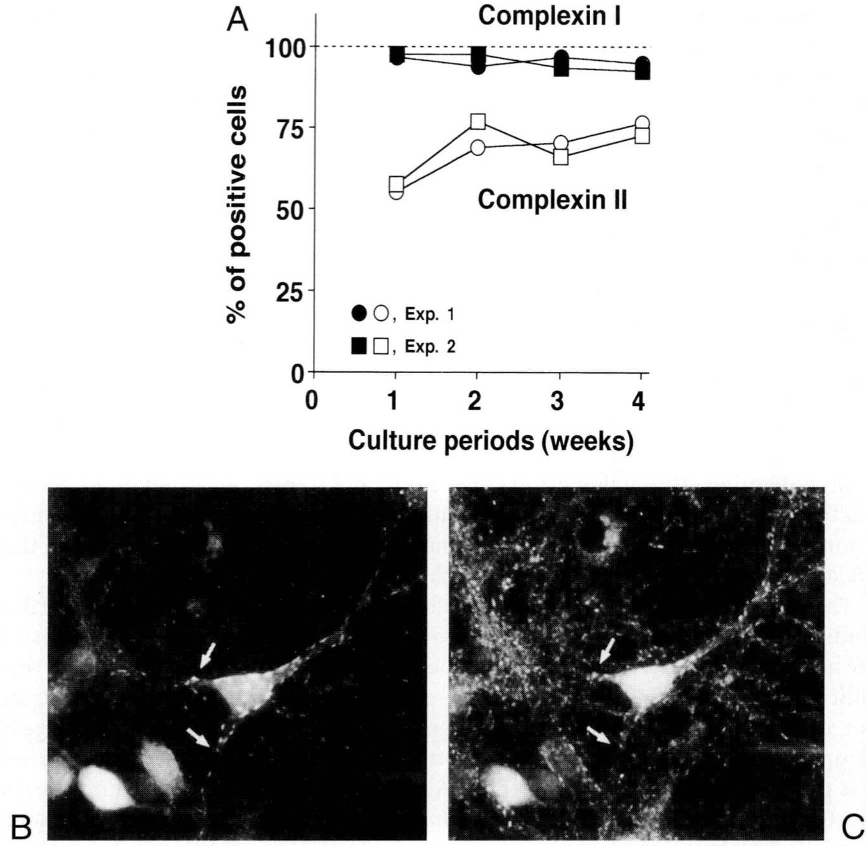

FIG. 2A–C. Differential distribution of complexin I and II in cultured hippocampal neurons. **A** Primary cultures of rat hippocampal neurons were fixed at indicated culture periods and triple-immunostained with the antibodies against complexin I, complexin II, and neuronal β-tubulin. Then, photographs were randomly taken from at least two different wells for cell counting. Numbers of neurons positive for each complexin isoform were counted and are expressed as percentages to that of total neurons positive for neuronal β-tubulin. **B** Rat hippocampal neurons matured in vitro (28 days in culture) were immunostained with anticomplexin I antibody (**B**) and anticomplexin II antibody (**C**). In mature neurons, complexin I and complexin II were sorted into distinct synapses, although they colocalized in some regions (*arrows*)

essential for neurotransmitter release such as synaptotagmin and synaptobrevin/VAMP-2 are accumulated in the synaptic releasing sites.

In contrast, neither complexin I nor complexin II was concentrated in synaptic region, but rather they were diffusely distributed throughout the cytoplasm of the cell body, dendrites, and axons and partly in the nucleus. In neurons cultured for 3–4 weeks, the spotlike staining of complexin I and II surrounding neuronal cell soma and neurites became prominent, as well as the cytoplasmic staining of the cell body

(Fig. 2B,C). Double-label immunofluorescence revealed that complexins and synaptotagmin I were colocalized in a large number of spots, indicating that complexins accumulate into the presynaptic nerve terminals as synaptic connections mature (data not shown). From the in vitro experiments using recombinant presynaptic proteins, complexin and αSNAP were shown to compete for binding to syntaxin in the SNARE complex [3]. Injection of recombinant complexin II in presynaptic neurons of *Aplysia* buccal ganglia markedly inhibited neurotransmitter release [4]. An opposite effect, i.e., a transient increase of evoked acetylcholine release, was observed after anticomplexin II antibody injections. Furthermore, the reversion of the decreasing effect of complexin II on acetylcholine release by αSNAP is in agreement with in vitro experiments demonstrating a competition between complexin II and αSNAP for binding to the SNARE complex [3] and shows that such a competition may indeed take place in vivo. All these results suggested that complexins may be not essential for exocytosis of synaptic vesicles and may have some modulatory roles in neurotransmitter release from the mature presynaptic terminals.

Quite interestingly, even when most of the neurons matured in vitro expressed both complexin I and II, complexin isoforms tended to be sorted into different synaptic terminals. In neurons shown in Fig. 2B,C, complexin I was primarily localized in synapses that surrounded neuronal cell soma and arising thick neurites, while complexin II was concentrated in the synaptic terminals attached to the networks of thin neurites. Some synapses contained both complexin I and II (arrows in Fig. 2B,C), indicating that sorting of complexin isoforms is not exclusive. A typical neuron might possesses a large number of neurotransmitter releasing sites. If complexin I and II have different functional roles in presynaptic terminals, the properties of neuronal networks could be changed by modifying the distribution of complexin isoforms. Further studies including functional differences between complexin isoforms, regulatory mechanisms of association of complexins with other presynaptic proteins, and specific delivery mechanisms of complexin isoforms into nerve terminals are necessary for the elucidation of plasticity of synaptic transmission.

Acknowledgments. This work was funded in part by a research grant from the Human Frontier Science Program (RG-79/96), by grants-in-aid 07279105 for scientific research on priority areas on "Functional Development of Neural Circuits," the Ministry of Education, Science Sports and Culture of Japan; by a grant from the Japan Health Science Foundation; and by the Program for Promotion of Fundamental Studies in Health Sciences of the Organization for Drug ADR Relief, R & D Promotion, and Product Review of Japan.

References

1. Takahashi S, Yamamoto H, Matsuda Z, et al (1995) Identification of two highly homologous presynaptic proteins distinctly localized at the dendritic and somatic synapses. FEBS Lett 368:455–460
2. Ishizuka T, Saisu H, Odani S, et al (1995) Synaphin: a protein associated with the docking/fusion complex in presynaptic terminals. Biochem Biophys Res Commun 213:1107–1114
3. McMahon HT, Missler M, Li C, et al (1995) Complexins: cytosolic proteins that regulate SNAP receptor function. Cell 83:111–119

4. Ono S, Baux G, Sekiguchi M, et al (1998) Regulatory roles of complexins in neurotransmitter release from mature presynaptic nerve terminals. Eur J Neurosci 10:2143–2152
5. Basarsky TA, Parpura V, Haydon PG (1994) Hippocampal synaptogenesis in cell culture: developmental time course of synapse formation, calcium influx, and synaptic protein distribution. J Neurosci 14:6402–6411
6. Ogura A, Iijima T, Amano T, et al (1987) Optical monitoring of excitatory synaptic activity between cultured hippocampal neurons by a multi-site Ca^{2+} fluorometry. Neurosci Lett 78:69–74
7. Muramoto K, Ichikawa M, Kawahara M, et al (1993) Frequency of synchronous oscillations of neuronal activity increases during development and is correlated to the number of synapses in cultured cortical neuron networks. Neurosci Lett 163:163–165

SNARE Proteins and Neurotransmitter Release: A Functional View

George J. Augustine

A number of presynaptic proteins, including NSF, SNAPs, synaptobrevin, syntaxin and SNAP-25, form complexes in vitro. Such SNARE protein complexes have been hypothesized to mediate synaptic vesicle docking and fusion. We have exploited the squid "giant" synapse to evaluate the roles of these proteins in neurotransmitter release. Our strategy has been to microinject, into living presynaptic terminals, chemical probes that specifically target the function of individual SNARE proteins while simultaneously assessing the effects of these probes upon neurotransmitter release. With this approach we have found that all proteins of the SNARE complex—as well as regulators of SNARE complex formation, such as sec-1 and complexin—are important for neurotransmitter release. These findings provide strong support for the SNARE hypothesis of neurotransmitter release. Reagents that perturb SNARE proteins invariably prevent the fusion of synaptic vesicles but do not prevent vesicles from docking at the plasma membrane. These results are inconsistent with the hypothesis that SNARE complex formation is required for vesicle docking. We therefore instead propose that the function of SNARE complexes is to make docked vesicles competent to fuse. Disrupting SNAP and NSF also blocks the production of synaptic vesicles, suggesting that SNARE complexes may be involved the presynaptic endocytosis pathway. Perhaps this action is due to the requirement for SNARE-dependent fusion events in the endocytic pathway. In sum, though the specific functions of SNARE complex proteins remain to be elucidated, it is clear that they are key players in presynaptic membrane traffic.

Department of Neurobiology, Duke Medical Center, PO Box 3209, Durham, NC 27710, USA

A Functional Role of *neuronal-synaptobrevin* Protein in Synaptic Transmission Revealed in *Drosophila* Mutants

MOTOJIRO YOSHIHARA[1], ATSUSHI UEDA[1], DAWEI ZHANG[1], DAVID L. DEITCHER[2], THOMAS L. SCHWARZ[3], and YOSHIAKI KIDOKORO[1]

Summary. Synaptobrevin is a key constituent of the synaptic vesicle membrane. It has been reported that cleavage of neuronal synaptobrevin with tetanus toxin in the *Drosophila* synaptic terminal abolishes nerve-evoked synaptic currents whereas miniature excitatory synaptic currents (mESCs) remain. In electron microscopy, synaptic vesicles were found to be docked in the presynaptic terminal. In accord with these findings, in the *Drosophila* null mutant of the *neuronal-synaptobrevin* gene there were no nerve-evoked synaptic currents, but mESCs remained. Despite the inability of an action potential to trigger fusion, high K^+ saline increased the frequency of mESCs 4 to 17 fold in a Ca^{2+}-dependent fashion, and the rate of fusion approached 25% of that seen in wild-type synapses under the same condition. Ca^{2+} ionophore and black widow spider venom were also effective in increasing the mESC frequency. Tetanic stimulation also increased the frequency of mESCs, particularly toward the end of a train and after the train of stimuli. These effects of treatments that are likely to cause a sustained Ca^{2+} increase stood in marked contrast to the inability of an individual action potential to stimulate fusion. The results suggest the existence of two pathways of calcium-dependent transmitter release.

Key words. *Neuronal-synaptobrevin*—*Drosophila*—SNARE—Neuromuscular synapse—Transmitter release

Introduction

Synaptic transmission is caused by fusion of synaptic vesicles containing transmitter molecules to plasma membrane. The fusion is believed to be mediated by vesicle proteins (VAMP/synaptobrevin, synaptotagmin, etc.), plasma membrane proteins

[1] Institute for Behavioral Sciences, Gunma University School of Medicine, 3-39-22 Showa-machi, Maebashi, Gunma 371-8511, Japan
[2] Section of Neurobiology and Behavior, W125 Mudd Hall, Cornell University, Ithaca, NY 14853, USA
[3] Department of Molecular and Cellular Physiology, Stanford University Medical Center, Stanford, CA 94305-5426, USA

(syntaxin, SNAP-25, etc.), and cytoplasmic proteins (α-SNAP, NSF, etc.) [1–3]). All these proteins have been identified in *Drosophila* [4]. Thus, because of established powerful genetic techniques, the *Drosophila* neuromuscular synapse provides an ideal model system for studies of functions of the synaptic proteins. In this chapter, we briefly review the genetic approaches to the role of neuronal synaptobrevin in synaptic transmission using the *Drosophila* neuromuscular synapse as a model system.

Genes Necessary for Synaptic Vesicle Fusion in *Drosophila*

The SNARE hypothesis [5] proposes docking of vesicles to occur via interaction between a vesicle membrane protein, VAMP/synaptobrevin (v-SNARE, [6,7]), and plasma membrane proteins, syntaxin and SNAP-25 (t-SNAREs [8,9]). In *Drosophila*, two synaptobrevin genes have been identified. One is a ubiquitous type, *syb* [10], and the other is a neural type, *n-syb* (*neuronal-synaptobrevin*) [11]· On the other hand, it is believed that there is only one syntaxin gene in the *Drosophila* genome [12]. The mutants completely lacking the syntaxin gene were isolated by Hugo Bellen's group [12], who reported that both nerve-evoked synaptic currents and mESCs were abolished in the null mutant although synaptic vesicles are docked [12,13]. A homolog of SNAP-25 has been cloned in *Drosophila* [14], but no mutant of the SNAP-25 gene has been isolated.

Synaptotagmin, a candidate for Ca^{2+} sensor [15], was cloned in *Drosophila* [16] and mutations in the gene were isolated [17,18]. In these mutants, nerve-evoked synaptic currents were severely impaired [17–19], whereas the mESC frequency was higher than normal, suggesting that synaptotagmin may act as a negative regulator [18,19]. Although it is reported that the some synaptotagmin mutations cause alteration of Ca^{2+} sensitivity [20], more data are necessary to determine a precise role of synaptotagmin. *N*-Ethylmaleimide-sensitive factor (NSF) is a soluble ATPase associated with SNAREs [21]. An NSF homolog in *Drosophila* has been cloned, and the mutation of the gene was found to be an already known mutant, *comatose* [22,23]. In the mutant, the nerve-evoked synaptic current is impaired, and the delayed time course of the defect is consistent with the results from yeast experiments, which suggests that NSF plays a role in repriming before vesicle fusion [23,24]. A *Drosophila* homolog of α-SNAP, a soluble protein associated with SNAREs [25], has been cloned whereas no mutant of the α-SNAP gene has been isolated [22]. A homolog of *n*-Sec1/Munc-18 (rop), a putative regulator of the SNARE complex [26], has been cloned in *Drosophila*, and overexertion of rop protein inhibited nerve-evoked synaptic currents, suggesting its role as a negative regulator [27].

Function of Neuronal Synaptobrevin Revealed by Transformants Expressing Tetanus Toxin

Tetanus toxin is known to cleave and inactivate neuronal synaptobrevin protein [28]. Sweeney et al. [29] have introduced the gene of tetanus toxin into *Drosophila* to make tetanus toxin protein expressed in the presynaptic terminal. They have shown that no nerve-evoked synaptic currents were observed but that miniature excitatory synaptie currents (mESCs) remained in the tetanus toxin transformants in which neuronal synaptobrevin protein was cleaved. The frequency of spontaneous synaptic currents was reduced in the transformant embryos to approximately half of that of the wild type. With electron microscopy, synaptic vesicles are found docked normally in this transformant. Although their experiment was elegant, some uncertainties remained; namely, complete cleavage of the molecule was not certain, cleaved fragments might have aberrant effects, and the possibility of side effect of tetanus toxin on other molecules cannot be ruled out. To approach more directly to the function of neuronal synaptobrevin in synaptic transmission, mutants completely lacking neuronal synaptobrevin protein were generated and synaptic transmission was examined in embryonic neuromuscular synapses using the patch-clamp technique.

Neuronal Synaptobrevin Mutants

Deitcher et al. [30] have isolated several mutants with a defect in the *n-syb* locus. The molecular defects of the isolated alleles were determined. One allele, named *ΔF33B*, was null for a certainty because the most of the coding region was deleted. *F33-R* is another severe allele, although it is not guaranteed null because it is caused by insertional mutation in a 5′-noncoding region.

Nerve stimulation did not evoke synaptic currents in homozygous mutants of both *ΔF33B* and *F33-R*, consistent with the previously reported phenotype of tetanus toxin transformant [29,13] (Fig. 1 [30]). An elevation of the Ca^{2+} concentration to 6 mM in the external solution did not restore the excitatory synaptic current, nor did a potassium channel blocker, 4-aminopyridine (4-AP), which is known to enhance synaptic transmission in *Drosophila* larvae [31], restore the excitatory synaptic current.

On the other hand, mESCs were observed as was reported in tetanus toxin-expressing transformants [29,13] (Fig. 2 [30]). However, the frequency was reduced by 75% compared with a control strain, line 34, which is an original strain of all synaptobrevin mutants. High K^+ saline increased the mESC frequency in the presence of external Ca^{2+}, suggesting that voltage-gated influx of Ca^{2+} is causing the observed increase of the mESC frequency. The relation between the mESC frequency and Ca^{2+} concentration shifted toward higher Ca^{2+} concentrations compared with control animals (Yoshihara et al., unpublished observation). Similarly, the mESC frequency in *n-syb* null mutants could be increased by a Ca^{2+} ionophore, A23187, and with black widow spider venom (Yoshihara et al., unpublished observation). Thus, it appears that Ca^{2+}-dependent spontaneous fusion of synaptic vesicles remains in *n-syb* null mutants.

Tetanic stimulation induced asynchronous release of transmitter in the *n-syb* null mutant, *ΔF33B* (Yoshihara et al., unpublished observation). This asynchronous release occurred more frequently toward the end of tetanus and continued to occur after its termination. Furthermore, this did not happen in the absence of Ca^{2+}, suggesting that voltage-gated Ca^{2+} channels are playing a role for this phenomenon.

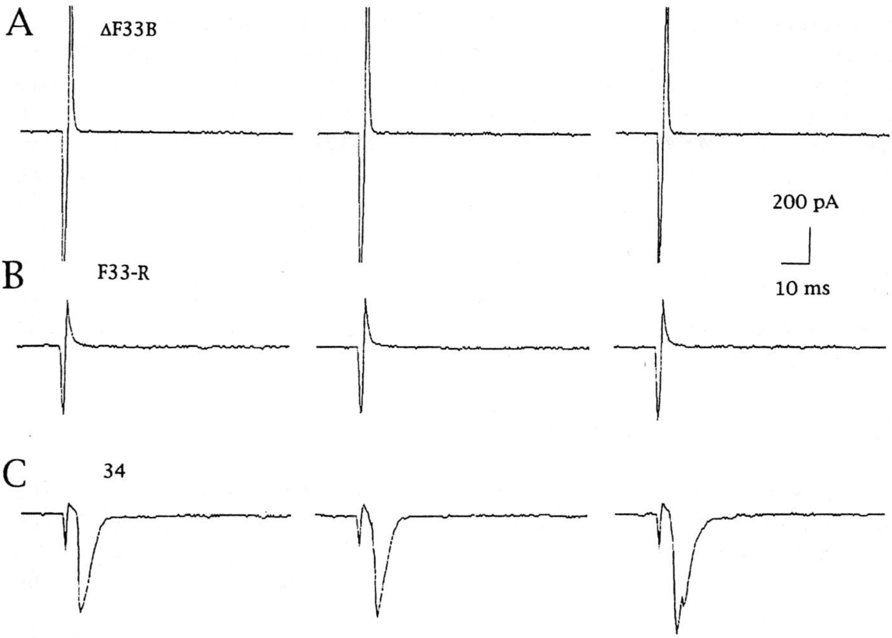

FIG. 1A–C. Nerve-evoked synaptic currents are absent from the neuromuscular junctions of *n-syb* mutants stimulated at 0.3 Hz. Evoked currents are lacking in *n-syb* null mutant *n-syb*$^{\Delta F33B}$ (**A**) and in the mutant *n-syb*$^{F33\text{-}R}$ (**B**) but present in the parental control, line 34 (**C**). The external solution contained 2 mM Ca^{2+} for lines *n-syb*$^{\Delta F33B}$ and in *n-syb*$^{F33\text{-}R}$ and 0.5 mM Ca^{2+} for line 34. The ability of the stimulus to activate the motoneuron was confirmed by observing the effects of tetanic stimuli, as described by Yoshihara et al. (in preparation). (From [30], with permission)

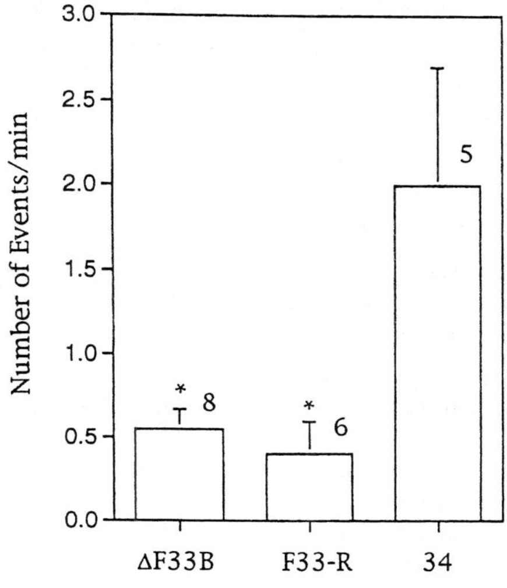

FIG. 2. Frequency of miniature synaptic currents in *n-syb*$^{\Delta F33B}$, *n-syb*$^{F33\text{-}R}$, and line 34. Error bars are the standard error of the mean. *Asterisks* denote statistical differences from line 34 at $P < 0.05$ by the ANOVA test. Numbers indicate the number of cells examined. Miniature synaptic currents were recorded in the presence of 3 μM TTX (tetrodotoxin) in 0.5 mM Ca^{2+} saline. (From [30], with permission)

Conclusion

In the *n-syb* null mutant, the synchronous excitatory synaptic current was not evoked by nerve stimulation whereas fusion of synaptic vesicles in response to Ca^{2+} influx was partly retained. The results suggest the existence of two pathways of calcium-dependent transmitter release. Only one pathway, which is neuronal synaptobrevin dependent, can cause synchronous release to nerve stimulus.

References

1. Südhof TC (1995) The synaptic vesicle cycle: a cascade of protein-protein interactions. Nature (Lond) 375:645–653
2. Zucker RS (1996) Exocytosis: a molecular and physiological perspective. Neuron 17:1049–1055
3. Bock JB, Scheller RH (1997) A fusion of new ideas. Nature (Lond) 387:133–135
4. Littleton JT, Bellen HJ (1995) Presynaptic proteins involved in exocytosis in *Drosophila melanogaster*: a genetic analysis. Invertebr Neurosci 1:3–13
5. Söllner T, Whiteheart SW, Brunner M, et al (1993) SNAP receptors implicated in vesicle targeting and fusion. Nature (Lond) 362:318–324
6. Trimble WS, Cowan DM, Scheller RH (1988) VAMP-1: a synaptic vesicle-associated protein. Proc Natl Acad Sci USA 85:4538–4542
7. Baumert M, Maycox PR, Navone F, et al (1989) Synaptobrevin: an integral membrane protein of 18 000 daltons present in small synaptic vesicles of rat brain. EMBO J 8:379–384
8. Bennet MK, Calakos N, Scheller RH (1992) Syntaxin: a synaptic protein implicated in docking of synaptic vesicles at presynaptic active zones. Science 257:255–259
9. Oyler GA, Higgins GA, Hart RA, et al (1989) The identification of a novel synaptosomal-associated protein, SNAP-25, differentially expressed by neuronal subpopulations. J Cell Biol 109:3039–3052
10. Chin AC, Burgess RW, Wong BR, et al (1993) Differential expression of transcripts from *syb*, a *Drosophila melanogaster* gene encoding VAMP (synaptobrevin) that is abundant in non-neuronal cells. Gene 131:175–181
11. DiAntonio A, Burgess RW, Chin AC, et al (1993) Identification and characterization of *Drosophila* genes for synaptic vesicle proteins. J Neurosci 13:4924–4935
12. Schulze KL, Broadie K, Perin MS, et al (1995) Genetic and electrophysiological studies of *Drosophila* syntaxin-1A demonstrate its role in nonneuronal secretion and neurotransmission. Cell 80:311–320
13. Broadie K, Prokop A, Bellen H, et al (1995) Syntaxin and synaptobrevin function downstream of vesicle docking in *Drosophila*. Neuron 15:663–673
14. Risinger C, Blomqvist AG, Lundell I, et al (1993) Evolutionary conservation of synaptosome-associated protein 25 kDa (SNAP-25) shown by *Drosophila* and *Torpedo* cDNA clones. J Biol Chem 268:24408–24414
15. Perin MS, Fried VA, Mignery GA, et al (1990) Phospholipid binding by a synaptic vesicle protein homologous to the regulatory region of protein kinase C. Nature (Lond) 345:260–263
16. Perin MS, Johnston PA, Özcelik T, et al (1991) Structural and functional conservation of synaptotagmin (p65) in *Drosophila* and humans. J Biol Chem 266:615–622
17. DiAntonio A, Parfitt KD, Schwarz TL (1993) Synaptic transmission persists in *synaptotagmin* mutants of *Drosophila*. Cell 73:1281–1290
18. Littleton JT, Stern M, Schulze K, et al (1993) Mutational analysis of *Drosophila* synaptotagmin demonstrates its essential role in Ca^{2+}-activated neurotransmitter release. Cell 74:1125–1134

19. Broadie K, Bellen HJ, DiAntonio, et al (1994) Absence of synaptotagmin disrupts excitation-secretion coupling during synaptic transmission. Proc Natl Acad Sci USA 91:10727–10731
20. Littleton JT, Stern M, Perin M, et al (1994) Calcium dependence of neurotransmitter release and rate of spontaneous vesicle fusions are altered in *Drosophila* synaptotagmin mutants. Proc Natl Acad Sci USA 91:10888–10892
21. Block MR, Glick BS, Wilcox DA, et al (1988) Purification of an *N*-ethylmaleimide-sensitive protein catalyzing vesicular transport. Proc Natl Acad Sci USA 85:7852–7856
22. Ordway RW, Pallanck L, Ganetzky B (1994) Neurally expressed *Drosophila* genes encoding homologs of the NSF and SNAP secretory proteins. Proc Natl Acad Sci USA 91:5715–5719
23. Pallanck L, Ordway RW, Ganetzky B (1995) A *Drosophila* NSF mutant. Nature (Lond) 376:25
24. Morgan A (1996) Classic clues to NSF function. Nature (Lond) 382:680
25. Clay DO, Griff IC, Rothman JE (1990) SNAPs, a family of NSF attachment proteins involved in intracellular membrane fusion in animals and yeast. Cell 61:709–721
26. Hata Y, Slaughter CA, Südhof TC (1993) Synaptic vesicle fusion complex contains *unc-18* homologue bound to syntaxin. Nature (Lond) 366:347–351
27. Schulze KL, Littleton JT, Salzberg A, et al (1994) rop, a *Drosophila* homolog of yeast Sec1 and vertebrate n-Sec1/Munc-18 proteins, is a negative regulator of neurotransmitter release in vivo. Neuron 13:1099–1108
28. Schiavo G, Benfenati F, Poulain B, et al (1992) Tetanus and botulinum-B neurotoxins block neurotransmitter release by proteolytic cleavage of synaptobrevin. Nature (Lond) 359:832–835
29. Sweeney ST, Broadie K, Keane J, et al (1995) Targeted expression of tetanus toxin light chain in *Drosophila* specifically eliminates synaptic transmission and causes behavioral defects. Neuron 14:341–351
30. Deitcher DL, Ueda A, Stewart A, et al (1998) Distinct requirements for evoked and spontaneous release of neurotransmitter are revealed by mutations in the *Drosophila* gene, *neuronal-synaptobrevin*. J Neurosci 18:2028–2039
31. Jan YN, Jan LY, Dennis MJ (1977) Two mutations of synaptic transmission in *Drosophila*. Proc R Soc Lond B 198:87–108

Enzyme-Based Real-Time Detection of Synaptic Glutamate Release Modulated by GABA and Neurotrophins in Cultured Rat Neurons

Keiichi Torimitsu[1], Ryoji Kurita[2], and Osamu Niwa[1]

Summary. Glutamate is well known as an important excitatory neurotransmitter for synaptic transmission in brain nerve cells. Although its real-time detection is extremely important for brain function analysis, no methods are capable of high temporal and spatial resolutions. Our method using enzyme-based electrochemical detection can meet this requirement and allow us to measure a transient glutamate release with nanomolar (nM) resolution. In this chapter, we describe this method in more detail and report the effect of neurotrophins on glutamate release. The method is highly selective for possible intereferents including ascorbic acid, and is capable of measuring glutamate distribution near a cell. Synaptic glutamate release could be measured when KCl and GABA were applied. A brief application of NGF induced a Ca-dependent transient release, which is considered to be related to synaptic activity. Its real-time capability and high sensitivity gives our method the potential to be a powerful tool for understanding the neural functions of the brain.

Key words. Cortical cell—Glutamate—GABA—NGF—Electrochemical detection—Os polyvinylpyridine

Introduction

Glutamate, an excitatory neurotransmitter, is essential for synaptic transmission in brain nerve cells. Although various methods [1–4] have been reported for measuring glutamate release, no method could provide a continuous measurement of synaptic release because of its transience and the small amount. Microdialysis-sampled liquid chromatography [2], which is commonly used for in vivo experiments or an enzyme detector [5], is not suitable for measuring such extremely low quantities and highly transient phenomena. Although carbon fibers [3] may have a higher spatial resolution, their sensing performance and reproducibility are unstable. We therefore developed a novel method using glutamate oxidase and a glassy carbon electrode [6]. The method is based on the electrochemical reaction via an enzymatic reaction of

[1]NTT Basic Research Laboratories and [2]NTT Advanced Technology, 3-1 Morinosato Wakamiya, Atsugi, Kanagawa 243-0198, Japan

glutamate oxidase (GluOx) [7] and Os polyvinylpyridine containing horseradish peroxidase (HRP) [8].

γ-Amino-butyric acid (GABA), an inhibitory neurotransmitter, has been reported to have depolarizing and neurotrophic effects on embryonic neurons [9,10]. Neurotrophins are also suggested to be related to neurotransmitter release [11]. We used the method to investigate the effect of GABA and neurotrophins on the synaptic glutamate release of cultured rat brain neurons.

Materials and Methods

Cell Culture and Slices

The cortices taken from E18 Wistar rats were cultured according to a previously described method [6]. Briefly, they were dissociated with 0.02% papaine (Boehringer, Mannheim, Penzberg, Germany) and plated on culture dishes precoated with laminine and poly-D-lysine. The following culture medium was used in 10% CO_2 at 37°C for maintaining the cells: 5% heat-inactivated fetal bovine serum (Hyclone, Logan, UT), and 5% heat-inactivated horse serum (Gibco, Grand Island, NY), 2.5 μg/ml insulin (Sigma, St. Louis, MO), 50 U/ml penicillin, and 50 μg/ml streptomycin (Sigma) in Dulbecco's Modified Eagles Medium (DMEM) (Gibco). Hippocampal slices (300-μm thickness) taken from P2 rats were also used. Measurements were carried out in a medium containing (in mM) 148 NaCl, 2.8 KCl, 2 $MgCl_2$, 2 $CaCl_2$, 10 glucose in 10 Hepes, pH 7.4.

On-Line Glutamate Measurement and HPLC Measurement

The medium was drawn with a glass microcapillary and transported continuously to the detector with a reverse-drive dual syringe pump (102; CMA Microdialysis, Stockholm, Sweden) at 4 μl/min. The detector has two enzyme layers consisting of L-GluOx (Yamasa Shoyu, Choshi, Japan) and HRP containing Os polyvinylpyridine formed on 6-mm glassy carbon (GC) electrodes, which are connected to an amperometric potentiostat (LC-4C; Bioanalytical Systems, West Lafayette, IN). We used the conventional method of *o*-phthalaldehyde (OPA)/mercaptoethanol (MCE) derivatization with an HPLC (BAS 200B; Bioanalytical Systems) and employed a CMA 200 autosampler and fluorescence detector (CMA 280) to confirm the glutamate measurement.

Results and Discussion

Figure 1 shows a block diagram of the detection method. The operating potential of the GC electrode was 0 mV vs. AgCl. The detector had a low detection limit of about 10 nM glutamate and temporal and spatial resolutions of seconds and micrometers, respectively.

As described in our previous paper [6], the detector has a highly specific response to glutamate, which was confirmed by HPLC measurement. The generating cathodic current of the detector exhibited a linear dependence on the glutamate concentration (Fig. 2a). Endogenous H_2O_2 did not affect the measurement. Examples of the KCl and

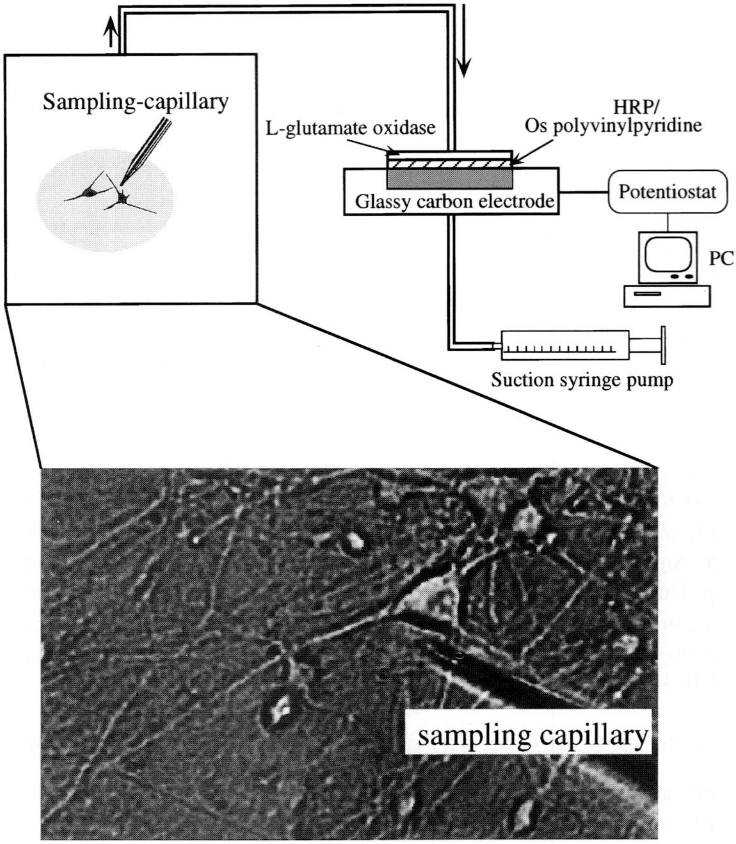

FIG. 1. Electrochemical method for detecting glutamate: block diagram of the detection method with an example of a sampling picture

FIG. 2a–c. Calibration curve and glutamate specificity. **a** Calibration curve of the generating current of the detector versus glutamate concentration. The generating current of the detector showed a linear response to the glutamate concentration over the whole measurement range. Although the detection limit depends on the suction flow rate, it is about 10^{-8} M in this system. Data were taken at 16 μl/min. Means ± SEM. **b** Induced glutamate release from rat cortical neurons. Transient glutamate release was observed by the addition of K. As removal of Ca from the external medium resulted in no transient change in glutamate release, this induced response is Ca dependent and synaptic. The addition of GABA (25 μM) induced a transient glutamate release; GABA modulation of the glutamate release is Ca dependent and blocked by bicuculline (25 μM). **c** Selectivity of the detection method. Each substrate concentration was 10 μM. The relative response is expressed as a percentage compared to the glutamate response (**a**, **b** are reproduced with permission of Rapid Science Publishers Ltd. from [6]: Torimitsu K., Niwa O. 1997. Real-time detection of GABA-induced synaptic glutamate release in cultured rat cortex. Neuro Report 8:1353–1358)

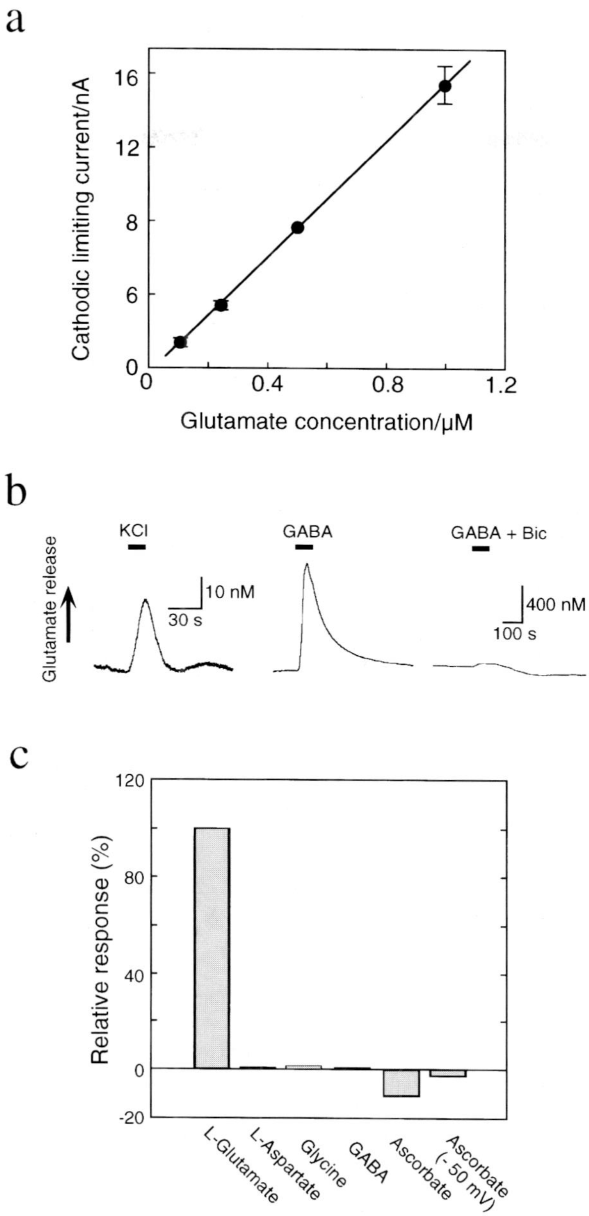
a
Cathodic limiting current/nA
0
6
8
12
16
0
0.4
0.8
1.2
Glutamate concentration/μM
b
Glutamate release
KCl
10 nM
30 s
GABA
GABA + Bic
400 nM
100 s
c
Relative response (%)
120
80
40
0
-20
L-Glutamate
L-Aspartate
Glycine
GABA
Ascorbate
Ascorbate
(- 50 mV)

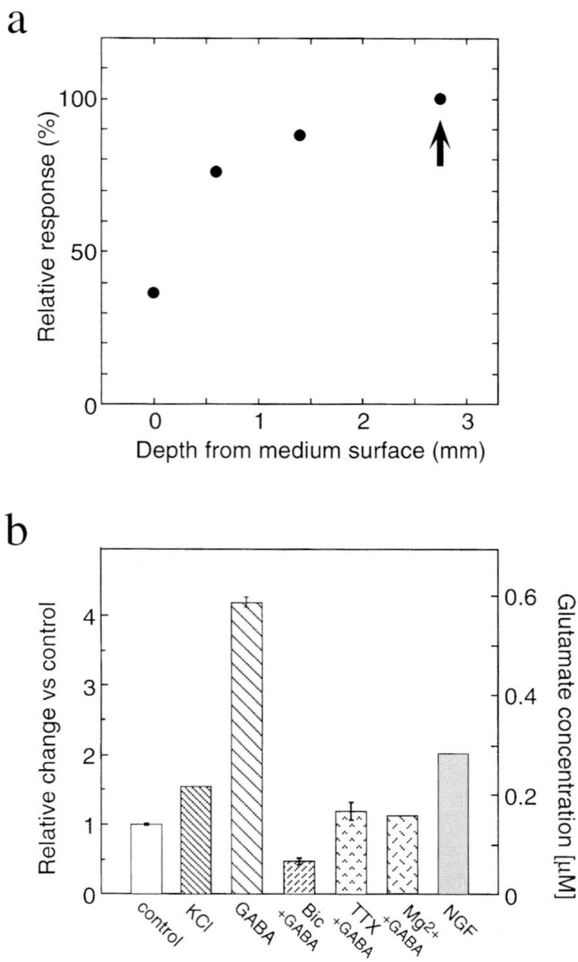

Fig. 3. **a** Glutamate concentration distribution according to the relative reponses vs. the concentration near the cell (*arrow*) in the culture dish. The glutamate concentrations were measured depending on the depth from the medium surface using the detector. The medium was changed and left for about 30 min without circulation (flow) before measurement. **b** HPLC measurement of glutamate response with the conventional method of *o*-phthalaldehyde/mercaptoethanol (OPA/MCE) derivatization when 30 µl of the medium was sampled 5 min after stimulation. The 10-µl sample was reacted with 10 µl OPA/MCE for 150 s at 10°C before injection. The experimental conditions were the same as in Fig. 2. Bicuculline (25–50 µM), TTX (100 µM), $MgCl_2$ (10 mM), and NGF (20 ng/ml) were used together with GABA. A capillary column (C-18 ODS, 5 µm, monomeric, 150 mm × 1.0 mm i.d.) and a precolumn (100 mm × 1.0 mm i.d.) were used at about 2700 psi with 0.1 M phosphate buffer (pH 6.0) containing 0.1 mM EDTA, 10% acetonitrile, and 3% tetrahydrofuran (THF)

GABA response are shown in Fig. 2b. The GABA response was blocked by the $GABA_A$ receptor antagonist bicuculline. As no release was detected in a Ca-free medium, this transient release was thought to be synaptic. The glutamate selectivity of the detector was at least 100 fold higher than various other amino acids (Fig. 2c). Although L-ascorbic acid (Asc) is an important interferent whose effect must be carefully considered, it poses no serious problem in this case. Because the current caused by Asc is anodic while that exhibited by the glutamate is cathodic, it is easy to distinguish the glutamate signal from the acid signal.

The detector could measure the distribution of the ambient glutamate in the culture dish (Fig. 3a). The concentration was higher near the cell (about 0.15 μM). A washing-out procedure undertaken with fresh medium indicated a transient decrease with a slow recovery in the glutamate concentration (data not shown).

HPLC measurements under the same experimental conditions indicated that the addition of GABA induced a glutamate release as in the detector experiment (Fig. 3b). The release was largely blocked by bicuculline. The addition of tetrodotoxin (TTX) or Mg^{2+} with GABA blocked the glutamate release but not that of the basal glutamate. Therefore, on the basis of these results, we believe the GABA-induced response to be synaptic. A brief application of neurotrophins, such as nerve growth factor (NGF) or brain-derived neurotrophic factor (BDNF), increased the glutamate release. Preliminary data on the NGF effect are shown.

Conclusion

Our enzyme-based real-time method of glutamate detection allows us to measure a transient glutamate release with a nanometer resolution and a high spatial resolution. With this method, GABA and NGF were found to be related to synaptic activities inducing a transient glutamate release in cultured rat cortical cells. This synaptic modulation suggests a close interaction between neurotrophic factors and neurotransmitter release. The detection method we described here is extremely effective for investigating such transient glutamate responses because of its real-time capability and high selectivity and sensitivity. Thus, this method could be a powerful tool for understanding neural functions in the brain.

Acknowledgments. The authors thank Dr. N. Matsumoto and Dr. A. Kawana for useful discussions and comments. The authors also thank Dr. J. Fukuda for helpful suggestions and Ms. Y. Furukawa for helping with the culture.

References

1. Wang LDL, Boyarsky LL (1979) Release of L-glutamate from the excitatory nerve terminals of the crayfish neuromuscular junction. Life Sci 24:1011–1014
2. Kissinger P (1991) Microdialysis and liquid chromatography. In: Robinson T, Justice J Jr (eds) Microdialysis in neuroscience. Elsevier, Amsterdam, pp 103–115
3. Hu Y, Mitchell KM, Albahadily FN, et al (1994) Direct measurement of glutamate release in the brain using a dual enzyme-based electrochemical sensor. Brain Res 659:117–125

4. Zilkha E, Obrenovitch TP, Koshy A, et al (1995) Extracellular glutamate: on-line monitoring using microdialysis coupled to enzyme-amperometric analysis. J Neurosci Methods 60:1–9
5. Niwa O, Torimitsu K, Morita M, et al (1996) Concentration of extracellular L-glutamate released from cultured nerve cells measured with a small-volume online sensor. Anal Chem 68:1865–1870
6. Torimitsu K, Niwa O (1997) Real-time detection of GABA-induced synaptic glutamate release in cultured rat cortex. NeuroReport 8:1353–1358
7. Kusakabe H, Midorikawa Y, Fujishima T, et al (1983) Purification and properties of a new enzyme, L-glutamate oxidase, from *Streptomyces* sp. X-119-6 grown on wheat bran. Agric Biol Chem 47:1323–1328
8. Yang L, Janle E, Huang T, et al (1995) Applications of "wired" peroxidase electrodes for peroxide determination in liquid chromatography coupled to oxidase immobilized enzyme reactor. Anal Chem 67:1326–1331
9. Obata K, Oide M, Tanaka H (1978) Excitatory and inhibitory actions of GABA and glycine on embryonic chick spinal neurons in culture. Brain Res 144:179–184
10. Ben-Ari Y, Khazipov R, Leinekugel X, et al (1997) $GABA_A$, NMDA and AMPA receptors: a developmentally regulated "ménage à trois". Trends Neurosci 20:523–529
11. Lauder JM (1993) Neurotransmitters as growth regulatory signals: role of receptors and second messengers. Trends Neurosci 16:233–240

Analysis of the Regulatory Mechanisms of Synaptic Exocytosis Using an Autapse of Cultured Rat Hippocampal Neuron

MAKOTO OSANAI[1,3], MAKOTO TAKADA[2], TOMONORI FUJIWARA[1], KIMIO AKAGAWA[1], and KAZUHIKO YAMAGUCHI[1]

Summary. Regulating mechanisms of synaptic exocytosis in the rat hippocampal neuron were investigated using autapse and whole-cell patch-clamp methods. First, we analyzed effects of anti-HPC-1/syntaxin 1A antibody on synaptic transmission. Intracellular application of the antibody enhanced the amplitude of the autaptic excitatory postsynaptic current (EPSC), while it did not change the asynchronous EPSC amplitude, indicating that the antibody increased neurotransmitter release from the presynaptic terminal. This result suggests a suppressive role for HPC-1/syntaxin 1A in exocytosis at the mammalian central synapse. Second, to elucidate the modulatory mechanism of the synaptic exocytosis, the dependence of EPSC amplitude and paired pulse facilitation (PPF) on extracellular Ca^{2+} concentration was analyzed using the autapse of the cultured rat dentate gyrus neuron. The Ca^{2+} dependency of the EPSC amplitude was well described by the multiple binding model with Hill coefficient as 3.7 and K as 0.97 mM. The Ca^{2+} dependency of PPF was also described by modifying this model, which suggested that PPF decreased as the release probability (p) increased. Using these models, we analyzed the mechanism of EPSC enhancement and concomitant PPF reduction by forskolin, suggesting that forskolin increased release probability.

Key words. HPC-1—Syntaxin 1A—Autapse—Paired pulse facilitation—Release probability

[1] Department of Physiology, Kyorin University School of Medicine, 6-20-2 Shinkawa, Mitaka, Tokyo 181, Japan

[2] Department of Physiology, Saitama Medical School, Moroyama, Saitama 350-04, Japan

[3] Present address: "Protecting the Brain" CREST, Japan Science and Technology Corporation, Department of Pharmacology, Faculty of Medicine, Tokyo Medical and Dental University, 1-5-45 Yushima, Bunkyo-ku, Tokyo 113-8519, Japan

Introduction

Synaptic exocytosis is characterized by its short latency from Ca^{2+} entry [1,2]. Although the mechanism for regulation by Ca^{2+} has not been clarified, interaction between synaptotagmin, a major Ca^{2+} sensor protein [3], and HPC-1/syntaxin 1A, a neuronal membrane protein [4,5], seem to play important roles in synaptic exocytosis [5,6]. To investigate roles for HPC-1/syntaxin 1A and Ca^{2+} in synaptic exocytosis in mammalian central neurons, we have examined effects of HPC-1/syntaxin 1A downregulation and extracellular Ca^{2+} concentration ($[Ca^{2+}]_e$) on the synaptic transmission of the rat hippocampal autapse [7] in culture. Results suggest that HPC-1/syntaxin 1A is involved in the suppressive process of exocytosis and that release probability depends on $[Ca^{2+}]_e$ in a cooperative manner.

Materials and Methods

The methods of microculture of the rat whole hippocampal or dentate neuron and preparation methods of anti-HPC-1/syntaxin 1A IgG have been described elsewhere [8,9]. Autaptic electrical activities were recorded using the whole-cell patch-clamp technique [8]. The bath solution consisted of (in mM): NaCl, 127; KCl, 4; $NaHCO_3$, 25; NaH_2PO_4, 1.2; $CaCl_2$, 2.5; $MgCl_2$, 1; glucose, 17. Picrotoxin (30–50 μM, Sigma, St. Louis, MO) and DL-AP5 (30–50 μM, Sigma) were also added. $CaCl_2$ was varied from 0.5 to 4 mM and osmolality was adjusted with sucrose. The intrapipette solution consisted of (in mM): K-gluconate, 129; KCl, 30; $MgCl_2$, 2; EGTA, 1; $CaCl_2$, 0.25; Na_2ATP, 3; GTP, 0.3; Hepes, 5 [8]. The pH was adjusted to 7.2. The calculated free Ca^{2+} concentration was about 50 nM.

Results

Under the whole-cell configuration, depolarizing voltage steps applied to the single neuron grown on the glial island elicited a voltage-dependent inward current and an autaptic excitatory postsynaptic current (EPSC) (Fig. 1a). Infusion of the polyclonal antibody against rat HPC-1/syntaxin 1A (IgG, 0.3 $mg\,ml^{-1}$) using a whole-cell patch pipette caused an increase in the amplitude of the autaptic EPSC (Fig. 1a, lower; 136% ± 9%, $n = 8$, $P < 0.05$). Preimmune IgG (0.3 $mg\,ml^{-1}$) did not increase the EPSC amplitude (Fig. 1a, upper). Amplitude distribution of an asynchronous miniature EPSC following the evoked EPSC was compared in each cell between the beginning phase, the period in which evoked EPSC amplitude was smaller than 120% (Fig. 1b, upper), and the peak phase, the period in which evoked EPSC amplitude was greater than 140% (Fig. 1b, lower). No difference was detected in the amplitude distribution of the asynchronous EPSCs between those recorded in the beginning phase and those recorded at the peak phase (in each of five cells; $P > 0.1$), which suggested that the increase in the evoked EPSC amplitude was attributable to the presynaptic mechanism. To investigate whether release probability was affected by the antibody against HPC-1/syntaxin 1A, the frequency of asynchronous EPSC was analyzed between the

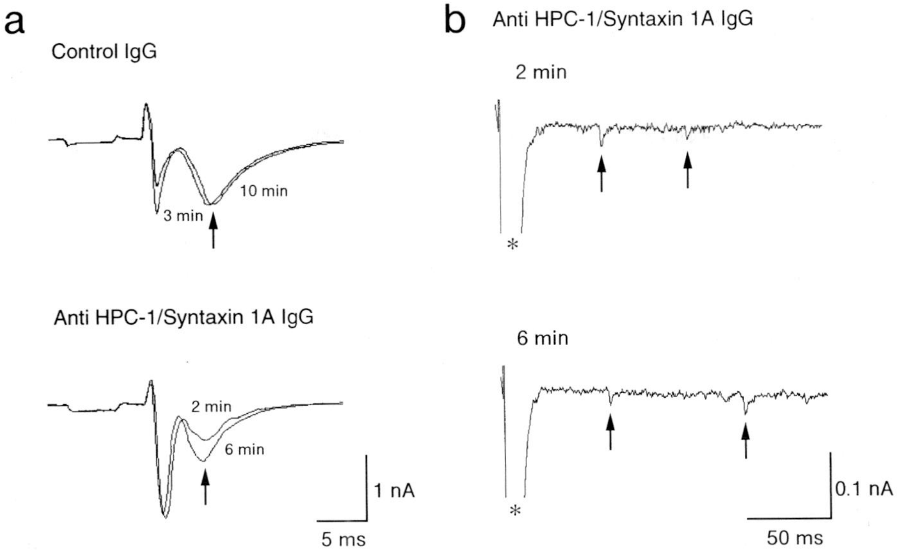

FIG. 1. **a** Effects of antibody against HPC-1/syntaxin 1A on the autaptic excitatory postsynaptic current (EPSC). Amplitude of the evoked EPSC (*arrow*) was enhanced by antibody (*lower*), but not by preimmune IgG (*upper*). **b** Asynchronous EPSCs (*arrows*) recorded by patch-pipette containing antibody against HPC-1/syntaxin 1A during the beginning phase (*upper*, 2 min after start of whole-cell recording) and the peak phase (*lower*, 6 min). *Asterisk*, evoked EPSC (truncated)

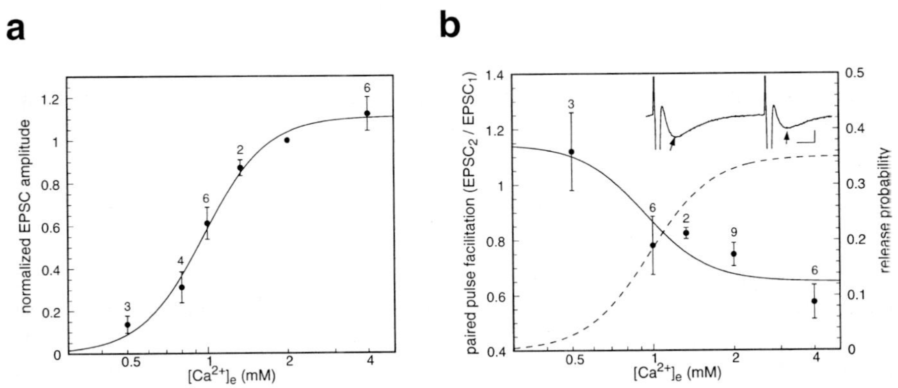

FIG. 2. **a** Relationship between extracellular Ca^{2+} concentration ($[Ca^{2+}]_e$) and EPSC amplitudes (*closed circles*). Each value was normalized by EPSC amplitude at $[Ca^{2+}]_e = 2\,mM$. The data were fitted by Eq. 1 (*solid line*; $n = 3.7$, $K = 0.97\,mM$). **b** Relationship between $[Ca^{2+}]_e$ and paired pulse facilitation (PPF; *closed circles*). The best-fit curves of PPF (*solid line*) and release probability (*dashed line*) were obtained using Eqs. 3 and 4 ($\Delta = 0.038$). *Inset*: EPSC trace (*arrow*) evoked by paired pulse ($[Ca^{2+}]_e = 2\,mM$); 100 pA, 5 ms

beginning and the peak phase of the evoked EPSC. The frequency of asynchronous EPSC appearing within a period of 200 ms following each evoked EPSC was 9.0 ± 2.4 Hz or 7.8 ± 1.7 Hz in the beginning or at the peak phase, respectively ($n = 5$), and the difference was not significant.

The dependence of EPSC amplitude on $[Ca^{2+}]_e$ was analyzed by using the following Hill equation (Fig. 2a):

$$EPSC = \frac{A[Ca^{2+}]^n}{K^n + [Ca^{2+}]^n} \tag{1}$$

The $[Ca^{2+}]_e$ dependence of the measured EPSC amplitude was well fitted by this model with $n = 3.7$ and $K = 0.97$ mM. If recruitment of the synaptic vesicle did not occur during the interstimulus interval (30 ms), amplitude of the first and second EPSC ($EPSC_1$, $EPSC_2$), and paired pulse facilitation (PPF) [10] would be as follows:

$$EPSC_1 \propto Np, \quad EPSC_2 \propto (N - Np)p' \tag{2}$$

$$PPF = EPSC_2/EPSC_1 = p'(1 - p)/p \tag{3}$$

where N is the total number of releasable vesicles and p and p' are the release probability at the first and second stimulus, respectively. Using Eqs. 1 and 2, and assuming that facilitation is caused by the residual Ca^{2+} ($\Delta[Ca^{2+}]$) [11], the release probability p and p' are as follows:

$$p = \frac{B[Ca^{2+}]^n}{K^n + [Ca^{2+}]^n}, \quad p' = \frac{B[Ca^{2+}]^n(1 + \Delta)^n}{K^n + [Ca^{2+}]^n(1 + \Delta)^n} \tag{4}$$

where n and K were obtained from Eq. 1 ($n = 3.7$, $K = 0.97$). Using this model, $[Ca^{2+}]_e$ dependence of release probability was estimated (Fig. 2b). As a result, PPF decreased as the release probability increased.

We also examined the mechanism(s) of EPSC enhancement by forskolin (data not shown) [12]. Forskolin enhanced EPSC amplitude while it reduced PPF. According to these models, EPSC enhancement by forskolin was suggested to be attributable to increase in release probability.

Discussion

Our results showed that intracellular administration of IgG against HPC-1/syntaxin 1A caused an increase in transmitter release, but did not change the frequency of asynchronous EPSC. Increase in the transmitter release by antibody against HPC-1/syntaxin 1A is contrary to the result in cultured rat superior cervical ganglion neuron [13]. The simplest explanation is that HPC-1/syntaxin 1A could have at least two different functional domains, a facilitatory domain [14] and a suppressing domain for synaptic exocytosis, and that the antibody in this study might recognize mainly the suppressing domain. Another possibility is that there is a region-specific chemical modification of HPC-1/syntaxin 1A in the hippocampal autapse, and this modification might decrease the accessibility of the antibody to the facilitatory domain of HPC-1/syntaxin 1A. Because antibody against HPC-1/syntaxin 1A enhanced evoked release without increasing asynchronous release, the underlying mechanisms for these two types of release could be different.

The present model (Eqs. 2, 3) is a primary approximation of the relationship between PPF and release probability in which recruitment of the synaptic vesicle is omitted, as the PPF ratio was not changed with interval range from 30 to 200 ms in most of the cells. This model would provide an useful tool to analyze whether changes in transmitter release at the presynaptic terminal would result from changes in release probability or in pool size of releasable vesicles.

References

1. Katz B, Miledi R (1967) A study of synaptic transmission in the absence of nerve impulses. J Physiol (Camb) 192:407–436
2. Llinás R, Steinberg IZ, Walton K (1981) Relationship between presynaptic calcium current and postsynaptic potential in squid giant synapse. Biophys J 33:323–352
3. Geppert M, Goda Y, Hammer RE, et al (1994) Synaptotagmin I: a major Ca^{2+} sensor for transmitter release at a central synapse. Cell 79:717–727
4. Inoue A, Obata K, Akagawa K (1992) Cloning and sequence analysis of cDNA for a neuronal cell membrane antigen HPC-1. J Biol Chem 267:10613–10619
5. Bennett MK, Calakos N, Scheller RH (1992) Syntaxin: a synaptic protein implicated in docking of synaptic vesicles at presynaptic active zones. Science 257:255–259
6. Südohof TC, Rizo J (1996) Synaptotagmins: C_2-domain proteins that regulate membrane traffic. Neuron 17:379–388
7. Segal MM, Furshpan EJ (1990) Epileptiform activity in microcultures containing small numbers of hippocampal neurons. J Neurophysiol (Bethesda) 64:1390–1399
8. Yamaguchi K, Takada M, Fujimori K, et al (1997) Enhancement of synaptic transmission by HPC-1 antibody in the cultured hippocampal neuron. Neuroreport 8:3641–3644
9. Baranes D, López-García JC, Chen M, et al (1996) Reconstitution of the hippocampal mossy fiber and associational-commissural pathway in a novel dissociated cell culture system. Proc Natl Acad Sci USA 93:4706–4711
10. Zucker RS (1989) Short-term synaptic plasticity. Annu Rev Neurosci 12:13–31
11. Katz B, Miledi R (1968) The role of calcium in neuromuscular facilitation. J Physiol (Camb) 195:481–492
12. Weisskopf MG, Castillo PE, Zalutsky RA, et al (1994) Mediation of hippocampal mossy fiber long-term potentiation by cyclic AMP. Science 265:1878–1882
13. Mochida S, Saisu H, Kobayashi H, et al (1995) Impairment of syntaxin by botulinum neurotoxin C_1 or antibodies inhibits acetylcholine release but not Ca^{2+} channel activity. Neuroscience 65:905–915
14. Kee Y, Lin RC, Hsu S, et al (1995) Distinct domains of syntaxin are required for synaptic vesicle fusion complex formation and dissociation. Neuron 14:991–998

Part 8
Neural Regeneration

Neuronal Survival and Axonal Regeneration After Axotomy of Cerebellar Purkinje Cells

Constantino Sotelo and Isabelle Dusart

Summary. Cerebellar Purkinje cells are extremely resistant to axotomy but lack the ability to spontaneously regenerate their axons. The resistance could depend on paracrine secretion of neurotrophins and the fast up-regulation of their receptor $p75^{NTR}$ after axotomy. By using organotypic cultures, the age and environmental factors implicated in Purkinje cell survival and axonal regeneration have been studied. In cultures taken from 1- to 5-day-old animals (P1–P5), Purkinje cells do not survive the in vitro conditions, whereas they survive when cultured before or after these ages, up to P10. Mature Purkinje cells do not regenerate their axons even when confronted with permissive environments. Fetal Purkinje cells regenerate their axons and can massively invade cocultured mature cerebella. Thus, the failure of mature Purkinje cell axons to regenerate is the result of maturational changes intrinsic to these neurons rather than of extrinsic environmental factors.

Key words. Purkinje cell axotomy—Neuronal survival—Axonal regeneration—Cerebellar cocultures

Introduction

Since the studies of Cajal [1], it has been known that mammalian adult neurons respond differently to axotomy according to the location of their axons in the peripheral or central nervous system (CNS). Peripheral neurons, if they survive axotomy, are able to regenerate; central neurons, however, have a very limited capacity for axonal regeneration. This difference in behavior has been interpreted as the consequence of the different environment encountered by the severed axons in the two situations. Aguayo and collaborators [2], by grafting bridges of peripheral nerves into injured CNS, showed that many types of neurons respond to the grafts by growing long axons into the peripheral nerves. Most central neurons have, therefore, the capacity to set up a program for axonal elongation when confronted with a permissive environment. The failure of central neurons to regenerate their axons would therefore appear to be

Neuromorphologie: Développement Evolution, INSERM U-106, Hôpital de la Salpétrière, 47 bld de l'Hôpital, 75651 Paris, Cédex 13, France the result of the nonpermissive environment offered by the lesioned CNS. However, not all central neurons have the intrinsic capacity for axonal regrowth [3]. Therefore, axonal regeneration in the CNS seems to depend on the interplay between favorable environmental cues and intrinsic properties of the axotomized neurons.

The following review summarizes work carried out in our laboratory [4–6] aimed at determining the capacity of Purkinje cells (PCs) to survive and regenerate their axons at various ages and in various environments.

Purkinje Cells are Extremely Resistant to Axotomy

The axons of PCs, the pivotal elements of the cerebellar cortex, enter the white matter axis of each folium, on the way toward their main postsynaptic targets, the deep cerebellar nuclei. The transection of the folia axotomizes all PCs dorsal to the lesion: some of them are axotomized very close to the emergence of their axons (even before the exit of the recurrent collateral), while in others the axotomy occurs far away from their cell bodies (about 1 mm). In the adult rat cerebellum, independently of a short or long axotomy, most if not all PCs survive without apparent somatic changes but with important axonic alterations [4]: the distal segments of the central stumps survived bearing large terminal varicosities. With the exception of sprouts of late appearance, neither axonal regeneration nor early terminal sprouting was observed. Moreover, PC axons exhibited hypertrophy of the recurrent collateral system and proximal axons, as well as torpedoes.

Axotomized Purkinje Cells Upregulate the Expression of the Neurotrophin Receptor p75 ($p75^{NTR}$)

$p75^{NTR}$ is highly expressed in rat PCs from P6 to P10, when it reaches its peak of expression, but by P15 it almost disappears [7]. In adult cerebellum, a low level of $p75^{NTR}$ expression can be detected only in those PCs coexpressing zebrin I [5]. However, 1–3 days after axotomy all axotomized PCs, independently of their zebrin I expression, are intensely stained with the anti-$p75^{NTR}$ antibody. This upregulation is maintained at high levels during the first week post lesion and decreases sharply thereafter, followed by a long period (up to 1 year) in which a faint immunostaining of all lesioned PCs persists [5].

This $p75^{NTR}$ upregulation could contribute to the PC resistance to axotomy. $p75^{NTR}$ binds all neurotrophins with similar low affinity (see the chapter by Y.A. Barde, this volume). It may be a functional receptor, and its high expression suggests that high-affinity receptors present in PCs (trkB, trkC) (see references in [8]) may also be upregulated (this hypothesis is now being tested in the laboratory). Neurotrophins (BDNF and NT3) are synthesized by granule cells (see references in [8]) and, if they are secreted into the extracellular space, they could act in a paracrine fashion, allowing the survival of lesioned PCs. Although still far from being proven, this hypothesis is supported by the common observation of surviving PCs with short axonal stumps, no recurrent collaterals, and apparently devoid of efferent synaptic connections [4].

An In Vitro System to Study Purkinje Cell Regeneration

The precise geometric arrangement of cerebellar elements makes possible the use of cerebellar slices, containing PCs and their postsynaptic neurons, for organotypic cultures. When taken from P10 cerebella, the organotypic cultures resemble mature, in vivo cerebellum offering an optimal system to study PC regeneration: (i) most PCs, unlike those in cultures taken from P2–P5 cerebella, survive for 2 or more weeks in vitro; (ii) PCs have finished their axonal growth and synapse on deep cerebellar neurons, allowing us to distinguish true regeneration from axonal growth; and (iii) the glial composition in the cultures mimics that of adult lesioned cerebellum (oligodendrocytes and myelinated axons, and astrogliosis and reactive microglia, are present) [6]. Cerebellar slices, P10 + 7 days in vitro (DIV), were divided with a glass knife into a dorsal two-thirds, containing the axotomized PCs, and a ventral third with intact PCs and some deep cerebellar neurons. The two parts were kept in contact for 7 DIV. Axotomized PCs have a very high survival rate but, despite the lack of a lesion cavity, their axons do not regenerate, and the ends of the amputated proximal stumps form a continuous front of terminal varicosities. Thus, the in vitro system corroborates the high resistance of these neurons to axotomy and their lack of spontaneous regeneration.

An advantage of the in vitro system is that the local cellular milieu of the terminal varicosities can easily be modified. We used two well-known permissive environments to promote regeneration of apposed, axotomized PCs: (i) slices of sciatic nerve, and (ii) the ventral part of slices of embryonic cerebellum (taken from rat embryos aged 17–21 days; E17–E21). After 7 or more DIV, neither of the two environments was able to promote regeneration. Thus, we corroborated previous work from our laboratory [3] in which the regenerative capacity of PCs was compared to that of inferior olivary neurons (the origin of climbing fibers), after filling the lesion cavity with embryonic cerebellum. Climbing fibers, confronted with this permissive environment, regenerate, invade the transplant, and synapse on grafted PCs whereas host PC axons remain unchanged. The different responses of these two classes of axons emphasizes the fact that a regeneration-promoting environment, even if necessary, is not by itself sufficient. The reasons for the different axonal behavior remain elusive, but it is of interest to recall that inferior olivary neurons, unlike PCs, are susceptible to axotomy (they follow a process of retrograde degeneration, and many of them die). Thus, retrograde changes seem to be implicated in the initiation of the program for axonal elongation.

Purkinje Cell Axons Can Regenerate but this Process is Age Dependent

To determine whether PCs have a regeneration-permissive critical period, we used a murine coculture setup in which immature wild-type PC axons (dorsal portions of E17–E18 CaBP+/+ cerebellar slices) were confronted with ventral portions of mature (P10 + 7 DIV) cerebellar slices, taken from a mouse line carrying a targeted disruption of the calbindin gene (CaBP−/−) [9]. In these cocultures, immature axotomized PC axons regenerate. They fasciculate and extensively invade the mature slices without any apparent selectivity for permissive environment (they can grow in the myelinated white matter axes of the folia).

The in vitro study has shown that during their development PCs pass through three sequential stages: (i) an early stage of axonal elongation in which these neurons survive the culture conditions and the axotomy and in addition are capable of massive regeneration; (ii) an intermediate stage of intense synaptogenesis with their target neurons and dendritic remodeling, in which stage PCs are extremely vulnerable and do not survive in our in vitro conditions; and (iii) a final stage of neuronal maturation in which the cells behave as adult PCs, with strong resistance to axotomy but lack of axonal regeneration. It can be suggested that synaptogenesis switches off, in these neurons, the genetic program for axonal elongation, impairing their regeneration.

References

1. Ramón y Cajal S (1928) Degeneration and regeneration of the nervous system. Hafner, New York, pp 597–630
2. David S, Aguayo AJ (1981) Axonal elongation into peripheral nervous system "bridges" after central nervous system injury in adult rats. Science 214:931–933
3. Rossi F, Jankovski A, Sotelo C (1995) Differential regenerative response of Purkinje cell and inferior olivary axons confronted with embryonic grafts: environmental cues versus intrinsic neuronal determinants. J Comp Neurol 359:663–677
4. Dusart I, Sotelo C (1994) Lack of Purkinje cell loss in adult rat cerebellum following protracted axotomy: degenerative changes and regenerative attempts of the severed axons. J Comp Neurol 347:211–232
5. Dusart I, Morel MP, Sotelo C (1994) Parasagittal compartmentation of adult rat Purkinje cells expressing the low-affinity nerve growth factor receptor: changes of pattern expression after a traumatic lesion. Neuroscience 63:351–356
6. Dusart I, Airaksinen MS, Sotelo C (1997) Purkinje cell survival and axonal regeneration are age dependent: an in vitro study. J Neurosci 17:3710–3726
7 Yan Q, Johnson EM Jr (1988) An immunohistochemical study of the nerve growth factor receptor in developing rats. J Neurosci 8:348–3498
8. Furukawa S, Sugihara Y, Iwasaki F (1998) Brain-derived neurotrophic factor-like immunoreactivity in the adult rat central nervous system predominantly distributed in neurons with substantial amounts of brain-derived neurotrophic factor messenger RNA or responsiveness to brain-derived neurotrophic factor. Neuroscience 82:653–670
9. Airaksinen MS, Eilers J, Garaschuk O, et al (1997) Ataxia and altered calcium transients in mice with targeted null mutation of calbindin-D28K gene. Proc Natl Acad Sci USA 94:1488–1493

Structural and Functional Repair of the Corticospinal Tract in the Adult Rat by Transplantation of Olfactory Ensheathing Cells

YING LI, PAULINE M. FIELD, and GEOFFREY RAISMAN

Summary. We made unilateral lesions destroying the upper cervical corticospinal tract in adult rats. We prepared a suspension of ensheathing cells cultured from adult rat olfactory bulb and injected it into the lesion site. We observed that this induced unbranched, elongative growth of the cut corticospinal axons, which grew through the transplant and continued uninterruptedly into the denervated caudal host tract. In a series of functional studies we established that rats with complete corticospital tract transections and no transplanted cells were unable to acquire the ability to use the forepaw on the lesioned side for directed reaching. However, rats in which we had transplanted cells that had formed a continuous bridge across the lesion were able to acquire directed forepaw reaching on the operated side.

Key words. Spinal cord—Repair—Function—Fiber tracts—Regeneration

Introduction

The failure of damaged nerve fibers to regenerate in the brain and spinal cord is the underlying cause of the permanent disabilities experienced by patients after accidents such as spinal injury and some of the more devastating types of stroke. The main motor pathway in the spinal cord is the corticospinal tract (CST). Schwann cells mediate the regeneration of peripheral nerves. Transplants of peripheral nerves [1] or Schwann cells cultured from peripheral nerves [2–4] have been used in an attempt to induce reparative growth of axons in the adult central nervous system (CNS). However, although these transplants are able to stimulate the growth of cut central axons, their ability to restore connections is still not established. In our own material we have observed that injections of Schwann cells greatly stimulate the growth of cut CST axons, but the sprouts were highly tortuous and branched profusely within the transplant region, and there was no indication that any substantial number of regenerating axons had left the transplant to reenter the host pathways [5].

The Norman and Sadie Lee Research Centre, Division of Neurobiology, National Institute for Medical Research, MRC, London, U.K.

We have sought a method of repair that would increase the ability of the regenerating CST axons to reenter the CNS. In the normal adult olfactory system the primary neurosensory cells are continuously replaced [6–9], and their newly formed axons are able to enter the olfactory bulb and form functional connections there. The pathway from the olfactory epithelium to the central targets is made up of specialized ensheathing cells [10–13]. Unlike Schwann cells, which stop at the dorsal root entry zone, the processes of the olfactory ensheathing cells (OECs) enter the CNS, where they are integrated into the sheaths of the olfactory glomeruli. Thus, they accompany the olfactory axons all the way to their synaptic terminations in the central neuropil [14]. In an attempt to use this property of the OECs to obtain more effective regeneration of the cut adult CST, we now describe the effects of intraspinal transplantation of OECs cultured from the adult olfactory bulb [15].

Results

The corticospinal tract was destroyed in adult rats by electrolytic lesions placed in the medioventral part of the dorsal columns on one side between the first and second cervical segments [5]. The lesions damaged an area of tissue about 0.5 mm in cross section and 1 mm in a rostrocaudal direction. Apart from the area at the tip of the electrode, there was no cavitation, and with time, the small central cystic area of the lesions shrank and became infiltrated with cells.

We obtained OECs from syngeneic adult olfactory bulbs by a simplified version of the method of Ramón-Cueto and Nieto-Sampedro [16], and maintained the cells in culture for 14–17 days [12]. In a series of animals OECs were injected into the lesion site immediately after lesioning. After 6 days we labeled the CST axons by anterograde labeling with biotin dextran from the medullary pyramid. We observed that the shafts of the cut CST axons were thickened and then extended as single, fine, unbranched regenerating processes with a few small varicosities close to a single, tapering, advancing tip. Many had already crossed the transplant, and their tips could be seen in the caudal part of the host CST. At longer survivals, progressive accumulation of regenerating CST axons formed parallel bundles of straight, unbranched axons.

The transplanted OECs could be recognized by the expression of p75 low-affinity neurotrophin receptor [12]. They infiltrated the lesion area, and extended beyond it for as much as 10 mm caudally. In electron photo micrographs the lesion area contained elongated pale cells and dark, Schwann-like cells. Both types were aligned with the longitudinal axis of the tract. Later, the Schwann-like cells established a peripheral-type myelinating relationship with individual CST axons. The regenerating axons became myelinated by oligodendrocytes after reentry into the caudal host CST.

Function

In a control group of five unoperated normal rats, we examined a directed forepaw reaching function (DFR) [17]. During 10 successive daily sessions, food pellets were presented until the rat had made 50 reaches through an aperture ("free" tests). Three rats showed varying degrees of preference for the right paw and two for the left. When the preferred paw was restrained by wrapping the forearm with a light sleeve ("restrained" tests), all five normal rats readily made 50 reaches per day for 5 successive days with the nonrestrained paw.

Of a group of 21 rats at 2–3 months after unilateral lesions of the CST, 14 showed a varying degree of paw preference in "free" tests and switched to the nonpreferred paw in "restrained" tests (i.e., DFR was present). Subsequent histology showed that in 13 of the 14 cases, the lesion had spared 20%–50% of the cross-sectional area of the CST. The 14th rat had only 698 surviving of a normal total of about 50 000 CST axons (EM).

The other group, consisting of 7 of the 14 lesioned rats, used only one paw, and even when this paw was restrained, they never carried out DFR with the other paw (50 daily failed attempts over a period of 5 days). Serial coronal semithin sections showed that in all 7 the CST had been completely destroyed on the side of the nonreaching paw. Thus, DFR depends on the presence of intact fibers in the CST ($P < 0.0001$; Fisher's exact probability test). As few as 1%–2% of the CST fibers may be sufficient for DFR.

We examined a group of seven rats at 2–3 months after OECs had been transplanted into the lesion sites immediately after lesioning. Four of these seven rats used both paws in the "free" tests, and when the preferred paw was restrained, they all used the unrestrained paw (50 daily reaches over 5 days; i.e., DFR was present, indistinguishable from the normal, unoperated controls). The histology showed that all four rats had complete unilateral destruction of the CST, and the transplants had formed a continuous column of OECs bridging the entire length of the lesion. The remaining three lesioned rats with OEC transplants used only one paw in the "free" tests, and never reached through the aperture with the other paw in "free" or "restrained" tests. The histology showed that these three rats also had complete unilateral destruction of the CST on the side of the nonreaching paw. In two of these three rats there were no OECs in the lesioned CST, and in the third OECs were present, but only in the caudal part of the CST lesion; i.e., they did not bridge the full rostrocaudal extent of the defect in the CST. Thus, the presence of DFR was correlated ($P = 0.029$; Fisher's exact probability test) with the presence of a continuous bridge of OECs and axons across the lesion.

Delayed Repair

In a subsequent series of experiments we placed unilateral lesions in the CST and monitored the pattern of DFR for several weeks after operation. In a first set of four cases we found one animal that had no loss of DFR on the operated side, and subsequent histology showed that the lesion was incomplete. In the remaining three animals, DFR was totally abolished on the operated side for the tested period of 5 weeks. Histology showed that all three had complete unilateral lesions of the CST. These three animals then received OEC transplants on the operated side at 5 weeks after lesion. Two of these manifested no improvement of DFR, and histology showed that the transplants had failed to form complete bridging columns. The third showed a recovery of DFR starting at 6 days after transplantation (which is about the time that the anatomical study had started to show axon regeneration). DFR reappeared first on restrained tests, and then on free tests, reached complete recovery at about 10 days, and was thereafter maintained for the entire posttransplant testing period of 4 weeks.

Discussion

The reentry of the regenerating CST axons into the CNS is the major difference between our previous grafts of Schwann cells [5] and the present grafts of OECs. In the normal adult olfactory system, the glial cells associated with the primary neurosensory fibers [18] permit continuous growth of axons into the CNS [6–9], and after cutting the olfactory nerves, new axons are able to reenter the CNS and grow back to their terminal sites into the olfactory bulb [19,20]. This result differs from the effects of cutting the dorsal roots of spinal nerves, where the fibers are ensheathed by Schwann cells and regenerating axon sprouts cannot reenter the CNS [21]. Ramón-Cueto and Nieto-Sampedro [22] demonstrated that cultured OECs transplanted into a cut dorsal root provide a bridge that enables the regenerating dorsal root axons to enter the dorsal horn of the spinal cord. In contrast to our present findings with the cut CST, however, the ingrowing fibers only grew in gray matter regions invaded by the OECs and did not enter the white matter tracts.

Like transplanted Schwann cells [23], transplants of OECs seem to suppress scar formation by the host astrocytes. Thus, although glial fibrillary acidic protein was upregulated in the host tract astrocytic processes immediately in contact with the transplanted OECs, there was no long-term formation of a thick, reduplicated scar of the type that we observed in comparably sized lesions in the same part of the CST without transplants [24]. Like Schwann cells, OECs are able to myelinate central axons [11], which is all the more striking because their normal axons, in the olfactory system, are totally unmyelinated [25]. A similar, normally unexpressed capacity to myelinate axons is found in the Schwann cells of the cervical sympathetic trunk [26].

OEC transplants, however, differ from Schwann cells in two important ways. First, the growing axons induced by OECs are unbranched, straight, and aligned with the longitudinal axis of the host tract. Second, the regrowing axon tips readily traverse the caudal transplant boundary and reenter the host CST. Like the axons, the OECs are also strictly aligned. In contrast to the restriction of injected Schwann cells to the vicinity of the transplant site [5,23], the considerable caudal migration (of up to 10 mm) we observed with the transplanted OECs demonstrates their superior ability to migrate and incorporate into the host CST and thereby induce the regenerating axon tips to reenter the caudal part of the host tract.

Further observations are needed to confirm what is the minimum number of CST fibers to maintain function, but the observations from rat L3 with a partial lesion (and no transplant) suggest that sparing of as few as 1% of the CST fibers may be sufficient for paw reaching. The improvement of function in this animal only occurred after 2 days of testing with the preferred paw restrained. Because this was 3 months after the operation, it suggests that recovery of function may not be a spontaneous event, but may require training; i.e., it is a learned phenomenon, dependent on activity [27,28]. It is possibly mediated by sprouting or reorganization of connections by the surviving fibers [29–31]. The ability of remodeled circuitry to "learn" a new pattern of function [32–34] may be important in designing strategies for future human spinal cord repair.

There have been previous reports of regeneration of CST fibers after intracerebral transplantation of hybridoma cells secreting an antibody against a central myelin-associated molecule inhibitory to axon growth [35–37]. The axons were described as highly branched, and their course was deflected by extensive cyst formation at the lesion site, so that they did not immediately reenter the distal CST and descended in abnormal locations. Further work is needed to elucidate the relationship to the present observations, where we found cyst formation to be minimal and the regenerating axons were unbranched, maintained their original position and alignment, and directly reentered the caudal CST, where they became at once myelinated by oligodendrocytes.

Our observations of a complete loss of directed forepaw reaching provided us with a precise and consistent measure of CST lesions and their repair. The full pattern of functional consequences of a selective lesion of the CST fibers in the upper cervical dorsal columns (e.g., in terms of hindlimb function) remains to be elucidated. At this stage, therefore, it is difficult to make a direct comparison with previously published data that have concentrated on restoration of hindlimb function [37–39], and have been based on much larger lesions which were placed at the mid- or lower thoracic levels, and destroyed many other constituents of the ascending and descending spinal pathways, sometimes also removing a large segment of the cord itself [38].

Our observations on discrete lesions of the CST at the upper cervical level show that directed forepaw reaching is totally and permanently abolished, and is restored in animals in which the lesion of the CST is bridged by a continuous column of OECs containing regenerating CST axons. The finding that transplanted cultured OECs are able to induce functionally useful regeneration of adult axons in adult rat corticospinal tract lesions opens up the prospect that in future they may also have a role in treating the more complex lesions caused by human spinal cord injury. Obtaining OECs from adult donors raises the important possibility of auto-transplantation for clinical situations.

Acknowledgments. We thank Dr. Daqing Li, who provided invaluable consultation and collaboration. Yewande Ajayi adapted the method for culturing the OECs. This work was supported by the Medical Research Council, the British Neurological Research Trust, the International Spinal Research Trust, the Barnwood House Trust, and Smith's Charity.

References

1. Vidal-Sanz M, Bray GM, Villegas-Pérez MP, Thanos S, Aguayo AJ (1987) Axonal regeneration and synapse formation in the superior colliculus by retinal ganglion cells in the adult rat. J Neurosci 7:2894–2909
2. Neuberger TJ, Cornbrooks CJ, Kromer LF (1992) Effects of delayed transplantation of cultured Schwann cells on axonal regeneration from central nervous system cholinergic neurons. J Comp Neurol 315:16–33
3. Montero-Menei CN, Pouplard-Barthelaix A, Gumpel M, Baron-Van Evercooren A (1992) Pure Schwann cell suspension grafts promote regeneration of the lesioned septo-hippocampal cholinergic pathway. Brain Res 570:198–208
4. Kuhlengel KR, Bunge MB, Bunge RP, Burton H (1990) Implantation of cultured sensory neurons and Schwann cells into lesioned neonatal rat spinal cord. II. Implant characteristics and examination of corticospinal tract growth. J Comp Neurol 293:74–91
5. Li Y, Raisman G (1994) Schwann cells induce sprouting in motor and sensory axons in the adult rat spinal cord. J Neurosci 14:4050–4063

6. Moulton DG (1974) Dynamics of cell populations in the olfactory epithelium. Ann NY Acad Sci 237:52–61
7. Barber PC, Raisman G (1978) Cell division in the vomeronasal organ of the adult mouse. Brain Res 141:57–66
8. Wilson KCP, Raisman G (1981) Estimation of numbers of vomeronasal synapses in the glomerular layer of the accessory olfactory bulb of the mouse at different ages. Brain Res 205:245–253
9. Graziadei PPC, Montigraziadei GA (1979) Neurogenesis and neuron regeneration in the olfactory system of mammals. I. Morphological aspects of differentiation and structural organization of the olfactory sensory neurons. J Neurocytol 8:1–18
10. Barber PC, Lindsay RM (1982) Schwann cells of the olfactory nerves contain glial fibrillary acidic protein and resemble astrocytes. Neuroscience 7:3077–3090
11. Devon R, Doucette R (1992) Olfactory ensheathing cells myelinate dorsal root ganglion neurites. Brain Res 589:175–179
12. Ramón-Cueto A, Valverde F (1995) Olfactory bulb ensheathing glia: a unique cell type with axonal growth-promoting properties. Glia 14:163–173
13. Franceschini IA, Barnett SC (1996) Low-affinity NGF-receptor and E-N-CAM expression define two types of olfactory nerve ensheathing cells that share a common lineage. Dev Biol 173:327–343
14. Raisman G (1985) Specialized neuroglial arrangement may explain the capacity of vomeronasal axons to reinnervate central neurons. Neuroscience 14:237–254
15. Li Y, Field PM, Raisman G (1997) Repair of adult rat corticospinal tract by transplants of olfactory ensheathing cells. Science 277:2000–2002
16. Ramón-Cueto A, Nieto-Sampedro M (1992) Glial cells from adult rat olfactory bulb: immunocytochemical properties of pure cultures of ensheathing cells. Neuroscience 47:213–220
17. Whishaw IQ, O'Connor RB, Dunnett SB (1986) The contributions of motor cortex, nigrostriatal dopamine and caudate-putamen to skilled forelimb use in the rat. Brain 109:805–843
18. Blanes T (1898) Sobre algunes puntos dudosos de la estructura del bulbo olfactorio. Revista Trimestral Micrográfica 3:99–127
19. Barber PC, Raisman G (1978) Replacement of receptor neurones after section of the vomeronasal nerves in the adult mouse. Brain Res 147:297–313
20. Graziadei PPC, Montigraziadei GA (1980) Neurogenesis and neuron regeneration in the olfactory system of mammals. 3. Deafferentation and reinnervation of the olfactory bulb following section of the fila-olfactoria in rat. J Neurocytol 9:145–162
21. Bignami A, Chi NH, Dahl D (1984) Regenerating dorsal roots and the nerve entry zone: an immunofluorescence study with neurofilament and laminin antisera. Exp Neurol 85:426–436
22. Ramón-Cueto A, Nieto-Sampedro M (1994) Regeneration into the spinal cord of transected dorsal root axons is promoted by ensheathing glia transplants. Exp Neurol 127:232–244
23. Li Y, Raisman G (1997) Integration of transplanted cultured Schwann cells into the long myelinated fibre tracts of the adult spinal cord. Exp Neurol 145:397–411
24. Li Y, Raisman G (1995) Sprouts from cut corticospinal axons persist in the presence of astrocytic scarring in long-term lesions of the adult rat spinal cord. Exp Neurol 134:102–111
25. Doucette JR (1991) PNS-CNS transition zone of the first cranial nerve. J Comp Neurol 312:451–466
26. Voyvodic JT (1989) Target size regulates calibre and myelination of sympathetic axons. Nature (Lond) 342:430–433
27. Goodman CS, Shatz CJ (1993) Developmental mechanisms that generate precise patterns of neuronal connectivity. Cell 72(suppl):77–98

28. Wernig A, Müller S, Nanassy A, Cagol E (1995) Laufband therapy based on "rules of spinal locomotion" is effective in spinal cord injured persons. Eur J Neurosci 7:823–829
29. Harris RM, Little JW, Goldstein B (1994) Spared descending pathways mediate locomotor recovery after subtotal spinal cord injury. Neurosci Lett 180:37–40
30. Eidelberg E, Straehley D, Erspamer R, Watkins CJ (1977) Relationship between residual hindlimb-assisted locomotion and surviving axons after incomplete spinal cord injuries. Exp Neurol 56:312–322
31. Aoki M, Fujito Y, Satomi H, Kurosawa Y, Kasaba T (1986) The possible role of collateral sprouting in the functional restoration of corticospinal connections after spinal hemisection. Neurosci Res 3:617–627
32. Wang X, Merzenich MM, Sameshima K, Jenkins WM (1995) Remodelling of hand representation in adult cortex determined by timing of tactile stimulation. Nature (Lond) 378:71–75
33. Florence SL, Jain N, Pospichal MW, Beck PD, Sly DL, Kaas JH (1996) Central reorganization of sensory pathways following peripheral nerve regeneration in fetal monkeys. Nature (Lond) 381:69–71
34. Manger PR, Woods TM, Jones EG (1996) Plasticity of the somatosensory cortical map in macaque monkeys after chronic partial amputation of a digit. Proc R Soc Lond (Biol) 263:933–939
35. Schnell L, Schwab ME (1990) Axonal regeneration in the rat spinal cord produced by an antibody against myelin-associated neurite growth inhibitors. Nature (Lond) 343:269–272
36. Schnell L, Schneider R, Kolbeck R, Barde Y-A, Schwab ME (1994) Neurotrophin-3 enhances sprouting of corticospinal tract during development and after adult spinal cord lesion. Nature (Lond) 367:170–173
37. Bregman BS, Kunkel-Bagden E, Schnell L, Dai HN, Gao D, Schwab ME (1995) Recovery from spinal cord injury mediated by antibodies to neurite growth inhibitors. Neuron 15:498–501
38. Cheng H, Cao Y, Olson L (1996) Spinal cord repair in adult paraplegic rats: partial restoration of hind limb function. Science 273:510–513
39. Kalderon N, Fuks Z (1996) Severed corticospinal axons recover electrophysiologic control of muscle activity after x-ray therapy in lesioned adult spinal cord. Proc Natl Acad Sci USA 93:11185–11190

Neuron–Glia Interactions in the Injured PNS and CNS

A.J. Aguayo, L. Aigner, H. Friedman, S. Singel, and G.M. Bray

Neuron-glia interactions mediated by surface bound and diffusable molecules appear to play an important role in neuronal survival and axonal growth. In the peripheral nervous system, Schwann cells that differentiate towards the myelinating phenotype express specific neurotrophins and cytokines in response to neuronal injury. Conversely, in certain neuropathies where myelination is deficient (e.g. PMP 22 mutations) such sheath cells fail to upregulate neurotrophin expression. Such a lack of trophic support may lead to the neuronal degeneration that is often associated with long standing demyelination.

In the injured central nervous system of mature mammals, the substitution of CNS glia by Schwann cells is associated with lengthy axonal regeneration and functional reconnection with distant target cells.

Finally, the neuroprotective role of non-neuronal cells in the CNS can be enhanced both in vitro and in vivo by manipulating gene expression. Müller cells in the mammalian retina infected with an Ad.BDNF viral vector secrete BDNF into the retina and transiently rescue retinal ganglion cells from axotomy-induced apoptosis. The short lasting effects of various forms of neurotrophin application suggests that additional requirements may need to be met to assure the long term viability of injured nerve cells.

Centre for Research in Neuroscience, McGill University, Montreal General Hospital Research Institute, 1650 Cedar Avenue, Montreal, Quebec H3G 1A4, Canada

Peripheral Nerve Regeneration

Chizuka Ide[1], Akira Mizoguchi[1], Tohru Morihara[1], Shigeru Kawano[1], Etsuko Fujimoto[2], Takuya Hayakashi[1], Takashi Tsujihara[1], Mikihiro Shirasu[1], Kazushi Kimura[1], and Masaaki Kitada[1]

Summary. In previous studies, we have demonstrated that Schwann cell basal laminae remained intact in the form of tubes after Schwann cells were killed by freeze-treatment, and that regenerating axons grew as vigorously through such Schwann cell basal lamina tubes as in Schwann cell columns. It has also been shown that Schwann cell basal lamina tubes can be used for allogeneic nerve graft. The current study examined the effects of basic fibroblast growth factor (bFGF) on the growth of regenerating axons through acellular Schwann cell basal lamina tubes. A solution containing bFGF was applied locally to rat saphenous nerve that had been treated by freezing in situ, and a gelatin hydrogel containing bFGF was applied to the surface of freeze-treated allografts (5 cm long) in canine common peroneal nerve. Exogenously applied bFGF enhanced the growth of regenerating axons through the long allogeneic acellular graft as well as through the autogeneic acellular saphenous nerve. This result indicates that Schwann cell basal lamina tubes treated by bFGF can be used for clinical application. Immunohistochemistry demonstrated localization of bFGF on the outer surface of the basal laminae, indicating that Schwann cell basal laminae function not only as mechanical scaffolds, but also as trophic sources attracting regenerating axons in vivo.

Key words. Schwann cell basal lamina—bFGF—Immunohistochemistry—Allograft—Peripheral nerve

Introduction

We have shown that Schwann cell basal laminae are effective conduits for the growth of regenerating axons [1–3]. Schwann cells were killed by freeze-treatment and the myelin sheaths were damaged and removed by macrophages, while basal laminae remained intact in the form of tubes. Because these freeze-treated nerves contain no living cells, they can be called an "acellular nerve segment." It was shown that acellular

[1] Department of Anatomy and Neurobiology, Kyoto University Graduate School of Medicine, Yoshidakonoe-cho, Sakyo-ku, Kyoto 606-8501, Japan
[2] Seishin School of Nursing, 78-73 Yoshinari, Kande-cho, Nishi-ku, Kobe 651-2301, Japan

nerve segments can be successfully used for allografting without immunosuppressants in the rat [4], rabbit [5], and monkey [6]. However, as the Schwann cell basal laminae gradually disintegrate, regenerating nerve growth should be accelerated so that regenerated axons extend the entire length of the long allograft before the basal laminae are lost.

In the present study, we examined whether nerve trophic factors such as basic fibroblast growth factor (bFGF) can promote the growth of regenerating axons in autogeneic as well as allogeneic acellular nerve segments in vivo [7,8]. bFGF was applied to the grafts before suturing the skin during surgery. Early nerve regeneration in the autograft was examined in the rat and long-term regeneration in the allograft was examined in the dog. In addition, the location of bFGF was examined by immunohistochemistry.

Materials and Methods

Autograft

The rat saphenous nerve was crushed at midthigh by ligation with a thread. After 1 week, the nerve segment distal to the ligation was frozen several times in situ to kill Schwann cells, and the ligation was then released, allowing regenerating axons to extend distally through the basal lamina tubes. The nerve was immersed in situ for 1 h in 0.2 ml bFGF solution containing 10 μg/ml bFGF in PBS. Nerve regeneration was examined 2 and 5 days after bFGF administration. Axons were visualized by anti-PGP9.5 immunohistochemistry. The length of regenerating axons was traced on several sections, then the tracings were superimposed [9].

Allograft

Allogeneic nerve segments were obtained from the sciatic nerve of dead beagle dogs, and after treatment by freezing and thawing these were transplanted to a 3-cm gap made in the common peroneal nerve of beagle dogs. bFGF (166.7 μg/500 μl) was impregnated in gelatin hydrogels [10] and applied to the surface of the grafts before suturing the skin. At 1 and 3 months after grafting, nerve regeneration was examined by light and electron microscopy. Electromyographs were recorded in the peroneus longus muscle.

Results

In the rat, regenerating axons extended up to 1.5 mm at 2 days and 5.9 mm at 5 days after administration. The corresponding control values were 0.7 mm and 4.3 mm, respectively. Electron microscopy showed that regenerating axons extended through basal lamina tubes, and that immature Schwann cells migrated along regenerating axons, always behind the growth cones. Immunoelectron microscopy demonstrated that exogenous bFGF was trapped in the proteoglycan on the outer surface of basal lamina (Fig. 1).

In the dog, the number of myelinated axons was 12.1 per $10^4 \mu m^2$ compared to the control value of 0.9 per $10^4 \mu m^2$ at the level 1.5 cm from the distal suture, 3 months after administration. Immunofluorescence histochemistry for neurofilaments showed

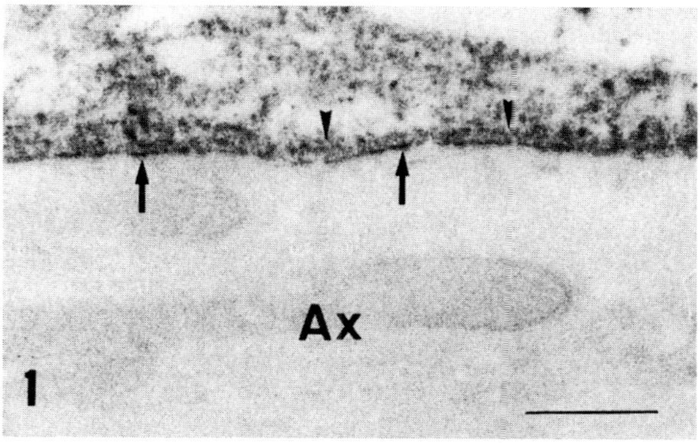

FIG. 1. Immunohistochemistry of basic fibroblast growth factor (bFGF). Basic FGF is present on the outer surface of the Schwann cell basal lamina (*arrowheads*), and on the plasma membrane (*arrows*) of the regenerating axon (*Ax*). *Bar* 1 μm

a 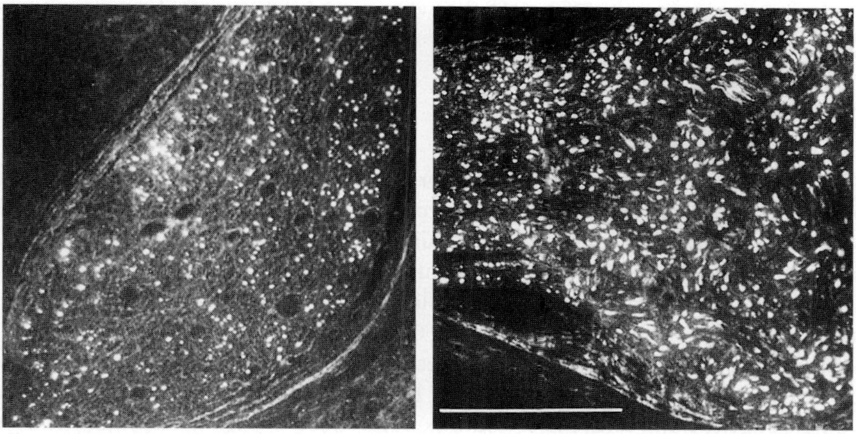 b

FIG. 2a,b. Regenerating axons 1.5 cm distant from the distal suture of dog common peroneal nerve, by neurofilament immunohistochemistry. **a** Control. **b** bFGF treatment. *Bar* 100 μm

an abundance of axon profiles in the bFGF-treated allografts (Fig. 2). Electron microscopy showed that regenerating axons extended in a group through basal lamina tubes, which were retained 3 months after grafting (Fig. 3). Electromyography showed M waves in the peroneus longus muscle in both the control and bFGF-administered grafts 3 months after grafting.

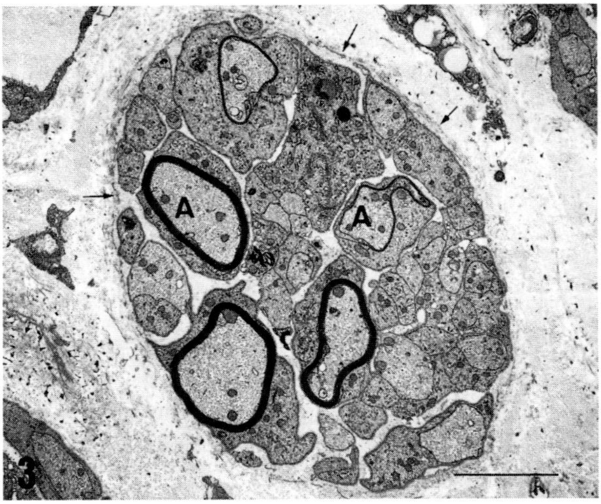

FIG. 3. Electron micrograph of group of regenerating axons (*A*) through the Schwann cell basal lamina tube (*arrows*). *Bar* 3 μm

Discussion

This study demonstrated that bFGF can promote the growth of regenerating axons through the basal lamina tubes of both autogeneic and allogeneic grafts in vivo. This experimental model is suitable for examining the effects of trophic factors on nerve regeneration in vivo because the trophic influences from living Schwann cells can be avoided in the acellular graft. The basal laminae have proteoglycans at the outer surface [11], and it can be said that the bFGF is trapped in the proteoglycans of the basal laminae and taken up by bFGF receptors on regenerating axons. Initially, growth cones would be the primary site for bFGF uptake. After axons are surrounded by migrating Schwann cells, bFGF would exert its effect through the Schwann cells on the axons.

The results of this study indicate that an allogeneic acellular nerve segment can be used for bridging a gap up to 5 cm in length, and that nerve regeneration is accelerated by administration of bFGF to the graft. Thus, acellular nerve segments might be used clinically as allografts without immunosuppressants.

References

1. Ide C, Tohyama K, Yokota R, et al (1983) Schwann cell basal lamina and nerve regeneration. Brain Res 288:61–75
2. Ide C, Osawa T, Tohyama K (1990) Nerve regeneration through allogeneic nerve grafts, with special reference to the role of the Schwann cell basal lamina. Prog Neurobiol 34:1–38
3. Ide C (1996) Peripheral nerve regeneration. Neurosci Res 25:101–121
4. Osawa T, Tohyama K, Ide C (1990) Allogeneic nerve grafts in the rat, with special reference to the role of Schwann cell basal laminae in nerve regeneration. J Neurocytol 19:833–849

5. Tohyama K, Ide C, Osawa T (1990) Nerve regeneration through the cryo-injured allogeneic graft in the rabbit. Acta Neuropathol 80:138–144
6. Tajima K, Tohyama K, Ide C, et al (1991) Regeneration through nerve allografts in cynomolgus monkey (*Macaca fasciculata*). J Bone Joint Surg 73A:172–185
7. Baird A, Böhlen P (1990) Fibroblast growth factors. In: Sporn MB, Roberts AB (eds) Handbook of experimental pharmacology, vol 95/I. Peptide growth factors and their receptors. Springer, Berlin, pp 369–418
8. Unsicker K, Reichert-Preibsch H, Wewetzer K (1992) Stimulation of neuron survival by basic FGF and CNTF is a direct effect and not mediated by non-neuronal cells: evidence from single cell cultures. Dev Brain Res 65:285–288
9. Fujimoto E, Mizoguchi K, Hanada K, et al (1997) Basic fibroblast growth factor promotes extension of regenerating axons of peripheral nerve. In vivo experiments using a Schwann cell basal lamina tube model. J Neurocytol 26:511–528
10. Tabata Y, Hijikata S, Ikada Y (1994) Enhanced vascularization and tissue granulation by basic fibroblast growth factor impregnated in gelatin hydrogels. J Controlled Release 31:189–199
11. Timpl R, Dziadek M (1986) Structure, development and molecular pathology of basement membranes. Int Rev Exp Pathol 29:1–112

Spinal Cord Repair in Adult Rats: Restoration of Corticospinal and Rubrospinal Tracts to Almost Normal Extent

S. Kawaguchi[1], K. Kojima[1], T. Hasebe[1], S. Kikukawa[1], M. Murata[1], and Y. Iwashita[2]

Under favorable conditions, regeneration of spinal cord pathways occurs spontaneously in neonatal and infant rats but not in adults. In adults, however, regeneration is inducible by manipulation of the lesion. Nevertheless, the extent of regeneration in adult rats in various endeavors to date is very restricted in both the amount and extent of axons. To overcome the restriction is of crucial importance if marked functional recovery after spinal cord injury is to be expected. The present experiments aimed at finding how to overcome the restriction and restore the injured projections to almost normal extent.

In Wistar rats aged 2–3 months, the spinal cord was overhemisected under ether anesthesia at the level of T10–T11. By this procedure the corticospinal tract was completely transected bilaterally and the rubrospinal tract unilaterally. Then, embryonic homologous structures dissected from E13–E14 embryos were grafted in the lesion. Control animals were left ungrafted. The corticospinal and rubrospinal tracts in these animals were examined by anterograde and retrograde labeling methods.

In the grafted rats but not in the controls, both tracts regenerated. In the most successful cases, gliosis in the lesion was very sparse, if any, and the regenerated projections were similar to normal ones in various aspects: the amount, extension, path, formation of a compact bundle, and termination. The difference from normal was additional aberrant projections. In the less successful cases, more significant gliosis was seen in the lesion, and the regenerated fibers crossed the lesion to descend a relatively short distance dispersedly.

Injured spinal cord pathways in adult rats are restorable to an almost normal extent. It appears unlikely that neurite growth-inhibitory activity of oligodendrocytes and CNS myelin or intrinsic deficiency of regrowth potential is responsible for the restriction of regeneration of mammalian CNS projections. It is worth noting that the regenerated fibers descended the length of the spinal cord without restriction so long as they coursed normally. It appears likely that the restriction is caused by mismatching of axons and their local environment, which contains positional and guidance cues for axons. Grafting of embryonic homologous structures would minimize the perturbation of such cues in the lesion by suppression of gliosis.

[1] Department of Integrative Brain Science, Graduate School of Medicine, Kyoto University, Kyoto 606, Japan

[2] Department of Orthopedic Surgery, Ehime University School of Medicine, Ehime 791-02, Japan

Transplants and Neurotrophins Increase Regeneration and Recovery of Function after Spinal Cord Injury

Barbara S. Bregman

Regenerative growth of axotomized central nervous system (CNS) pathways after spinal cord lesions is increased by the application of each of the following interventions individually: transplants of fetal spinal cord tissue, exogenous application of neurotrophic factors, and blocking antibodies to myelin-associated neurite growth inhibitors. In recent experiments, we have examined the extent to which these interventions may act in concert to increase regeneration and recovery of function. After spinal cord lesions in adult rats, neurotrophic factors increase the extent of axonal regrowth in the presence of a transplant. We find that transplants and neurotrophic factors (independently) increase c-Jun expression in axotomized CNS neurons, and prevent the lesion-induced atrophy of these neurons as well. Together, they have an additive influence on the axotomized neurons. An additional series of experiments was carried out in order to determine whether regenerating axons can bridge the gap created by a complete spinal cord transection, and to ascertain whether immediate and delayed intervention have similar effects on the regrowth. We transected the spinal cords of adult Sprague-Dawley rats at the mid-thoracic level. Fetal spinal cord tissue (embryonic day 14, E14) was then placed into the lesion site, either immediately (acute injury group), or after a 2 week delay (chronic injury group). In both groups, an Alzet osmotic pump was inserted subcutaneously at the time of transplantation, and oriented so as to deliver neurotrophin-3 continuously and directly over the transplant. All the pumps were removed after 2 weeks. The animals were sacrificed 2 to 4 weeks following transplantation, and their spinal cords analyzed by immunocytochemistry and by neuroanatomical tracing to determine the extent of regrowth of supraspinal axons. In both groups, numerous serotonergic fibers grew into the transplant, and were visible after both a 2 week and a 4 week survival time following transplantation. Serotonergic fibers also regrew caudal to the lesion in animals from both groups. We conclude that following spinal cord transection in adult rats, transplantation and neurotrophin administration can support the regeneration of supraspinal axons. Such axons can bridge the gap created by a spinal cord transection after either acute or chronic spinal cord injury. Recovery of locomotor function was

Department of Cell Biology, Division of Neurobiology, Georgetown University Medical Center, Washington, D.C. 20007, USA examined after spinal cord transection. The extent of recovery of locomotor function was greatest in animals with the combination of transplants and neurotrophins, and was signifcantly greater than that in transection only animals. Taken together, these findings suggest that particular interventions may target different cellular mechanisms in the neuronal response to injury leading to an additive influence on axonal regeneration and recovery of function.

Supported by NIH grants NS 19259 and 27054 from NINDS.

Defective Herpes Simplex Virus Vectors for the Study of Neural Regeneration

Takahito Yazaki[1,2], Michio Hoshi[1,2], Satoshi Terao[1,2], Yoko Nakai[1], Lynn Yoshida[1], Takeshi Kawase[2], and Keiichi Uyemura[1]

Summary. Defective herpes simplex virus (HSV) vectors offer unique advantages over other systems. Our vector genome consists of multiple copies of a plasmid-based amplicon, with a human cytomegalovirus promoter and functional genes. Helper functions were provided by an HSV1 mutant incapable of replication at physiological temperatures. Multiple functions of the neural cell adhesion molecule L1 have been reported including cell–cell interactions, neurite elongation, axonal fasciculation, cell migration, and myelination. L1 plays important roles in neural development and neuronal regeneration in the peripheral nervous system (PNS). We used defective herpes simplex virus vectors to express human or rat L1 in cultured primary rat cortical astrocytes or Schwann cells. Rat cerebellar neurons cocultured on a feeder layer of L1-expressing astrocytes demonstrated increased migration and neurite extension compared with neurons cocultured on lacZ-expressing astrocytes. Rat dorsal root ganglion (DRG) neurons cocultured on a feeder layer of L1-deficient Schwann cells by infection of antisense L1- expressing defective HSV vector showed decreased neurite extension. Because this vector system can be used to confer phenotypic changes in primary nervous system cells, it will be useful for in vitro and in vivo studies of neural regenerative sprouting and plasticity in the central nervous system (CNS).

Key words. Herpes simplex virus—L1—Central nervous system—Regeneration—Gene therapy

Introduction

Neurons of the adult mammalian CNS usually fail to regenerate following injury [1]. However, in the developing nervous system, cell–cell and cell–substrate interactions are associated with neurogenesis, neuronal cell migration, neurite elongation, fasciculation of axons, and finally the formation of correct neural networks. Direct cell–cell

Departments of [1]Physiology and [2]Neurosurgery, School of Medicine, Keio University, 35 Shinanomachi, Shinjuku-ku, Tokyo 160-8582, Japan adhesions are regulated by two adhesive systems: a calcium-dependent adhesive system, mediated by the cadherin family [2], and a calcium-independent system, mediated by the immunoglobulin superfamily [3].

The neural cell adhesion molecule L1 is a member of the immunoglobulin superfamily. Expression of L1 is restricted to post-mitotic neurons, pre- and non-myelinating Schwann cells, and proliferating epithelial progenitor cells of crypts in the intestinal tract [4–7]. L1 contains a short, highly conserved cytoplasmic tail, a hydrophobic membrane-spanning domain and a large extracellular region consisting of six immunoglobin-like domains and five fibronectin type III homologous repeats [8]. Important roles for L1 during the development of the neocortex [9], cerebellar cortex [10], and peripheral nervous system (PNS) [11] have been reported. Neurons extend their axons on purified L1 or on L1-transformed cells [12,13] via a homophilic interaction between L1 molecules. Antibodies to L1 inhibit neurite outgrowth and granular cell migration in culture [12,14,15]. These interactions may also be important during neuronal regeneration in the PNS because L1 expression is upregurated in non-myelinating Schwann cells and neurons after peripheral nerve injury [16], during which nerve regeneration can occur. L1 is not expressed in adult astrocytes in the CNS, and expression in juvenile astrocytes can only be induced by nerve growth factor. We hypothesized that the lack of L1 expression in adult astrocytes could be partially responsible for the lack of regenerative capacity in the adult CNS and that novel expression of L1 in astrocytes could promote neuronal regeneration.

Defective HSV vectors are an efficient means for introducing an amplified gene of interest into postmitotic cells in culture and into the adult brain in vivo [17–23]. Among the advantages of defective herpes simplex virus (HSV) vectors over other systems are the following. (i) The amplicon plasmid is easily manipulated in vitro. (ii) Large-size inserts can be accommodated, including multiple transcription units. (iii) Transcription units are amplified because the defective genome (~153 kb) contains tandem repeats of the amplicon plasmid. (iv) The vector does not spread transneuronally because of a total lack of autonomous replication. (v) Alterations in the physiology of infected cells are confined to those mediated by genes on the amplicon and not other viral genes. (vi) Gene regulation is not affected by surrounding chromosomal sequences because the defective genome is maintained episomally. (vii) The recipient cells need not be mitotically active.

We have investigated the ability of L1 to modify the behavior of neurons using primary cortical astrocytes infected with defective HSV vectors containing L1 cDNA. We present here the evidence of L1 expression in infected primary astrocytes and demonstrate that (1) the homophilic interaction of L1 between neurons and astrocytes can promote neurite outgrowth and neural cell migration and (2) decreased expression of L1 in Schwann cells inhibits neurite outgrowth.

Neurite Extension of Neurons on the Infected Astrocytes or Schwann Cells

Expression of L1 in the infected astrocytes was confirmed by immunocytochemistry. To determine the physiological activity of this novel expression of L1, we investigated whether human or rat L1-infected astrocytes can stimulate the elongation of neurites

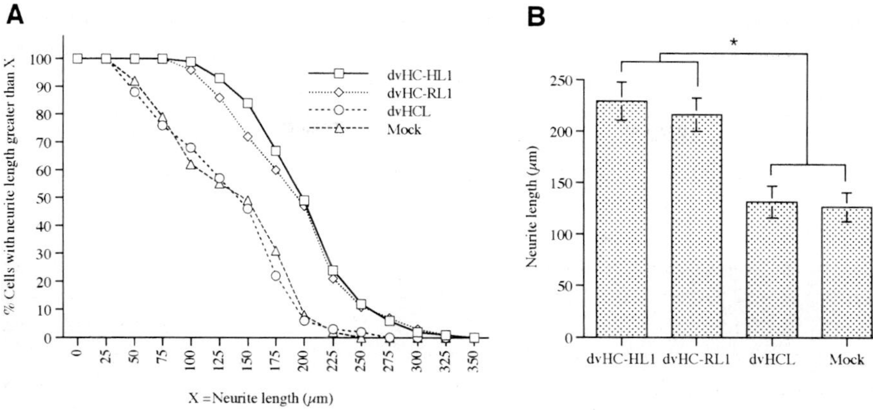

FIG. 1A,B. Distribution (A) and mean (B) of neurite lengths of cerebellar granule cells grown on astrocytes infected with dvHC-HL1, dvHC-RL1, or dvHCL, or mock infected; 100 neurites were measured at random 12h after plating in each group. Data are mean of triplicate wells. *, $P < 0.005$, unpaired *t*-test

and migration of cerebellar neurons when used as a feeder cell layer. To assess neurite outgrowth quantitatively, dissociated cerebellar neurons from rats (postnatal day 3) were plated on astrocytes infected with dvHC-HL1, dvHC-RL1, dvHCL, or mock infected. Twelve hours after plating, the cells were fixed and neurite lengths were measured. Unlike cultures grown on astrocytes infected with dvHCL or mock infected, cerebellar neurons grown on astrocytes infected with dvHC-HL1 or dvHC-RL1 displayed remarkable neurite outgrowth promotion (Fig. 1A). The neurite lengths of neurons plated on astrocytes infected with dvHC-HL1 or dvHC-RL1 were markedly longer than those of neurons plated on the astrocytes infected with dvHCL or mock infected (Fig. 1). Infection of astrocytes with dvHCL did not alter the inherent ability of the astrocytes to promote neurite outgrowth. By 12h after plating, a significant fraction of neurons (48%) grown on L1-expressing astrocytes displayed neurite lengths greater than 200mm, whereas few neurons (7%) grown on lacZ-expressing astrocytes were 200mm in length. The mean neurite length of cerebellar neurons grown on human or rat L1-expressing astrocytes was significantly greater ($P < 0.005$ by unpaired *t*-test) than that of neurons grown on L1-nonexpressing astrocytes (Fig. 1B). In contrast, the L1-nonexpressing Schwann cells by the infection of antisense L1-expressing defective HSV vectors demonstrated decreased neurite outgrowth of dorsal root ganglion (DRG) neurons (data not shown).

Cell Migration of Cerebellar Neurons Grown on the Astrocytes

To quantify the ability of L1-expressing cells to promote neuronal migration, we used a coculture system developed by Asou et al. [12] that measures the disaggregation of cerebellar neuron reaggregates. The morphology of reaggregates after plating is similar to that illustrated in Fig. 2 (upper, **A**). In this coculture system, migration of neurons grown on the lacZ-expressing astrocytes (infected with dvHCL) or mock-infected astrocytes was not observed until 12 h after plating (Fig. 2, lower). In contrast, reaggregated neurons coultured with L1-expressing astrocytes (infected with dvHC-HL1 or dvHC-RL1) began migrating outward from the original reaggregates within 4 h after they were plated, with the number of migrating cells increasing linearly with time

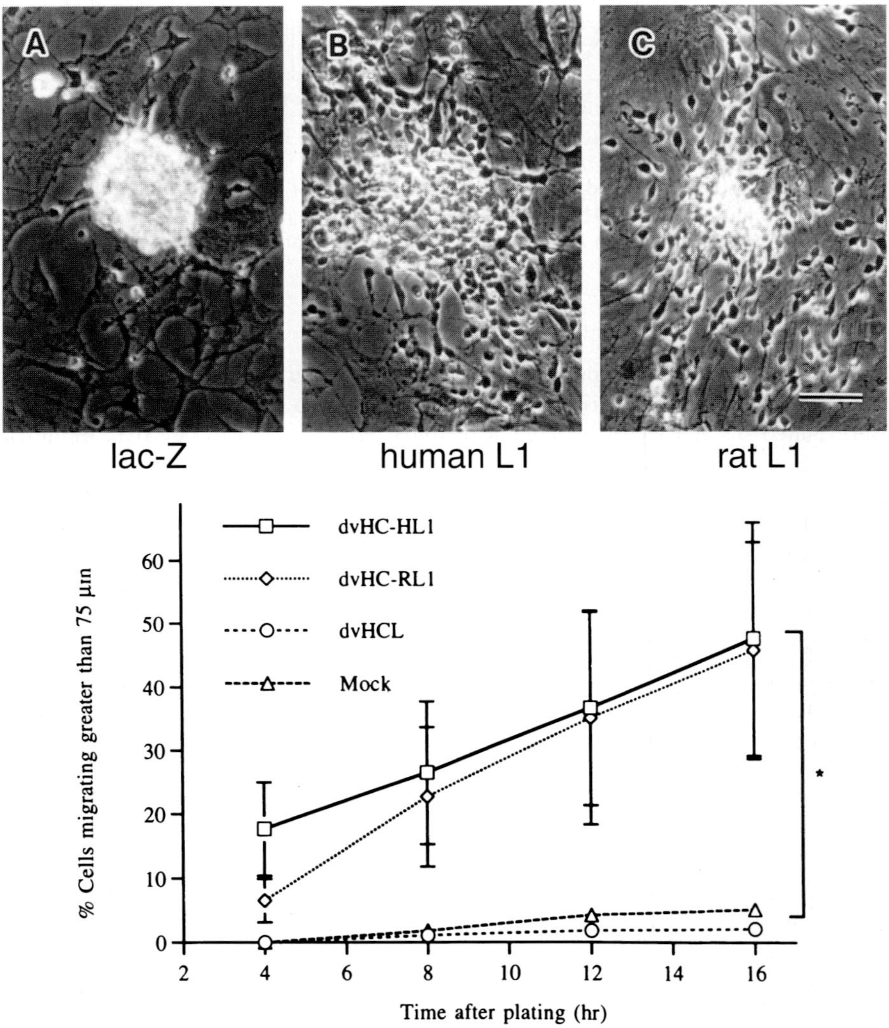

FIG. 2. *Upper*: cell migration of cerebellar granule cells grown on astrocytes infected with dvHCL (**A**), dvHC-HL1 (**B**), or dvHC-RL1 (**C**) 12 h after plating. *Bar* 30.4 μm. *Lower*: time-course analysis of cell migration. Data are mean of triplicate wells. *, $P < 0.001$ at 12 and 16 h, unpaired t-test. The number of countable cells per aggregate in the mock- and dvHCL-infected wells is about 50% the number of countable cells in the disaggregates. Therefore the % cells migrating in the mock- and dvHCL-infected wells is actually about 50% less than indicated

(Fig. 2, lower). By 16h after plating, more than 40% of neurons cocultured with L1-expressing astrocytes had migrated more than 75mm, whereas less than 5% of neurons cocultured with L1-nonexpressing astrocytes had migrated this distance (Fig. 2, lower). There is a significant difference in migration between L1-expressing and L1-nonexpressing astrocytes ($P < 0.001$ at 12 and 16h, unpaired *t*-test).

Conclusion

Direct injection of defective HSV vectors into the brain and subsequent gene expression has been reported. Our results in vitro demonstrate the feasibility of this approach to alter neuronal regeneration. Infection of the CNS with L1 containing defective HSV vectors should lead to expression in both neurons and glial cells. Increased L1 expression in neurons and novel expression in astrocytes should result in homophilic interactions between both cell types and subsequent increased regeneration after CNS damage. Gene transfer techniques that increase the expression of L1 or related molecules merit further study in the promotion of neural regeneration following CNS injury.

References

1. Bahr M, Bonhoeffer F (1994) Perspectives on axonal regeneration in the mammalian CNS. Trends Neurosci 17:473–479
2. Takeichi M (1988) The cadherins: cell-cell adhesion molecules controling animal morphogenesis. Development (Camb) 102:639–655
3. Edelman GM (1985) Cell adhesion and the molecular processes of morphogenesis. Annu Rev Biochem 54:135–169
4. Stallcup WB, Arner L, Levine J (1983) Antiserum against the PC12 cell defines cell surface antigens specific for neurons and Schwann cells. J Neurosci 4:2354–2368
5. Mirsky R, Jessen KR, Schachner M, et al (1986) Distribution of the adhesion molecules N-CAM and L1 on peripheral neurons and glial in adult rats. J Neurocytol 15:799–815
6. Thor G, Probstmeier R, Schachner M (1987) Characterization of the cell adhesion molecules L1, N-CAM and J1 in the mouse intestine. EMBO J 6:2581–2586
7. Miragall F, Kadmon G, Husmann M, et al (1988) Expression of cell adhesion molecules in the olfactory system of the adult mouse: presence of the embryonic form of N-CAM. Dev Biol 129:516–531
8. Moos M, Tacke R, Scherer H, et al (1985) Neural cell adhesion molecule L1 as a member of the immunoglobulin superfamily with binding domains similar to fibronectin. Nature (Lond) 334:701–703
9. Fushiki S, Schachner M (1986) Immunocytological localization of cell adhesion molecules L1 and NCAM and the shared carbohydrate epitope L2 during development of the mouse neocortex. Dev Brain Res 24:153–167
10. Chuong C-M, Crossin KL, Edelman GM (1987) Sequential expression and differential function of multiple adhesion molecules during the formation of cerebellar cortical layer. J Cell Biol 104:331–342
11. Miura M, Kobayashi M, Asou H, et al (1991) Molecular cloning of cDNA encoding the rat neural cell adhesion molecule L1: the L1 isoforms in the cytoplasmic region are produced by differential splicing. FEBS Lett 289:91–95
12. Asou H, Miura M, Kobayashi M, et al (1992) The cell adhesion molecule L1 has a specific role in neural cell migration. Neuroreport 3:481–484
13. Miura M, Asou H, Kobayashi M, et al (1992) Functional expression of a full-length cDNA coding for rat neural cell adhesion molecule L1 mediates homophilic intercellular adhesion and migration of cerebellar neurons. J Biol Chem 267:10752–10758

14. Fischer G, Kunemund V, Schachner M (1986) Neurite outgrowth patterns in cerebellar microexplant cultures are affected by antibodies to the cell surface glycoprotein L1. J Neurosci 6:605–612
15. Linder J, Rathjen FG, Schachner M (1983) L1 mono- and poly- clonal antibodies modify cell migration in early postnatal mouse cerebellum. Nature (Lond) 305:427–430
16. Seilheimer B, Schachner M (1987) Regulation of neural cell adhesion molecula expression on cultured mouse Schwann cells by nerve growth factor. EMBO J 6:1611–1616
17. Kaplitt MG, Pfaus JG, Kleopolos ST, et al (1991) Expression of a functional foreign gene in adult mammalian brain following in vivo transfer via herpes simplex virus type 1 defective viral vector. Mol Cell Neurosci 2:320–330
18. Ho DY, Mocarski ES, Sapolsky RM (1993) Altering central nervous system physiology with a defective herpes simplex virus vector expressing the glucose transporter gene. Proc Natl Acad Sci USA 90:3655–3659
19. Bergold PJ, Casaccia-Bonnefil P, Xiu-Liu Z, et al (1993) Transsynaptic neuronal loss induced in hippocampal slice cultures by a herpes simplex virus vector expressing the GluR6 subunit of the kainate recepter. Proc Natl Acad Sci USA 90:6165–6169
20. Federoff HJ, Geschwind MD, Geller AI, et al (1992) Expression of NGF in vivo, from a defective HSV-1 vector, prevents effects of axotonomy on sympathetic ganglia. Proc Natl Acad Sci USA 89:1636–1640
21. Bahr BA, Neve RL, Sharp J, et al (1994) Rapid and stable gene expression in hippocampal slice cultures from a defective HSV-1 vector. Mol Brain Res 26:277–285
22. During MJ, Naegele JR, O'Malley KL, et al (1994) Long-term behavioral recovery in Parkinsonian rats by an HSV vector expressing tyrosine hydroxylase. Science 266:1399–1403
23. Wood MJA, Byrnes AP, Kaplitt MG, Pfaff DW, et al (1994) Specific patterns of defective HSV-1 gene transfer in the adult central nervous system: implications for gene targeting. Exp Neurol 130:127–140

Localization of K^+ Channel Complex on Demyelinated Axons

Hiroko Baba[1], Kensuke Nakahira[1], Tetsushi Kagawa[1], Yoshiro Inoue[2], and Kazuhiro Ikenaka[1]

Summary. The characteristic localization of voltage-gated ion channels is crucial for the propagation of saltatory conductions in myelinated nerves. The humoral factor from oligodendrocytes has been recently reported to induce clustering of Na^+ channels at the nodes of Ranvier. However, the molecular mechanism for K^+ channel clustering at the juxtaparanodal region is still uncertain. The transgenic mouse line generated by overexpression of proteolipid protein gene, the 4e line, causes extensive demyelination in the heterozygote in the later stage of life. Because axonal degeneration is not significant in the 4e heterozygote, this is a useful animal model to understand the functional significance of myelin on the axons. In the present study, the heterozygote was used to examine how K^+ channel distribution was affected by demyelination. The axons in various ages of the heterozygotes were examined by double immunostaining using antibodies to K^+ channels and myelin proteins. The spinal cord from younger animals showed no significant demyelination and regular juxtaparanodal channel clustering. However, this characteristic immunoreactivity disappeared in demyelinating older animals, suggesting that the channel reorganization occurred after chronic demyelination, which might have affected the electric excitabilities. Taken together with the absence of channel clusters in the hypomyelinating mouse, the presence of a normal myelin sheath is required to form and to maintain the K^+ channel clustering.

Key words. Demyelination—K^+ channel—Na^+ channel—Channel clustering—hypomyelinating mutant

Introduction

The characteristic localization of ion channels is crucial for the propagation of saltatory conduction in myelinated nerves. For instance, voltage-gated Na^+ channels are located at the nodes of Ranvier while voltage-gated K^+ channels are mainly found

[1] National Institute for Physiological Sciences, Okazaki National Research Institute, 38 Nishigonaka, Myodaiji-cho, Okazaki, Aichi 444-8585, Japan
[2] Department of Anatomy, Hokkaido University School of Medicine, Sapporo 060-8638, Japan

at juxtaparanodal regions. These specific distributions of ion channels in myelinated axons and the electric shield by the myelin sheath are both critical for the fast conduction velocities associated with these axons. Changing of channel localization during the myelination process represents dynamic interactions between neuron and glia.

A humoral factor secreted by oligodendrocytes has recently been reported to induce clustering of Na^+ channels at the nodes of Ranvier [1]. However, the molecular mechanisms for K^+ channel clustering as well as the roles of oligodendrocytes on this clustering are still uncertain. In some of the hypomyelinating mutants, K^+ channel localization is altered and the expression levels of the channels are upregulated [2], suggesting that oligodendrocyte or myelin itself plays some important role in K^+ channel clustering.

The transgenic mouse line, the 4e line, which has been generated by overexpression of the myelin proteolipid protein (PLP) gene, causes severe hypomyelination in homozygotes and extensive demyelination in heterozygotes [3]. Histological analysis of the heterozygote showed that the myelin structure was formed normally during development, then striking demyelination occured abruptly at a later stage of life [4]. Because axonal degeneration is not significant in the heterozygote, this is a useful animal model to understand the functional significance of myelin on the axons in the central nervous system (CNS).

To clarify whether the myelin sheath itself can influence channel distribution, CNS axons from the heterozygotes, which spontaneously demyelinate, and from the jimpy mouse, which has severe hypomyelination, were used to examine how K^+ channel distribution was affected by chronic demyelination or by hypomyelination.

Materials and Methods

Materials

The genetic types of transgenic and jimpy mice were determined by the polymerase chain reaction (PCR) method using specific primers. For immunohistological studies, anti-K^+ channel subtype antibodies (Kv1.1, Kv1.2, and Kvβ2) were kindly provided by Dr. Trimmer (SUNY at Stony Brook, NY, USA) [5]. Monoclonal antibodies against myelin basic protein (MBP) and neurofilament 160-kDa (NF-M) were purchased from Boehringer-Mannheim (Mannheim, Germany) and Amersham (Buckinghamshire, England), respectively.

Immunohistological Studies

The heterozygotes (4e/−) and jimpy mice were perfused with 4% PFA/0.1 M PBS. The optic nerve and spinal cord were postfixed, cryoprotected with 30% sucrose/0.1 M PBS, and embedded in O.C.T. Compound (Miles, Elkhart, IN, USA). Frozen sections (12 μm) were treated with 1.0% Triton X-100/0.1 M PBS for 1 h at room temperature (RT). Sections were further rinsed with PBS, and were incubated with blocking solution (0.5% skim milk/0.1% Triton X-100/PBS) for 1 h at RT. The slides were rinsed with PBS and incubated with first antibodies (Abs) diluted with blocking solution overnight at 4°C. Fluorescein-conjugated second Abs were overlaid and incubated for 30 min at RT. The slides were examined with a BX-50 optical microscope (Olympus,

Tokyo, Japan). For peroxidase-based staining, biotinylated second Abs were used and visualized using an Elite ABC kit (Vector, Burlingame, CA, USA) according to the manufacturer's instructions.

Results

K^+ Channels, Kv1.1, 1.2, and β2-Subunit, are Localized at the Juxtaparanodal Regions of Spinal Cord Axons

To identify the localization of various K^+ channel subunits in the myelinated axons, cryostat sections from 21-day-old ICR mouse spinal cord were immunostained with antibodies against three different K^+ channel subunits (Fig. 1). Kv1.1 and Kv1.2 were found to be located in the juxtaparanodal regions in myelinated axons, which was further confirmed by double-immunostaining methods using anti-NF-M or anti-MBP antibodies (data not shown). The immunostaining pattern of the Kvβ2 subunit, which binds to the N-terminus of α-subunits, was similar to those of Kv1.1 and 1.2, suggesting that the β2-subunit of K^+ channels also localizes in the same region.

Localization of K^+ Channels Is Altered in Demyelinating Axons

The localization of the K^+ channel Kvβ2-subunit in the spinal cord axons was examined in two different ages of the heterozygotes and wild-type controls (wild). The spinal cord from 4.5-month-old heterozygotes showed no significant demyelination, and juxtaparanodal localization of Kvβ2-subunit immunoreactivity was similar to

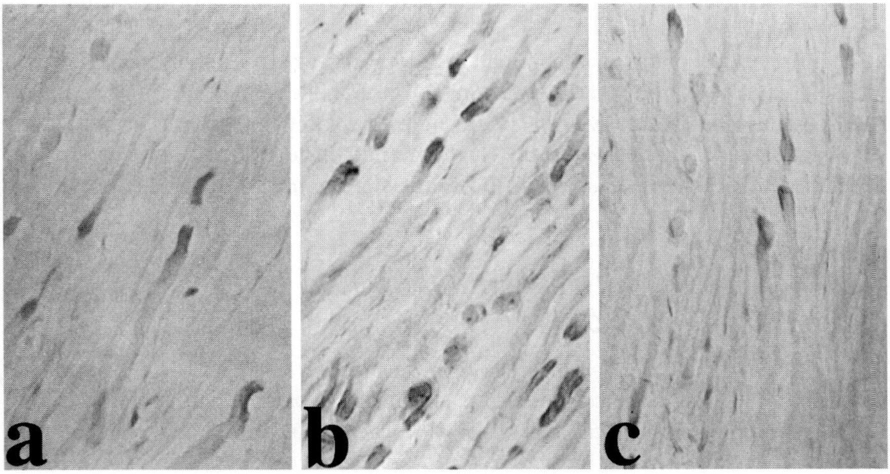

FIG. 1a–c. Localization of K^+ channel subunits in the myelinated axons. The cryostat sections from ICR mouse spinal cord (P21) were immunostained with antibodies against three different K^+ channel subunits. Kv1.1 (**a**) and Kv1.2 (**b**) were found to be located in the juxtaparanodal regions of myelinated axons. The immunostaining patterns of these three subtypes were all similar, suggesting that the Kvβ2 subunit of K^+ channels (**c**) was also located in the same region

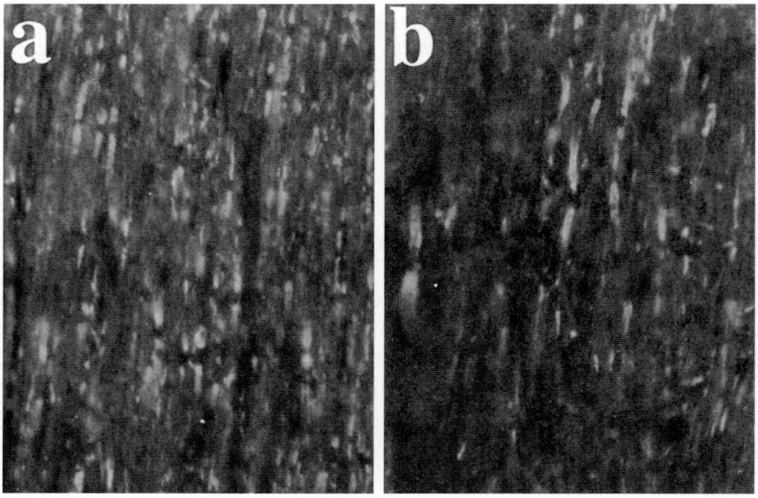

FIG. 2a,b. Localization of the K^+ channel Kvβ2 subunit in demyelinating axons. Spinal cords from 12-month-old transgenic heterozygotes (4e/−) and wild-type control (wild) were immunostained with Kvβ2 antibody. The characteristic immunoreactivity, as shown in the wild type (**a**), decreased significantly in the demyelinated axons (**b**)

that of the wild type. However, this characteristic immunoreactivity disappeared in demyelinating spinal cord from 12-month-old heterozygotes (Fig. 2b) but not from age-matched controls (Fig. 2a), suggesting reorganization of K^+ channels on demyelinated axons.

The similar observation was obtained in demyelinating optic nerves from 8-month-old heterozygotes. In the wild-type control, the characteristic clustering patterns of the K^+ channel were observed in the myelinated region of the optic nerve. On the other hand, the K^+ channel was diffusely distributed throughout the optic nerve in the heterozygous optic nerve in which the axons were almost totally naked (data not shown).

Localization of K^+ Channels Is Altered in Hypomyelinating Axons

The spinal cord from the 21-day-old jimpy mouse and its littermate were examined to determine if K^+ channel localization was affected in hypomyelinating conditions. In the normal control, the K^+ channel already localized at juxtaparanodal regions, while it was distributed diffusely in the jimpy mouse spinal cord where only some of the axons were myelinated. The homogenous and strong staining pattern in jimpy spinal cord was similar to that of the shiverer mouse, in which the expression level of the K^+ channel was upregulated [2]. Even in the jimpy mouse, the characteristic channel localization appeared in some axons, which were wrapped by myelin, suggesting that

K^+ channel clustering is well correlated with the presence of the myelin sheath (data not shown).

Discussion

The characteristic immunoreactivity in the juxtaparanodal region was changed in demyelinating spinal cord and optic nerve as the heterozygote animals grew older, suggesting redistribution of K^+ channels on chronically demyelinated axons. Although it is still controversial [6,7], a similar change in the localization of Na^+ channels was also reported previously [8]. Therefore the dynamic relocalization of these channels occurs following chronic demyelination to compensate electric excitabilities. After chronic demyelination, channel proteins may be localized more diffusely on the axonal surface to partially retain electric excitabilities. This delocalization effect may reflect a compensatory reorganization for impulse conduction to occur in these demyelinating axons, and also these electric inpulses may induce remyelination by oligodendrocytes. However, if the channels are delocalized too diffusely, the channel densities may be too low to retain normal channel functions and, in this case, conduction block may occur in these axons, which may eventually be degenerated.

Present studies using demyelinating and hypomyelinating mouse models demonstrated that the presence of the myelin sheath with the axons is required to maintain the specific localization of K^+ channels. This result is in contrast to the previous report on Na^+ channel clustering, which occurs in cultured neurons in oligodendrocyte-conditioned medium [1]. Taken together, it can be speculated that the humoral factor secreted from oligodendrocyte facilitates Na^+ channel clustering with intervals similar to those of nodes of Ranvier, and oligodendrocyte processes then find a place to myelinate between these clusters. Finally after the myelin sheaths are formed and some signals have been sent to the axons, K^+ channels may form clusters in the juxtaparanodal regions.

Recently, several proteins such as paranodin [9] have been reported to be concentrated in the paranodal region. It is important to investigate the interaction of these proteins with K^+ channels to understand the mechanism of K^+ channel clustering in this region.

References

1. Kaplan MR, Meyer-Franke A, Lambert S, et al (1997) Induction of sodium channel clustering by oligodendrocytes. Nature (Lond) 386:724–728
2. Wang H, Allen ML, Grigg JJ, et al (1995) Hypomyelination alters K^+ channel expression in mouse mutants shiverer and Trembler. Neuron 15:1337–1347
3. Kagawa T, Ikenaka K, Inoue Y, et al (1994) Glial cell degeneration and hypomyelination caused by overexpression of myelin proteolipid protein gene. Neuron 13:427–442
4. Inoue Y, Kagawa T, Matsumura Y, et al (1996) Cell death of oligodendrocytes or demyelination induced by overexpression of proteolipid protein depending on expressed gene dosage. Neurosci Res 25:161–172
5. Rhodes KJ, Monaghan MM, Barrezueta NX, et al (1996) Voltage-gated K^+ channel β subunits: expression and distribution of Kvβ1 and Kvβ2 in adult rat brain. J Neurosci 16:4846–4860

6. Deerinck TJ, Levinson SR, Bennett GV, et al (1997) Clustering of voltage-sensitive sodium channels on axons is independent of direct Schwann cell contact in the dystrophic mouse. J Neurosci 17:5080–5088
7. Rasband MN, Trimmer JS, Schwarz TL, et al (1998) Potassium channel distribution, clustering, and function in remyelinating rat axons. J Neurosci 18:36–47
8. Black JA, Felts P, Smith KJ, et al (1991) Distribution of sodium channels in chronically demyelinated spinal cord axons: immuno-ultrastructural localization and electrophysiological observations. Brain Res 544:59–70
9. Menegoz M, Gaspar P, Le Bert M, et al (1997) Paranodin, a glycoprotein of neuronal paranodal membranes. Neuron 19:319–331

Treatment of Metastatic Brain Tumors with a Replication-Competent Multimutated Herpes Simplex Virus-1

MASAHIRO TODA[1,2,3], KEIICHI UYEMURA[2], TAKESHI KAWASE[3], SAMUEL D. RABKIN[1], and ROBERT L. MARTUZA[1]

Summary. To investigate the therapeutic efficacy of a replication-competent mutant of herpes simplex virus type-1, G207, for the treatment of metastatic brain tumors, we developed two experimental models: the first, an intracerebral xenograft tumor model with human breast adenocarcinoma in athymic mice, and the second, a syngeneic tumor model with murine colon adenocarcinoma in BALB/c mice. In athymic mice harboring intracerebral MDA-MB-435 human breast adenocarcinoma cells, a single intraneoplastic inoculation of G207 caused tumor growth inhibition and prolonged survival. In addition, we have provided experimental evidence for one of the underlying hypotheses, that replication and the consequent spread of G207 within tumor cells are responsible for the inhibition of tumor growth seen in vivo. In the syngeneic bilateral subcutaneous tumor model, intratumoral inoculation of G207 into CT26 murine colon tumors caused a significant reduction in growth of both the infected tumors and the contralateral non-inoculated tumors. Intratumoral inoculation with G207 induced tumor-specific cytotoxic T lymphocytes that recognized a major histocompatibility complex class I-restricted peptide which is a dominant tumor antigen. Therefore, G207 is able to inhibit local tumor growth by direct cytotoxic viral replication and to induce a systemic antitumor immune response to inhibit metastatic tumors.

Key words. Brain metastasis—Herpes simplex virus—Replication-competent—Intratumoral inoculation—Antitumor immunity

Introduction

The incidence of brain metastases is rising, partly because of improved systemic therapies for primary tumors [1–3]. Metastatic brain tumors are present at multiple sites in approximately half the cases, and often cannot be surgically removed because

[1]Georgetown Brain Tumor Center, Department of Neurosurgery, Georgetown University Medical Center, 3800 Reservoir Road, NW, Washington, D.C. 20007, USA
Departments of [2]Physiology and [3]Neurosurgery, School of Medicine, Keio University, 35 Shinanomachi, Shinjuku-ku, Tokyo 160-8582, Japan

of their location in or near critical brain structures [4,5]. Current treatments have failed to improve the survival of these patients, and metastases to the brain frequently result in severe and debilitating neurological complications [2,6]. There is a need to develop novel strategies for the treatment of metastatic brain tumors.

We have developed a multigene mutant of herpes simplex virus type-1 (HSV-1), G207, with deletions in the γ34.5 gene and a *lac Z* gene insertion in the ICP6 gene [7]. G207 replicates in dividing cells, undergoing a lytic infection with consequent cell death; however, growth in nondividing cells is highly attenuated, targeting viral spread to the tumor. A number of antiviral drugs (i.e., acyclovir) are available that would allow unforeseen viral spread to be treated, as opposed to many viruses for which no antiviral therapies exist. More importantly, G207 is nonneuropathogenic, causing no detectable disease in mice and non-human primates [7]. In this chapter we have demonstrated the therapeutic efficacy of G207 in two different metastatic tumor models, an intracerebral xenograft tumor model in athymic mice and a syngeneic bilateral subcutaneous tumor model in BALB/c mice.

Intracerebral Xenograft Tumor Model

Brain tumors represent a dividing cell population within an essentially nondividing population of support cells and terminally differentiated neurons. Thus, in the context of brain tumor therapy, one approach is to use viruses whose replication is restricted to dividing cells [8]. G207 is particularly well suited to this approach, because the virus replicates selectively in dividing cells, undergoing a lytic infection with consequent cell death, but is incapable of lytic infection in normal brain cells.

We tested the susceptibility of four human breast cancer cell lines, MDA-MB-231 [9], MDA-MB-435 [10], MCF-7 [11], and T47D [12] against G207 in vitro. G207 destroyed efficiently three of four human malignant breast cancer cell lines. MDA-MB-435 was most susceptible and MDA-MB-231 least susceptible to replication by G207 [13]. In athymic mice harboring subcutaneous or intracerebral MDA-MB-435 cells, a single intraneoplastic inoculation of G207 caused tumor growth inhibition or prolonged survival of mice. In contrast, G207 had minimal effects on MDA-MB-231 tumor growth or survival in the intracerebral tumor model. In histological studies, sections from mice treated with G207 showed that the virus replication is restricted to tumor cells in vivo and does not occur in the surrounding brain tissue.

The therapeutic efficacy of G207 in two different in vivo tumor models (subcutaneous and intracerebral) correlated well with the susceptibility of the tumor cells to viral replication in vitro. These results further suggest that there are likely to be human breast cancer tumors that may be relatively resistant to infection by G207. Many tumors contain cells of different lineage, so it is also possible that some cells within a breast tumor may be sensitive to viral therapy while others would be resistant. Therefore, pretreatment testing of tumor cells in vitro may allow the selection of patients appropriate for this form of therapy. The cause of these differences in susceptibility, which could result from differences in viral entry, viral replication, shutoff of protein synthesis, or interferon expression, is important. Identification of the cause of these differences would allow design of HSV vectors to overcome the lack of susceptibility of some tumor cells.

Syngeneic Bilateral Subcutaneous Tumor Model

Strategies of cancer vaccination to destroy metastatic tumors or prevent metastases have focused on the use of killed tumor cells or lysates delivered in combination with adjuvants or cytokines [14]. A major difficulty in the therapeutic use of "cancer vaccines" has been the need to obtain and culture a patient's autologous tumor cells for manipulation in vitro, including transduction with cytokine genes, irradiation, and subsequent vaccination, or the need to identify and purify a specific tumor antigen. The ability to elicit antitumor immunity by in situ inoculation of a tumor would overcome such major difficulties.

A nonmutated nonamer peptide derived from the envelop protein (gp70) of an endogenous ecotropic murine leukemia provirus (MuLV), *env-1*, has been identified as the immunodominant major histocompatibility complex (MHC) class I-restricted antigen for the murine colorectal carcinoma CT26 [15]. Adoptive transfer of peptide-specific cytotoxic T-lymphocyte (CTL) lines was able to cure established subcutaneous CT26 tumors, indicating the correlation between induction of tumor-specific CTL and an antitumor effect. The poorly immunogenic carcinoma cell line CT26, expressing an identified tumor antigen, was used to investigate the efficacy of attenuated replication-competent HSV-1 mutant G207 as an "in situ cancer vaccine."

In the syngeneic bilateral subcutaneous tumor model, intratumoral inoculation of G207 into established CT26 tumors caused a significant reduction in growth of both the infected tumors and the contralateral noninoculated tumors [16]. Seven days after the virus inoculation, lacZ expression from G207 was detected by X-gal histochemistry in the inoculated tumor, but not the noninoculated tumor. The antitumor effect on the noninoculated contralateral tumor depended on intratumoral inoculation of G207, because intradermal inoculation of G207 in the right flanks of mice with established unilateral tumors in the left flanks had no effect on tumor growth. Treatment of established CT26 subcutaneous tumors in athymic mice with similar doses of G207 did not reduce tumor growth.

CT26 cells express the MuLV *env* gene product while the normal tissues of BALB/c mice do not, and thus this viral antigen, gp70, can serve as a potential tumor rejection antigen for the immune system [15]. Intradermal vaccination with G207 virus or intratumoral inoculation of mock-extract was not able to enhance the activation of specific T cells against CT26 tumors. In contrast, intratumoral inoculation of G207 induced an antigenic peptide-specific CTL response [17]. This result suggests that "in situ cancer vaccination" with G207 could overcome potential mechanisms of tolerance to endogenous antigen expression and inhibit the growth of established noninoculated tumors.

It is unclear what mechanism(s) is involved in the generation of specific antitumor immunity after HSV infection of tumor cells. It has recently been shown that tumor antigens can be transferred to host bone marrow-derived antigen-presenting cells before MHC class I presentation to CD8+ T cells [18]. Local HSV infection of the tumor might induce circulating precursors to differentiate into antigen presenting cells. A subset of macrophages can process exogenous antigens by the same MHC class I pathway as endogenous antigens and present them to CD8+ T cell clones [18,19]. Viral destruction of tumor cells might release tumor antigens that are then picked up by antigen-presenting cells, carried to the draining lymph nodes and presented to T cells.

Conclusion

A replication-competent multigene mutant of HSV-1, G207, inhibits local tumor growth by direct cytotoxic viral replication. As an "in situ cancer vaccine," it induces a systemic antitumor immune response to inhibit metastatic tumors by intratumoral viral inoculation.

References

1. Patchell RA (1991) Brain metastases. Neurol Clin 9:817–824
2. Penar PL, Wilson JT (1994) Cost and survival analysis of metastatic cerebral tumors treated by resection and radiation. Neurosurgery (Baltim) 34:888–893
3. Zimm S, Wampler GL, Stablein D, et al (1981) Intracerebral metastases in solid-tumor patients: natural history and results of treatment. Cancer (Phila) 48:384–394
4. Walker AE, Robins M, Weinfeld FD (1985) Epidemiology of brain tumors: the national survey of intracranial neoplasms. Neurology 35:219–226
5. Delattre JY, Krol G, Thaler HT, et al (1988) Distribution of brain metastasis. Arch Neurol 45:741–744
6. Bindal RK, Sawaya R, Leavens ME, et al (1993) Surgical treatment of multiple brain metastases. J Neurosurg 79:210–216
7. Mineta T, Rabkin SD, Yazaki T, et al (1995) Attenuated multi-mutated herpes simplex virus-1 for the treatment of malignant gliomas. Nat Med 1:938–943
8. Hunter W, Rabkin S, Martuza R (1995) Brain tumor therapy using genetically engineered replication-competent virus. In: Kaplitt MG, Loewy AD (eds) Viral vectors. Academic Press, San Diego, pp 259–274
9. Cailleau R, Young R, Olive M, et al (1974) Breast tumor cell lines from pleural effusions. J Natl Cancer Inst 53:661–674
10. Cailleau R, Olive M, Crucigar QV (1978) Long term human breast carcinoma cell lines of metastatic origin: preliminary characterization. In Vitro 14:911–915
11. Soule HD, Vazquez J, Long A, et al (1973) A human cell line from a pleural effusion derived from a breast carcinoma. J Natl Cancer Inst 51:1409–1413
12. Keydar I, Chen L, Karby S, et al (1979) Establishment and characterization of a cell line of human breast carcinoma origin. Eur J Cancer 15:659–670
13. Toda M, Rabkin SD, Martuza RL: Treatment of human breast cancer in a brain metastatic model by G207, a replication-competent multimutated herpes simplex virus 1. Hum Gene Ther (in press)
14. Dranoff G, Mulligan RC (1995) Gene transfer as cancer therapy. Adv Immunol 58:417–454
15. Huang AYC, Gulden PH, Woods AS, et al (1996) The immunodominant major histocompatibility complex class I-restricted antigen of a murine colon tumor derives from an endogenous retroviral gene product. Proc Natl Acad Sci USA 93:9730–9735
16. Toda M, Rabkin SD, Martuza RL (1997) Intratumoral inoculation of a replication-competent herpes simplex virus, G207, induces an antitumor immune response. Proc Am Assoc Cancer Res 38:1176

17. Toda M, Martuza RL, Kojima H, et al (1998) In situ cancer vaccination: a defective IL-12/replication-competent herpes simplex virus vector combination induces local and systemic antitumor activity. J Immunol 160:4457–4464
18. Kovacsovics-Bankowski M, Rock KL (1995) A phagosome-to-cytosol pathway for exogenous antigens presented on MHC class I molecules. Science 267:243–246
19. Reis e Sousa C, Germain RN (1995) Major histocompatibility complex class I presentation of peptides derived from soluble exogenous antigen by a subset of cells engaged in phagocytosis. J Exp Med 182:841–851

Reactive Astrocytes and Mossy Fiber Sprouting in Rat Hippocampus Induced by Kainic Acid Injection

CHUNXI HAN, WANG-JIE YU, HIROAKI ASOU, and KEIICHI UYEMURA

Summary. Administration of kainic acid (KA), a glutamate receptor agonist to the rat, induced neuronal degeneration, reactive gliosis, and sprouting of mossy fibers in the hippocampus. These changes play an important role in the adaptation of the nervous system to injury. The molecular mechanisms underlying them are poorly understood. Using the rat model of temporal lobe epilepsy induced by KA, we examined, particularly in the rat hipocampus, the relationship between several proteins and morphological changes. At 4 days after KA treatment, neuronal death occurred especially in CA1, where characteristic reactive astrocytes appeared. The reactive astrocytes contained all the glial fibrillary acidic protein (GFAP), vimentin, and the polysialylated neural cell adhesion molecule (NCAM-H). The sproutings of mossy fibers were observed only in the inner molecular layer of dentate gyrus, accompanied by NCAM-H-positive reactive astrocytes in the molecular layer. In the light of the known functional properties of NCAM-H, we suggest that reactive astrocytes by means of expression of NCAM-H may play a promoting role in the sprouting of mossy fibers in the hippocampus following the systemic injection of KA.

Key words. Kainic acid—Neural cell adhesion molecule—Hippocampus—Reactive astrocyte—Neuronal sprouting

Introduction

It has been reported that injection of kainic acid (KA) produced neurological and pathological features associated with human temporal epilepsy, neuronal cell death, and sprouting of mossy fibers. We have focused on the expression of polysialylated neural cell adhesion molecule (NCAM-H) in reactive astrocytes in rat hippocampus

---

Department of Physiology, School of Medicine, Keio University, 35 Shinanomachi, Shinjuku-ku, Tokyo 160-8582, Japan after injection of KA. We have examined expression of specific markers, such as NCAM-H and glial fibrillary acidic protein (GFAP), and how these relate to morphological changes such as neuronal death and sprouting in the hippocampus of the epileptic adult rat.

Neuronal Cell Death of Hippocampal Neurons by KA Injection

Selective neuronal cell death was observed in hippocampus by systemic KA (12.5 mg/kg) injection. At 4 days after KA administration, the death of neurons was consistently observed in the pyramidal cell layer of the hippocampus. At 28 days, loss of pyramidal neurons was the most extensive in CA1, followed by neuronal damage in the hilus of the dentate gyrus. The most remarkable occurrence was that of neuronal death in CA1 region, followed by the hilus of the dentate gyrus, which agreed well with previous reports [1,2]. On the other hand, neuronal death was less prominent in the CA3 region and was not detectable in CA2 and the dentate gyrus. When KA was directly injected into the lateral ventricle or hippocampus, remarkable neuronal death was reported in CA3 [3], indicating a difference in the major lesion caused by the method of injection. KA-1, a subunit of the KA receptor, is selectively expressed in CA3 pyramidal cells with mossy fiber terminals and granular cells in the dentate gyrus [4]. Therefore, in the case of injection of KA in ventricle or hippocampus, a high concentration of KA may induce directly the cell death of CA3 pyramidal cells. On the other hand, a low level of KA delivered by systemic injection may bind to the receptors in CA3 and dentate gyrus, resulting in abnormal release of glutamate and induction of cell death in CA1 [5]. When KA was administered systemically after the lesion of CA3, which is a major input to CA1, cell death in CA1 was not observed. Systemic administration of a large quantity of KA induced cell death in CA3 by overexcitation, but not in CA1, because CA1 neurons may be protected by decrease of input to CA1 from CA3 [3].

Appearance of GFAP-Positive Reactive Astrocytes by KA Injection

In the normal rat hippocampus, GFAP-positive astrocytes were found in the stratum lacunosum and hilus, in contrast to weak staining in other layers. The astrocytes were characterized by small cell bodies and short processes in the stratum lacunosum, but by long processes and light staining in the stratum oriens, stratum radiatum, and molecular layer, which findings are in good agreement with the previous report of Dalby et al. [6]. Reactive astrocytes that were positive for both GFAP and vimentin were first observed at 4 days after KA treatment in the stratum oriens and radiatum in the vicinity of CA1 and increased strikingly in all layers of hippocampus, especially in the molecular layer and hilus at 14 days and in the stratum oriens, radiatum, and hilus at 28 days, remained at this level until 90 days, and decreased thereafter. It is interesting that both GFAP- and vimentin-positive reactive astrocytes were also observed in the hilus and molecular layer of the dentate gyrus.

Appearance of NCAM-H in Reactive Astrocytes by KA Treatment

Distinct NCAM-H-positive cells were found only in the basal side of the dentate gyrus in normal adult rats, which did not correlate with GFAP staining. These cells corresponded to the newly generated neurons as previously reported by Seki [7]. After KA administration, some of the NCAM-H-positive cells appeared in the hilus and in stratum oriens and radiatum under the field of CA1. In addition, significant NCAM-H-positive cells were observed in the molecular layer of dentate gyrus, before and coincident with sproutings of mossy fibers. These stainings agreed well with GFAP staining, indicating clearly that reactive astrocytes expressed NCAM-H (Fig. 1).

By western blotting of proteins from the hippocampus of adult normal rats, a band of 200 kDa was significantly stained by anti-NCAM-H antibody, suggesting that this is a neuronal NCAM-H. By KA treatment, NCAM-H reactivity increased remarkably in similar molecular weight in the hippocampus. Immunostaining was clearly decreased by neuraminidase treatment, which specifically removes the polysialic acids from NCAM-H. The results indicate that after KA treatment reactive astrocytes expressed similar NCAM-H molecules as newly generated neurons in rat hippocampus.

NCAM is a cell adhesion protein of the immunoglobulin superfamily for neuro-neuronal and neuro-glial interactions [8–10]. Among the three kinds of NCAM isoforms, the 180-kDa isoform is mainly expressed in neurons while the 140-kDa and 120-kDa isoforms are found in both neurons and glial cells. In embryonic neurons, NCAM-H is expressed region- and stage specifically; it has carbohydrate chains with a high content of the unusual α-2,8-linked polysialic acids [9]. Removal of polysialic acids from the molecules on the cell surface by neuraminidase increased cell adhesion, aggregation, and axonal fasciculation. The polysialic acids in NCAM-H play roles in branching of axons, neurite outgrowth, and maintenance of intercellular spaces [7]. Expression of NCAM-H is observed not only in embryonic neurons, but also in special neural regions in the adult, such as the olfactory system and dentate gyrus in hippocampus [7]. In these regions, where neurons are actively renewed, NCAM-H may contribute to neurite outgrowth, pathfinding, and finally circuit formation.

Astrocytes occasionally express NCAM but not NCAM-H. In the hippocampus, KA treatment induced morphological and functional changes of astrocytes to reactive astrocytes [5,11]. Some of the reactive astrocytes in CA1 and the dentate gyrus expressed NCAM-H by KA treatment, but the astrocytes did not express NCAM-H by kindling stimulation, which did not induce glial reaction [12]. While the physiological significance of NCAM-H expression in astrocytes is not known in detail, it is intriguing that the astrocytes were mainly observed in the neuronal death-dominant CA1 region and the sprouting-rich dentate gyrus.

In general, damage to neurons induces activation of microglia, then the activated microglias secrete IL1, which may play a role in transforming astrocytes to reactive astrocytes [13,14]. Only some of the reactive astrocytes in specific regions of the hippocampus expressed NCAM-H, while in the other regions reactive astrocytes never expressed it. It is probable that NCAM-H plays a role in active sproutings, which are also characteristic in the hippocampus.

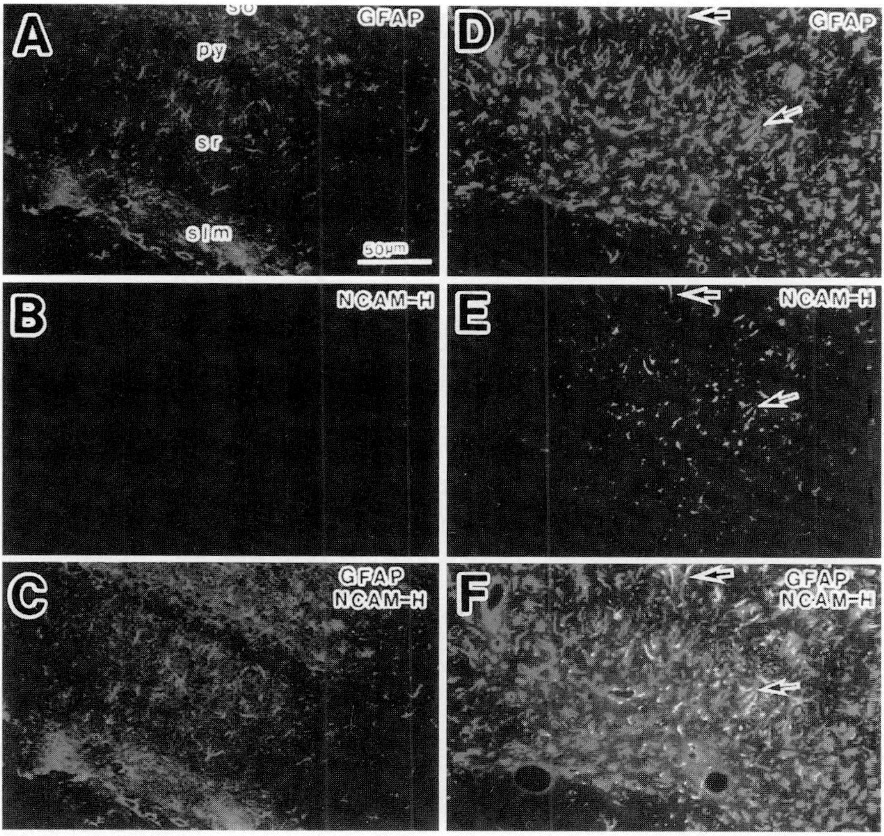

FIG. 1A–F. Double immunostaining of glial fibrillary acidic protein (GFAP) and polysialylated cell adhesion molecule (NCAM-H) in the CA1 region of hippocampus in control rats (**A–C**) and in rats at 28 days (**D–F**) after kianic acid (KA) administration: GFAP staining (**A,D**), NCAM-H staining (**B,E**), and double staining (**C,F**). While NCAM-H immunoreactivity was negative in CA1 of control (**B**), the NCAM-H-positive astrocytes were remarkably increased in the stratum oriens and stratum radiatium after KA treatment (**E,F**; *arrows*). The staining indicated clearly that reactive astrocytes expressed NCAM-H. *so*, stratum oriens; *py*, pyramidal cell layer; *sr*, stratum radiatum; *slm*, stratum lacunosum moleculare. *Bars* (**A–F**) 50 μm

Reactive Astrocytes and Mossy Fiber Sprouting

By modified Timm's staining, the sprouting of mossy fibers was first observed in the molecular layer of the dentate gyrus at 14 days after KA administration. The sprouting reached the greatest extent at 28 days, and maintained this level thereafter. The mossy fiber sproutings were observed in 2 of 9 rats (22%) at 14 days and in 7 of 11 rats (63%) at 28 days, in agreement with the results reported by Cronin et al. [15]. The dramatic increase of reactive astrocytes in the molecular layer of the dentate gyrus suggests its close relationship to synaptic remodeling of mossy fibers. Increase of GFAP-positive reactive astrocytes in rats with sprouting was much greater compared to that in rats without sprouting (Fig. 2).

Mossy fibers are axons of granular cells in the dentate gyrus and form glutaminergic synapses to dendrites of pyramidal cells in CA3 and of neurons in the hilus of the dentate gyrus [16]. The mossy fiber terminals are clearly stained by Timm's method, a sensitive reaction to zinc ions. With KA treatment, sprouting of mossy fibers occurred in the whole inner molecular layer surrounding the dentate gyrus. Electron microscopy examination revealed that the sprouting axons newly formed synapses to dendrites of granular cells [17,18]. Increase of mossy fiber terminals in CA3 and elongation of the fiber to CA2–1 were also reported. However, the mechanism of sprouting is not yet known. By removal of inputs to molecular layer of dentate gyrus, such as cutting of the perforant pathway or entorhinal area, mossy fibers showed active sprouting [19]. Therefore, mossy fiber sprouting observed in

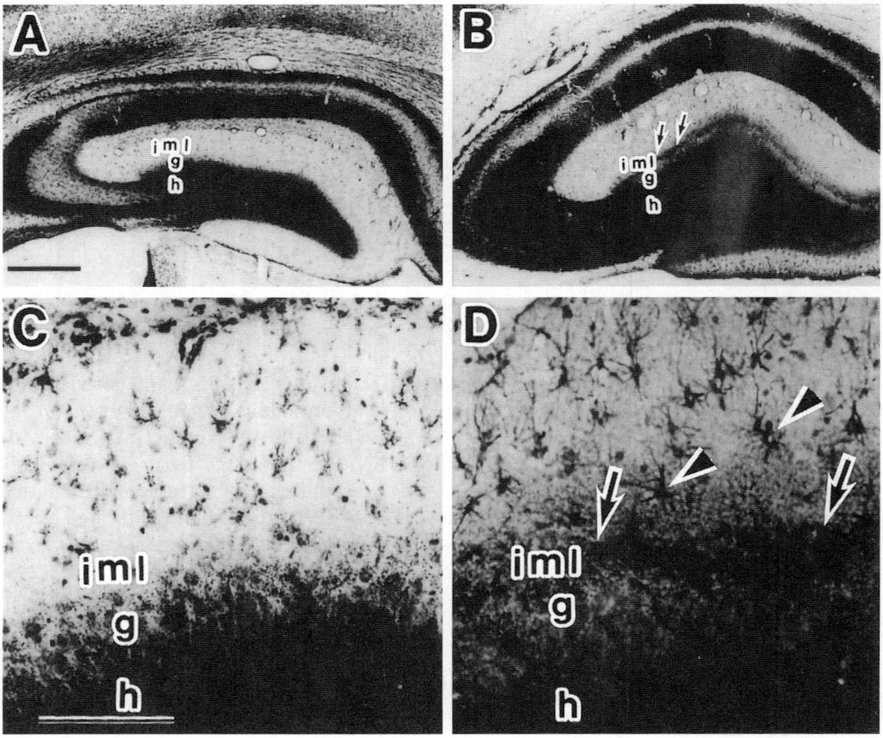

FIG. 2. Timm's and GFAP staining 28 days after KA administration. **C** and **D** are magnifications of **A** and **B**. In several cases, intensive aberrant Timm's staining was observed in the inner molecular layer of the dentate gyrus (**B**,**D**; *arrows*), while in other cases such sproutings were not observed (**A**,**C**). In **D**, the increase of GFAP-positive reactive astrocytes (*arrowheads*) was more remarkable than that in C. *iml*, inner molecular layer; *g*, granule cell layer; *h*, hilus. *Bars* 500 μm (**A**,**B**); 50 μm (**C**,**D**)

epilepsy may be caused by neuronal cell damage in CA1 and the hilus of the dentate gyrus, which form synapses to dendrites of granular cells in the inner molecular cell layer [17,18,20].

Concerning to the relationship among neuronal cell death, mossy fiber sprouting and synaptic remodeling, various changes have been reported, such as immediate-early genes, neurotrophins, cytoskeletal proteins, glial cells, and cell adhesion molecules [12,13,21–25]. In these studies, we have developed a modified Timm's method for examination of mossy fiber sprouting and reactive astrocytes. To detect zinc by Timm's staining, pretreatment is necessary to perfuse tissues with a high concentration of sodium sulfate. Therefore, it is difficult to use the same tissues for immunostaining. We used 0.1% sodium sulfate, instead of the usual 0.5%–1.0%, in 4% paraformaldehyde for perfusion and fixation and developed a double staining method with Timm's staining and immunostaining.

In the case of active sprouting of mossy fibers after KA stimulation, the number of GFAP-positive astrocytes increased more compared to those without sprouting. Astrocytes in the hilus of the dentate gyrus were reported to be tenascin positive, NCAM positive, and proliferative, while those in the molecular layer were tenascin negative, NCAM weakly positive, and not proliferative. These astrocytes may play a cooperative role for mossy fiber sprouting [5]. The evidence that KA induces NCAM-H-positive astrocytes in the hilus and the molecular layer of the dentate gyrus strongly suggests that the reactive astrocytes play a role in sprouting and synaptic remodeling by mossy fibers.

References

1. Ben-Ari Y (1985) Limbic seizure and brain damage produced by kainic acid: mechanisms and relevance to human temporal lobe epilepsy. Neuroscience 14:375–403
2. Lothman EW, Collins RC (1981) Kainic acid induced limbic seizures: metabolic, behavioral, electroencephalographic and neuropathological correlates. Brain Res 218:299–318
3. Balchen T, Berg M, Diemer NH (1993) A paradox after systemic kainate injection in rats: lesser damage of hippocampal CA1 neurons after higher doses. Neurosci Lett 163:151–154
4. Wisden W, Seeburg PH (1993) A complex mosaic of high-affinity kainate receptors in rat brain. J Neurosci 13:3582–3598
5. Represa A, Niquet J, Pollard H, et al (1995) Cell death, gliosis, and synaptic remodeling in the hippocampus of epileptic rats. J Neurobiol 26:413–425
6. Dalby NO, Rondouin G, Lerner-Natoli M (1995) Increase in GAP-43 and GFAP immunoreactivity in the rat hippocampus subsequent to perforant path kindling. J Neurosci Res 41:613–619
7. Seki T, Arai Y (1993) Distribution and possible roles of the highly polysialylated neural cell adhesion molecule (NCAM-H) in the developing and adult central nervous system. Neurosci Res 17:265–290
8. Edelman GM (1986) Cell adhesion molecules in neural histogenesis. Annu Rev Physiol 48:417–430
9. Rutishauser U, Acheson A, Hall AK (1988) The neural cell adhesion molecule (NCAM) as a regulator of cell-cell interactions. Science 240:53–57
10. Rutishauser U, Jessell TM (1988) Cell adhesion molecules in vertebrate neural development. Physiol Rev 68:819–857

11. Niquet J, Ben-Ari Y, Represa A (1994) Glial reaction after seizure induced hippocampal lesion: immunohistochemical characterization of proliferating glial cells. J Neurocytol 23:641–656
12. Le Gal La Salle G, Rougon G, Valin A (1992) The embryonic form of neural cell surface molecule (E-NCAM) in the rat hippocampus and its reexpression on glial cells following kainic acid-induced status epilepticus. J Neurosci 12:872–882
13. Fagan AM, Gage FH (1990) Cholinergic sprouting in the hippocampus: a proposed role for IL-1. Exp Neurol 110:105–120
14. Ide CF, Scripter JL, Coltman BW, et al (1996) Cellular and molecular correlates to plasticity during recovery from injury in the developing mammalian brain. Prog Brain Res 108:365–377
15. Cronin J, Dudek EF (1988) Chronic seizures and collateral sprouting of dentate mossy fibers after kainic acid treatment in rats. Brain Res 474:181–184
16. Brown TH, Zador AM (1990) Hippocampus. In: Shepherd GM (ed) The synaptic organization of the brain, 3rd edn. Oxford University, New York, pp 346–388
17. Babb TL, Kupfer WR, Pretorius JK, et al (1991) Synaptic reorganization by mossy fibers in human epileptic fascia dentata. Neuroscience 42:351–363
18. Babb TL, Pretorius JK, Mello LE, et al (1992) Synaptic reorganization in epileptic human and rat kainate hippocampus may contribute to feedback and feedforward excitation. Epilepsy Res 9:193–202
19. Nadler JV, Perry BW, Cotman CW (1980) Selective reinnervation of hippocampal area CA1 and the fascia dentata after destruction of CA3-CA4 afferents with kainic acid. Brain Res 182:1–9
20. Houser CR, Miyashiro JE, Swartz BE, et al (1990) Altered patterns of dynorphin immunoreactivity suggest mossy fiber reorganization in human hippocampal epilepsy. J Neurosci 10:267–282
21. Gage FH, Olejniczak P, Armstrong DM (1988) Astrocytes are important for sprouting in the septohippocampal circuit. Exp Neurol 102:2–13
22. Khrestchatisky M, Ferhat L, Charton G, et al (1995) Molecular correlates between reactive and developmental plasticity in the rat hippocampus. J Neurobiol 26:426–436
23. Pollard H, Khrestchatisky M, Moreau J (1994) Correlation between reactive sprouting and microtubule protein expression in epileptic hippocampus. Neuroscience 61:773–787
24. Seki T, Arai Y (1993) Highly polysialylated neural cell adhesion molecule (NCAM-H) is expressed by newly generated granule cells in the dentate gyrus of the adult rat. J Neurosci 13:2351–2358
25. Strauss S, Otten U, Joggerst B, et al (1994) Increased levels of nerve growth factor (NGF) protein and mRNA and reactive gliosis following kainic acid injection into the rat striatum. Neurosci Lett 168:193–196

Part 9
Human Genome Project

Human Genome Project: Current Status, Keio Strategy, and Prospects

Nobuyoshi Shimizu

Summary. Through the human genome project (HGP) at Keio University, we were able to find a number of new genes including SIM1 (single-minded), SIM2, MNB (minibrain), KNP1 (keio novel protein), KNP3, APECED/AIRE, MYOC (myocilin), RAO (retina amine oxidase), and ISLR (immunoglobulin superfamily containing LR domain). Also, we determined the genomic sequence of immunoglobulin light chain λ (IGL) gene cluster of more than 1 megabase (Mb) and identified 69 IGLV genes, 7 IGLC genes, and 12 non-Ig genes/pseudogenes. Moreover, we established the genomic structures of previously known genes such as HCS, ETS2, TMEM1, PWP2, and PFKL. It should be noted that SIM2 and MNB are candidate genes for mental retardation in Down syndrome, that APECED/AIRE is a gene responsible for the autoimmune disease APECED, and that MYOC is identical to TIGR, which is a pathogenic gene for primary open angle glaucoma.

Key words. Human genome project—Down syndrome—Autoimmune disease—Glaucoma

Current Status of Human Genome Project (HGP)

Since its conception in 1991, the HGP has made rapid progress and has accomplished its initial goal, namely, mapping of the human genome. To date, substantial numbers of STS (sequence tagged site) markers have been located in the genetic and physical maps of individual human chromosomes [1]. Among 5000 Mendelian inherited disorders, more than 800 disease-causing genes have been isolated by positional cloning and candidate gene approaches [2]. Moreover, several thousands of biologically important genes have been cloned and characterized [3]. Sequence information on more than 250 000 cDNAs and ESTs (expressed sequence tags) has been deposited in the database, and most of these have been made publicly available [4]. Various computer databases including GDB have been established, providing service through the Internet with international collaboration [5].

Department of Molecular Biology, School of Medicine, Keio University, 35 Shinanomachi, Shinjuku-ku, Tokyo 160-8582, Japan

Despite the overwhelmingly successful progress in the past, there are still enormous tasks to reach the final goal, sequencing of the entire human genome consisting of 3000 Mb as well as complete understanding of the estimated 100 000 human genes. Now, HGP has entered a new challenging era of genomic DNA sequencing.

Keio Strategy and Outcome

We have begun genomic DNA sequencing of human chromosomes 21 and 22, each of which consists of 50 Mb, through the Keio strategy using highly advanced technologies and automated instruments (Fig. 1). Our initial targets are four regions of chromosome 21 and three regions of chromosome 22. These regions are known to be involved in important medical and biological problems including Down syndrome, epilepsy, autoimmune disease, immune response, bipolar affective disorder, and human embryogenesis [6–8].

We are using a shotgun sequencing strategy in combination with primer walking and nested deletion methods to completely determine the sequence of DNA clones in the BAC/cosmid contigs [9,10]. We have so far determined more than 3 Mb of DNA sequence with accuracy >99.99%. Genomic sequence was subjected to homology search and analysis for protein coding potential using GRAIL, HEXON, and GENSCAN (Fig. 1). Putative exon sequences were used as probes to screen corresponding cDNA clones or as primers to amplify cDNA fragments by PCR. Nucleotide sequence of these cDNA clones allowed us to find new genes. These genes include SIM2 (single-minded), MNB (minibrain), KNP1 (keio novel protein), KNP3, KNP5, APECED/AIRE (autoimmune regulator), and others [9–17] (Table 1A).

It should be noted that SIM2, encoding a PAS family transcription factor, and MNB, encoding a dual specific protein kinase, are candidate genes for mental retardation in Down syndrome, and that APECED/AIRE, encoding a novel transcription factor with 3 LxxLL motifs, is a gene responsible for the autoimmune disease APECED. The KNP3 protein has seven putative transmembrane domains that probably constitute a Ca^{2+} channel. Also, we established the genomic structures of previously known genes such as HCS (holocarboxylase synthetase), ETS2 (v-*ets* oncogene), TMEM1 (transmembrane protein 1), PWP2 (periodic tryptophan protein 2), and PFKL (phosphofructokinase, liver) [17,18] (Fig. 2, Table 1C). Moreover, we determined the genomic organization of immunoglobulin light chain λ (IGL) gene cluster of more than 1 Mb and identified 69 IGLV genes, 7 IGLC genes, and 12 non-Ig genes/pseudogenes including a protein phosphatase and a DNA topoisomerase [8,9] (Table 1B).

As a separate line of work, we developed a subtracted cDNA library to enrich retina-specific mRNAs and applied differential hybridization to isolate useful cDNA clones [19]. This analysis resulted in the isolation of several new genes including MYOC (myocilin), RAO (retina amine oxidase), and ISLR (immunoglobulin superfamily containing LR domain) [20–23] (Table 1C). It should be noted that MYOC was identical to TIGR, which was found to be a pathogenic gene for primary open angle glaucoma. Also, RAO was proposed to be a candidate gene for hereditary ocular disease. The sequence information generated at Keio is available through a Web site (http://www-alis.tokyo.jst.go.jp) of JST as well as in DDBJ/EMBL/GenBank databases.

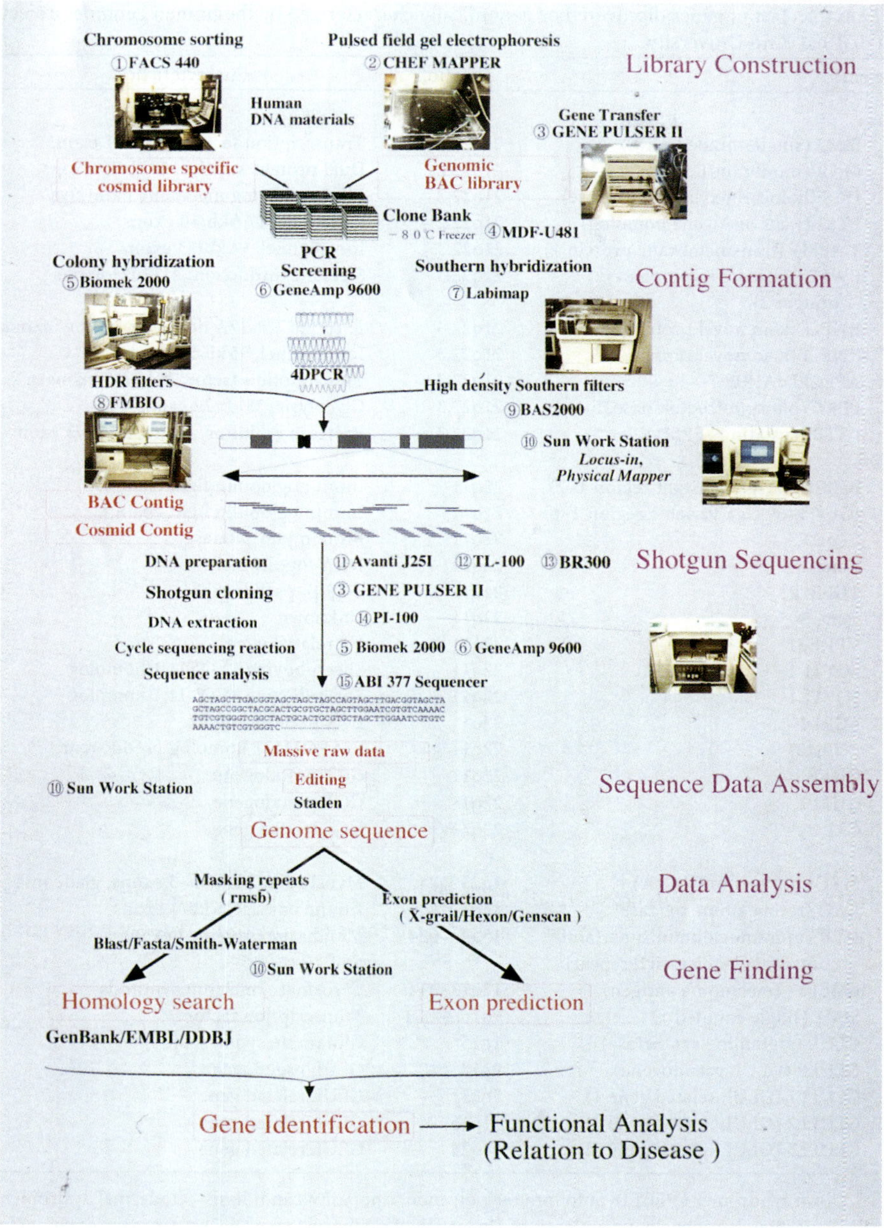

FIG. 1. Strategy of mapping and sequencing of the human genome at Keio

TABLE 1. List of genes discovered or genomically characterized by the human genome project (HGP) at Keio University

Genes	Location	Characteristics
A		
SIM2 (single-minded 2/DS)	21q22.2	Transcription factor, 55 kb/11 exons
MNB (minibrain/DS)	21q22.2	Dual protein kinase, 150 kb/13 exons
HCS (holocarboxylase synthetase)	21q22.2	Biotin-utilizing metabolic pathways
ETS2 (v-*ets* oncogene homolog)	21q22.2	Oncogene, 17.6 kb/10 exons
TMEM1 (transmembrane protein 1)	21q22.3	Ion channel, 94 kb/23 exons
PWP2 (periodic tryptophan protein 2)	21q22.3	Signal transduction, 24 kb/21 exons
KNP-1 (keio novel protein 1)	21q22.3	*E. coli* SCRP-27A homolog, 12 kb/7 exons
KNP-3 (keio novel protein 3)	21q22.3	Ion channel, 95 kb/32 exons
APECED/AIRE	21q22.3	Transcription factor, 13 kb/15 exons
PFKL (phosphofructokinase, liver)	21q22.3	Glycolysis, 28 kb/22 exons
CSTB (cystatin B/EPM1)	21q22.3	Protease inhibitor, EPM1, 2.5 kb/3 exons
B		
IGLC 1–7 (Igλ constant region 1–7)	22q11	Immunoglobulin light chain λ
IGLV 1–69 (Igλ variable region 1–6)	22q11	Immunoglobulin light chain λ
Gene 1	22q11	Protein phosphatase
Gene 2	22q11	DNA topoisomerase
TGFB1P1	22q11	TGF-β_1 pseudogene
Gene 3	22q11	Unknown
VPREB1	22q11	Vλ related gene
5OY11.1	22q11	Sheep/bovine 5′-OY11.1 homolog
3OY11.1	22q11	Sheep/bovine 3′-OY11.1 homolog
BCRL4	22q11	BCR pseudogene
POM121	22q11	Rat POM121 homolog pseudogene
GGTP1	22q11	GGT pseudogene
GGTP2	22q11	GGT pseudogene
Gλ1	22q11	J-Cλ related gene
C		
MYOC (myocilin/GLC1A)	1q23–q24	Myosin-like protein, 3 exons, glaucoma
RAO (retina amine oxidase)	17q21	Amine oxidase, 6 kb/4 exons
ISLR (immunoglobulin superfamily containing leucine-rich repeat)	15q23–q24	Membrane protein, 2 exons
pMEL17 (melanoma antigen)	12q13–q14	Tyrosinase, melanin synthesis
SIM1 (single-minded 1)	6q16.3–q21	Transcription factor
GLUL (glutamine synthetase)	1q25	Glutamine synthesis [25]
GLULP (GLUL pseudogene)	9p13	GLUL pseudogene
GLULL1 (GLUL related gene 1)	5q33	GLUL related gene
GLULL2 (GLUL related gene 2)	11p15	GLUL related gene
GLULL3 (GLUL related gene 3)	11q24	GLUL related gene

DS, Down syndrome; APECED, autoimmune polyendocrinopathy-candidiasis-ectodermal dystrophy; EPM1, progressive myoclonus epilepsy of Unverricht-Lundborg Type; GLC1A, primary open angle glaucoma (POAG).

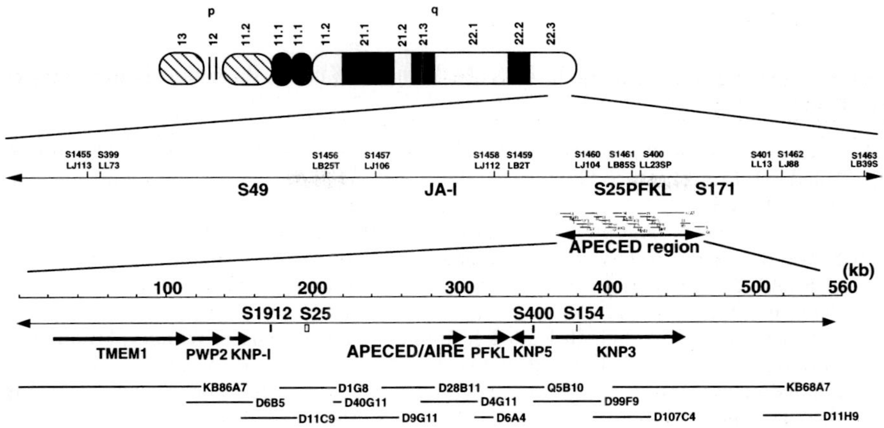

FIG. 2. Seven genes identified in the APECED region of chromosome 21q22.3

Prospects of HGP

The prospected goal of genomic DNA sequencing is the year 2005, and massive efforts are in progress. It is critical to accelerate the preparation of sequence-ready DNA contigs, and hence we have developed a novel method of DNA library screening called "Digital hybridization screening" [24]. In the coming few years, we will gain a much better understanding of the human genome, which contains a blueprint of our life. The sequence information will undoubtedly serve as an essential foundation of human biology, including neuroscience and brain research.

Acknowledgments. The author thanks laboratory members and collaborators who have made significant contributions to the HGP at Keio.

References

1. Hudson JJ, Stein LD, Gerety SS, et al (1995) An STS-based map of the human genome. Science 270:1945–1954
2. On-line Mendelian inheritance in man (OMIM). http://www3.ncbi.nih.gov/omim/
3. Schuler GD, Bognski MS, Stewart EA, et al (1996) A gene map of the human genome. Science 274:540–546
4. NCBI/Nentrez. http://www.ncbi.nlm.nih.gov/
5. Nucleic Acids Research, vol. 26, #1. Database Issue, January 1998
6. Osoegawa K, Susukida R, Okano S, Kudoh J, Minoshima S, Shimizu N, de Jong PJ, Groet J, Ives J, Lehrach H, Nizetic D, Soeda E (1996) An integrated Map with cosmid/PAC contigs of a 4-Mb Down syndrome critical region. Genomics 32(3):375–387
7. Kudoh J, Nagamine K, Asakawa S, Abe I, Kawasaki K, Maeda H, Tsujimoto S, Minoshima S, Ito F, Shimizu N (1997) Localization of 16 exons to a 450-kb region involved in the autoimmune polyglandular disease type I (APECED) on human chromosome 21q22.3. DNA Res 4(1):45–52
8. Kawasaki K, Minoshima S, Schooler K, Kudoh J, Asakawa S, de Jong PJ, Shimizu N (1995) The organization of the human immunoglobulin λ gene locus. Genome Res 5(2):125–135
9. Kawasaki K, Minoshima S, Mine E, Shibuya K, Shintani A, Schmeits JL, Wang J, Shimizu N (1997) One megabase sequence analysis of the human immunoglobulin λ gene locus. Genome Res 7(3):250–261

10. Asakawa S, Abe I, Kudoh Y, Kishi N, Wang Y, Kubota R, Kudoh J, Kawasaki K, Minoshima S, Shimizu N (1997) Human BAC library: construction and rapid screening. Gene 191(1):69–79
11. Chen H, Chrast R, Rossier C, Gos A, Antonarakis SE, Kudoh J, Yamaki A, Shindoh N, Maeda H, Minoshima S, Shimizu N (1995) Single-minded and Down syndrome? Nat Genet 10(1):9–10
12. Yamaki A, Noda S, Kudoh J, Shindoh N, Maeda H, Minoshima S, Kawasaki K, Shimizu Y, Shimizu N (1996) The mammalian single-minded (SIM) gene; mouse cDNA structure and diencephalic expression indicate a candidate gene for Down syndrome. Genomics 35(1):136–143
13. Chrast R, Scott HS, Chen H, Kudoh J, Rossier C, Minoshima S, Wang Y, Shimizu N, Antonarakis SE (1997) Cloning of two human homologs of the *Drosophila* single-minded gene SIM1 on chromosome 6q and SIM2 on 21q within the Down syndrome chromosomal region. Genome Res 7(6):615–624
14. Shindoh N, Kudoh J, Maeda H, Yamaki A, Minoshima S, Shimizu Y, Shimizu N (1996) Cloning of a human homolog of the *Drosophila* minibrain/rat dyrk gene from "the Down syndrome critical region" of chromosome 21. Biochem Biophys Res Commun 225(1):92–99
15. Nagamine K, Kudoh J, Minoshima S, Kawasaki K, Asakawa S, Ito F, Shimizu N (1996) Isolation of cDNA for a novel human protein KNP-I that is homologous to the *E. coli* SCRP-27A protein from the autoimmune polyglandular disease type I (APECED) region of chromosome 21q22.3. Biochem Biophys Res Commun 225(2):608–616
16. Nagamine K, Peterson P, Scott HS, Kudoh J, Minoshima S, Heino M, Krohn KJE, Lalioti MD, Mullis PE, Antonarakis SE, Kawasaki K, Asakawa S, Ito F, Shimizu N (1997) Positional cloning of the APECED gene. Nat Genet 17(4):393–398
17. Nagamine K, Kudoh J, Kawasaki K, Minoshima S, Asakawa S, Ito F, Shimizu N (1997) Genomic organization and complete nucleotide sequence of the TMEM1 gene on human chromosome 21q22.3. Biochem Biophys Res Commun 235(1):185–190
18. Nagamine K, Kudoh J, Minoshima S, Kawasaki K, Asakawa S, Ito F, Shimizu N (1997) Genomic organization and nucleotide sequence of the human PWP2 gene on chromosome 21. Genomics 42(3):528–531
19. Kubota R, Wang Y, Minoshima S, Kudoh J, Mashima Y, Oguchi Y, Shimizu N (1995) Mapping of the human gene for Pmel 17 (D12S53E) to chromosome 12q13-q14. Genomics 26(2):430–431
20. Kubota R, Noda S, Wang Y, Minoshima S, Asakawa S, Kudoh J, Mashima Y, Oguchi Y, Shimizu N (1997) A novel myosin-like protein (miocilin) expressed in connecting cilium of photoreceptor: molecular cloning, tissue expression and chromosomal mapping. Genomics 41(3):360–369
21. Kubota R, Kudoh J, Mashima Y, Asakawa S, Minoshima S, Hejtmancik JF, Oguchi Y, Shimizu N (1998) Genomic organization of the human myocilin gene (MYOC) responsible for primary open angle glaucoma. Biochem Biophys Res Commun 242:396–400
22. Imamura Y, Kubota R, Wang Y, Asakawa S, Kudoh J, Mashima Y, Oguchi Y, Shimizu N (1997) Human retina-specific amine oxidase (RAO): cDNA cloning, tissue expression and chromosomal mapping. Genomics 40(2):277–283
23. Nagasawa A, Kubota R, Imamura Y, Nagamine K, Wang Y, Asakawa S, Kudoh J, Minoshima S, Mashima Y, Oguchi Y, Shimizu N (1997) Cloning of the cDNA for a new member of the immunoglobulin superfamily (ISLR) containing leucine-rich repeat (LRR). Genomics 44(3):273–279
24. Asakawa S, Shimizu N (1998) High fidelity digital hybridization screening. Genomics 49(5):209–217
25. Wang Y, Kudoh J, Kubota R, Asakawa S, Minoshima S, Shimizu N (1996) Chromosomal mapping of a family of human glutamine synthetase genes: functional gene (GLUL) on 1q25, pseudogene (GLULP) on 9p13 and three related genes (GLULL1, GLULL2, GLULL3) on 5q33, 11p15 and 11p24. Genomics 37(2):195–199

Key Word Index

Keio University Symposia
for Life Science and Medicine 2

Springer
Tokyo
Berlin
Heidelberg
New York
Barcelona
Hong Kong
London
Milan
Paris
Singapore